THE BLUE-GREEN ALGAE

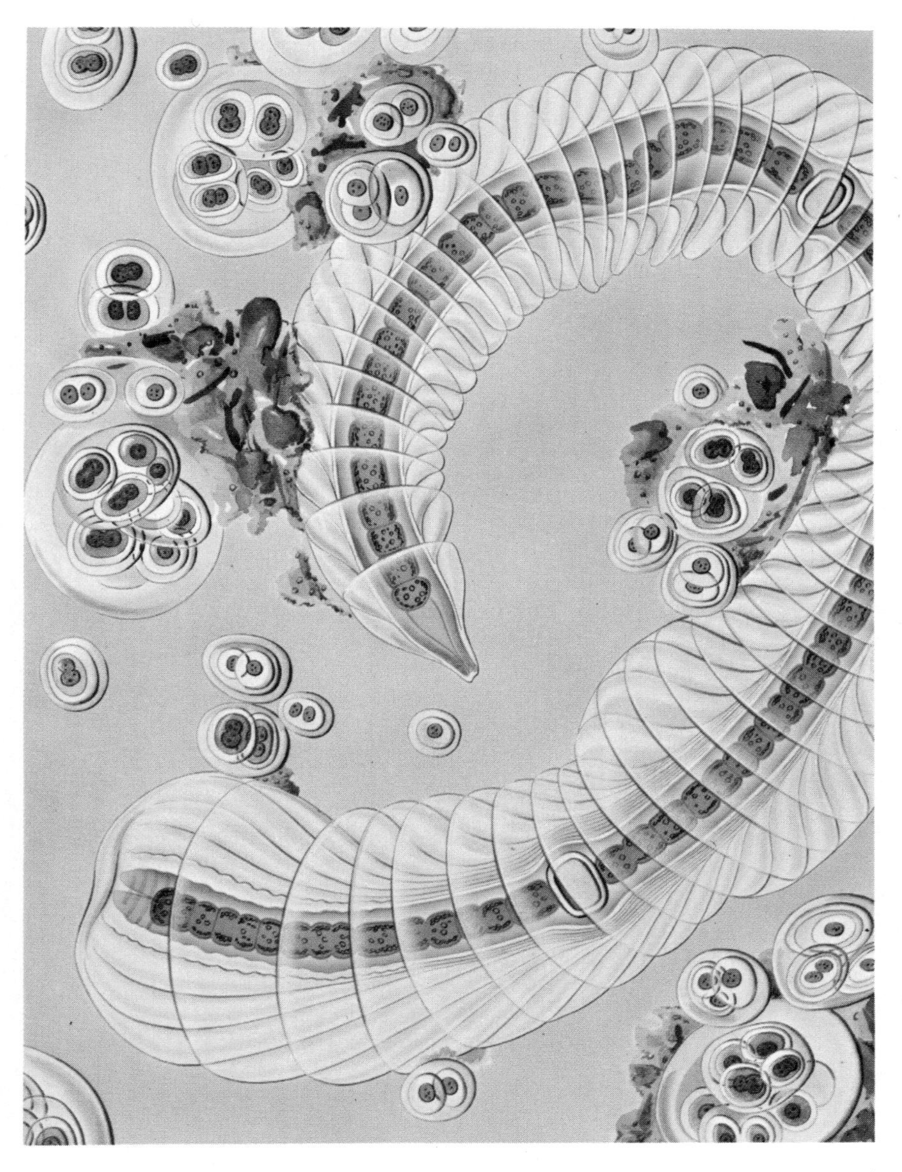

Blue-green algae from a wet rock face, Salen, Isle of Mull, Scotland; *Petalonema alatum* Berk. (filamentous with complex sheath), *Gloeocapsa alpina* Näg. emend. Brand (unicellular, violet sheath) and *Gloeocapsa montana* Kutz. (unicellular, colourless sheath), x 400. E. F. Wilkinson, after a water colour drawing by G. E. Fogg.

THE BLUE-GREEN ALGAE

G. E. FOGG, W. D. P. STEWART
P. FAY *and* A. E. WALSBY

1973

Academic Press - London *and* New York
A subsidiary of Harcourt Brace Jovanovich, Publishers

ACADEMIC PRESS INC. (LONDON) LTD.
24/28 Oval Road,
London NW1

United States Edition published by
ACADEMIC PRESS INC.
111 Fifth Avenue
New York, New York 10003

Second printing 1974

Library of Congress Catalog Card Number: 72–12269
ISBN: 0–12–261650–2

PRINTED IN GREAT BRITAIN BY
J. W. Arrowsmith Ltd., Bristol BS3 2NT

Preface

The blue-green algae, for a long time a disregarded group of micro-organisms, are now a fashionable subject for research. The possible reasons for this are various. One may be that current speculations about the origin and early evolution of life have reminded biologists that this is a group of undoubted antiquity. It has biochemical characteristics, on the one hand, which may be related to an existence on the primitive Earth and, on the other hand, it shows possible links with the green plants which are dominant today. Another reason may be that the electron microscope has revealed features of fine structure which confirm the long suspected kinship of these organisms with the bacteria. Mounting evidence that blue-green algae play an important, if unobtrusive, part in maintaining soil fertility and the often all too obvious fact that they are a nuisance in freshwater have also made their study of some economic importance.

For the four authors of this book this upsurge of interest has been exhilarating, as ways of solving apparently insoluble problems have presented themselves, facts have fallen into place and new, intriguing, questions have arisen. Nevertheless they are already conscious of the increasing mass of specialist literature and feel that now, before it becomes overwhelming, is the time to attempt the assembly of a unified picture of blue-green algae as living organisms. They hope that the result will be of use to students and research workers in various branches of botany, microbiology and biochemistry. On a more personal level this book is a tangible reminder of a most happy period of collaboration in the Department of Botany at Westfield College, University of London.

The authors' thanks are due to many people who have generously given their help in this enterprise. Successive versions of the text have been typed by Mrs. N. Last, Miss Marianne Joyce and Mrs. B. Cunningham. Dr. Hilda M. Lund has given us considerable help with the chapter on pathogens. We also thank Mr. Ray Mitchell and Miss Gail Alexander for help with photographs and figures. The names of those who have allowed the use of copyright illustrative material, thereby greatly enhancing the usefulness and appearance of the book, are too numerous to mention here but will be found in the appropriate places in the text. Lastly, the patience and forbearance of the publishers in the face of repeated delays on the part of the authors must be acknowledged.

February, 1973

G. E. FOGG
Department of Marine Biology,
University College of North Wales,
Marine Science Laboratories,
Menai Bridge, Anglesey,
North Wales.

W. D. P. STEWART
Department of Biological Sciences,
The University,
Dundee DD1 4HN,
Scotland.

P. FAY
Department of Botany,
Westfield College
(University of London),
London N.W.3,
England.

A. E. WALSBY
Department of Botany,
University of California,
Berkeley, California 94720,
U.S.A.

Contents

Chapter 1

Introduction

The micro-organisms known as blue-green algae form an unusually well-defined group. Blue-green algae are not always blue-green in colour but after a little experience one can recognize them instantly under the microscope because of the characteristically homogeneous appearance which the absence of membrane-bound organelles, such as nuclei and chloroplasts, gives to their protoplasm. In distinguishing the smaller species from bacteria, which also lack such organelles, there may be some difficulty. However, the blue-green algae usually contain the pigments chlorophyll *a* and biliproteins, by means of which they carry out photosynthesis with the production of oxygen, whereas bacteria, if photosynthetic, have other pigments and do not produce oxygen. Furthermore, blue-green algae, although they may show a characteristic gliding movement, never possess flagella as means of locomotion as some bacteria do. Only with some colourless species which have close morphological resemblances to blue-green algae but lack their photosynthetic apparatus, may there be uncertainty about taxonomic position.

Mentions of remarkable growths in fresh water, which we now know to have been planktonic blue-green algae, are found in early records such as those which Giraldus Cambrensis wrote in 1188. Shakespeare assumed that his audiences were familiar with such manifestations when he put into the mouth of Gratiano in the Merchant of Venice the words "There are a sort of men whose visages do cream and mantle like a standing pond". However, the first scientific descriptions of particular kinds of blue-green algae did not appear until the early nineteenth century. Roth between 1797 and 1806, distinguished the genus *Rivularia* and Vaucher in 1803 described *Oscillatoria* and *Nostoc*. Vaucher placed these blue-green algae in the animal kingdom but nevertheless their subsequent study was undertaken mainly by botanists. Harvey (1846) put the blue-green algae together with green algae in a group Chlorospermae but as biologists learnt more about the internal structure of cells and the significance of the photosynthetic pigments it was realized that there are important differences between the two kinds of organism. The blue-green algae were segregated as a distinct group, the Myxophyceae, by Stitzenberger in 1860. This name has priority but has not been universally accepted because it is liable to be confused with Myxophyta, which denotes the slime moulds. The name Cyanophyceae, introduced by Sachs in 1874, has usually been preferred as it conforms with the principle of naming algal groups according to their characteristic colours. In the present day classification of the algae, the basic pattern of which was crystallized by Pascher in 1914, the group is given the rank of a division of the plant kingdom under the name Cyanophyta, a term introduced by Smith in 1938 (a fuller account of the history of phycology is given by Smith, 1951).

In gross morphology and nutrition the blue-green algae have many features in common with the other algal groups and it is therefore convenient to deal with them as algae for many purposes. However, as long ago as 1853, Cohn drew attention to the similarities between blue-green algae and bacteria and in 1871/2 put them together in a division Schizophyta, giving the name Schizophyceae to the former and Schizomycetes to the latter. The name Schizophyceae has not been widely used but the affinity of the blue-green algae and bacteria has been substantiated by modern work. Indeed, some authorities (Stanier *et al.*, 1971) categorically state that blue-green algae are bacteria. This recognition of the affinities of the two groups has grown largely out of studies of fine structure. Cells of blue-green algae are small, generally being only a few micrometres in diameter, about a tenth of that of an average cell of a higher plant or animal. The details of subcellular structure are often near the limits of resolution of the light microscope. Although it is remarkable how much information was collected by the early workers, real progress scarcely began until, somewhat tardily, the electron microscope was used. Although much remains to be learnt it has become quite clear that the blue-green algae share with the bacteria a type of cellular organization known as prokaryotic. This stands in sharp contrast to the eukaryotic type of cell, with membrane-enclosed nucleus, shown by all other types of organism. Also certain bacteria and blue-green algae produce a unique type of gas-containing subcellular structure, the gas vesicle, which confers buoyancy on the algae. Classification of the blue-green algae alongside the bacteria in the kingdom Prokaryota is therefore a truer representation of their phylogenetic position than classification with the other algal groups, which together with higher plants and animals constitute the kingdom Eukaryota.

The interest of the prokaryotic state to the biologist scarcely needs emphasis. The basic chemical mechanisms on which life is based are generally similar in the Prokaryota and Eukaryota but they are carried out in somewhat different structures in the two kingdoms. The sequence of reactions in aerobic respiration, for example, is generally similar although in Eukaryota the enzymes concerned are contained in specific membrane-bounded organelles, the mitochondria, whereas in Prokaryota they are associated with simpler membranous structures. Comparison of the molecular architecture and efficiency of these two types of structure is of obvious interest. The relatively simple cell structure of the Prokaryota gives an impression of being primitive and its antiquity is indicated both by the evidence of fossils and biochemical considerations. Study of the Prokaryota may thus shed light on the origin of life on the one hand and the evolution of the Eukaryota on the other. Beyond this, the small size, lack of differentiation, ease of culture and handling, and rapid growth of certain Prokaryota has made them excellent experimental material for investigation of a great variety of general biological problems. One may call to mind here the enrichment of biochemistry by study of the tremendous variety of metabolic processes exhibited by the bacteria and the massive contribution which research utilizing *Escherichia coli* has made to molecular biology. In all this, however, work with blue-green algae has so far played little part. There are various reasons for this. Blue-green algae are often difficult to isolate and grow in culture and a belief, largely unfounded, that they grow extremely slowly was current at one time. Few of them are even indirectly pathogenic and, until recently, none has been considered of great importance either as being a nuisance or being useful. Thus, there has not been the urgency and economic incentive in their investigation that there has been with certain bacteria. Nevertheless it is important that the kind of investigations that have been carried out with bacteria should also be made with blue-green algae. Even if it were to turn out that their biochemical and genetical mechanisms closely parallel those of bacteria, a point of general biological importance would have been established, but there are certainly interesting differences that should be investigated. In some respects blue-green algae appear to possess different biochemical machinery to the bacteria for there are indications that they lack some of the mechanisms of control of enzyme synthesis possessed by bacteria. The photosynthetic mechanism of blue-green algae differs distinctly from that of bacteria both in the chemical nature of the pigments carrying out the photochemical reactions and in the involvement of oxygen liberation from water. Here, in fact, we have the type of photosynthetic machinery characteristic of the higher plants present in prokaryotic cells.

Another point is that the greater extent of cellular differentiation and morphological elaboration exhibited by the blue-green algae offers the possibility of extending molecular biology into fields which could not be tackled using bacteria as experimental material. Two kinds of differentiated cell are produced by blue-green algae: spores (or akinetes) and heterocysts. The former, although different from bacterial spores in their method of formation and in appearance, have, like them, the function of tiding the species over periods of adverse conditions. The function of heterocysts, empty-looking thick-walled cells, has for long baffled phycologists and earned these structures the title of "botanical enigmas" (Fritsch, 1951). The recent discovery of the specialized biochemical activities of the heterocyst indicates their probable function as that of nitrogen fixation but they still present many problems. Whatever their significance, their regular positioning in the filament seems to present a simple linear morphogenetic situation particularly suitable for experimental investigation.

The evolutionary potentialities of the prokaryotic cell are evidently strictly limited as compared with those of eukaryotes. This perhaps depends both on a limited capacity for genetic recombination and on the restrictions on specialization and organization of cells imposed by the non-compartmentalized type of cell structure. The blue-green algae include unicellular, colonial and filamentous forms and the most advanced type of morphological structure attained is the multiseriate branched filament. Within this range there has been evolution parallel to that in the eukaryotic algal classes. The broad classification of these forms into the five orders Chroococcales, Chamaesiphonales, Pleurocapsales, Nostocales and Stigonematales suggested by Fritsch (1942) is generally accepted although it is becoming apparent that at least the first of these is polyphyletic (Kenyon and Stanier, 1970). Stanier *et al.* (1971) have established on the basis of cell physiology, fatty acid composition and DNA base composition, that several distinct and widely separated groups can be discerned in the Chroococcales. They have demonstrated that it is possible to make such taxonomic distinctions only by investigations with organisms in pure culture, a statement which will no doubt prove to be also true of the other major groups of blue-green algae. It is not proposed to discuss the taxonomy of the blue-green algae in any detail in this book but the comment must be made that at the levels of genus and species the situation at present is chaotic. In taxonomic works such as those of Geitler (1932) and Desikachary (1959) distinction between species is frequently made on the basis of characters, such as trichome diameter and the form of the sheath, that may vary according to environmental conditions. Drouet has made a laudable attempt to reduce the number of species and systematize their classification into genera but has taken little account of experimental studies; perhaps, in reducing more than 1000 specific and subspecific taxa of the coccoid families to 30 (Drouet and Daily, 1956), he has gone too far in the direction of simplification. The taxonomic treatment of the Chroococcales by Stanier *et al.* (1971) provides a model for work on other orders of blue-green algae. No major taxonomic treatment has yet taken account of the fact that the presence or absence of heterocysts, which has often been used as an important diagnostic feature in filamentous forms, depends on the concentration of combined nitrogen in the medium as well as on the genotype and in turn has far-reaching repercussions on the general morphology of the alga. Perhaps the numerical taxonomy approach advocated by Whitton (1969) will ultimately prove to be one solution to these various problems. Meanwhile the nomenclature used by the individual authors whose papers are quoted will be followed in this book.

The ability of the blue-green algae to fix the free nitrogen of the atmosphere has attracted the attention of ecologists, physiologists and biochemists. At present it appears that this remarkable biochemical process is confined to Prokaryota. Towards the end of the nineteenth century, at a time when identification of the principal nitrogen-fixing agents was beginning, some microbiologists were convinced that blue-green algae were particularly important. Frank (1889) found nitrogen fixation associated with the growth of these algae on soil and Prantl in the same year reported on nitrogen fixation in cultures of *Nostoc*, but their experiments provided no proof that these, and not accompanying fungi or bacteria, were the organisms responsible. Three years later Schloesing and Laurent (1892) reported that two *Nostoc* species fixed nitrogen in impure culture whereas another blue-green alga, *Microcoleus vaginatus* did not. Beijerinck (1901) obtained copious growth of certain species of blue-green algae in media containing no source of combined nitrogen but, being unsuccessful in isolating the algae in pure culture, was unable to substantiate his conviction that they fixed free nitrogen. It was incidentally in the course of this work that he characterized the well known nitrogen-fixing bacterial genus *Azotobacter*. Axenic cultures of blue-green algae were eventually obtained by Pringsheim (1914) but he and his pupils Glade (1914) and Maertens (1914) failed to find evidence of nitrogen fixation by the species isolated. This finding is somewhat inexplicable since some of the species belonged to genera which we now know to be characterized by this property. It led to the belief that blue-green algae are not themselves capable of nitrogen fixation and that their ability to grow in media deficient in combined nitrogen depended on symbiotic association with nitrogen-fixing bacteria (see, for example, Jones, 1930). Conclusive proof that some species of blue-green algae are themselves able to fix nitrogen was provided by Drewes in 1928 but by then *Azotobacter*, *Clostridium* and the legume nodule symbiotic association had become established as the "classical" nitrogen-fixing systems and interest of biochemists and agriculturalists in the algae was slight. It was not realized that in combining capacities for photosynthesis, nitrogen fixation and growth under aerobic conditions the blue-green algae have a unique potential for contributing to productivity in a variety of agricultural and ecological situations and for many years their importance was generally discounted. Following the work of De (1939), who showed nitrogen-fixing blue-green algae to be abundant and widely distributed in Indian paddy fields, it became apparent that they could be of value in contributing to fertility in certain specialized tropical habitats. Only recently, when the development of techniques for *in situ* determinations has led to the demonstration that blue-green algae are the principal nitrogen-fixing agents in a variety of habitats and often surprisingly active in this respect, has their more general importance been recognized. At the same time work in the laboratory has revealed remarkable interrelations between nitrogen fixation, photosynthesis and the oxygen content of the environment so that a marked increase of interest in nitrogen-fixing algae as objects for physiological and biochemical research has resulted.

Blue-green algae have a world-wide distribution and the majority of species are cosmopolitan. *Nostoc commune*, for example, is a terrestrial species which is abundant alike in temperate, tropical and polar regions, on isolated islands as well as continents (Geitler, 1932). It is described as common in Britain (West and Fritsch, 1927), the Congo (Duvigneaud and Symoens, 1948), the Western Great Lakes area, United States (Prescott, 1951), India (Desikachary, 1959), Puerto Rico (Almodovar, 1963), and Russia (Shtina, 1964a), and is abundant on both Signy Island in the South Orkney group, Antarctica (Fogg and Stewart, 1968), and the tropical coral atoll of Aldabra in the Indian Ocean (Whitton, 1971). Such world-wide distribution is presumably related both to the antiquity of the group and to the high degree of resistance which most of its members show towards adverse conditions. A unicellular blue-green alga, *Gloeocapsa* sp., has been found to occur regularly in the air over England and although the concentrations recorded are low in comparison with those that are normal for pollen and fungal spores they indicate that air dispersal may be important for such forms (Gregory *et al.*, 1955). More recently Schlichting (1969) has found as many as 55 different species of blue-green algae in air in various parts of the United States. Many blue-green algae show great resistance to desiccation and high temperatures, surviving in air-dry conditions for as long as 65–107 years (see Fogg, 1969a), so that their dissemination in air-borne dust is likely. Resistance to desiccation is, however, correlated with dryness of the habitat in which the species grow (Hess, 1962) so that aquatic forms may not survive this type of transport. Blue-green algae, however, are readily dispersed by external carriage on a variety of aquatic insects, which are considered to be more effective dispersal agents than waterfowl. *Nostoc* sp. has been found in a viable condition in the hind gut of dragon flies (Stewart and Schlichting, 1966). Whatever the mechanism of dispersal may be, there is no doubt that blue-green algae are among the first organisms to establish themselves on new ground. Behre and Schwabe (1970) found seven species in the pioneer flora which had established itself by 1968 on the island of Surtsey, formed off Iceland by volcanic eruption in 1963.

In the range of habitats which they occupy the blue-green algae are rivalled only by the bacteria. On the one hand they predominate in the flora of arid deserts and are abundant on rocks and buildings exposed to the full intensity of tropical sunlight whilst, on the other, different species show massive growth in temperate fresh-water lakes. Blue-green algae are the dominant plant life in the frigid lakes of the Antarctic and, together with other prokaryotes and an anomalous eukaryotic alga, *Cyanidium caldarium*, are the only organisms which grow in hot springs of temperatures above 51°C. They are abundant on rocky seashores and some species grow in strong brine. They are generally common in soil. Only three principal limitations are apparent. Many species are obligately phototrophic, so that although some may be found in caves or at depth in soil, they usually grow only in lighted habitats. Secondly, although again there are exceptions, most blue-green algae prefer neutral or alkaline conditions. Thirdly, except for *Trichodesmium*, which often forms massive blooms in tropical seas, blue-green algae are generally absent from the open sea. Why this should be so is difficult to explain at present. The ability to withstand extreme conditions and the inability to grow well under acid conditions, both of which properties the blue-green algae share with most bacteria, evidently depend on some basic feature of the prokaryotic cell.

Blue-green algae form a substantial fraction of the biomass in several important types of habitat and from this point of view alone it is desirable that we should understand their activities if natural resources are to be conserved and used to best advantage. In addition, these organisms are of direct practical importance in several respects. Evidence is mounting that nitrogen-fixing blue-green algae make a major contribution to the fertility of paddy fields. Since in many eastern countries peasant farmers do not fertilize their fields in any way, it appears that blue-green algae may often permit a moderate rice harvest to be gathered when in their absence there would be only a poor one. Indeed, it does not seem unreasonable to suppose that many millions of people survive largely because of nitrogen fixation by blue-green algae. In underdeveloped countries an agent which fixes nitrogen *in situ* has many advantages over a nitrogenous fertilizer which must be transported and so research with the object of increasing the effectiveness of algal fixation is of considerable economic importance. Additionally, blue-green algae are of importance in reducing soil erosion, increasing the organic content of soils, and perhaps in producing growth factors for higher plants. The possibility of using blue-green algae directly as human food is an interesting one. The Chinese consider *Nostoc commune* as a delicacy and *Aphanothece sacrum*, *Nostoc verrucosum*, *N. commune* and *Brachytrichia quoyi* are listed by Watanabe (1970) as having been used as side dishes from ancient times in Japan. However, in certain parts of the world blue-green algae have been, or are, a staple article of diet. Bernal Diaz del Castillo who accompanied Cortez to Mexico described in 1521 how people living in the area where Mexico City now is ". . . sell some small cakes made from a sort of ooze which they get out of the great lake, which curdles and from this they make bread having a flavour something like cheese". This was probably a blue-green alga, perhaps *Spirulina* or *Arthrospira platensis*. In present times a related form is collected and eaten in the vicinity of Lake Chad (Pirie, 1969). The mass culture of such species, which is currently under investigation, may provide an economic means of producing a food supplement with high protein content (Marty and Busson, 1970). It is also suggested that blue-green algae may be a valuable supplement in the modern diet on account of the relatively high concentrations in them of the medicinally valuable γ-linolenic acid (Watanabe, 1970). *Anabaena cylindrica* is one of the richest known sources of vitamin B_{12} (Brown *et al.*, 1956).

It is paradoxical that while the underdeveloped countries need to encourage the growth of blue-green algae, the countries of the West are seeking means to control their growth. Blue-green algae are a nuisance in fresh waters in many ways. In reservoirs they clog filters and impart musty tastes to drinking water. By making swimming unpleasant and by producing offensive smells when they decay they detract from the amenity of lakes used for recreation. Decay may lead to deoxygenation of water and death of fish but some species also produce toxic substances which may kill water-fowl and cattle. Dense blooms of these nuisance algae are liable to occur in waters polluted with sewage effluents or receiving drainage from agricultural land and present an increasing problem. No satisfactory means of prevention is yet available so that study of the nutrition and growth of planktonic blue-green algae is a prerequisite for any programme of eutrophication control.

Thus the blue-green algae, which were probably predominant among the aboriginal organisms on the earth, are still quantitatively and qualitatively important in the biosphere. Study of these remarkable organisms is not only of particular intrinsic interest for biology but is also of economic importance.

Chapter 2

General features of form and structure

The morphologically simplest blue-green algae are unicellular and the most complex are branched multiseriate filaments, although even these show no more than two major types of differentiated cell. In this chapter the chief features of the group as seen under the light microscope will be considered and their taxonomic value discussed. Their appearance under the electron microscope is dealt with in Chapter 3 .

The light microscope shows a typical blue-green algal cell with a definite shape, surrounded by a firm cell wall and frequently embedded in a massive mucilaginous sheath (Geitler, 1936; Fritsch, 1945). The protoplast appears to be rather viscous, often packed with highly refractive granules of diverse size and differentiated into two main regions, a peripheral pigmented region called the chromatoplasm and a central colourless region termed the centroplasm, without a sharp boundary between them. The chromatoplasm contains the photosynthetic pigments whereas the centroplasm gives a positive reaction with the Feulgen reagent, indicating the presence of deoxyribosenucleic acids (DNA). Various more-or-less specific cytochemical reactions have been used to distinguish the many granular inclusions and to demonstrate the presence of starch-like polysaccharides, fat droplets, proteinaceous cyanophycin granules and metachromatic or volutin granules.

The sheath

In blue-green algae the sheath surrounds the colonies, unicells or filaments in a rather similar way to the capsular material of bacteria. It is often extensive in natural populations and may be seen readily by phase contrast microscopy or on negative staining with Indian ink (Fig. 2.1). Details of the chemistry of the sheath are given in Chapter 3 (p. 74).

Active growth seems necessary for sheath formation, a fact which may explain its sometimes poor development round spores and heterocysts (Fig. 2.1). It usually develops first as a colourless diffluent jelly and in

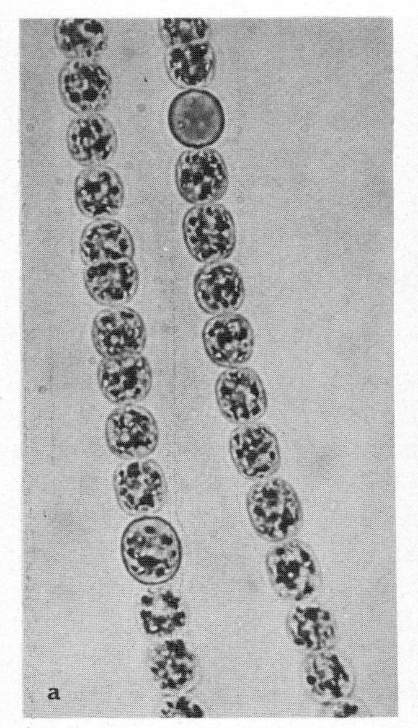

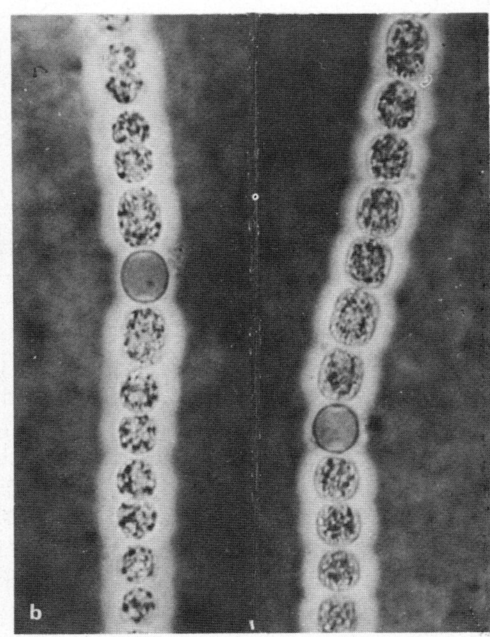

FIG. 2.1. Micrographs of *Anabaena flos-aquae* (×1100). In (b) the filaments are suspended in diluted Indian ink; the ink particles are excluded from the sheath (made up of submicroscopic microfibrils orientated normal to the cell surface) which appears as a halo around the cells. The sheath surrounding the heterocysts is much reduced. Photographs A. E. Walsby.

liquid culture is gradually dispersed in the medium. Sloughed-off sheath material probably accounts for a large proportion of the soluble polysaccharide found in cultures of blue-green algae. In terrestrial algae the sheath often becomes gel-like, solid, or lamellate and is frequently pigmented. The lamellations may be parallel to the main filament, as in *Lyngbya majuscula* (Fig. 2.3c), or divergent, as in *Petalonema* species (Fig. 2.7*a* and Frontispiece) and both types may occur in the same filament as in certain Scytonemataceae. In thickness sheaths vary from being almost undetectable as in *Aphanocapsa* (Fig. 2.2a) to thick and striate in the colonies of *Gloeocapsa* (Fig. 2.2c, e). Similarly in the filamentous algae very thin hyaline sheaths are characteristic of *Oscillatoria* (Fig. 2.3a), while the closely related genus *Phormidium* (Fig. 2.3b) often has diffluent sheaths which aggregate the filaments together to form an expanded substratum. *Lyngbya* (Fig. 2.3c) has firm distinct sheaths.

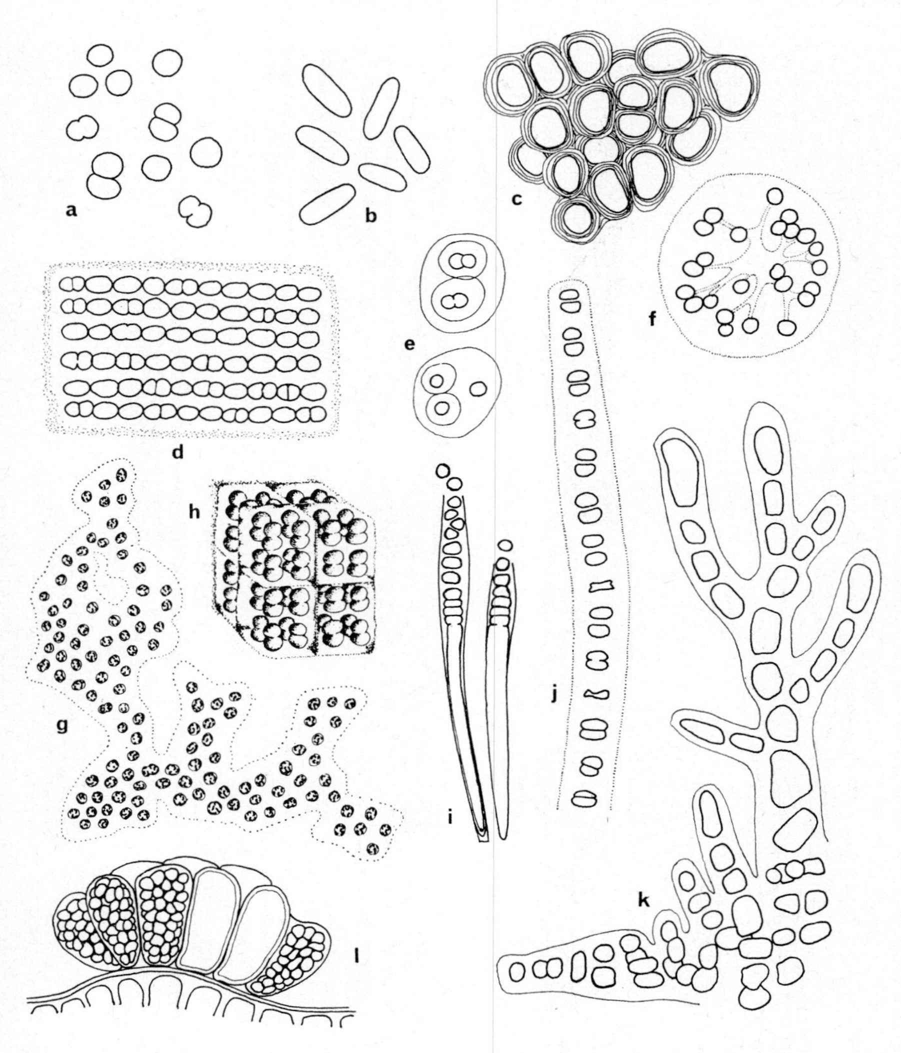

FIG. 2.2. Various unicellular, colonial and pseudo-parenchymatous blue-green algae. a, *Aphanocapsa* sp. (orig.) (×600); b, *Synechococcus* sp. (orig.) (×1000); c, *Gloeocapsa pleurocapsoides* Nck. (after Skuja) (×750); d, *Merismopedia convoluta* Bréb. (after Frémy) (×1000); e, *Gloeocapsa granosa* (Berk.) Kütz. (after Wille) (×400); f, *Gomphosphaeria lacustris* Chodat (after Geitler) (×1000); g. *Microcystis aeruginosa* (orig.) (×400); h, *Eucapsis alpina* Clements et Shantz (after Clements and Shantz) (500); i, *Chamaesiphon curvatus* Nordst. (after Geitler) (×500); j, *Johannesbaptistia pellucida* (Dickie) Taylor et Drouet (after Iyengar and Desikachary) (×750); k, *Hyella balani* (after Frémy) (×1000); l, *Dermocarpa prasina* (Reinsch) Bonn. et Thur. (after Geitler) (×1000).

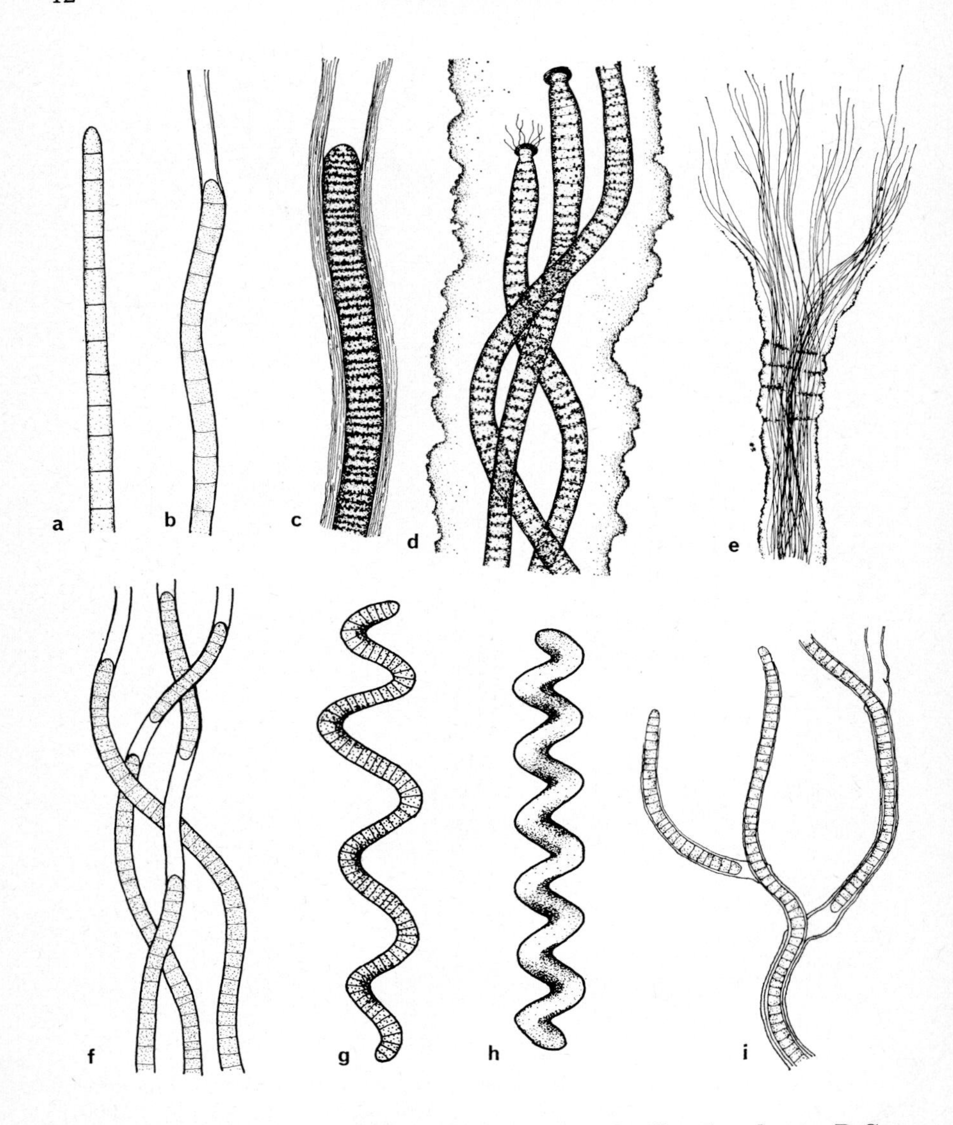

FIG 2.3. Some members of the Oscillatoriaceae. a, *Oscillatoria rubescens* D.C. ex. Gom. (after Desikachary) (×500); b, *Phormidium incrustatum* (Naeg.) Gom. (after Frémy) (×600); c, *Lyngbya majuscula* Harv. (after Frémy) (×170); d, *Hydrocoleum lyngbyaceum* Kütz. (after Frémy) (×300); e, *Sirocoleus kurzii* (Zell.) Gom. (after Frémy) (×200); f, *Symploca muscorum* Gom. (after Frémy) (×400); g, *Arthrospira platensis* v. *non-constricta* (Banerji) comb. nov. (after Banerji) (×600); h, *Spirulina gigantea* Schmidle (after Frémy) (×750); i, *Plectonema radiosum* (Schiederm.) Gom. (after Gom.) (×1300).

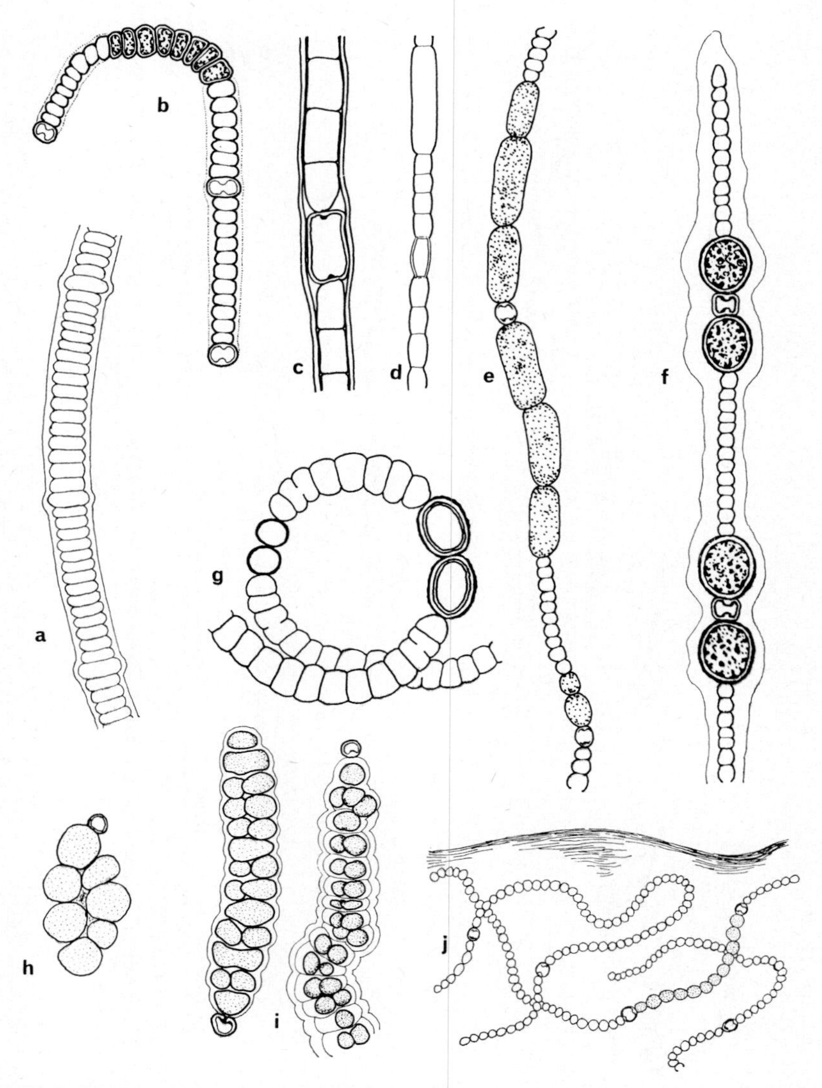

FIG. 2.4. Some members of the Nostocaceae. a, *Nodularia spumigena* Mert. var. *litorea* Born. et Flah. (after Bornet) (× 500); b, *Nodularia sphaerocarpa* Born. et Flah. (after Geitler) (× 500); c, *Aulosira fertilissima* var. *tenuis* Rao (× 400); d, *Aphanizomenon flos-aquae* Ralfs (after Geitler) (× 400); e. *Anabaena oscillarioides* Bory (after Frémy) (× 400); f, *Wollea bharadwajae* Singh (after Singh) (× 500); g, *Anabaenopsis arnoldii* Aptekarj var. *indica* Ramanathan (after Ramanathan) (× 300); h, *Chlorogloea fritschii* Mitra (orig.) (× 850); i, *Nostoc punctiforme* var. *populorum* (after Geitler) (× 1000); j, *Nostoc sphaericum* Vauch. (orig.) (× 300).

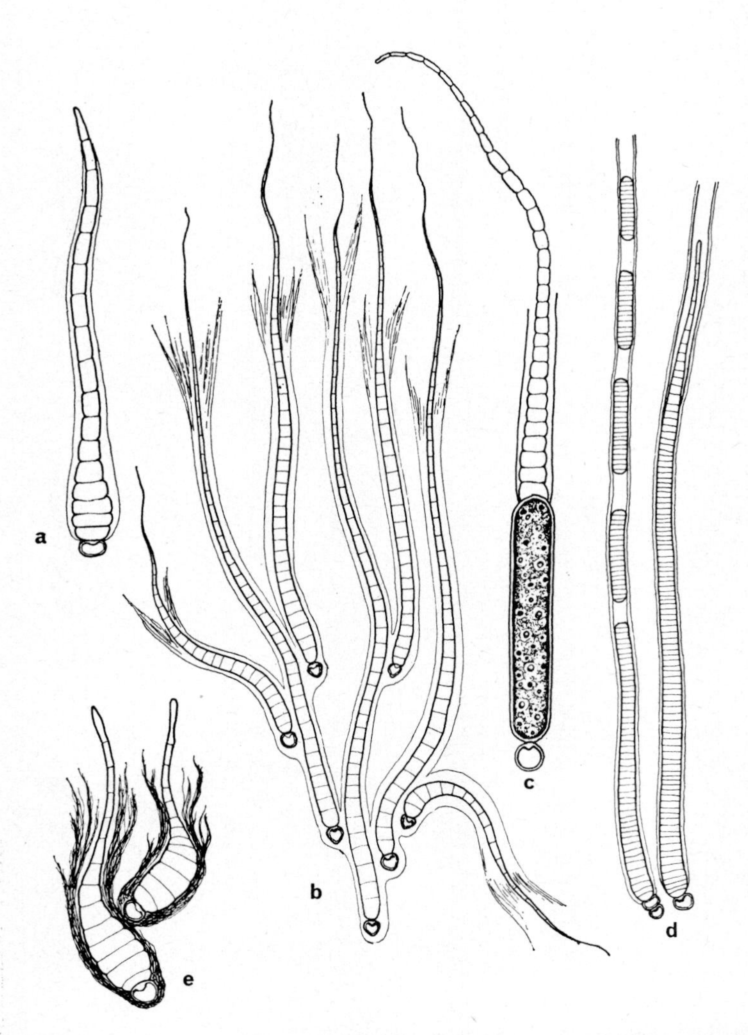

FIG. 2.5. Some Rivulariaceae. a, *Calothrix clavata* G. S. West (after Frémy) (×1000) ; b, *Rivularia nitida* Ag. (after Frémy) (×1000); c, *Gloeotrichia pisum* Thur. (after Frémy) (×900); d, *Calothrix confervicola* (Roth) Ag. (after Frémy) (×200); e, *Calothrix parietina* Thur. (after Frémy) (×200).

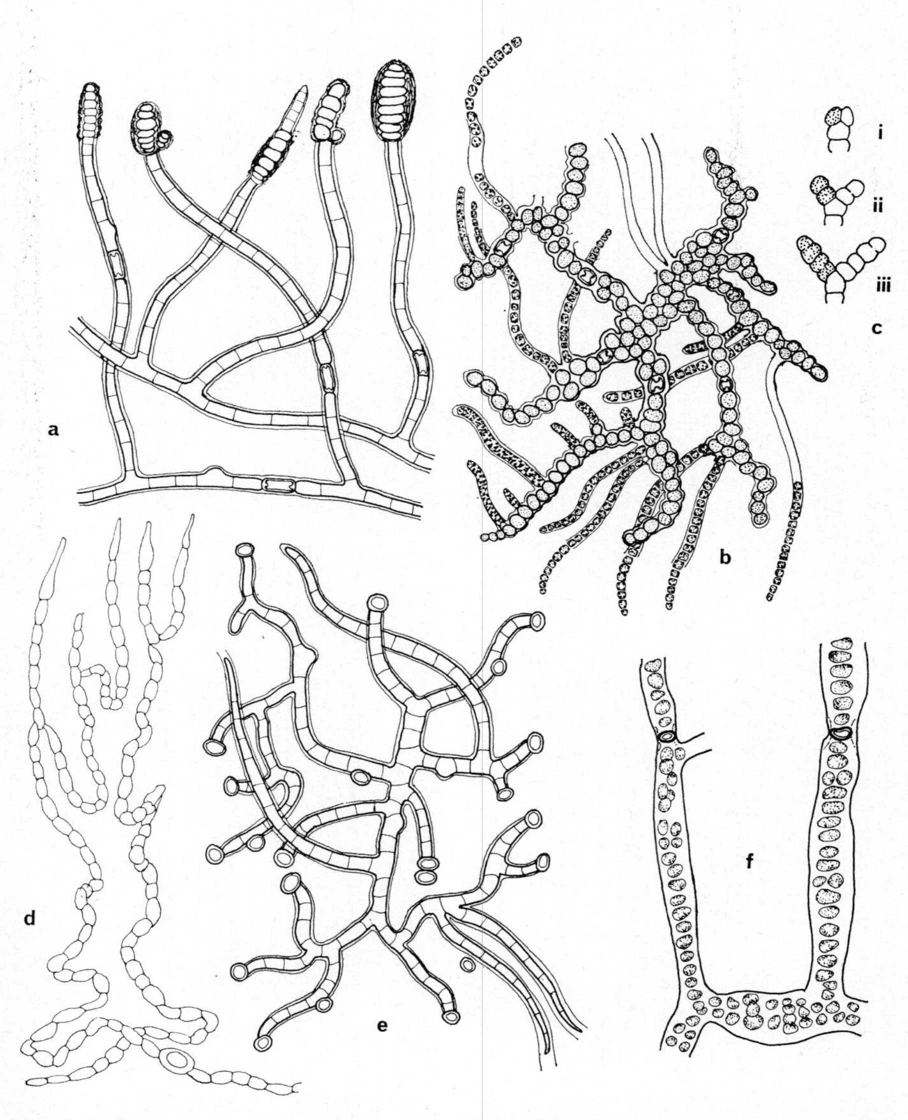

FIG. 2.6. Filamentous blue-green algae showing different types of branching. a, *Westiella lanosa* Frémy (after Frémy) (×150); b, *Fischerella muscicola* (Thur.) Gom. (after Frémy (×200); c, dichotomous branch formation in *Colteronema* sp. (after Frémy) (×300); d, *Brachytrichia balani* Born. et Flah. (after Ercegovic) (×400); e, *Mastigocoleus testarum* Lagerh. (after Frémy) (×200); f, *Stigonema dendroideum* Frémy (after Desikachary) (×130).

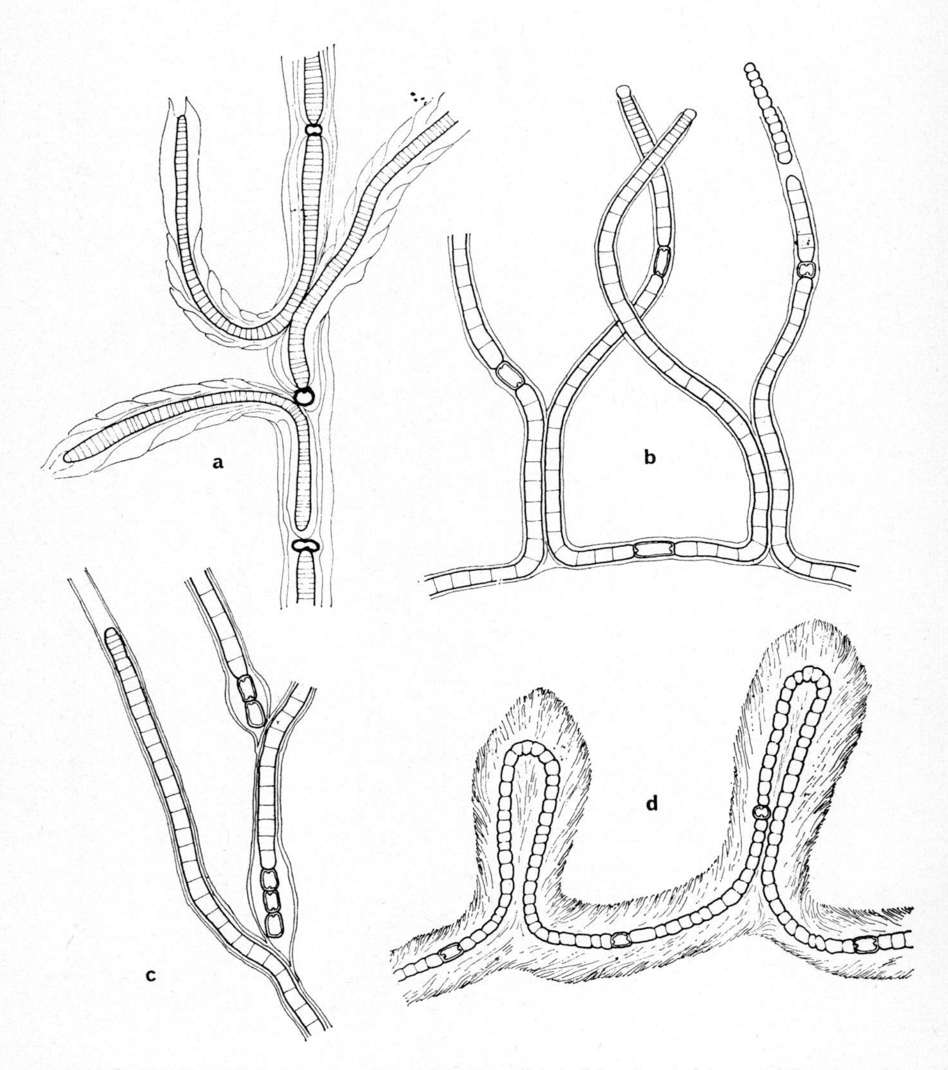

FIG. 2.7. Various Scytonemataceae showing (a–c) false branching. a, *Petalonema densum* (A.Br.) Migula (after Desikachary) (×200); b, *Scytonema hofmanni* Ag. (after Frémy) (×200); c, *Tolypothrix tenuis* Kütz. emend. Schmidt (after Frémy) (×180); d, *Scytonema crustaceum* Ag. var. *incrustans* (Kütz.) Born. et Flah. (after Frémy) (×100).

Sheaths often show brown, blue, red and yellow pigmentation (Frontispiece and Fig. 2.8—facing p. 256) which may be dependent on the environmental conditions under which the algae are growing. Algae often develop pigmented sheaths under high light intensities and colourless sheaths under low light intensities, although some never have coloured sheaths whatever the light intensity. Drouet (1968) who has investigated thousands of specimens concludes that red sheaths are most often found in algae from highly acid soils while blue sheaths are characteristic of algae of basic habitats. He also notes that yellow and brown sheaths are common in specimens from habitats of high salt content, particularly after the algae dry out. We have confirmed this in laboratory studies on *Calothrix scopulorum*. Whether or not the sheath stains with chlor-zinc-iodine solution was used as a taxonomic characteristic by Gomont (1892) and others, but Drouet (1962) found this to be unreliable. For example in studies on *Microcoleus vaginatus* and *M. lyngbyaceous* different results can be obtained in tests on different parts of the same filament let alone on separate specimens. Drouet (1964) also noted that in *Schizothrix arenaria* parts of the sheath infected with fungi stain blue with chlor-zinc-iodine while other parts of the sheath do not.

The trichome

In filamentous algae, the living structure within the sheath is referred to as the trichome. In all genera, including *Spirulina* (Fig. 2.3h), the trichome is divided into well-defined cells by cross walls which develop centripetally from the outer wall of the trichome.

In size the trichomes of filamentous algae vary from less than 1 μm in diameter (as in certain *Lyngbya* species) to more than 20 μm in diameter in others. In algae such as *Oscillatoria* spp. (Fig. 2.3a) there is little difference in width along the length of the filaments but in the genera *Rivularia* (Fig. 2.5b) and *Gloeotrichia* (Fig. 2.5c) the trichomes taper and the terminal cells become depleted of cell contents so that distinct hairs are produced. In certain species of *Scytonema* (Fig. 2.7b) tapering of the trichomes occurs although hairs are not produced and in the Stigonemataceae the trichomes in the main and side branches may differ markedly in width (Fig. 2.6b).

Branching

Branching is a characteristic feature of many species of filamentous heterocystous algae and is of two main types: *true branching*, in which the branch is continuous with the trichome from which it originates, and *false branching* in which the formation of the branch depends on a break in the original filament.

The most usual type of true branching is found in Stigonematales such as *Westiella lanosa* (Fig. 2.6a) where a cell of the main trichome divides longitudinally and by repeated division of one of the newly formed cells in one plane a branch is formed. Two other types of true branching occur. In *Colteronema* branching takes place when the apical cell of a trichome divides longitudinally and both of the cells formed continue to divide equally in a transverse fashion to form a dichotomous system (see Fig. 2.6c). The other type of true branching, found frequently in *Brachytrichia balani* is reverse V-branching, first studied by Bornet and Thuret (1880). Here the filament forms the shape of an inverted V, and from the base of this V a branch is produced (Fig. 2.6d). There are different views on the exact morphological changes which bring this about (see Desikachary, 1959).

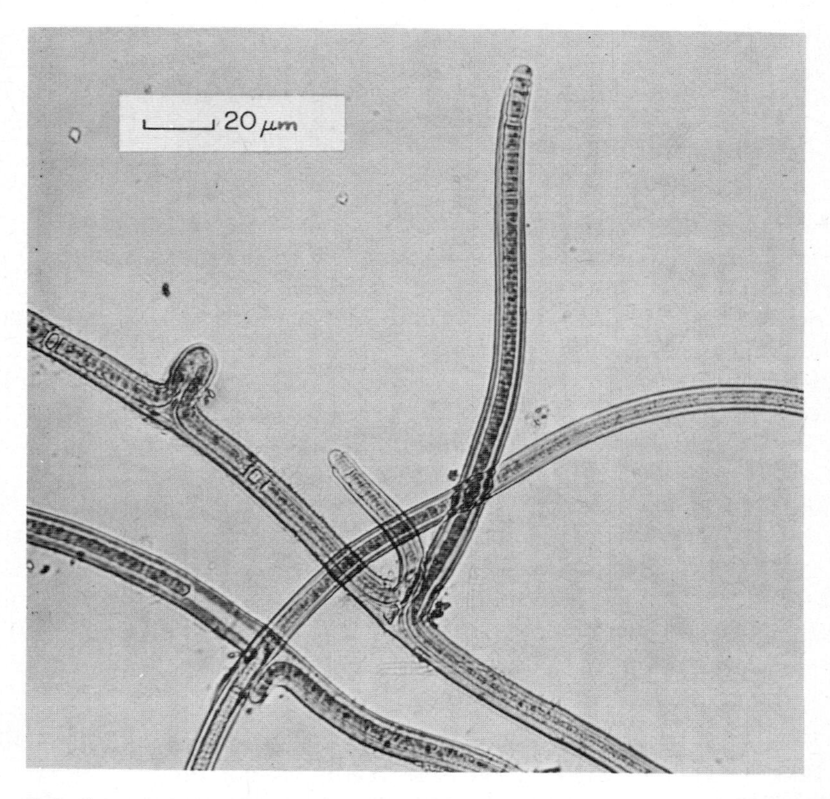

FIG. 2.9. *Scytonema* sp. showing loop formation and false branching. The branches may form by breakage of the loop, with the two ends then growing out separately. Photograph G. E. Fogg.

False branching results either when a cell dies or separation discs are formed in a filament and the part of the filament below continues to divide and breaks through the sheath to form a side-branch. The cell on the other side of the breakage may develop as a heterocyst. This type of branching is characteristic of *Petalonema*, *Tolypothrix* and *Scytonema*. In *Scytonema* in addition to the *Tolypothrix* type, branches may be formed as a result of loop formation (see Figs. 2.7, 2.9) Here the filament divides rapidly in one part and a loop is formed which breaks through the sheath. This then breaks in the middle and two false branches are formed. False and true branching may occur in the same filament as happens in *Hapalosiphon*.

The physiological and biochemical factors governing branching in blue-green algae require investigation. Branches are positively phototactic and their formation is regulated by the level of combined nitrogen in the medium; levels high enough to inhibit heterocyst formation also inhibit false branching. Branching may also be induced artificially by ultraviolet irradiation (Singh and Tiwari, 1969).

Heterocysts

Heterocysts are cells, with thick hyaline walls, characteristic of filamentous algae belonging to the orders, Nostocales (except the Oscillatoriaceae) and Stigonematales (see p. 29). They are derived from ordinary vegetative cells which usually (*Anabaena*, *Rivularia*, *Calothrix*), but not always (*Chlorogloea fritschii*, see footnote* and p. 250), become larger than adjacent cells. Under the light microscope they appear to have homogeneous cell contents but a complex membraneous system is revealed under the electron microscope (p. 222). They are usually yellow-green in colour due to the presence of chlorophyll *a* and carotenoid pigments and the loss of phycocyanin (see p. 228). In shape they generally resemble the cells from which they are derived and vary from 3–12 times longer than broad, as in *Anabaenopsis*, to 2–3 times shorter than broad, as in *Nodularia* (Fig. 2.4a and b).

In position, heterocysts may be terminal or intercalary. Intercalary heterocysts such as those found in *Anabaena* and *Aphanizomenon* have a

* *Chlorogloea fritschii* was isolated from Indian soil and described as *species nova* by Mitra in 1950 who referred it to the family Entophysalidaceae, order Chroococcales. However, it was later observed that the alga produces short filaments at an early stage in its life cycle and forms heterocysts when grown in a medium free of combined nitrogen (Fay *et al.*, 1964). It was therefore considered by Dr. J. W. G. Lund to be an anomalous species of the genus *Nostoc*. This assumption was confirmed by Schwabe and El Ayouty (1966) who renamed the alga *Nostoc fritschii* (Mitra) Schwabe (*combinatio nova*). However, because its growth is not like that of a *Nostoc*, Stanier, *et al.* (1971) consider that it is best placed in the genus *Chlorogloeopsis* Mitra *et* Pandey.

pore at each end through which there appears to be a protoplasmic connexion with the vegetative cell on either side. Terminal heterocysts occur at the ends of filaments. In *Mastigocoleus* they are formed terminally on short branches (Fig. 2.6e) and in *Gloeotrichia* they occur at the bases of the filaments (Fig. 2.5c). Terminal heterocysts have one pore only and what appear to be two-pored terminal heterocysts are probably intercalary cells from which the filament has become detached on one side. Three-pored heterocysts sometimes occur in *Brachytrichia balani* (Iyengar and Desikachary, 1953) and *Mastigocladus laminosus* (Venkataraman, 1957), each pore connecting with the basal cell of a vegetative branch. Recently Singh and Tiwari (1969) have shown that mutants of *Nostoc linckia* obtained by ultraviolet irradiation may also produce 3-pored heterocysts, each pore subtending a branch as in *Brachytrichia* and *Mastigocladus* (Fig. 2.10).

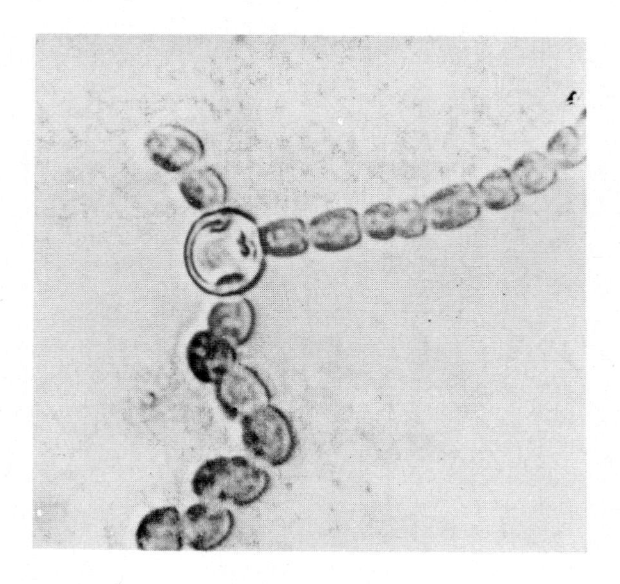

FIG. 2.10. Three-pored heterocyst in *Nostoc linckia* (×1000). By courtesy of R. N. Singh and D. N. Tiwari (1969) *Nature, Lond.* **221**, 64, Fig. 5.

Heterocysts usually occur singly at fairly regular intervals along the filament or in particular positions such as at the base of filaments in the Rivulariaceae. However, paired heterocysts are characteristic of *Anabaenopsis* and Singh and Tiwari (1969) have described *Nostoc* mutants in which chains of up to eight heterocysts regularly occur (Fig. 2.11).

The spacing of the heterocysts along algal filaments is regulated by environmental factors and is fairly constant under any one set of conditions.

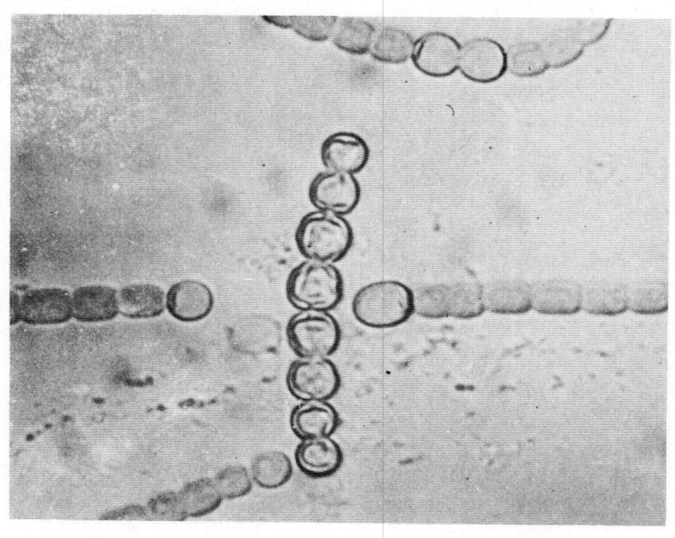

FIG. 2.11. A chain of eight heterocysts in a mutant of *Nostoc linckia* (×700). By courtesy of R. N. Singh and D. N. Tiwari (1969) *Nature, Lond.* **221**, 64, Fig. 4.

In algae such as *Anabaena*, *Nodularia* and *Nostoc*, heterocysts are dispersed regularly along the filament with new heterocysts forming exactly, or almost exactly, equidistantly between existing heterocysts. This is shown by the studies of Fritsch (1951) on *Anabaenopsis circularis* and of Singh and Srivastava (1968) on *Anabaena doliolum*. In *Anabaenopsis circularis* the mature filament consist of eight vegetative cells with a heterocyst at either end. The two cells in the centre of the filament then divide unequally and two proheterocysts (see p. 235) are formed side by side. The original filament breaks between these and each proheterocyst then develops into a mature heterocyst. As it does, so cell division of the vegetative cells between the two heterocysts occurs until a total of eight vegetative cells is formed. Thus, each new filament contains an old heterocyst at one end and a new heterocyst at the other. In *A. doliolum* the first heterocyst develops at the four-celled stage and as the filament divides heterocysts are developed 4, 10, 16 and 23 cells apart until eventually the number of vegetative cells between the heterocysts stabilizes at about 24.

The finding that the distribution of heterocysts along a filament can be regulated by environmental factors (see chapter 11) poses the question of how reliable the use of heterocysts is as a taxonomic character. Bornet and Flahault (1886–1888) used the presence or absence and position of heterocysts as a major criterion for separating both genera and species. On this basis a *Gloeotrichia* filament grown on high levels of combined nitrogen inhibiting the development of its heterocysts, would probably be put into the non-heterocystous genera *Homeothrix* or *Leptochaete*. These resemble *Gloeotrichia* filaments but have no basal heterocysts. The taxonomic significance of intercalary heterocysts is even more questionable; these can be eliminated completely in *Calothrix* grown on high levels of ammonium nitrogen. The absence of heterocysts prevents branching and the presence or absence of branches in *Calothrix* was regarded as a taxonomic character by Bornet and Flahault (1886–1888). The general conclusion from studies on *Tolypothrix tenuis* (Stein, 1963), *Calothrix scopulorum* (Stewart, unpublished), *C. membranacea* (Pearson and Kingsbury, 1966) and *Gloeotrichia* (Walsby, unpublished) is that heterocysts alone cannot be regarded as useful taxonomic criteria on which to separate genera or even species. This view is implicit in the work of Fan (1956) who, in a taxonomic revision of the genus *Calothrix* based mainly on herbarium material, reduced the twenty three species recognized by Bornet and Flahault to six. He did not consider the additional species described by workers after Bornet and Flahault.

THE ROLE OF THE HETEROCYST

Despite much effort the problem of the functions of heterocysts has not been resolved entirely and controversy continues. It is probable that heterocysts play several roles and, as emphasized by Stewart (1972b), "It would be unwise to treat in isolation the variety of hypotheses which have been forthcoming on their possible function, or to consider that their function in one particular metabolic process rules out their participation in another". Earlier views on the possible function of heterocysts have been considered by Fritsch (1951) and Wolk (1965b). Certain aspects of the problem will be discussed in detail in other chapters but it is convenient to consider some possible functions at this point.

(a) *Heterocysts and vegetative reproduction*

An early suggestion (Borzi, 1878) was that heterocysts promote vegetative reproduction. While it is true that heterocyst senescence promotes filament fragmentation, it is difficult to accept that a highly specialized structure such as the heterocyst should merely serve as a breaking point in vegetative reproduction (Fritsch, 1951). In addition, filaments of algae such as *Oscillatoria* fragment readily but these are without heterocysts. Fragmentation may in fact occur by the death of any intercalary vegetative cell during hormogonium production (see below p. 26) and the hypothesis provides no explanation for the function of the terminal heterocysts of *Cylindrospermum* and *Gloeotrichia* (Tischer, 1957).

(b) *Heterocysts as reproductive units*

Geitler (1921) first observed germination of heterocysts of blue-green algae and suggested that they may be archaic reproductive structures. Fritsch (1951) rejected the theory that heterocysts play an important role as reproductive structures on the basis that germinating heterocysts are seldom found in nature and that it would be unusual for an organism to develop two independent perennating structures in the form of akinetes and heterocysts. Heterocysts form most abundantly during active growth (Fogg, 1949), in a period when spore formation and germination is not only unusual but without apparent advantage to the species. However, Kale and Talpasayi (1969) record heterocyst germination in at least seven genera: viz. *Anabaena*, *Brachytrichia*, *Calothrix*, *Gloeotrichia*, *Nostoc*, *Rivularia* and *Tolypothrix*. Wolk (1965a), using *Anabaena cylindrica*, found that on the addition of 4 m mol l^{-1} of ammonium chloride and 50 m moles l^{-1} of glucose, 3 to 10% of the heterocysts germinated. More recently Singh and Tiwari (1970) reported germination of up to 83% of the heterocysts of what they considered to be a non-sporing mutant of *Gloeotrichia ghosei* when it was supplied with 0·001 M ammonium chloride. Germination was rare or absent in medium free from combined nitrogen ("nitrogen-free") and was uncommon in the presence of nitrate–nitrogen (0·005 M). These heterocysts germinated *in situ*—an unusual finding! Germination was preceded by an increase in size and then the protoplast divided into two, one new cell of which then developed as a heterocyst while the other continued to divide, giving rise eventually to a new filament. These workers, who also report 24% germination of the heterocysts of a mutant of *Nostoc linckia*, consider that in *Gloeotrichia* the heterocyst takes over the function of reproduction. Whether the ability of heterocysts to germinate is of any advantage in nature is doubtful because the levels of combined nitrogen necessary to induce germination are seldom, if ever, found there. Furthermore, heterocysts do not possess the typical features of spores. They lack reserve substances and are not completely enclosed by a protective coat. The septum between heterocyst and adjacent vegetative cell is a fragile structure, and when heterocysts are detached from the filament it breaks open (Fay and Lang, 1971). The contents of detached heterocysts rapidly disintegrate.

(c) *Heterocysts as organs of attachment*

De Puymaly (1957) suggested that a heterocyst may play a role as an attachment organ which anchors the filament to the substratum in a rather similar fashion to the rhizoids of other algae. This hypothesis has been developed by Allsopp (1968) who, from studies on *Scytonema javanicum*, agreed that it acted as an attachment organ but not in the same way as originally proposed by de Puymaly. He found that, on germination, the hormocysts of *Scytonema* which are formed by encystment of hormogonia (see p. 26) develop a median heterocyst. This acts as a means of attaching the germinating trichome to the sheath, which in turn remains attached to the substratum. The advantage of such a mechanism is questionable because algae without heterocysts are still capable of attaching themselves readily to the substratum. Also many planktonic algae, which do not require attachment, possess heterocysts.

(d) *Heterocysts as regulators of spore formation*

Carter (1856) was the first to suggest that heterocysts may function in the control of sporulation in blue-green algae and this hypothesis has been considered subsequently by Fritsch (1904, 1951), Geitler (1921) and Wolk (1966). The suggestion is based on two lines of evidence. Firstly, akinetes are found only in heterocystous algae and usually develop adjacent to the heterocysts, on either one or both sides (Fig. 2.4 e, f). In a few genera, such as *Nodularia*, they develop remote from the heterocysts (Fig. 2.4b). Secondly, Wolk (1966) observed, using *Anabaena cylindrica*, that if heterocysts are gently detached from the vegetative filaments, the cells adjacent to the heterocysts, which would normally develop into akinetes, no longer do so. Caution must be exercised in the interpretation of these findings because it is likely that detachment of the heterocysts results in some damage to the vegetative cells to which they were connected. Nevertheless, the result provides what is probably the most convincing evidence in support of the hypothesis of heterocysts as regulators of akinete formation. Fritsch (1951) suggested that this control may be exerted by the secretion of substances which stimulate akinete formation under certain conditions. The active substance could perhaps be ammonium–nitrogen or some other nitrogenous compound produced in the heterocysts.

The view that heterocysts regulate spore formation has not been accepted by all workers. Singh and Srivastava (1968) found that in *Anabaena doliolum* all vegetative cells are capable of developing into heterocysts and that nitrite–nitrogen inhibits sporulation without affecting heterocyst formation. They conclude therefore that heterocyst development and sporulation are probably independent morphogenetic effects. Wolk (1965b) has shown that spore development is affected by a variety of factors in addition to the presence of heterocysts (see p. 248).

(e) *Heterocysts as sites of nitrogen fixation*

This hypothesis was proposed by Fay *et al.* (1968) after consideration of a variety of diverse and at first sight unrelated morphological and physiological data. There can now be little doubt that heterocysts are sites of nitrogen fixation (see p. 190) and if we look on them as centres producing the combined nitrogen on which other cells are dependent, a ready explanation for many features of the morphology of blue-green algae is provided. The regulatory function of heterocysts in akinete formation already referred to is a possible example. Another is the polarity of the filament of *Gloeotrichia* and similar forms, in which the akinete, if produced, occurs next to the terminal heterocyst with a zone of actively dividing cells next, grading into narrow inactive hair cells. If an ammonium salt is supplied to *Gloeotrichia* the whip-like form is lost and the filaments appear uniform throughout (Fay *et al.*, 1968). Even in forms with uniformly spaced heterocysts such as *Anabaena* spp. there may be a detectable gradient of cell size, with minimum cell diameter mid-way between two heterocysts (Troitzkaya, 1924). The early appearance of heterocysts in akinete germlings of *Anabaena* spp. is also explicable if these structures are essential to provide combined nitrogen for growth.

Reproductive structures

Increase in cell number in the blue-green algae is usually by binary fission, but in addition there are cells which are involved specifically with reproduction and perennation.

1. AKINETES

In algae such as the Nostocaceae, Rivulariaceae and in some Stigonemataceae, for example, *Hapalosiphon* and *Fischerella*, vegetative cells may develop into perennating structures called spores or akinetes. Mature akinetes are usually much larger than ordinary vegetative cells and range in shape from spherical to several times longer than broad (Fig. 2.4 e, f). They have a thick wall which is often brown or yellow and is sometimes elaborate and sculptured as in *Cylindrospermum alatum*. The protoplasm has reduced amounts of photosynthetic pigments but accumulates large quantities of cyanophycin granules, which are proteinaceous reserves. The protoplasm of *Gloeotrichia* spores is banded because of the formation of incomplete septa within the cell.

The distribution of akinetes along the algal filament varies according to species. In algae such as *Gloeotrichia* akinetes always develop adjacent to the basal heterocysts (Fig. 2.5c) whereas in members of the Nostocaceae they may occur singly or in chains which originate next to (*Anabaena oscillarioides*, Fig. 2.4e) or remote from (*Nodularia sphaerocarpa*, Fig. 2.4b) the heterocysts. In old filaments of some Nostocaceae all the vegetative cells may develop into spores.

Under favourable environmental conditions germination of spores is usually rapid, although not all germinate in a similar fashion. The commonest type of germination is that in which part of the akinete wall gelatinizes, the new filament growing out through the gelatinized area, as in *Anabaena cylindrica* (Fig. 11.36). Sometimes, as in *Nostoc ellipsosporum*, the entire akinete coat gelatinizes. Germination may start within the spore before the cell contents are extruded. The contents of the spore develop either into a hormogonium (see below) or sometimes directly into a trichome. Heterocysts differentiate in both at an early stage.

Akinetes have been reported in a few non-filamentous blue-green algae but it is doubtful whether these structures are strictly comparable with those just described (see Fritsch, 1945).

2. HORMOGONIA AND HORMOCYSTS

Hormogonia (or hormogones) are characteristic of all truly filamentous blue-green algae and are short pieces of trichome which become detached from the parent filament and move away with a gliding motion, eventually developing into a separate filament. In simple filamentous algae such as *Oscillatoria* and *Cylindrospermum* the entire filament may break up, but in others such as *Fischerellopsis* hormogonia are produced at the tips of special branches. In *Rivularia* they are produced in an intercalary position, being released when the apical hair breaks off. Hormogone production in *Calothrix* is shown in Fig. 2.5d.

In some algae such as *Nostoc* species hormogonia are formed by direct fragmentation of the trichomes and rounding off of the end cells of the fragments. In other algae specialized separation discs, or necridia are involved in hormogone production. These are distinct cells which become biconcave in shape due to pressure from adjacent cells. The morphological changes are associated with lysis and dehydration of the discs, which first appear darker but become pale and colourless as lysis proceeds. The trichomes break at these points of lysis and hormogonia are produced (Kohl, 1903; Lotsy, 1907; Fuhs, 1958d; Lamont, 1969b). The way in which the separation discs and tears develop varies with the species. In *Microcoleus vaginatus* the tear is usually along junctional pores between the membrane of the separation disc and an adjacent cell (see p. 73). In *Oscillatoria chalybea* the tear is usually irregular; in *Schizothrix calcicola* both transcellular and intercellular breakages occur. It has been suggested that these discs may be an important source of algal extracellular products (Goryunova and Rzhanova, 1964) but they are obviously not the only source because algae without separation discs still produce extracellular products. For further information on separation discs the reader should consult Lamont (1969b).

Hormogonia production is not always dependent on the formation of separation discs. We have found in *Tolypothrix* for example that the end cells of developing hormogonia round off without cell lysis occurring.

Although hormogonia are usually liberated from the parent sheath they sometimes germinate *in situ* so that then there are, as in *Microcoleus*, numerous trichomes within the same sheath. Following germination the hormogonia usually develop an intercalary (*Scytonema*) or terminal (*Tolypothrix*) heterocyst at an early stage.

In some species such as *Westiella lanosa* (Fig. 2.6a) specialized structures called hormocysts are formed and are composed of a short row of highly granulated cells completely surrounded by a common condensed sheath. They may be produced terminally or in an intercalary position. They thus appear to be intermediate between hormogonia and akinetes. When they become detached from the main filament by breakage they may germinate at either or both ends (Allsopp, 1968).

3. EXOSPORES, ENDOSPORES AND NANNOCYTES

Several workers have reported on the production of various reproductive structures such as exospores, endospores, nannocytes and planococci in non-filamentous blue-green algae. These have been found in a few genera only and are considered to be units of reproduction rather than perennating structures.

Exospores, reported in *Chamaesiphon* and *Stichosiphon*, are budded off from the free ends of the filaments by transverse division (Fig. 2.2i). Endospores, in contrast to exospores, form by internal division of the protoplast. In species such as *Endoderma* two endospores are produced per cell but in *Dermocarpa* numerous endospores, each with a thin wall, may form in each cell (Fig. 2.2l). These are released by rupture or by gelatinization of the sporangium wall. Nannocytes, which occur in algae such as *Gloeocapsa* and *Microcystis*, are endospores formed by cell division without subsequent cell enlargement. Other reported reproductive structures are planococci. These are unicells exhibiting a slow creeping movement and occur in *Desmosiphon* (Borzi, 1916). Most of the reports of the occurrence of reproductive units in non-filamentous algae have come from early workers (see Geitler, 1932, 1936; Fritsch, 1945) and require further investigation in pure cultures to eliminate the possibility that the structures observed were parasites or other contaminating organisms.

Gross morphological form

Blue-green algae can be separated into unicellular, colonial and filamentous forms. The cells of unicellular algae are usually spherical

(*Chroococcus*), or cylindrical (*Synechococcus*) and each remains separate usually within a well-defined mucilaginous sheath. In others the cells remain aggregated after cell division to form distinct colonies. These may be irregular in shape, as in *Microcystis* (Fig. 2.2g), or symmetrical as in the cube-shaped colonies of *Eucapsis* (Fig. 2.2h), or the plate-like colonies of *Merismopedia* (Fig. 2.2d, Fig. 11.5). In others such as *Gomphosphaeria* the cells are aggregated within a mucilaginous mass which is attached to the substratum by a mucilaginous stalk (Fig. 2.2f). In *Chamaesiphon* the cells divide linearly to give a pseudo-filamentous form (Fig. 2.2i). Pseudo-parenchymatous colonies are represented by *Hyella* (Fig. 2.2k) and *Pleurocapsa* where the cells are aggregated into upright and prostrate systems. There may be division of labour in some pseudoparenchymatous algae in that specialized reproductive cells such as exospores (*Chamaesiphon*), endospores (*Dermocarpa*) or nannocytes (*Gloeocapsa* and *Aphanothece*) have been reported (see above).

Filamentous algae may have evolved from unicellular forms by the products of repeated divisions in one plane remaining in connexion. Drouet (1968) in his reclassification of the non-heterocystous filamentous algae (Oscillatoriaceae) considers that whether this happens or not may sometimes depend on environmental factors and that some unicellular forms and filamentous algae are simply ecophenes (ecological growth forms) of the same species. This view has been challenged in the case of the organism frequently called *Anacystis nidulans* and which Drouet (1968) places in the Oscillatoriaceae but which Allen and Stanier (1968a) consider to be truly unicellular.

The classification of the unicellular blue-green algae, according to Fritsch (1945) is shown in Table 2.I. It is based solely on the morphological characteristics of the group. Stanier *et al.* (1971) who have recently published a comprehensive account of unicellular blue-green algae, conclude, quite rightly, that structural characteristics alone are inadequate. They emphasize that all satisfactory work on the taxonomy and classification of blue-green algae should be based on studies using pure strains, and on physiological and biochemical, as well as morphological characteristics. They classified unicellular blue-green algae into seven typographical groups based on the plane, or successive planes of cell division (Table 2.II) and further characterized these using DNA base composition, fatty acid composition and various other physiological and biochemical criteria.

The following comments are concerned with some of the larger filamentous families:

(a) The Oscillatoriaceae is composed of algae with simple untapered, unbranched filaments and includes genera such as *Oscillatoria*, *Phormidium*

TABLE 2.I. The taxonomic classification of blue-green algae adopted by Fritsch (1945)

Order	General characteristics	Families	Representative genera
CHROOCOCCALES	Plants unicellular or colonial; colonies not showing polarity; multiplication by binary fission or by endospores; no heterocysts.	Chroococcaceae	*Chroococcus*, *Microcystis*
		Cyanochloridaceae[a]	*Placoma*, *Entophysalis*
		Entophysalidaceae	
CHAMAESIPHONALES	Plants unicellular or colonial; colonies showing distinct polarity; multiplication by endospores or exospores; no heterocysts	Dermocarpaceae	*Dermocarpa*
			Stichosiphon
		Chamaesiphonaceae	*Chamaesiphon*
		Endonemataceae	*Endonema*
		Siphononemataceae	*Siphononema*
PLEUROCAPSALES	Plants filamentous; multiplication by endospores; no heterocysts.	Pleurocapsaceae	*Pleurocapsa*
			Oncobyrsa
		Hyellaceae	*Hyella*, *Solentia*
NOSTOCALES	Plants filamentous but with no division into prostrate and upright filaments; multiplication by short motile filaments called hormogonia, with or without heterocysts.	Oscillatoriaceae	*Microcoleus*, *Oscillatoria*, *Phormidium*, *Spirulina*
		Nostocaceae	*Anabaena*, *Anabaenopsis*, *Nostoc*, *Wollea*
		Microchaetaceae	*Microchaete*,
		Rivulariaceae	*Calothrix*, *Dichothrix*, *Gloeotrichia*
		Scytonemataceae	*Scytonema*, *Tolypothrix*
		Brachytrichiaceae	*Brachytrichia*
STIGONEMATALES	Plants filamentous; showing distinct prostrate and upright systems; multiplication by hormogonia, or rarely by akinetes; heterocysts present.	Pulvinulariaceae	*Pulvinularia*
		Capsosiraceae	*Capsosira*
		Nostochopsidaceae	*Mastigocoleus*, *Nostochopsis*
		Loefgreniaceae	*Loefgrenia*
		Stigonemataceae	*Fischerella*, *Hapalosiphon*, *Westiella*

[a] It seems to us that the members placed by Fritsch in this family are not true blue-green algae.

TABLE 2.II. Key to the typological groups of unicellular blue-green algae (Stanier *et al.*, 1971)

A. Reproduction by repeated binary fission in a single plane, frequently resulting in the formation of short chains of cells.	Group I
1. Cells cylindrical, ellipsoidal, or spherical immediately after division; cells do not contain refractile polar granules.	Group IA
2. Cells cylindrical or ellipsoidal after division; cells always contain refractile polar granules.	Group IB
B. Reproduction by binary fission in two or three successive planes at right angles to one another; cells spherical.	
1. Cell divisions occur at regular intervals; never form parenchymatous masses of cells.	Group II
a. Cells not ensheathed; do not contain gas vacuoles.	Group IIA
b. Cells ensheathed; do not contain gas vacuoles.	Group IIB
c. Cells not ensheathed; contain gas vacuoles.	Group IIC
2. Cell division irregular; growth leads to the formation of parenchymatous, tightly packed masses of cells.	Group III

and *Lyngbya* (Figs 2.3a, b, c). There may be more than one trichome within a sheath, as for example in *Hydrocoleum lyngbyaceum* where there are few, or, as in *Sirocoleus kurzii* where there are many (Fig. 2.3d, e). In *Symploca* (Fig. 2.3f) the trichomes aggregate in bundles but not within a common sheath and in genera such as *Arthrospira* and *Spirulina* (Fig. 2.3g, h) the trichomes are helical in shape.

(b) The Nostocaceae also contains forms with untapered, unbranched filaments but these differ from the Oscillatoriaceae in that the algae are capable of producing heterocysts and akinetes. The genera and species are separated usually on the basis of size and shape of the cells, heterocysts, and spores, and on the distribution of the latter two cell types along the filament. Examples are species of *Nodularia*, *Aulosira*, *Aphanizomenon*, *Anabaena*, *Anabaenopsis*, *Wollea* and *Nostoc* (Fig. 2.4).

(c) Members of the Scytonemataceae have false branches and untapered heterocystous filaments, typical examples being species of *Scytonema* (Fig. 2.7b) and *Tolypothrix* (Fig. 2.7c). The genus *Plectonema* (Fig. 2.3i) which resembles typical Scytonemataceae in all characters except that it has no heterocysts was nevertheless placed in this family by Fritsch (1945) although Gomont (1892) considers it as a member of the Oscillatoriaceae. This is an example of a problematical genus which does not fit exactly into any one family. There are many forms like this in the blue-green algae and where they are placed ultimately depends on the weighting given by particular taxonomists to the various morphological characters.

(d) The Rivulariaceae comprises heterocystous species in which the filaments, which may, or may not, show false branching, are tapered and often terminate in a hair. The filaments have basal and sometimes intercalary heterocysts. *Rivularia*, *Gloeotrichia* and *Calothrix* are typical genera (Fig. 2.5).

(e) Genera placed in the Stigonemataceae and Nostochopsidaceae are morphologically the most complex Cyanophyceae. They show true branching and often there is differentiation between prostrate and upright filaments. In the Nostochopsidaceae the upright branches are only 1–4 cells long and terminate in a heterocyst (Fig. 2.6e). The filaments show cell-division predominantly in two but sometimes in three planes. Examples of this are found in the genera *Stigonema* (Figs. 2.6f) and *Westiella* (Fig. 2.6a).

Kenyon *et al.* (1972) carried out a comparative study of the fatty acids and lipids of 32 axenic strains of filamentous blue-green algae and assigned the strains examined to provisional typographical groups. They found three physiological characteristics: the ability to grow heterotrophically in the dark on glucose, the ability to grow on glucose in the light in the presence of DCMU (3-3,4-dichlorophenyl-1,1-dimethylurea)(10^{-5} M), and the ability of various non-heterocystous algae to fix nitrogen anaerobically, which could be used to sub-divide these groups. The classification adopted is shown in Table 2.III.

Classification of blue-green algae

To date, the classification of the blue-green algae into orders, families and species has been based, with the exception of the studies of Stanier *et al.* (1971) and Kenyon *et al.* (1972) discussed above, almost entirely on morphological characters of the sort just considered with some reference to the ecological distribution of the organisms. The divergence of opinion which has resulted from the use of these criteria is enormous. Nevertheless morphological characters are about the only features which can be used easily in ordinary practice. Routine biochemical tests such as those used with bacteria are seldom possible because of the difficulty in obtaining quickly pure growths of the algae under investigation. Major classifications which have been put forward, based on morphological characteristics, will now be considered briefly.

The earliest detailed attempt was made by Thuret (1875) who considered that there were two main groups: the Hormogonae, which contained all filamentous algae, and the Coccogonae, which contained the rest. This was followed by several classifications by other workers emphasizing

TABLE 2.III. Some physiological properties of filamentous blue-green algae (Kenyon *et al.*, 1972)

			Growth with 1% glucose		Nitrogenase synthesis	
	Typological subgroup	Strain number	in the dark	in the light with 10^{-5} M DCMU	Aerobic	Anaerobic
Strains not forming heterocysts						
I. *Oscillatoria* type						
	1	6304	−	−	−	−
		6401	−	−	−	−
	2	6407	+	+	−	+
		6412	+	+	−	+
		6506	+	+	−	+
		6602	+	+	−	+
II. *Lyngbya* type						
	1	6703	−	−	−	−
	2	6404	+	+	−	[a]
		6409	+	+	−	+
		7004	+	+	−	
		7104	+	+	−	
III. *Plectonema* type						
	1	6306	−	+	−	+
		6402	−	+	−	+
IV. *Spirulina* type						
	1	6313	−	−	−	−
Strains forming heterocysts						
I. *Anabaena* type						
	1	6302	+	+	+	
		6310	+	+	+	
		6720	+	+	+	
	2	6309	−	−	+	
		6314	−	−	+	
		6411	−	−	+	
		6705	−	−	+	
		6719	−	−	+	
II. *Calothrix* type						
	1	6303	−	−	+	
		7102	−	−	+	
	2	7101	+	+	+	
		7103	+	+	+	
III. *Microchaete* type						
	1	6305	−	+	+	
		6601	−	+	+	
IV. *Chlorogloeopsis* type						
	1	6718	+	+	+	

[a] Blank space = not determined.

different morphological characters. At one extreme there was the classification of Elenkin (1936) who considered that there were 12 orders and 47 families, and at the other extreme there is the classification of Drouet (1951) who recognizes no distinct orders and 8 families only. Frémy (1929, 1934) divided the non-hormogonial types into the orders Chroococcales and Chamaesiphonales, and placed all truly filamentous types in the Hormogonales. Geitler (1942) considered that there are four distinct orders: Chroococcales, Dermocarpales, Pleurocapsales and Hormogonales, while Fritsch (1943, 1944, 1945) adopted the classification outlined in Table 2.I. It may be noted that according to the International Code of Botanical Nomenclature the works of Bornet and Flahault (1886–1888) are to be used as the later starting point for the taxonomy of heterocystous algae and Gomont's 1892 treatise is to be used for the taxonomy of non-heterocystous algae. Many workers tend to be in sympathy with Drouet and Daily (1956) and Drouet (1968) who have drastically reduced the number of species in their taxonomic treatments. The work of Stanier *et al.* (1971) shows that he is not entirely correct in his delineation of genera and species, but there is little doubt that in reducing the number of taxa he has gone in the proper direction. The question is where to draw the line in the "lumping" of algal forms into simple species. A good example of the difficulty comes from studies on *Anacystis nidulans* (*Synechococcus* sp.).

Anacystis nidulans was the name given to a rod-shaped unicell obtained in pure culture by Kratz and Allen (see Kratz and Myers, 1955a). This organism was subsequently classified by Drouet and Daily (1956) into the filamentous genus *Phormidium* (see Allen and Stanier, 1968) and in his 1968 "Revision of the Oscillatoriaceae" Drouet placed it as a form of the filamentous *Schizothrix calcicola*. Allen and Stanier (1968) however, investigated cell division in *Anacystis nidulans* and, finding that the longest filaments obtained consisted only of four cells, suggested that the mode of growth and reproduction was not significantly different from that in rod-shaped unicellular bacteria. They concluded, "we cannot therefore accept the contention that they (*Anacystis*-like unicellular algae) are filamentous blue-green algae in the sense that an *Oscillatoria* with its many-celled structure and hormogonial mode of reproduction can be described as filamentous". However, more recently Kunisawa and Cohen-Bazire (1970) working in the same laboratory have obtained stable filamentous mutants of *Anacystis nidulans* and Ingram and Van Baalen (1970) similarly have obtained filamentous mutants of *Agmenellum quadruplicatum* on treatment with NTG (N-methyl-N′-nitro-N-nitrosoguanidine). These mutants have similar growth rates to the wild type *Agmenellum* and can commonly be induced on NTG treatment. Thus the question arises—should a filamentous mutant be regarded as a specialized ecophene of a unicellular alga? It may be noted that mutants of blue-green algae have been obtained usually with heavy doses of ultraviolet irradiation or with NTG, a particularly powerful mutagen, and that compared with eukaryotic algae, blue-green algae are resistant to irradiation (see p. 87). This is a typical example of the difficulties in classifying unicellular blue-green algae. Kenyon *et al.* (1972) have also drawn attention to the taxonomic confusion within the filamentous blue-green algae. Such studies emphasize the enormous amount of work still required in this area, and while recognizing only too clearly the inadequacies of the existing system, we feel that until more data of the sort being provided by Kenyon *et al.* become available, it would be premature for us, in this text, to abandon the conventional taxonomy of our predecessors.

Chapter 3

Cellular organization of blue-green algae

Classical light microscopy is largely unsatisfactory with such small organisms as blue-green algae and although much information was obtained by early cytologists, this initial progress was followed by controversy and confusion which persisted into the 1930s. Physiological and biochemical studies, on the other hand, provided information on photosynthesis, respiration and nitrogen fixation (see Fogg, 1956b) which suggested that these major metabolic processes did not differ markedly from those found in higher plants. What was of particular interest was that all these complex activities occurred in minute cells lacking the biochemical compartmentalization of higher plant cells. This apparent discrepancy between a high level of physiological and biochemical organization and a seemingly low level of structural differentiation was an important factor leading to contemporary studies on the fine structure of blue-green algae. Electron microscope investigations were first carried out by Niklowitz and Drews (1956, 1957) with *Phormidium uncinatum* and later with other filamentous blue-green algae. Their observations and conclusions aroused great interest and were soon followed by other significant contributions (Fuhs, 1958a; Hopwood and Glauert, 1960; Lefort, 1960a, b; Ris and Singh, 1961; Pankratz and Bowen, 1963; Wildon and Mercer, 1963a; Leak and Wilson, 1965; Lang, 1965). With increased interest and improved techniques of preservation, staining and embedding, based largely on methods devised originally for bacteria (Luft, 1956; Kellenberger *et al.*, 1958; Glauert and Glauert, 1958) there is now available a wealth of information on the sub-cellular organization of blue-green algae (see Lang, 1968).

Prokaryotic cellular organization

One of the first features which became apparent from electron microscope observations was that there were no true boundaries in the protoplast between chromatoplasm and centroplasm (Ris and Singh, 1961). The cellular organization was in fact different from that in other algae and

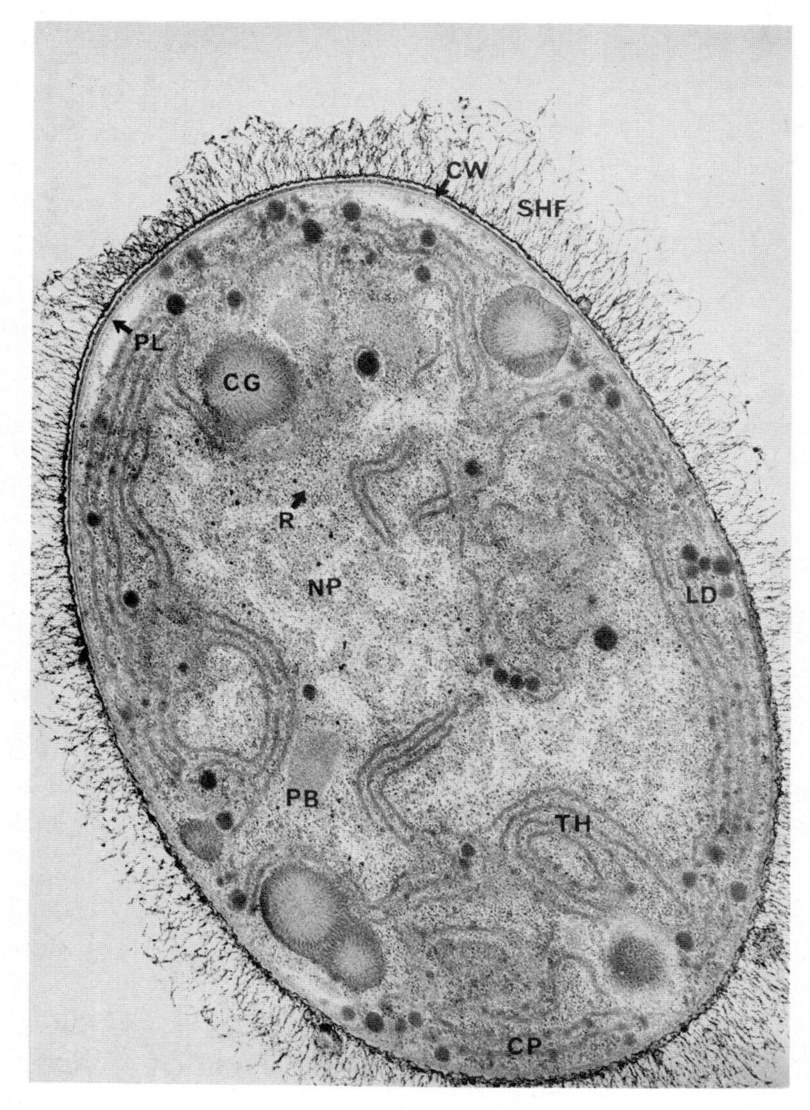

FIG. 3.1. Electron micrograph of a section of *Anabaena variabilis* illustrating the characteristic ultrastructure of the cell of a blue-green alga. CG, cyanophycin granule; CP, chromatoplasmic region; CW, cell wall; LD, lipid droplets; NP, nucleoplasmic region; PB, polyhedral body; PL, plasmalemma; R, ribosomes; SHF, sheath fibre; TH, thylakoids ($\times$33,000). By courtesy of L. V. Leak (1967), *J. Ultrastruct. Res.* **20**, 190, Fig. 6.

higher plants but resembled that of the bacteria closely in the absence of a membrane-bound nucleus and other membrane-bound organelles. The term *prokaryotic* coined by Chatton (1937) with what Stanier and van Niel (1962) called "singular prescience" is now used to denote this type of cell structure. It is accepted that the division into prokaryotes and eukaryotes is the most fundamental distinction existing in the plant kingdom (but see Bisset, 1973). The main features of a blue-green alga as seen under the electron microscope (Fig. 3.1) are: a prokaryotic cellular organization; an elaborate photosynthetic membrane system; a fibrillo-granular nucleoplasmic region; a diversity of characteristic cytoplasmic inclusions; a rigid multilayered cell wall; a fibrous sheath. In these cells cytoplasmic membrane-containing regions are in direct contact with regions containing DNA fibrils and ribosomes and thus the cell can be considered as a single physiological unit with a close biochemical relationship between subcellular structures of different function.

The plasma membrane

The physiological integrity of the cell is maintained by a thin *plasma membrane* or *plasmalemma* which appears as a typical tripartite "unit membrane" about 7 nm thick. It acts as a selective semipermeable membrane, a function well demonstrated when the cell wall is removed by the action of lysozyme leaving the plasmalemma intact (see p. 74; Biggins, 1967a; Vance and Ward, 1969). The metabolic activity of the released protoplasts can be maintained only in an osmotically balanced medium.

The structure of the membrane has been investigated by Jost (1965) using freeze-etching techniques. He showed that the plasma membrane is composed of globular substructures or "globuli" (Fig. 3.2) and estimated that about 2900 of these are present per square micron of membrane surface. These structures may be protein molecules with possible enzymatic functions in respiration and cell wall and sheath formation (Jost, 1965) although this has not been demonstrated.

Folds of the plasma membrane often protrude into the cytoplasm (Pankratz and Bowen, 1963; Jost, 1965; Smith and Peat, 1967b; Peat and Whitton, 1968; Allen, 1968c). Whether true pinocytosis, i.e. local invagination of the plasmalemma followed by separation and dispersal of the vesicle so formed within the cytoplasm, occurs is uncertain because of the complex arrangement of the membranes within the cells. It is also uncertain whether such protrusions are the main source of new thylakoid membranes (see below) or not. Jost (1965) suggested from studies on *Oscillatoria rubescens* that thylakoids disintegrate in senescent cells and are

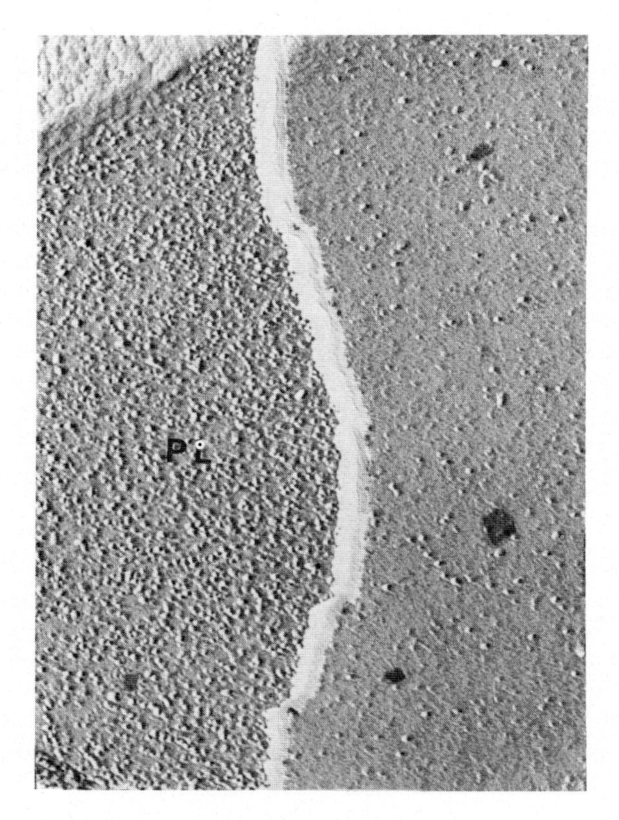

Fig. 3.2. Image of plasmalemma (PL) of *Oscillatoria rubescens* prepared by freeze etching and showing globular ultrastructure. Electron micrograph (× 75,000). By courtesy of M. Jost (1965) *Arch. Mikrobiol.* **50**, 211, Fig. 8.

newly formed in the hormogonia by protrusion of the plasmalemma and secondary invagination. Division of existing thylakoids during septum formation (Pankratz and Bowen, 1963), splitting and fusion of thylakoids (Hall and Claus, 1962) and proliferation of thylakoids by growth (Chapman and Salton, 1962; Lang, 1968) have been suggested as alternative or additional mechanisms. Unusual membranous structures, produced by the invagination of the plasmalemma and the photosynthetic lamellae, and *nidulans* (*Synechococcus* sp.) by Echlin (1964b) and Allen (1972). These structures, termed "lamellosomes" or "mesosomes" may be concerned in respiration and cell division as in bacteria (Fitz-James, 1960).

The photosynthetic apparatus

1. THE PHOTOSYNTHETIC MEMBRANE SYSTEM

The most extensive cellular structures seen with the electron microscope in the blue-green algae are the flattened vesicles called *thylakoids* (Menke, 1961) which ramify throughout the outer regions of the cell. They may be arranged either peripherally in parallel concentric shells, or less regularly when they appear to form a three-dimensional anastomosing system of lamellae traversing most of the cytoplasm. These arrangements may be of some taxonomic significance.

Thylakoids contain the photosynthetic pigments (Calvin and Lynch, 1952; Shatkin, 1960). They are the sites of photosynthesis and their

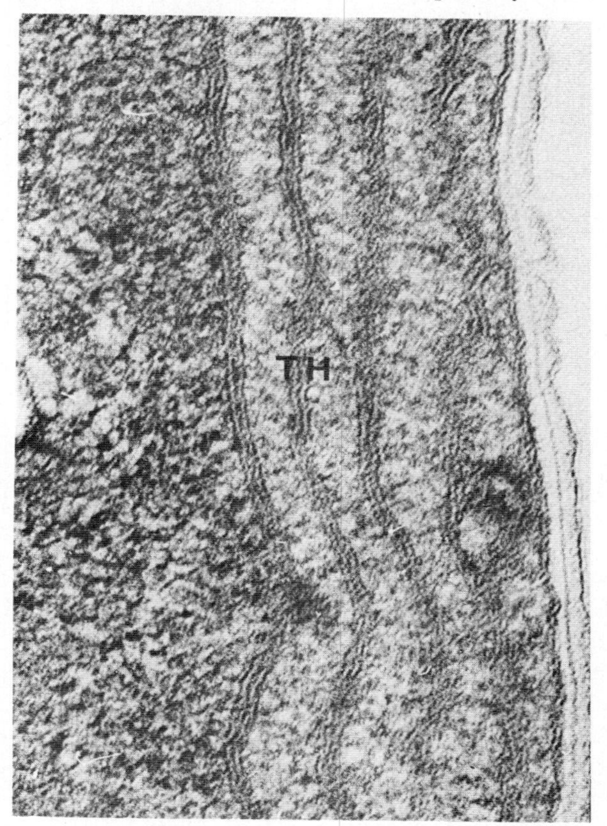

FIG. 3.3. Portion of longitudinal section of *Synechococcus lividus* showing thylakoids (TH) with closely apposed membranes which exhibit a tripartite substructure. Electron micrograph (×140,000). By courtesy of M. R. Edwards, D. S. Berns, W. C. Ghiorse and S. C. Holt. (1968) *J. Phycol.* **4**, 283, Fig. 11.

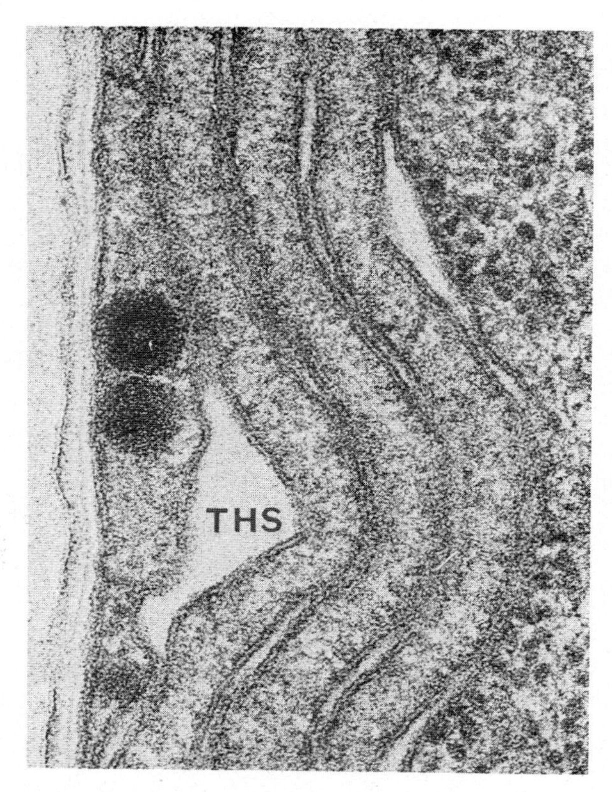

FIG. 3.4. Part of longitudinal section of *Synechococcus lividus* showing the thylakoid membranes partially separated by an intrathylakoidal space (THS). Electron micrograph (× 160,000). By courtesy of M. R. Edwards, D. S. Berns, W. C. Ghiorse and S. C. Holt. (1968) *J. Phycol.* **4**, 283, Fig. 10.

ability to carry out photophosphorylation (Petrack and Lipman, 1961), photosynthetic oxygen evolution (Cox *et al.*, 1964) and the Hill reaction (Susor and Krogman, 1964) have all been demonstrated in cell-free extracts. The amount of chlorophyll present in *Anacystis nidulans* has been shown to be directly proportional to the amount of thylakoid membrane present (Allen, 1968b). Thylakoids also occur in dark-grown cells of *Chlorogloea fritschii* and evidence from metabolic studies using cell-free extracts of photosynthetic lamellae suggests that they are probably sites of respiratory activity as well (Fay and Cox, 1966; Webster and Hackett, 1966; Smith *et al.*, 1967; Biggins, 1969). Bisalputra *et al.* (1969) have shown from electron microscope studies that when *Nostoc sphaericum* is incubated in the presence of potassium tellurite and succinate in the dark,

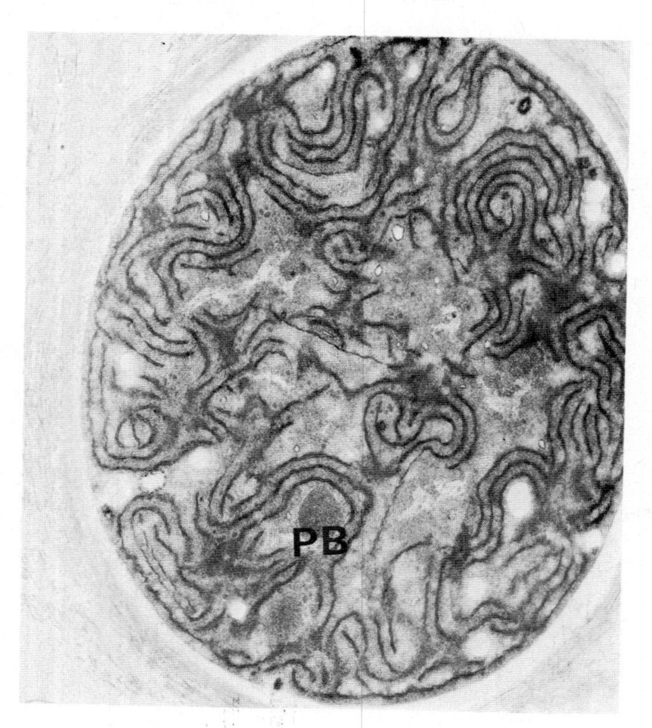

FIG. 3.5. Section of a cell of *Chlorogloea fritschii* grown at low light intensity (200–600 lux). Electron micrograph (×20,700). By courtesy of D. L. Findley, P. L. Walne and R. W. Holton (1970) *J. Phycol.* **6**, 182, Fig. 5.

tellurite reduction occurs on the surface of the thylakoids, another line of evidence which supports the idea of their respiratory function (see p. 169).

Structurally the thylakoid vesicles appear tripartite, having a typical "unit membrane" structure (Fig. 3.3). The membranes are usually closely apposed to each other in the flattened vesicles, without an internal lumen, and form a lamella about 14 nm thick. Sometimes the membranes appear to be slightly separated by a narrow intrathylakoidal space (Fig. 3.4). The lumen may appear dilated in aged cells (Echlin, 1964a) or if the alga is grown under unfavourable conditions.

Stacks of thylakoids separated by spaces as narrow as 16 nm have occasionally been seen in blue-green algae (Ris and Singh, 1961; Pankratz and Bowen, 1963). The number and arrangement of the thylakoids seem to vary according to the physiological condition and stage of development of the alga. This is shown clearly in *Chlorogloea fritschii* (Peat and Whitton, 1967; Findley *et al.*, 1970). In young filaments the thylakoids run mainly

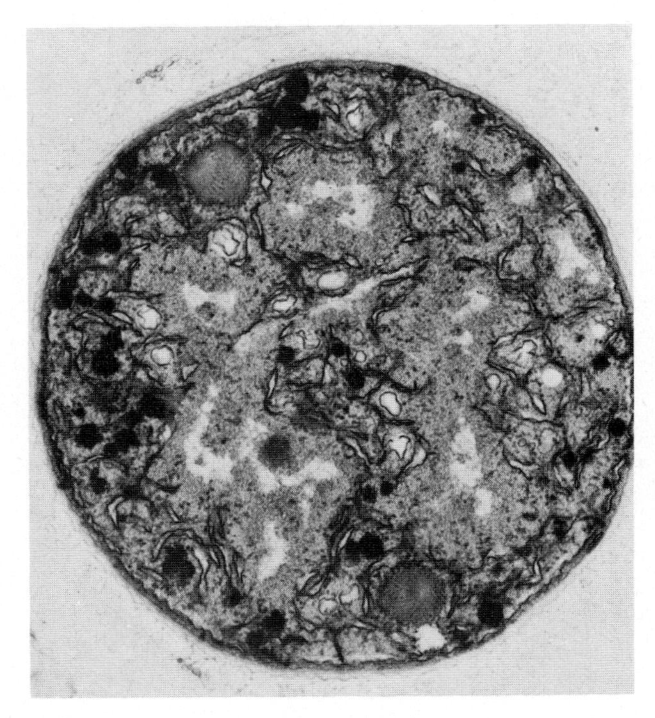

FIG. 3.6. Section of a cell of *Chlorogloea fritschii* grown at high light intensity (7000 lux). Electron micrograph (×18,400). By courtesy of D. L. Findley, P. L. Walne and R. W. Holton (1970) *J. Phycol.* **6**, 182, Fig. 9.

parallel to the cell wall, especially in algae grown at low light intensity (Fig. 3.5). At higher light intensities and during the later stages of cell development they become scattered throughout the outer part of the cytoplasm (Fig. 3.6). In cells grown heterotrophically in the dark the thylakoids are distributed evenly throughout the cytoplasm (Fig. 3.7). A similar inverse relationship between thylakoid content, or pigment concentration, and light intensity occurs in *Anacystis nidulans* (Allen, 1968b). In aged cultures striking changes in thylakoid structure are observed. "Grana"-like stacked arrays of tightly packed thylakoids have been recorded in *Symploca muscorum* by Pankratz and Bowen (1963) and paracrystalline thylakoid lattices similar to the "prolamellar bodies" of chloroplasts of etiolated plants have been seen in an *Anabaena* species (Fig. 3.8; Lang and Rae, 1967) where they form a cubic lattice of interconnected tubules. Although this is so far the only record of the presence of such structures in vegetative cells, tubular structures are common

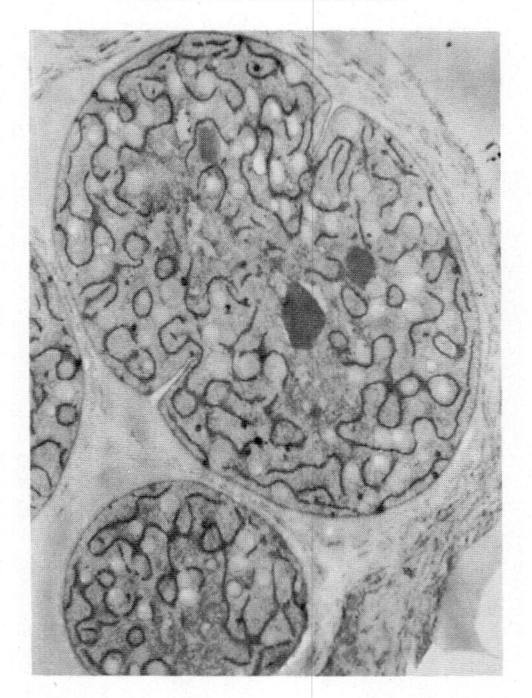

FIG. 3.7. Section of *Chlorogloea fritschii* grown in the dark with sucrose. Electron micrograph (× 9500) by courtesy of A. Peat and B. A. Whitton (1967) *Arch. Mikrobiol.* **57**, 155, Fig. 10.

in heterocysts (see p. 227) and their occurrence lends additional support to the view that thylakoids in blue-green algae and in chloroplasts are homologous structures.

2. CHEMISTRY OF THE PHOTOSYNTHETIC LAMELLAR SYSTEM

The tripartite fixation image of these membranes supports the notion that, like other membranes, they comprise a lipid bimolecular layer sandwiched between two layers of protein. Freeze-etching studies on the thylakoids of *Oscillatoria rubescens* by Jost (1965) also show the presence of particles on the fracture faces which separate the two lipid layers (Branton, 1966). By analogy with particles seen in other membrane systems these are probably protein macromolecules or aggregates of proteins which occur in the lipid layer. Although some direct analyses have been made on isolated photosynthetic membranes of blue-green algae (e.g. Calvin and Lynch,

1952), for the most part our knowledge of their composition depends on the assumption that components of thylakoids studied extensively in higher plant systems, if detected in extracts of blue-green algal cells, will also be located on their thylakoid systems. In discussing the probable

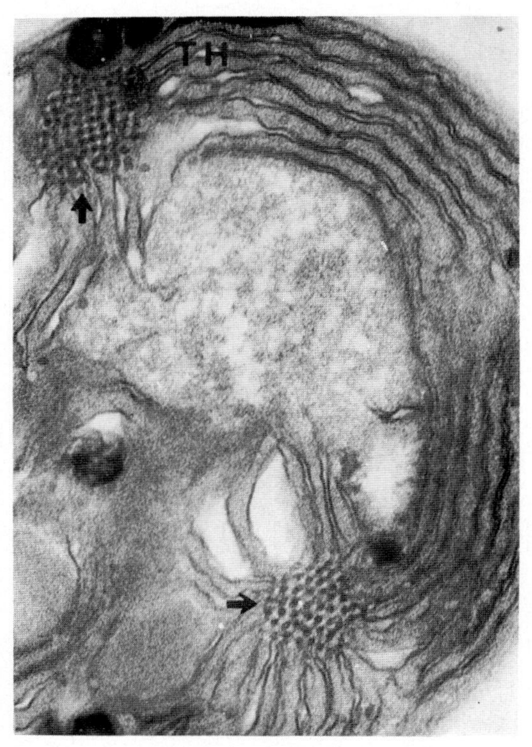

FIG. 3.8. Section of an aged cell of *Anabaena* sp. illustrating thylakoid lattices (arrows) continuous with thylakoids (TH). Electron micrograph (×40,400). Courtesy of N. J. Lang and P. M. M. Rae (1967) *Protoplasma* **64**, 67, Fig. 5.

content of the "quantasome", the supposed morphological manifestation of the physiological photosynthetic unit, Park and Biggins (1964) listed the known components of photosynthetic lamellae from higher plants. The principal classes of compounds were as follows: (a) Lipids: chlorophylls, carotenoids, quinones, phospholipids, diglycerides, sulpholipids and possibly sterols; (b) Proteins: cytochromes, other iron- and copper-containing proteins, and other uncharacterized proteins. Following this precedent, the chemistry of probable components of the photosynthetic membranes system in blue-green algae is now briefly described.

(a) *Lipids*

Photosynthetic lamellae probably have a greater variety of different lipid components than any other membrane system, these substances being involved in electron transport, light gathering, light shielding and perhaps also in the process of oxygen evolution. The lipid components fall into the following categories:

(i) *Chlorophyll.* Unlike eukaryotic algae, blue-green algae possess only one kind of chlorophyll, chlorophyll *a*, the structure of which is as follows:

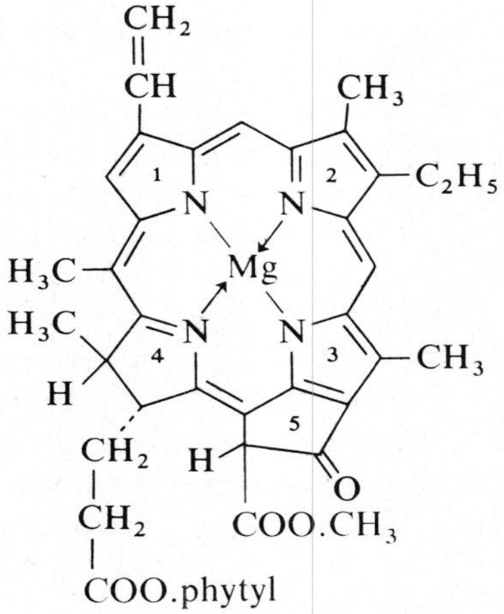

CH_2
CH
CH_3
1 2 C_2H_5
N N
H_3C Mg
H_3C N N
4 3 CH_3
H
5
CH_2 H
O
CH_2 $COO.CH_3$
COO.phytyl

The molecule may, however, exist in different forms determined by the chemical environment and distinguishable by their absorption spectra. The principal forms are C*a* 670, C*a* 680 and P700 (Smith and French, 1963), the numbers designating the wavelengths, in nm, of the major red-absorbing band. The proportions of the different forms may change with different conditions of illumination and supply of nutrients (see Öquist, 1971). According to Thornber (1969), 70% of the chlorophyll *a* in *Phormidium luridum* is bound to protein. He demonstrated a protein complex which would dissociate into subunits of 35,000 MW, each containing five chlorophyll molecules.

(ii) *Carotenoids.* Carotenoids are yellow, orange or red pigmented compounds whose basic structure is that of a tetraterpene, a conjugated chain of eight isoprene subunits. The units at one or both ends of the chain may be cyclized, forming 6-membered rings as in β-carotene:

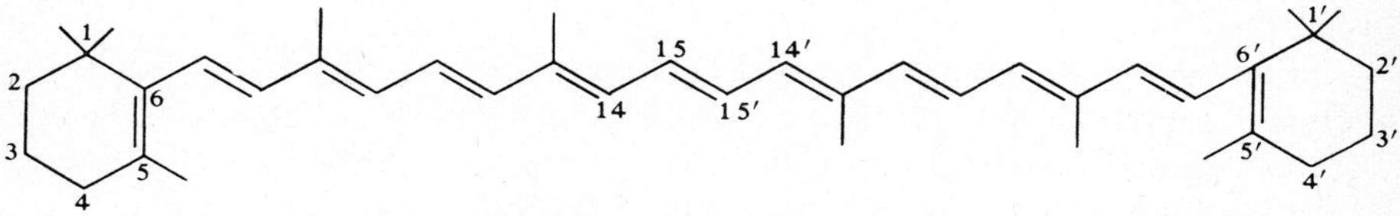

There are two basic categories of carotenoid: carotenes, which are hydrocarbons, and xanthophylls which are oxygen-containing derivatives. Both groups comprise lipophilic substances which are highly soluble and easily extracted in acetone, chloroform-methanol and ether.

Table 3.I lists the carotenoids that have been described in various blue-green algae. β-carotene appears to be universal and is often the principal carotenoid and only carotene present (Healey, 1968), though flavacene occurs in some species (Hertzberg and Liaaen-Jensen, 1967) and Goodwin (1965) also lists ε-carotene as occurring in this group of algae. Another isomer, α-carotene, has also been reported to occur in small quantities but as the material was obtained from field samples which may have contained other algae, such reports should be treated with some caution (see Fogg, 1956b); Stransky and Hager (1970) state categorically that blue-green algae are not able to make α-carotene or its derivatives.

Much of the definite work on the chemistry of cyanophycean xanthophylls has been done by Hertzberg and Liaaen-Jensen (1966 a, b, 1967, 1969 a, b) who determined the structures of myxoxanthophyll and oscillaxanthin, two xanthophylls which appear to be specific to blue-green algae. Many of the other xanthophylls they contain have a wide distribution, although myxoxanthin, identical with echinenone, occurs otherwise only in certain animal tissues. The synonymy of other carotenoids is also given in Table 3.I. This list is not complete, as there are a number of other carotenoids (see Healey, 1968; Stransky and Hager, 1970) the identities of which have yet to be determined, occurring in blue-green algae.

Carotenoids have, perhaps, two principal functions. The first is that they transfer light energy, which they trap, to photosystem I of photosynthesis. According to Goedheer (1969) the efficiency of the transfer from β-carotene to chlorophyll *a* is high, approaching 100%, though he was unable to obtain any evidence that the xanthophylls present in the system he investigated were able to mediate such a transfer. Perhaps they, like α-carotene, are more important in providing light shielding and preventing photo-oxidation of the other photosynthetic pigments (see Krinsky, 1966).

(iii) *The diglyceride lipids*. There are four major groups of diglyceride lipids in blue-green algae, which are also the predominant lipids in the chloroplasts of higher plants. These lipids, whose structures are shown below, are mono- and di-galactosyl diglycerides, sulphoquinovosyl diglyceride, and phosphatidyl glycerol (Nichols, 1970). They can perhaps be considered to be the photosynthetic lipids.

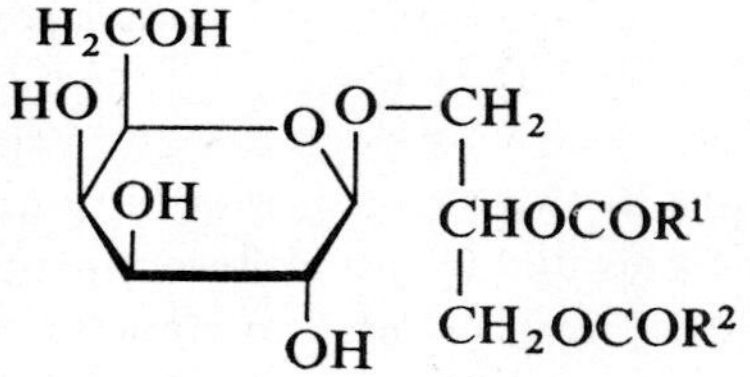

(I) Monogalactosyl diglyceride [β-D-galactosyl-(1-1′)–2′, 3′-diacyl-D-glycerol]

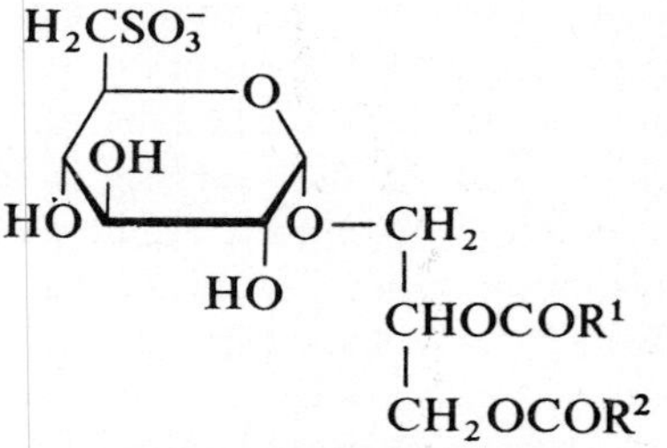

(III) Sulphoquinovosyl diglyceride [6-Sulpho-α-D-quinovosyl-(1-1′)-2′, 3′-diacyl-D-glycerol]

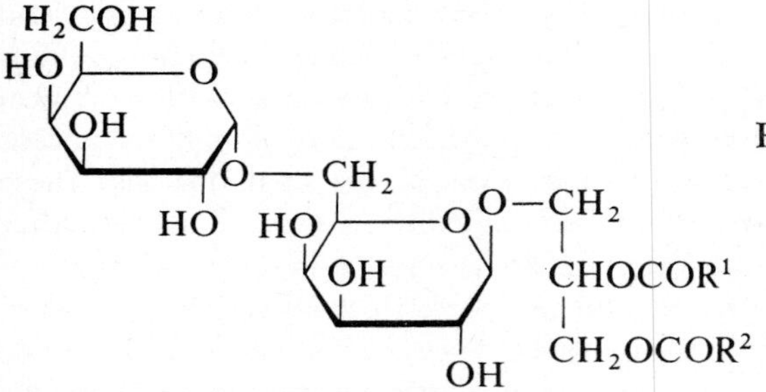

(II) Digalactosyl diglyceride [α-D-galactosyl-(1-6)-β-D-galactosyl-(1-1′)-2′, 3′-diacyl-D-glycerol]

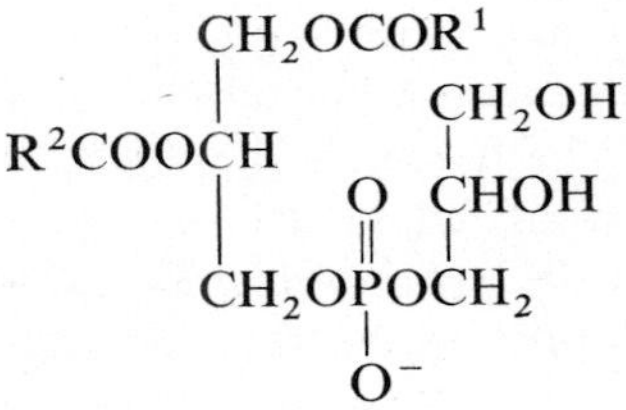

(IV) Phosphatidyl glycerol

Within these four groups occurs a large number of compounds distinguished by the different fatty acids (whose hydrocarbon chains are denoted by "R" in the structural formulae) attached to carbons 1 and 2 of the glycerides. The distribution of these fatty acids (released by saponification of the diglycerides) has itself proved to be of considerable phylogenetic interest. Preliminary surveys seemed at first to indicate that polyunsaturated fatty acids (i.e. fatty acids having two or more double bonds in the hydrocarbon chain) which are the principal components of higher plant diglycerides, were present in all blue-green algae; this, and their absence from photosynthetic bacteria, led Erwin and Bloch (1963) to suggest that they might be involved in the oxygen-evolving process in photosynthesis. This theory lost credence, however, when Holton *et al.* (1964) reported that the unicellular blue-green alga *Anacystis nidulans*, now grouped in the genus *Synechococcus* by Stanier *et al.*, (1971), contained only saturated and mono-unsaturated acids. An extensive survey of 34 strains of unicellular forms has now shown that this is a common feature of this group, though more than a third of those investigated are able to produce di- and tri-enoic acids (Kenyon, 1972; Kenyon and Stanier, 1970).

Amongst the filamentous forms analysed by Kenyon *et al.* (1972) a similar degree of variation is found, though here the trend is towards a greater degree of unsaturation. Of the 32 strains investigated, only two had mono-, and five strains had di-enoic acids as the highest degree of saturation.

The results of these surveys seem to indicate that the degree of saturation is a legacy of certain metabolic pathways inherited by the different groups whose phylogenetic affinities are marked by certain other metabolic and morphological characteristics. If these lipids are intimately involved in the photosynthetic process then it must be concluded that the types of fatty acids they contain are unimportant to the roles they play. We can now find a reciprocal correlation between oxygen-producing photosynthesis and lipid composition only in the digalactosyl diglycerides, for monogalactosyl diglycerides and sulpholipids have been recorded in some, and phosphatidyl glycerol in all, photosynthetic bacteria (Nichols, 1970; Kenyon and Stanier, 1970). It is interesting that digalactosyl diglycerides, together with the other three groups of photosynthetic lipids, are absent from heterocysts and particularly so in view of the evidence that these cells have photochemical activity but do not evolve oxygen. Their peculiar lipid composition, and its implications, are discussed below (p. 230).

(iv) *Quinones*. Quinones are substances in which two hydrogen atoms of the benzene nucleus are replaced by two oxygen atoms attached by double covalent bonds. Most naturally occurring quinones are *p*-quinones with the oxygens located on opposite sides of the ring (*o*-quinones being less stable structures). This class of structures is generally considered to be important in the electron transport systems of respiration and photosynthesis. The first report of quinones in blue-green algae appears to be that by Lester and Crane (1959) who found both vitamin K and another quinone (now known to be a plastoquinone) in *Anacystis nidulans*. In the same organism, Henninger *et al.* (1965) found three different quinones which were subsequently characterized by Allen *et al.* (1967) as plastoquinone A (2,3-dimethyl-5-nonaprenyl-1,4-benzoquinone), vitamin K_1 (2-methyl-3-phytyl-1,4-napthoquinone) the structure of which is:

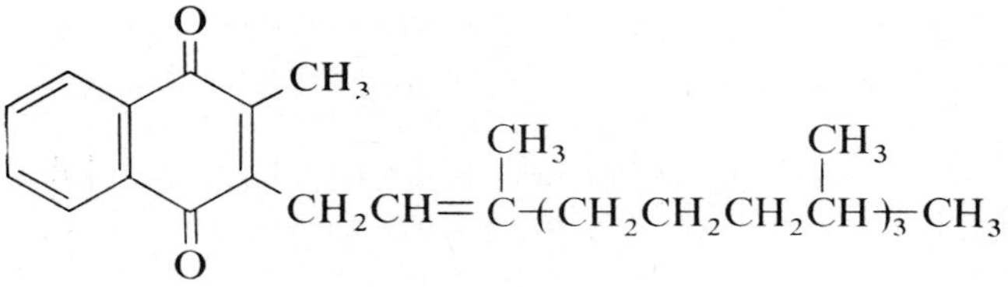

$$CH_3 \quad CH_2CH{=}\overset{CH_3}{\overset{|}{C}}\!\left(CH_2CH_2CH_2\overset{CH_3}{\overset{|}{C}H}\right)_3CH_3$$

and a monohydroxy analogue of vitamin K_1. Carr and his co-workers (see Carr and Hallaway, 1966; Carr *et al.*, 1967) found the first two of these three compounds in four filamentous blue-green algae, which also contained α-tocopherol. This compound, which has been implicated in oxidation-reduction reactions in photosynthesis and the protection against peroxidative destruction of membranes (see Henninger *et al.*, 1965), was not found in *Anacystis nidulans* though it has been found in another unicellular species (DaSilva and Jensen, 1971). Ubiquinone, which is thought to be involved in electron transfer systems of respiration in other organisms, has not been found in any blue-green algae.

(v) *Sterols.* It was at one time thought that sterols were absent from blue-green algae (Carter *et al.*, 1939; Levin and Bloch, 1964) but it has now been shown that such compounds are present in these organisms, though harsh methods are needed to extract them. Reitz and Hamilton (1964) were first to demonstrate their presence in *Anacystis nidulans* and *Fremyella diplosiphon.* The two sterols isolated, cholesterol and α-sitosterol, have also since been demonstrated in *Spirulina maxima* by Martinez Nadal (1971). De Souza and Nes (1968) also demonstrated small amounts of cholesterol, and larger quantities of unsaturated 24-ethylcholesterols in *Phormidium luridum.* Phytol and squalene were present in addition. Carr and Craig (1970) reported that sterols are present in *Anabaena variabilis* and that they appear on centrifugal fractionation to be associated with the photosynthetic lamellae.

(b) *Protein*

The figures of Park and Biggins (1964) show that protein accounts for approximately half of the photosynthetic material, but much less is known about the identity of the proteins involved. It is known that certain classes of proteins are involved in electron transport, namely the cytochromes, plastocyanin and the terminal electron acceptor, ferredoxin. Because electron transfer cannot occur over large distances (>5 nm certainly) these proteins must at some point be in close proximity to the membrane, if not forming an integral part of it.

In the case of these three classes of proteins evidence that they can be washed free from lamella systems suggests that the association is a loose one, but their chemistry is nevertheless considered here, briefly.

(i) *Ferredoxins.* Ferredoxins are proteins containing non-haem iron which, functioning as a link between hydrogenases and different electron donors and acceptors in both photosynthetic and non-photosynthetic systems, are capable of reducing the hydrogen ion to molecular hydrogen. They have the lowest redox potential of any substance characterized in photosynthetic systems and are thought to be the primary electron acceptors from photosystem I (see Fig. 8.1). They are small molecules, by protein standards, having a molecular weight of between 6000 and 13,000, and are distinguished by their characteristic absorption spectra which show marked changes on oxidation and reduction.

Arnon (1965) was first to demonstrate ferredoxin in a blue-green alga, a *Nostoc* sp., and he commented on its resemblances to ferredoxins of higher plants. Similar ferredoxins have been demonstrated in *Anabaena variabilis* (Susor and Krogmann, 1966) and *Anacystis nidulans* (Yamanaka *et al.*, 1969). In the latter the molecule contains two atoms of non-haem iron and one of labile sulphur, and it has a molecular weight of about 10,000.

(ii) *Cytochromes*. Cytochromes are pigmented proteins involved in electron transfer, which have a prosthetic haem group as the active site, this example being from a *b*-cytochrome:

CH_2
||
CH_3 CH
N
H_3C — — CH_3
N---Fe^{++}---N
$^{-}OOC—H_2C—H_2C$ — — $CH=CH_2$
N
CH_2 CH_3
|
CH_2
|
COO^{-}

Different cytochromes are distinguished by the position and intensity of their absorption maxima when in the reduced state. They have different protein moieties and different classes (*a*, *b*, *c*) have haem groups of different compositions; a consequence of this is that they have different redox potentials which determine their positions in electron transport chains.

Katoh (1959) was first to isolate a cytochrome, of the *c*-type, from a blue-green alga, *Tolypothrix tenuis*. Holton and Myers (1963, 1967a) isolated three cytochromes from *Anacystis nidulans*, $c_{(549)}$ (properly designated "cytochrome $c_{(549)}$ (*Anacystis nidulans*)", a strongly autoxidizable compound and perhaps a terminal oxidase in respiration, $c_{(552)}$, and

$c_{(554)}$, possibly analogous to photosynthetic *f*-type cytochromes (Holton and Myers 1967b) (see Fig. 8.2). Susor and Krogmann (1966) described a similar cytochrome to $c_{(554)}$ (*Anacystis nidulans*) in *Anabaena variabilis*, later referred to as a cytochrome f (Lightbody and Krogmann, 1967). In each case the cytochromes from broken cells were obtained by a mild washing procedure.

(iii) *Plastocyanin.* Plastocyanin is a copper-containing protein which is found in photosynthetic tissue, where it exhibits physiological redox properties. It appears to act in the electron transport pathway between plastoquinone and photosystem I see (Fig. 8.1) and the studies of Levine (1968) on *Chlamydomonas* mutants indicate that it acts in parallel with cytochrome *f*.

Lightbody and Krogmann (1967) isolated plastocyanin from *Anabaena variabilis*, as a blue copper protein having an absorption peak at 597 nm. They showed that it was required for a number of partial reactions in the photosynthetic electron transfer chain, requirements which could also be met by cytochrome *f*. Like the various cytochromes, plastocyanin can be removed from photosynthetic membranes simply by washing.

3. LOCATION OF THE PHOTOCHEMICAL SYSTEMS

There is some information concerning the location of photochemical systems within the thylakoid membranes. Membrane fragments released from the cells of *Anabaena variabilis* by sonic oscillation yielded two fractions when treated with the detergent Triton X-100 (Ogawa *et al.*, 1968, 1969): one was heavy and blue-pigmented, the other was lighter and orange in colour (Fig. 3.9). 80% of the chlorophyll was present in the heavy fragment, which also had a higher ratio of β-carotene to total xanthophylls. On the other hand, the concentration of total carotenoids, lipid and protein was higher in the light fragments (see p. 148). In the electron microscope the heavy fragments appeared as flat discs with many small particles about 10 nm in diameter scattered on their surface, while the light fragments consisted of small membranes with 5-6 nm diameter particles. Comparing these data with the observations on freeze-etched preparations from *Oscillatoria rubescens* (Jost, 1965) it is tempting to speculate that the thylakoid membrane was split in half by the detergent treatment. Fluorescence measurements indicate that the two fragments represent two different components of the photosynthetic apparatus in *Anabaena variabilis*, the heavy fragments being related to photosystem I while the light fragments are associated with photosystem II.

TABLE 3.I Carotenoids of blue-green algae

Carotenoid	Structure
CAROTENES	
β-carotene[a,b]	see p. 46
flavacene	probably identical with mutachrome, a rearranged furanoid oxidation product of β-carotene[i]
XANTHOPHYLLS	
echinenone[c,d]	4-keto-β-carotene
= calorhodin[a,m]	
= aphanin[c,h,o,p]	
= myxoxanthin[c,l]	
	3′-hydroxy-4-keto-β-carotene[n]
	4-hydroxy-4′-keto-β-carotene[n]
cryptoxanthin[g,q]	3-hydroxy-β-carotene
isocryptoxanthin[e]	4-hydroxy-β-carotene
canthaxanthin[d,h]	4,4′-diketo-β-carotene
= aphanicin[h,l]	
lutein[b,f]	3,3′ dihydroxy-α-carotene
zeaxanthin[b,l]	3,3′ dihydroxy-β-carotene
caloxanthin[n]	3,3′-dihydroxy-5-hydro-7-dehydro-β-carotene
nostoxanthin[n]	3,3′-dihydroxy-5,5′-dihydro-7,7′-didehydro-β-carotene
oscillaxanthin[b,k,l]	1,1′-dihydroxy-2,2′-di-β-L-rhamnosyl-1,2,1′,2′-tetrahydro-3,4,3′,4′-tetradehydrolycopene[k]
myxoxanthophyll[b,j]	1′,2′-dihydro-3′4′-didehydro-3,1′-dihydroxy-γ-carotene-2′-yl-glycoside (mainly rhamnose) [= 2′-O-rhamnosyl-myxol][d]
4-keto-myxoxanthophyll	2′-0-rhamnosyl-4-keto-myxol[d]

[a]Fogg (1956b)
[b]Goodwin (1965)
[c]Goodwin and Taha (1951)
[d]Halfen and Francis (1972)
[e]Healey (1968)
[f]Heilbron and Lythgoe (1936)
[g]Hertzberg and Liaaen-Jensen (1966a)
[h]Hertzberg and Liaaen-Jensen (1966b)
[i]Hertzberg and Liaaen-Jensen (1967)
[j]Hertzberg and Liaaen-Jensen (1969a)
[k]Hertzberg and Liaaen-Jensen (1969b)
[l]Karrer and Rutschmann (1944)
[m]Kylin (1943)
[n]Stransky and Hager (1970)
[o]Tischer (1938)
[p]Tischer (1939)
[q]Tischer (1958)

4. PHYCOBILIPROTEIN GRANULES ("PHYCOBILISOMES")

The accessory pigments *c*-phycocyanin, *c*-allophycocyanin and *c*-phycoerythrin are usually present in high concentration in the cells of blue-green algae, constituting as much as 40% of their total soluble protein

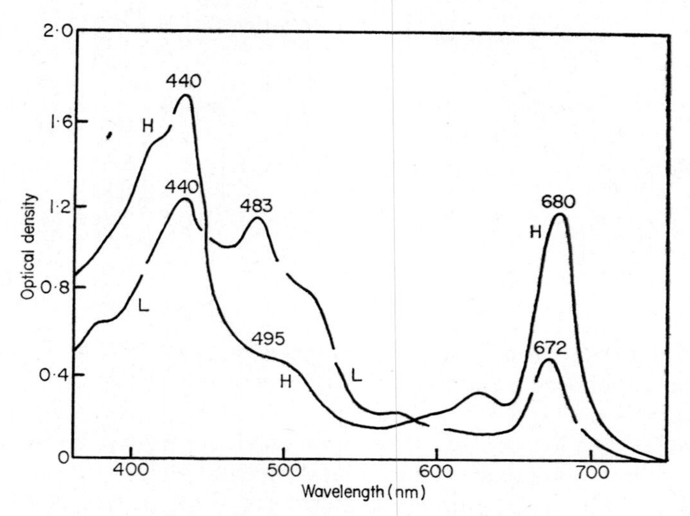

FIG. 3.9. Absorption spectra of the heavy (H) and light (L) thylakoid fragments of *Anabaena variabilis*, prepared by using 0·75 and 0·2% Triton X–100, respectively. Note the differences in the absorption at 483, 495 and 520 nm wavelengths (related to carotenoids) and at 672 and 680 nm wavelengths (associated with chlorophyll). After T. Ogawa, L. P. Vernon and H. H. Mollenhauer (1969) *Biochim. biophys. Acta* **172**, 216, Fig. 2.

(Fay, 1969b). Chemically, phycobiliproteins comprise a pigmented prosthetic group (bilin) attached to a protein (biliprotein). Bilins, so called because they occur in the bile juices of animals, are open-chain tetrapyrroles (conjugated chains of four pyrrole rings). They may be formed by oxidation of porphyrins, which are closed-ring structures composed of four pyrrole nuclei, such as occur in haemoglobin and chlorophyll. The probable structures are:

HOOC COOH
CH_3 CH_2 CH_2 CH_3
H CH CH_3 CH_2 CH_2 CH_3 CH_3 CH_2
H_3C
A B C D
O N N N N O
H H H H H H H

Phycocyanobilin

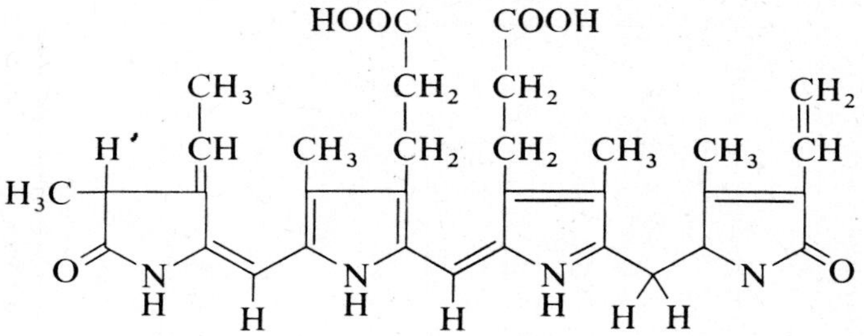

Phycoerythrobilin

These structures are according to Chapman *et al.* (1967). Schram and Kroes (1971) have proposed a slightly different structure for phycocyanobilin.

Recently Glazer and Cohen-Bazire (1971) have shown that the phycobiliprotein "monomers" are themselves made up to two distinguishable protein subunits designated α and β, of molecular weight in the region of 15–22,000, and each carrying chromatophores. This proves to be true of all three phycobiliproteins found in blue-green algae (see Table 3.II), with the exception that there are two species of the β-subunit in allophycocyanin of *Aphanocapsa*.

TABLE 3.II. Properties of the phycobiliprotein subunits isolated from *Aphanocapsa* sp. (from Glazer and Cohen-Bazire, 1971)

			Subunit composition		
Protein	λ_{max}		*Mol. wt.*	*Colour*	*Fluorescence*
1. Phyococyanin	620	α	16,600	purple	strong red
		β	20,200	violet	red
2. Allophycocyanin	650	α	16,000	violet grey	very weak red
		β_1	17,000	violet grey	
		β_2	17,900	violet grey	
3. Phycoerythrin	565	α	20,000	pink	orange
		β	22,000	pink	reddish

Glazer *et al.* (1971) investigated the immunology of various phycobilins extracted from a wide variety of unicellular and filamentous blue-green algae and red algae. They found that allophycocyanins from each of the algae investigated were immunologically identical, indicating the presence of a strong determinant common to the protein in each organism. The same proved to be true with *c*-phycocyanins and also with phycoerythrins. However, the three spectroscopically distinct phycobilins themselves showed no cross reactions, even when they were isolated from the same organism. This, and other evidence, demonstrated that the common determinants of particular phycobilins in different algal groups must reside in the biliproteins rather than in the bilins, which are identical in phycocyanins and allophycocyanins (Siegelman *et al.*, 1966) and would consequently be cross-reactive.

In contrast to the surprising degree of cross-reactivity by phycobilins in the blue-green and red algae, investigations with the phycoerythrin of a *Cryptomonas* species indicate that the phycobilins of the Cryptophyceae are quite distinct. This difference is significant in relation to the fact that these pigments do not form phycobilisomes in cryptophytes, but are contained in the intrathylakoidal spaces (Gantt *et al.*, 1971).

The role of these pigments in the photosynthesis of blue-green algae is discussed later (p.145). Here, it is sufficient to note that transfer of absorbed light energy to other components of the photochemical system occurs with high efficiency, indicating that the interacting pigment molecules are in fairly close association. Theoretically a proximity of between 0·7–1 nm and not more than 5 nm must be assumed for resonance transfer of excitation energy to be efficient (Arnold and Oppenheimer, 1950; Brody and Vatter, 1959). However, phycobiliproteins are readily released from lysing cells. They can be washed completely from thylakoid fragments leaving the chlorophyll and carotenoids in the membranes (Calvin and Lynch, 1952; Shatkin, 1960). This indicates that the phycobiliprotein pigments, although in close proximity to the chlorophyll molecules, are not integral parts of the thylakoid membranes. Some workers (Bergeron, 1963) suggested that the pigment is located between the thylakoids, and others believed that phycocyanin is an integral part of the lamella, or that it is present in the lumen between the thylakoid membranes.

The first experimental evidence indicating the possible location of phycobiliprotein came from studies on the plastid lamellae of the unicellular red alga, *Porphyridium cruentum* (Gantt and Conti, 1966). Granules about 35 nm in diameter arranged in regular arrays on the outer surface of the thylakoid membranes were detected but only when cells were prefixed with glutaraldehyde. Similar granules were not seen in broken cells from which phycobilins had been released. These granules may represent macromolecular aggregations of phycobiliproteins and they were therefore termed *phycobilisomes*. Phycobilisome-like granules had been noted, but only later recognized as such, in the endosymbiont of *Glaucocystis nostochinearum* (Lefort, 1965). They have also been found in *Tolypothrix tenuis* and *Fremyella diplosiphon* (Gantt and Conti, 1969) and *Synechococcus lividus* (Edwards and Gantt, 1971; Figs 3.10 and 3.11). The arrangement of granules on the thylakoid membranes is similar to that seen in *Porphyridium* with the rows of granules alternating on the membrane surfaces of

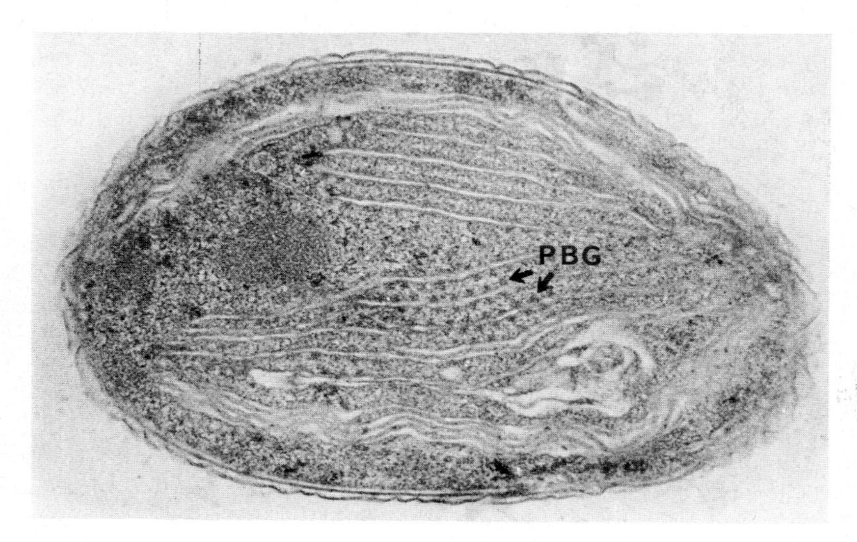

FIG. 3.10. Section of a filamentous marine blue-green alga showing arrays of phycobilisome granules (PBG) in the interthylakoidal space. Electron micrograph (×60,000). By courtesy of E. Gantt and S. F. Conti (1969) *J. Bacteriol.* **97**, 1486, Fig. 4.

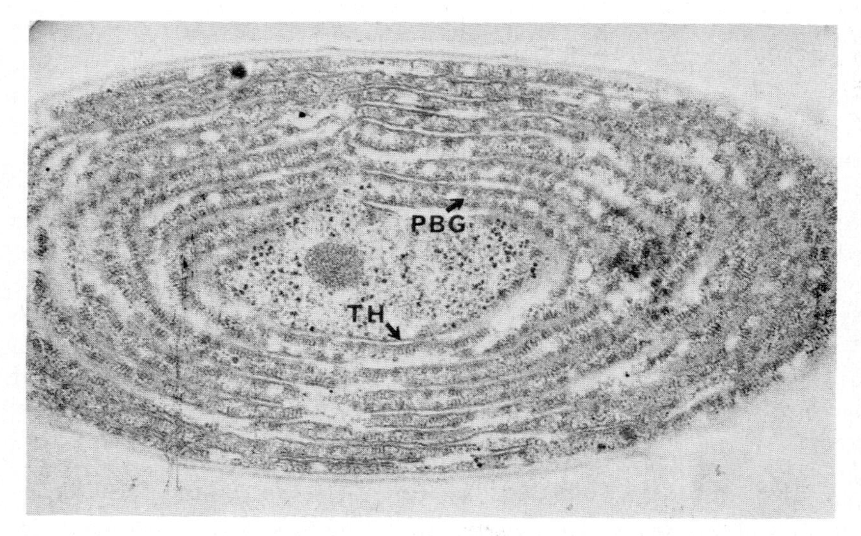

FIG. 3.11 Slightly oblique section of *Synechococcus lividus* displaying various planes of section of the concentrically arranged thylakoids (TH) and of the attached phycobilisome granules (PBG). Electron micrograph (×42,000). By courtesy of M. R. Edwards and E. Gantt (1971) *J. Cell Biol.* **50**, 896, Fig. 1.

adjacent thylakoids (Fig. 3.12). The glutaraldehyde treatment appears to stabilize the otherwise labile association between the chlorophyll-containing thylakoids and phycobilisomes (Cohen-Bazire and Lefort-Tran, 1970). Hallier and Park (1969) have demonstrated that glutaraldehyde fixation does not inhibit photosystem I and II reactions in *Anacystis nidulans*.

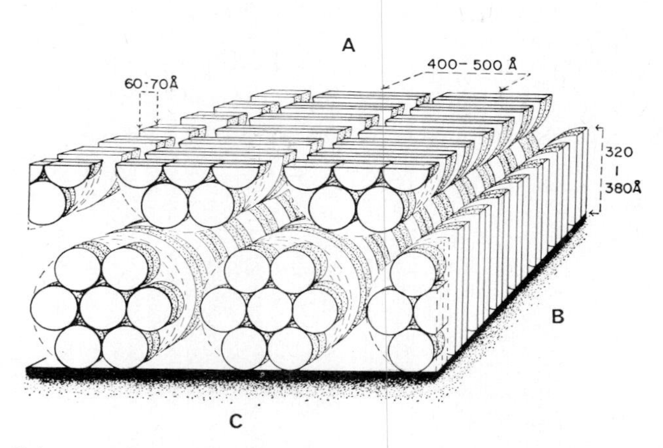

FIG. 3.12. Schematic three-dimensional representation of phycobilisomes in relation to the thylakoid membrane. The disc-shaped units appear as short dense lines in a grazing section parallel to the membrane (A), and in a longitudinal cut passing through the membrane (B); they show a round broad face in cross section (C) which also reveals the arrangement of subunits. By courtesy of M. R. Edwards and E. Gantt (1971) *J. Cell Biol.* **50**, 896–900.

Most interesting are the studies by Berns and Edwards (1965) on the ultrastructural aspects of *c*-phycocyanin extracted from *Plectonema calothricoides*. With the aid of negative contrasting electron microscopy they showed that phycocyanin is present in four different structural units, i.e. in the form of monomer, trimer, hexamer and dodecamer, the equilibrium between these forms being affected by pH. The monomer appears rod shaped 7 nm long and 1·5–2 nm in diameter when unfolded, but *in vivo* is probably present in a folded conformation. The ring-shaped hexamer, with an overall diameter of 13 nm and a molecular weight of 200,000, is probably the predominant species *in vivo*. In this structure the tetrapyrrole chromophores are probably oriented in a random array. Such an orientation would promote radiant energy transfer in any direction. It was postulated that the chlorophyll molecule may be associated with each monomer unit in the hexamer as this would favour efficient energy transfer. However, the evidence is not yet sufficient for further speculation.

The nucleoplasmic region

The presence of DNA in the cells of blue-green algae was demonstrated in early studies (Poljansky and Petruschovsky, 1929; Spearing, 1937; Fogg, 1951a) in which light and ultraviolet microscopy were applied in combination with cytochemical techniques. These investigations failed to demonstrate convincingly either the presence of an organized nucleus or of distinct chromatic structures resembling plant or animal chromosomes. Nor was there any convincing indication of a recurrent process corresponding to the mitotic cycle in eukaryotic cells. Nevertheless, attempts were made quite recently to correlate structures giving DNA reaction with

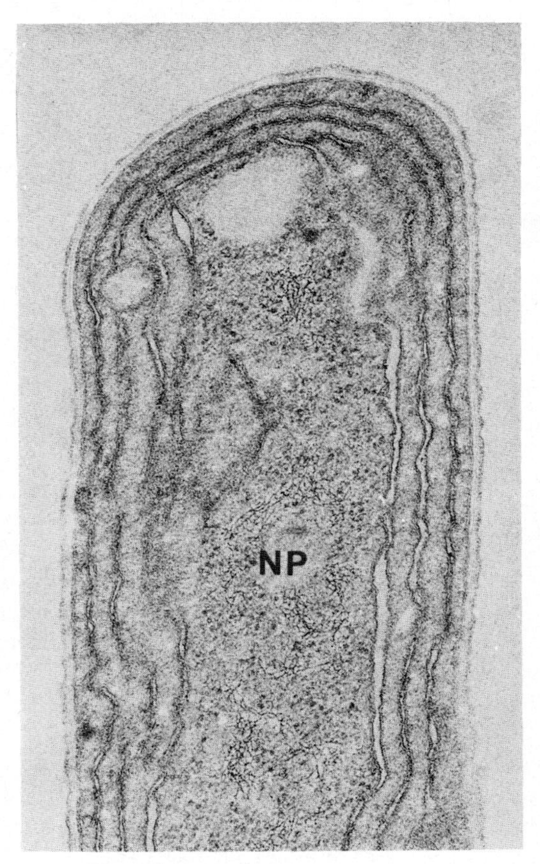

FIG. 3.13. Part of a longitudinal section of *Synechococcus lividus* showing DNA fibrils and ribsomes within the nucleoplasmic region (NP). Electron micrograph ($\times 60{,}000$). By courtesy of M. R. Edwards, D. S. Berns, W. C. Ghiose and S. C. Holt (1968). *J. Phycol.* **4**, 283, Fig. 3.

"chromatic elements" and to compare various stained patterns with the stages of a mitotic cycle (Leak and Wilson, 1960; Hofstein and Pearson, 1965).

An early electron microscope examination of *Phormidium* (Drews and Niklowitz, 1956) detected only granular inclusions within the electron-transparent region of the cytoplasm. Using improved techniques of fixation Hopwood and Glauert (1960) showed that the Feulgen-positive material is dispersed in the form of fine fibrils 5–7 nm thick. It was suggested that these fibrils correspond to DNA macromolecules. If this is correct the distribution of DNA in blue-green algae is similar to that found in bacteria (Kellenberger *et al.*, 1958; Ris and Singh, 1961). The genetic material in prokaryotic cells seems to be organized as a three-dimensional network of anastomosing fibrils (Pankratz and Bowen, 1963; Fig. 3.13). This pattern is slightly altered by fixation procedures which affect the degree of aggregation of the finest DNA fibrils, measuring 2–3 nm in diameter Leak, 1967a). The fibrils are not associated with histone type proteins similar to those found in eukaryotic chromosomes (Ris and Singh, 1961; Fuhs, 1964).

Evidence for the location of DNA has been provided by the Feulgen test, by u.v. fluorescence of acridine orange and by autoradiographic techniques following the incorporation of ^{3}H-labelled thymidine into rapidly growing cells of an *Anabaena* species (Leak, 1965). Silver grain formation is associated with the central and peripheral nucleoplasmic regions (Fig. 3.14). Cells previously treated with the enzyme desoxyribonuclease or with trichloro-acetic acid (both known to extract DNA) did not produce the reactions characteristic for DNA.

Ribosomes

The most common cytoplasmic elements in blue-green algae are electron-dense granules 10–15nm in diameter which can be easily detected by electron microscopy especially after fixation with osmium tetroxide and staining with uranyl acetate (Ris and Singh, 1961). They are concentrated mainly in the central electron-transparent regions of the cytoplasm, in close contact with DNA fibrils, but they can be observed throughout the cytoplasm in areas not occupied by thylakoids (Fig. 3.15). These granules resemble in size and appearance ribosomes of bacteria (although bacterial ribosomes are usually not associated with DNA fibrils). No ribosome-like granules are seen in the inter-thylakoidal regions (Edwards *et al.*, 1968).

The distribution of RNA corresponds to the distribution of these granules (Ris and Singh, 1961) as shown by staining with pyronine before and after treatment with the enzyme ribonuclease. Ribonuclease-treated cells of *Pseudanabaena catenata* and of *Oscillatoria amoena* did not show any ribosome-like granules (Fuhs, 1963). In freeze-etched preparations of

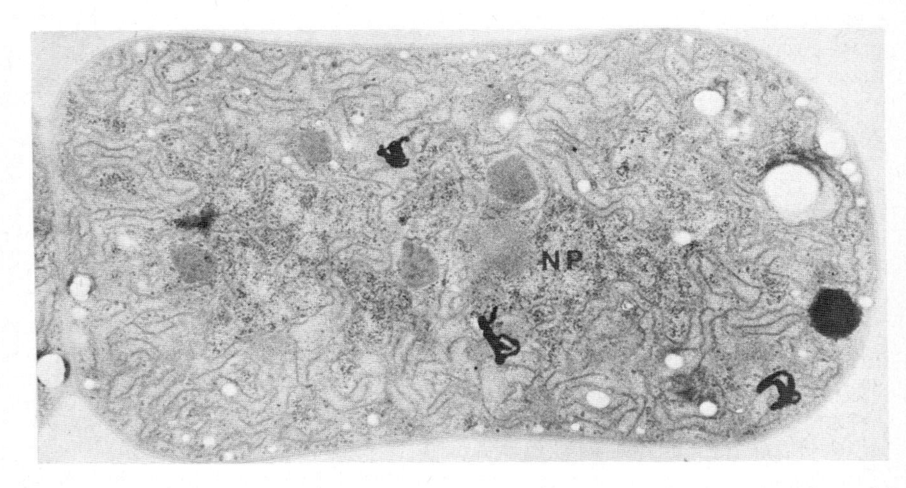

FIG. 3.14. Electron microscopic autoradiograph of a longitudinal section through a ^{3}H-thymidine-labelled *Anabaena* cell. Silver grains appear over the nucleoplasmic region (NP) (× 25,200). By courtesy of L. V. Leak (1965), *J. Ultrastruct. Res.* **12**, 135, Fig. 5.

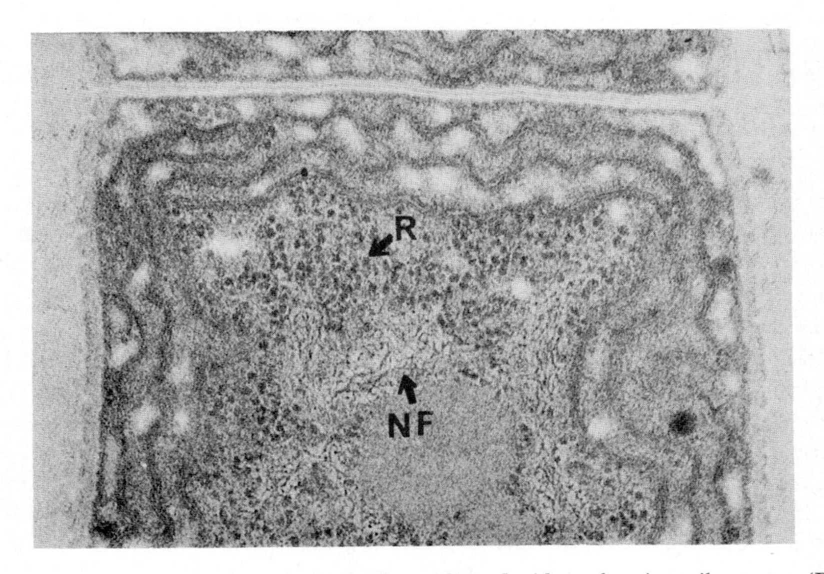

FIG. 3.15. Longitudinal section of *Phormidium luridum* showing ribosomes (R) in close contact with DNA fibrils (NF) in the nucleoplasmic area. Electron micrograph (× 80,000). Courtesy of M. R. Edwards (unpublished).

Oscillatoria rubescens ribosomes have been frequently seen arranged in rows of up to 15 (Fig. 3.16). These structures may correspond to polyribosomes (Jost, 1965).

FIG. 3.16. Image of the nucleoplasm of *Oscillatoria rubescens* prepared by freeze-etching, and displaying polyribosome-like arrays (arrows). Electron micrograph (×60,000). By courtesy of M. Jost (1965) *Arch. Microbiol.* **50**, 211, Fig, 42.

Ribosomes from cell homogenates of *Anacystis nidulans*, *Anabaena cylindrica* (Taylor and Storck, 1964), *Anabaena variabilis* (Craig and Carr, 1967, 1968) and *Phormidium luridum* (Vasconcelos and Bogorad, 1971) possess similar sedimentation characteristics to those from bacteria such as *E. coli*. Both correspond to the 70s (= Svedberg sedimentation coefficient) type of ribosome, also present in chloroplasts, and differ from the 80s type of ribosome characteristic of the cytoplasm of eukaryotic cells. Cleavage of blue-green algal ribosomes to form 50s and 30s sub-units also follows the bacterial pattern and the molecular weights of the rRNAs of various species (0·55–0·56 and 0·68–0·69 million daltons) resemble those of *E. coli* rRNA's (0·56 and 1·07 million daltons) (see Loening, 1968; Howland and Ramus, 1971; Vasconcelos and Bogorad, 1971). On the other hand Vasconcelos and Bogorad (1971) have shown that the ribosomal proteins of *P. luridum* differ from those of *E. coli* and from higher plant chloroplasts.

Cytoplasmic inclusions

1. POLYGLUCAN GRANULES

The presence of a glycogen-type photosynthetic product was earlier assumed on the evidence of the brown coloration of the "chromatoplasm" on treatment with iodine (Fritsch, 1945). Ris and Singh (1961) observed 25nm particles in the space between the thylakoids in many species. These granules showed a special affinity for the lead hydroxide used for post-staining (Fig. 3.17). Similar granules were called "α-granules" by Pankratz

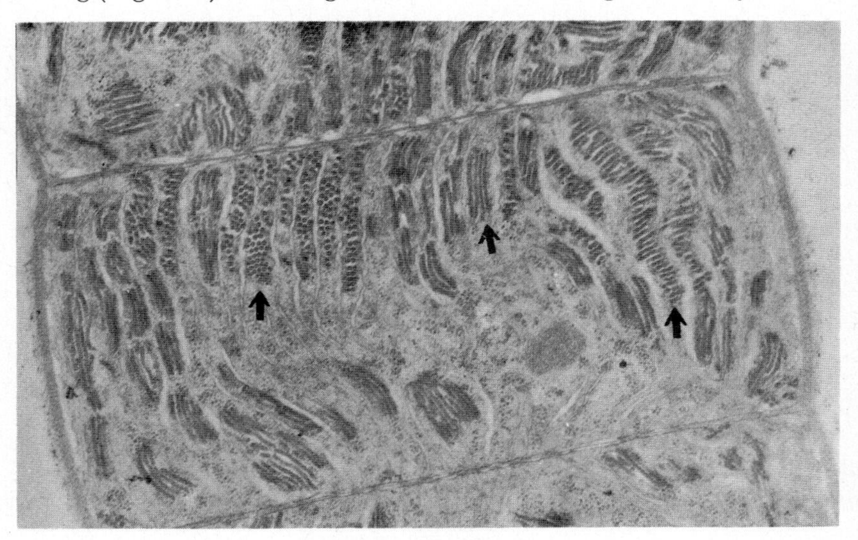

FIG. 3.17 Longitudinal section of *Oscillatoria chalybea* showing stacks of rod-shaped polyglucan granules (arrows) deposited in the interthylakoidal space. Electron micrograph (× 17,000). By courtesy of M. Giesy (1964) *Am. J. Bot.* **4**, 388, Fig. 1.

and Bowen (1963). The granules react with periodic acid Schiff's (PAS) reagent for carbohydrate but the intensity of the reaction greatly depends on the age and the conditions of the material (Giesy, 1964). The reaction does not occur and granules are absent in cells previously exposed to diastase (Fuhs, 1963).

A massive accumulation of polyglucan granules has been observed in heterotrophically grown cells of *Nostoc muscorum* (Wildon and Mercer, 1963a) and *Chlorogloea fritschii* (Peat and Whitton, 1967). In freeze-etched preparations of *Oscillatoria rubescens* polyglucan granules appear as densely packed rods ("botuli") with a striated substructure. In developing hormogonia these granules gradually disappear from the cytoplasm (Jost 1965).

Hough *et al.* (1952) extracted a reserve polysaccharide from a species of *Oscillatoria*, which is presumably the component of polyglucan granules. It yielded only D-glucose on hydrolysis and analysis indicated a cold water-soluble compound similar to amylopectin, with an average chain length 14–16 residues. Chao and Bowen (1971) have isolated polyglucan granules from *Nostoc muscorum*. They measure about 31 × 65 nm and appear to be composed of two main parts and several small sub-units. Chemical analysis shows the presence of highly branched polyglucosyl units with short side chains, indicating a similarity to animal glycogen. Periodate oxidation suggests the average chain length is 13 glucosyl units. Fredrick (1951) demonstrated an enzyme in *Oscillatoria princeps* which is capable of synthesizing a similar polysaccharide, with an average chain length of 14–16 residues, from glucose-1-phosphate, and has more recently (Fredrick, 1971) demonstrated the presence of branching enzymes, which are capable of introducing branches in polyglucoside chains, in several blue-green algae.

The studies of Tsusue and her collaborators have shown that blue-green algae also produce oligosaccharides which may be specific to this group of organisms. In species of *Anabaena*, *Anabaenopsis*, *Calothrix*, *Nostoc* and *Tolypothrix*, Tsusue and Fujita (1964) found a series of oligosaccharides having one glucose residue followed by an increasing number of fructosyl units. Tsusue and Yamakawa (1965) tentatively proposed fructofuranosyl-$(2\rightarrow4)_n$-fructofuranosyl-$(2\rightarrow1)$-α-glucopyranoside as the formula for the series. These oligosaccharides are, however, soluble substances and are unlikely themselves to be components of any histologically recognizable structure.

2. LIPID DROPLETS

Spherical osmiophilic granules of about 30–90 nm in diameter are common cytoplasmic inclusions in blue-green algae (Drews and Niklowitz, 1956; Pankratz and Bowen, 1963). They are scattered among the thylakoids but are more frequent near the cell surface (Fig. 3.18). They resemble the lipid droplets ("plastoglobuli") of chloroplasts but usually have a denser periphery and a less dense core (Lang, 1968). They are probably reserve products but little is known about their metabolic relations.

3. CYANOPHYCIN (STRUCTURED) GRANULES

Cells of blue-green algae frequently appear under the light microscope to be packed with large conspicuous granules of two principal types. Those

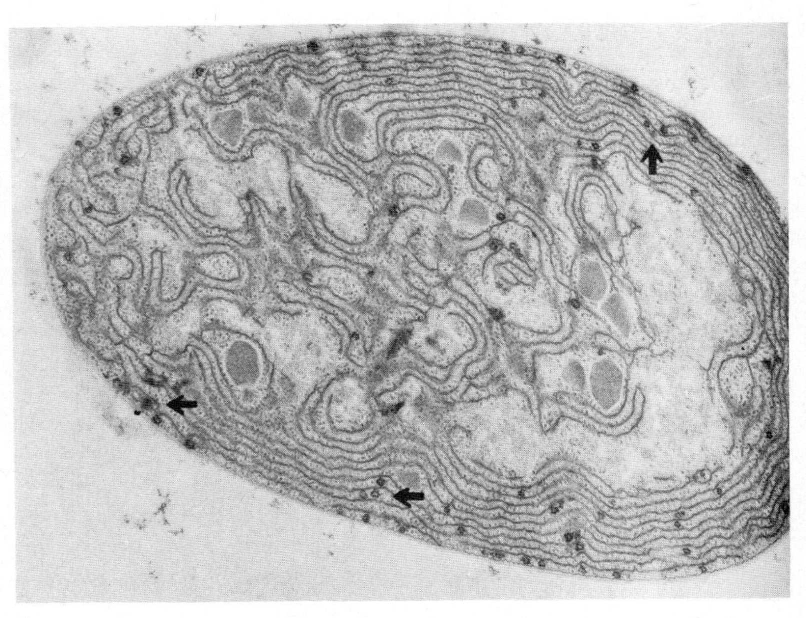

FIG. 3.18. Section of a vegetative cell of *Anabaena azollae* showing lipid droplets (arrows) mostly in the peripheral region. Electron micrograph (×10,000). By courtesy of N. J. Lang (1965) *J. Phycol.* **1**, 127, Fig. 1.

which do not stain with methylene blue or toluidine blue but which stain with carmine or neutral red were called "cyanophycin" granules (Fritsch, 1945). Fogg (1951a) used the Sakaguchi test, which is specific for arginine, an amino acid which occurs in blue-green algal proteins, to demonstrate that the latter were protein-containing granules. They are absent from younger cells but are abundant in older cells (Tischer, 1957) and in spores (Miller and Lang, 1968).

Drews and Niklowitz (1956), who carried out electron microscopy on *Phormidium*, *Oscillatoria*, *Anabaena* and *Cylindrospermum* species, first described them as "structured granules". According to their observations the granules reduce various tetrazolium salts and selectively absorb the dye Janus green and for this reason were considered as "mitochondrial equivalents". The latter supposition has been questioned by other workers (Ris and Singh, 1961; Fuhs, 1963) and has not been confirmed in critical studies (Bisalputra *et al.*, 1969; Brown and Bisalputra, 1969).

Structured granules have an irregular shape and may measure up to 500 nm in diameter (Fig. 3.19). Their appearance varies according to the duration and type of fixation, temperature and electron bombardment during examination under the electron microscope (Lang and Fisher,

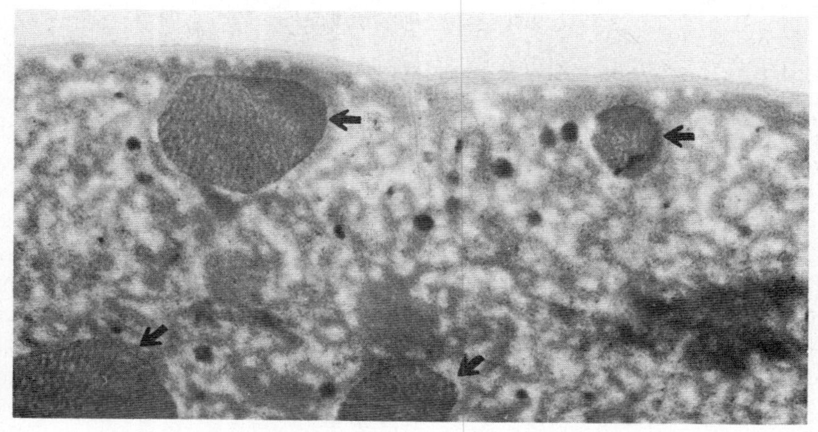

FIG. 3.19. Portion of a dividing cell of *Anabaena* sp. displaying "structured" cyanophycin granules (arrows). Electron micrograph (×29,400). By courtesy of N. J. Lang and K. A. Fisher (1969) *Arch. Mikrobiol.* **67**, 173, Fig. 5.

1969). Under certain conditions of fixation and poststaining they appear as a mass of tightly packed undulating membranes (Ris and Singh, 1961) and it has been suggested (Brown and Bisalputra, 1969) that they may form from thylakoid membranes and serve as a reserve of membrane material. However, this possibility is rendered most unlikely by the results of chemical analyses recently performed on cyanophycin granules isolated from *Anabaena cylindrica* by Simon (1971). Simon was unable to detect any pigments, lipids, or significant quantities of carbohydrate in his preparations. An unusual type of protein, containing only the two amino acids, arginine (explaining Fogg's (1951a) observations) and aspartic acid, accounted for 98% of the dry weight of the granules. The protein exists as a population of heterogeneous molecular weights, from 25,000 to 100,000. Considering its composition it seems unlikely that it would have any metabolic function, and Simon has suggested that it provides a store of combined nitrogen, which seems likely in view of the fact that arginine has four nitrogen atoms per molecule.

Stewart (1972a) has reported that while cyanophycin granules develop abundantly in cultures supplied with high levels of combined nitrogen, they are rare in nitrogen-deficient cultures.

The identity of cyanophycin granules and structured granules has been confirmed by electron microscope studies on the isolated structures by Lang *et al.* (1972).

4. POLYPHOSPHATE BODIES

Structures originally termed "metachromatic" or volutin granules (Fritsch, 1945) can be seen clearly under the light microscope in vegetative cells of blue-green algae. Unlike "cyanophycin" granules they stain with methylene blue, changing its colour to red, hence the name metachromatic. They are small and scarce in young cultures of *Anabaena cylindrica* but become large and prominent in old cultures. Tischer (1957) showed that such granules do not form when the algae are grown in phosphorus-deficient medium. When cells containing them are extracted with trichloroacetic acid no metachromatic staining is observed and phosphate can be precipitated from the extract with barium acetate (Talpasayi, 1963). Earlier suggestions that metachromatic granules are at least partly organic

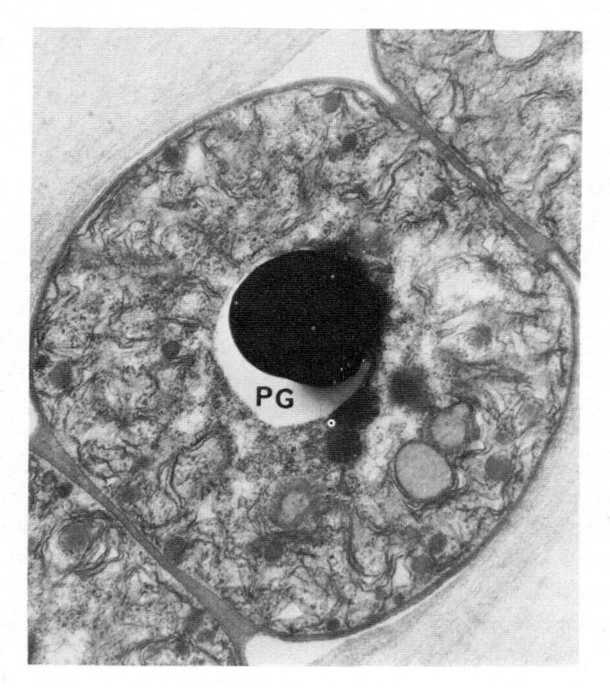

FIG. 3.20. Longitudinal section of a cell of *Nostoc pruniforme* with a large centrally located polyphosphate granule (PG). Electron micrograph (× 24,000). By courtesy of T. E. Jensen (1968) *Arch. Mikrobiol.* **62**, 144, Fig. 3.

and may contain DNA fibrils and ribonucleoprotein were not confirmed in an examination of *Oscillatoria amoena* by Fuhs (1958b) and these structures are now usually referred to as polyphosphate bodies.

Polyphosphate bodies are not easily preserved by conventional methods of fixation and often disintegrate under an intense electron beam, or turn into so-called "alveolar bodies" (Hall and Claus, 1965; Edwards *et al.*, 1968). Empty areas in electron micrographs of the cytoplasm generally indicate the removal of polyphosphate. However, if well preserved, polyphosphate granules appear as fairly large (100–500 nm in diameter) and electron dense inclusions (Drews and Niklowitz, 1957; Pankratz and Bowen, 1963; Jensen, 1968, 1969; Stewart and Alexander, 1971; Fig. 3.20). Phosphate deposition follows a characteristic pattern in *Plectonema boryanum* where the granule is preformed as a porous structure of medium electron density. Polyphosphate is deposited first in the surrounding area of the cytoplasm, and gradually penetrates into the porous structure until the latter becomes a dense electron-opaque body (Jensen, 1969; Fig. 3.21).

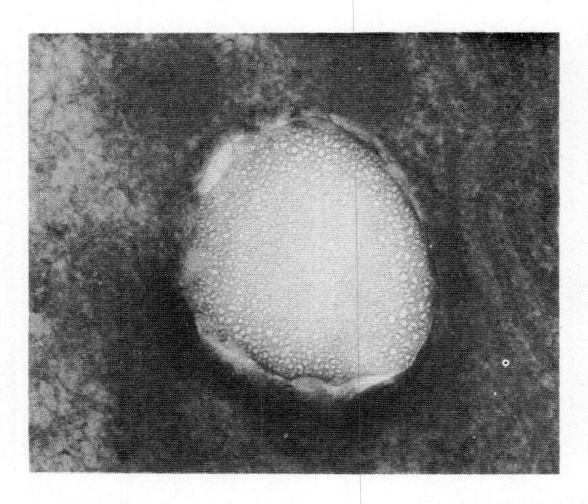

FIG. 3.21. Section of a developing polyphosphate granule of *Plectonema boryanum* displaying a porous substructure. Electron micrograph (× 100,000). By courtesy of T. E. Jensen (1969) *Arch. Mikrobiol.* **67**, 328, Fig. 7.

5. POLYHEDRAL BODIES

Electron microscopy has revealed another type of large cytoplasmic inclusion about 200–300 nm in diameter which has a distinct polygonal profile and a medium electron density. They were first observed in *Nostoc pruniforme* by Jensen and Bowen (1961) and called "polyhedral bodies" and since then have been found commonly in various blue-green algae,

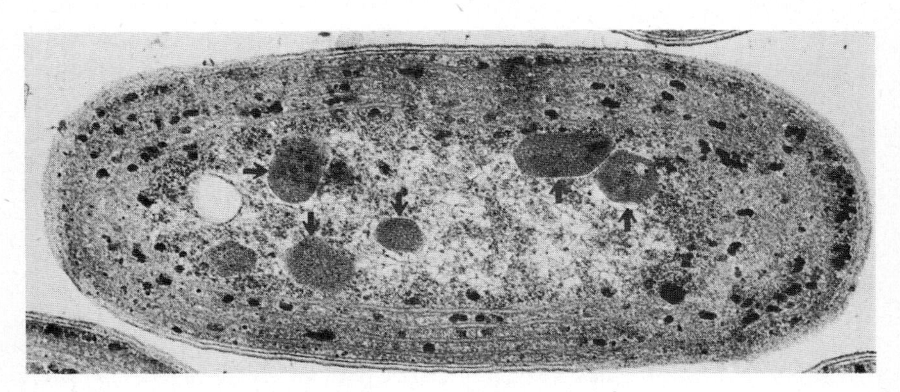

FIG. 3.22 Longitudinal section of *Anacystis nidulans* (*Synechococcus* sp.) showing several polyhedral bodies (arrows) in the nucleoplasmic region. Electron micrograph (×44,000). By courtesy of E. Gantt and S. F. Conti (1969) *J. Bacteriol.* **97**, 1486, Fig. 6.

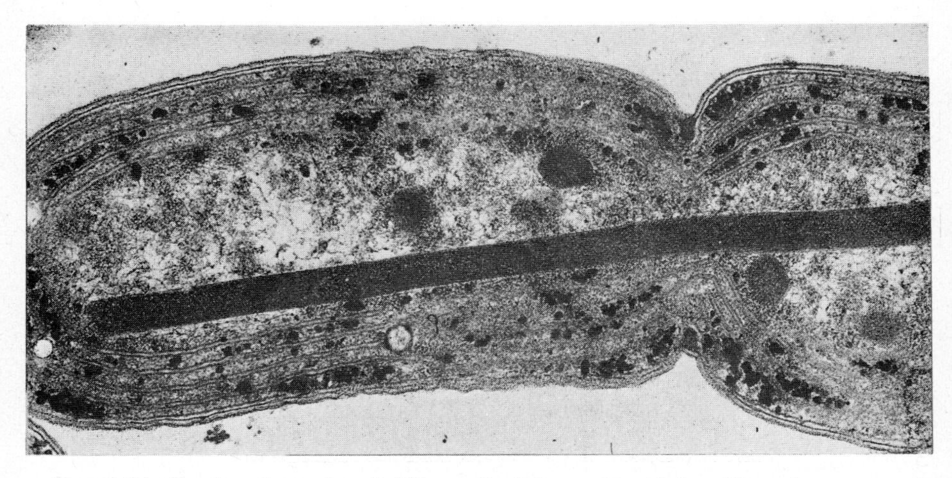

FIG. 3.23. Section through a dividing cell of *Anacystis nidulans* (*Synechococcus* sp.) with a long rod-shaped polyhedral body extending across the potential daughter cells. Electron micrograph (×40,000). By courtesy of E. Gantt and S. F. Conti (1969) *J. Bacteriol.* **97**, 1486, Fig. 7.

where they occur in the fibrous nucleoplasmic region (Fig. 3.22). Occasionally these bodies are surrounded by a dark narrow line about 3 nm thick. In *Anacystis nidulans* polyhedral bodies have been found as long, straight, rigid rods, often spanning the length of two cells and preventing the completion of cell division (Gantt and Conti, 1969; Fig. 3.23). They show a highly ordered assembly of identical subunits which results in the crystalline appearance of the bodies (Edwards *et al.*, 1968; Gantt and Conti, 1969; Fig. 3.24).

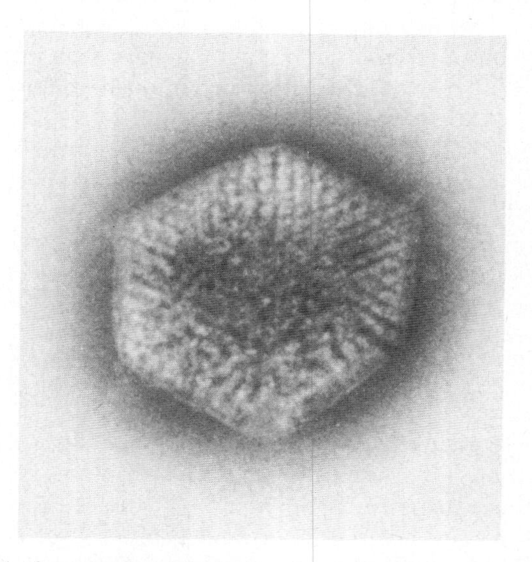

FIG. 3.24. Negatively stained polyhedral body of *Anacystis nidulans* (*Synechococcus* sp.) revealing crystalline substructure with regular arrangement of subunits. Electron micrograph (× 250,000). By courtesy of E. Gantt and S. F. Conti (1969) *J. Bacteriol.* **97**, 1486, Fig. 9.

Large basophilic structures similar to polyhedral bodies were regarded as nuclear equivalents of blue-green algae (Hall and Claus, 1962, 1965) but in view of the lack of experimental evidence these suggestions have been regarded with reservation (Wildon and Mercer, 1963a; Edwards *et al.*, 1968). Unfortunately nothing is so far known about the composition and function of polyhedral bodies but their consistent presence in the nucleoplasmic region favours the idea (Edwards *et al.*, 1968) of a role in the storage of ribonucleoprotein.

6. UNUSUAL INCLUSIONS

Many bacteria contain as a food reserve the compound poly-β-hydroxybutyric acid (PHB) and Carr (1966) has demonstrated that this product also occurs in the blue-green alga *Chlorogloea* (*Chlorogloeopsis*) *fritschii*. Jensen and Sicko (1971a) found that in addition to the various granules described in the foregoing section, there appeared in thin sections of this alga spherical or elongate, slightly electron-dense, granules of diameter varying from 100–800 nm. Surrounded by a membrane approximately 3 nm thick, they resembled structures which have been identified as PHB granules in a number of bacteria in which this reserve material occurs. These granules may be more common in blue-green algae than hitherto supposed as they are difficult to preserve with conventional techniques of electron microscopy.

Jensen and Bowen (1970) in an extensive electron microscope survey of blue-green algae detected several unusual inclusions. These include membranous "stacks", spheres and "scrolls", clusters of small "spheroids" in the interthylakoidal space, and a variety of filamentous and crystalline arrays. Most of these structures were observed only occasionally. They consider that at least some of these unusual structures are produced on virus infection (see p. 285).

Surface structures

The surface structures surrounding blue-green algae may be scarcely detectable but more often the cells are enclosed in an extensive structure which is sometimes many times the volume of the cell proper.

These structures are heterogeneous and composed of two main layers. Adjoining the plasma membrane is the *cell wall* ("inner investment" or "Hautschicht"), which may be encompassed by the sheath ("Wandschicht"). Fritsch (1945) deplored the fact that most workers had failed to distinguish between these two components. Early electron microscope studies did not improve on this situation and perhaps added to the confusion. Lang (1968) has attempted to clarify the position as follows: "Usage of terms such as cell envelope, cell boundary or inner investment should now be replaced by the term *cell wall*, designating the multilayered structure external to the *plasmalemma* and internal to the *sheath* (if present). One morphological feature which may serve to separate the outermost layer of the cell wall and innermost layer of the sheath is the globular substructure of the former and unglobular (i.e. fibrous) structure of the latter".

1. CELL WALL

The presence of a cell wall in blue-green algae was demonstrated by early workers (Gomont, 1888; Geitler, 1936) who soaked various blue-green algae in strong chromic acid, a treatment which dissolved away the protoplast and left behind the cell wall which could then be examined by the light microscope. This rather drastic treatment incidentally demonstrates the resistance of the surface structures to extreme chemical conditions.

The wall of a blue-green alga appears as a multilayered structure under the electron microscope. Some of these structures were observed by early workers (Niklowitz and Drews, 1956; Pankratz and Bowen, 1963) but it was Jost (1965), using replicas obtained from freeze-etched preparations of *Oscillatoria rubescens*, who first clearly demonstrated the presence of four distinct layers, each about 10 nm thick, between the plasma membrane and the sheath (Fig. 3.25). These four layers are designated as L_I, L_{II}, L_{III}, and L_{IV}, in an outward order. All have a globular substructure. The

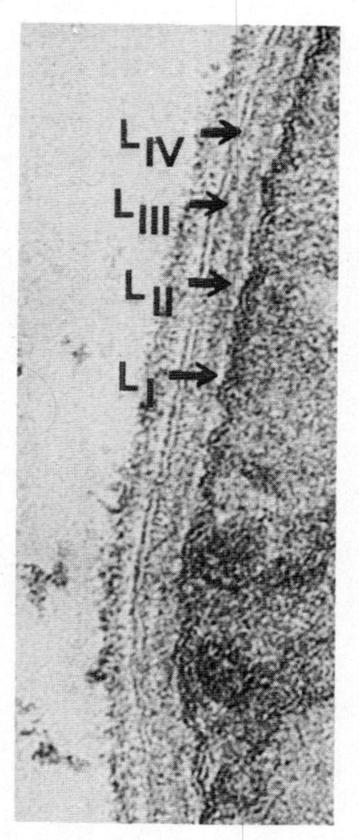

FIG. 3.25. Portion of a section of *Oscillatoria rubescens* showing the construction of the cell wall with four wall layers (L_I–L_{IV}). Electron micrograph (×110,000). By courtesy of M. Jost (1965) *Arch. Mikrobiol.* **50**, 211, Fig. 2.

innermost L_I layer is electron transparent after chemical fixation and contains about 500 granules, each about 7 nm in diameter, per square micron. Jost postulated that this layer is composed of a mucopolymer present in the form of fine fibrils and disposed in a reticulum. Others (Lamont, 1969a) associate the next, L_{II}, layer with the mucopolymer. Evidence for this was obtained by Jensen and Sicko (1971b) in an electron microscope study which showed that treatment with lysozyme resulted in the removal of the L_{II} layer in a *Cylindrospermum* species. L_{II} is electron dense after permanganate fixation while L_{III} is electron transparent. L_{IV}, which again appears electron dense, shows a unit membrane-like substructure with two electron opaque components separated by a less electron dense middle component (Edwards *et al.*, 1968; Lamont, 1969a). This outermost layer often shows a wavy appearance (Fig. 3.26) but this seems to be an artifact

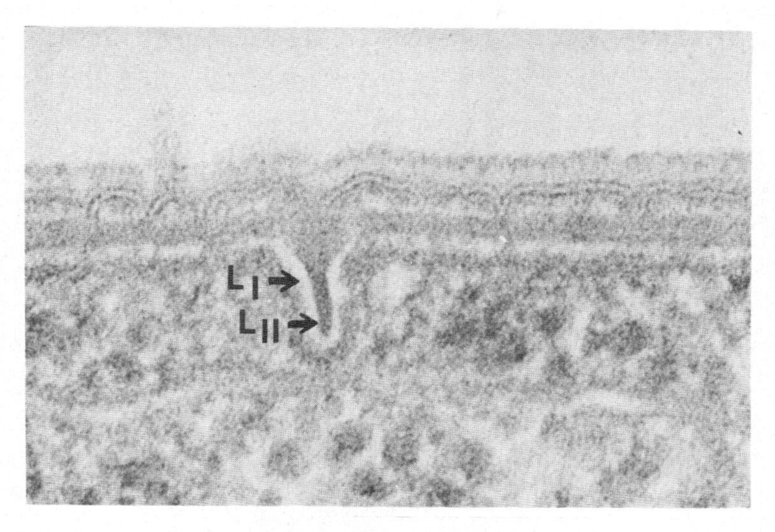

FIG. 3.26. Portion of a longitudinal section of *Lyngbya* sp. illustrating initial stage of septum formation by the ingrowth of two inner cell wall layers (L_I and L_{II}). Electron micrograph (× 210,000). By courtesy of H. C. Lamont (1969) *J. Bacteriol.* **97**, 350, Fig. 4.

and does not occur after freeze-etching (Jost, 1965). There is little evidence available about the chemical nature of the individual layers of the cell wall. Septa are thinner than the external walls. They form in the process of cell division and appear as annular ingrowths. This is discussed in more detail on p. 219.

Intercellular contact between adjacent cells of the filament is probably maintained by means of fine, plasmodesma-like structures (about 15 nm wide) which traverse the septum and connect the plasma membranes of the two cells (Drawert and Metzner, 1956; Pankratz and Bowen, 1963; Peat and Whitton, 1968; Fig. 3.27). Similar structures are present in the septum between vegetative cells and heterocysts (Wildon and Mercer, 1963b) and have been termed "microplasmodesmata" by Lang and Fay (1971). In *Anabaenopsis* species it has been calculated that there may be up to 4000 microplasmodesmata between two vegetative cells (Peat and Whitton, 1968).

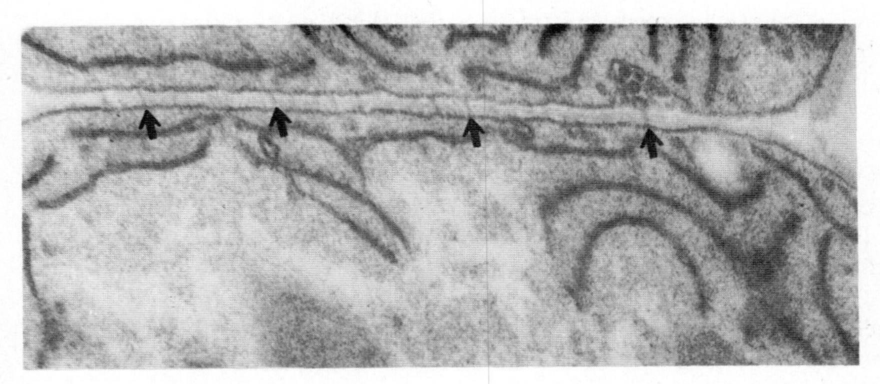

FIG. 3.27. Section through the cross wall between vegetative cells of *Anabaenopsis* sp. displaying intercytoplasmic connexions called "microplasmodesmata" (arrows). Electron micrograph (×75,000). By courtesy of A. Peat and B. A. Whitton (1968) *Arch. Mikrobiol.* **63**, 170, Fig. 3.

Other pore-like structures have been detected in the longitudinal walls of various blue-green algae (Metzner, 1955; Drawert and Metzner, 1956, 1958; Hagedorn, 1960; Pankratz and Bowen, 1963). Some are perhaps more properly described as pits or depressions and are larger than junctional pores (Schultz, 1955; Ris and Singh, 1961). The latter are located at both sides of the junction of the septum with the longitudinal wall and are regularly spaced at 18–25 nm intervals. They are actually narrow channels, about 18 nm long and 10 nm in diameter and extend between the plasmalemma and the L_{III} layer of the cell wall (Lamont, 1969a; Fig. 3.28).

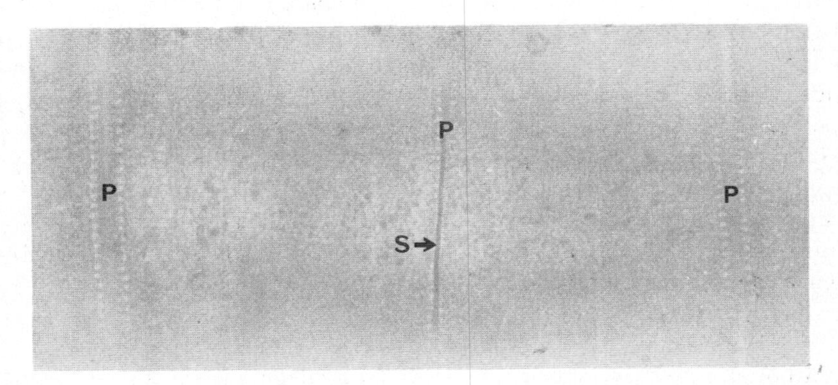

FIG. 3.28. Grazing longitudinal section along the trichome of *Lyngbya* sp. showing rows of junctional pores (P) flanking the cross walls (S) around the circumference of the trichome. Electron micrograph (× 54,000). By courtesy of H. C. Lamont (1969) *J. Bacteriol.* **97**, 350, Fig. 4.

The suggestion, however attractive, that these pores represent channels through which mucilage is excreted, has not so far been supported by direct experimental evidence (see p. 115).

The junction between the cross wall and longitudinal wall might be a weak point because of the thinning of the electron-dense L_{II} layer of the longitudinal wall. Van Baalen and Brown (1969) have suggested that this weakness could be responsible for the ready lysis of the marine planktonic blue-green alga *Trichodesmium erythraeum* and for the difficulty in culturing this interesting alga.

Chemical analysis of cell wall fractions provides more evidence of a close relationship between bacteria and blue-green algae. Most significant is the presence in blue-green algae of mucopolymer components similar to those found in gram-negative bacteria. Cyanophyceae are sensitive to penicillin (see p. 88) and when *Phormidium uncinatum* is grown in the presence of this antibiotic, local defects are seen in the mucopolymer. Analysis of cell wall fractions of species of widely different taxonomic position show that in the lysozyme-soluble or acid-hydrolysable fraction, mucopolymer is the major component and amounts to as much as 50% of the wall fraction (Frank *et al.*, 1962; Höcht *et al.*, 1965; Drews and Meyer, 1964). The purified mucopolymer is composed of five typical building blocks: two amino sugars (muramic acid and glucosamine) and three amino acids (α-ϵ-diaminopimelic acid, glutamic acid and alanine) present in a molecular ratio of about 1 : 1 : 1 : 1 : 2, a similar ratio to that found in *E. coli*. Frank *et al.* (1962) recorded muramic acid, glucosamine, diaminopimelic acid glutamic acid and alanine in the molar ratios of 0·63 : 1·00 : 1·00 : 0·94 : 2·00 in the cell walls of *Phormidium uncinatum*. Drews and Meyer (1964) looked at the cell walls of *Anacystis nidulans* and *Chlorogloea fritschii* and reported the same components (the first two as N-acetyl derivatives) in the ratios of 1·47 : 1·34 : 1·0 : 1·52 : 2·4, with another eight amino acids in smaller proportions (between 0·1 and 0·4). With few exceptions (see Stanier and van Niel, 1962) diaminopimelic acid can be regarded as a structural wall component specific to prokaryotic organisms. A lipopolysaccharide, composed of 30–30% lipid and about 60% carbohydrate has been extracted from *Anacystis nidulans* in 45% phenol (Weise *et al.*, 1970). This polysaccharide is again similar to the bacterial (*E. coli*) lipopolysaccharide and its main carbohydrate component is mannose.

2. SHEATH

Although most blue-green algae have a mucilaginous sheath, its extent, consistency, pigmentation and structure varies a great deal as seen under the light microscope. Outside the L_{IV} layer may appear various types of encapsulating material. These are all referred to as sheaths but it is not certain that they are homologous in either structure or origin. The fibrous substructure was first shown by Frey-Wyssling and Stecher (1954) but their suggestion that the submicroscopic network is composed of cellulose has not been confirmed by chemical investigations (Dunn and Wolk, 1970). The fibres are embedded in an amorphous matrix and show a typical orientation which varies according to species (Metzner, 1955; Figs 3.29, 3.30). Lamont (1969a) proposed that the microfibrils are oriented by shear in the direction of the flow of mucilage relative to the surface of the trichome (see p. 119). He supported his idea with observations on two species of Oscillatoriaceae, *Oscillatoria chalybea* and a *Lyngbya* species, which differ from each other in direction of rotation during gliding movement. The fibrils lay roughly parallel to the stream line, and thus their orientation seemed to be related to the motion.

Sheaths of blue-green algae stain readily with basic dyes such as ruthenium red and alcian blue (Fritsch, 1945; Leak, 1967b; Tuffery, 1969) and were hence considered to be composed of pectic acids and acid mucopolysaccharides. This has been confirmed in a detailed analysis of the sheath material by Dunn and Wolk (1970). When filaments of *Microcoleus vaginatus* are exposed to the action of hemicellulase this results in the removal of the sheath, prolonged action causing the fragmentation of the filaments (Durrell and Shields, 1961). Lysozyme has no effect on the sheath (Fuhs, 1958c).

The distinction between sheaths and mucilages *per se* is not one that has been properly made by any author. Perhaps the term mucilage should be retained for components which occur in the sheath but which are not firmly attached to the cell, embracing the substances left in a trail by gliding organisms (see p. 113) and those which form the matrix which holds together colonial forms, such as species of *Nostoc* and *Microcystis*. There have been few attempts to analyse the chemical structures of mucilages, which appear to be highly complex polysaccharides comprising a number of different sugar and uronic acid residues. Kylin (1943) detected galactose, glucose and pentose in the hydrolysis products of a mucilage obtained from the marine blue-green alga *Calothrix scopulorum*. Hough *et al.* (1952) obtained a complex polysaccharide from the mucilage of a species of *Nostoc*, containing a large proportion of hexuronic acids, galactose and D-xylose, with smaller quantities of glucose and other, unidentified, sugars. Ankel and Tischer (1969) have demonstrated the presence in *Anabaena flos-aquae* strain A-37 of the enzyme UDP-glucuronate 4-epimerase which may be involved in the synthesis of such polymers. Another polysaccharide was obtained from cultures of *Anabaena cylindrica* by Bishop *et al.* (1954). Acid hydrolysis gave glucose, xylose, galactose, rhamnose and arabinose in the approximate molar ratios of 5 : 4 : 1 : 1 : 1, with a further 4 molecules

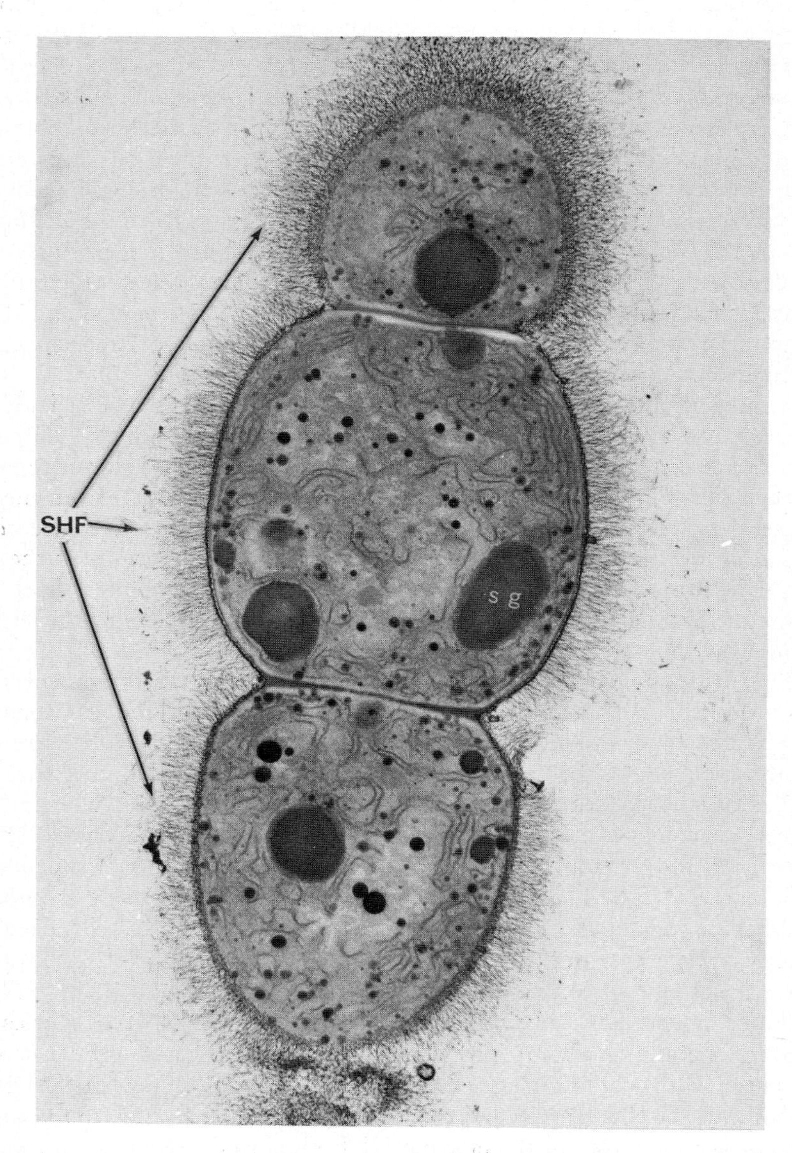

FIG. 3.29. Longitudinal section through *Anabaena* sp. illustrating the arrangement of sheath microfibrils (SHF). Structured granules (sg) are also shown. Electron micrograph (×22,000). By courtesy of L. V. Leak (1967) *J. Ultrastruct. Res.* **21**, 61, Fig. 4.

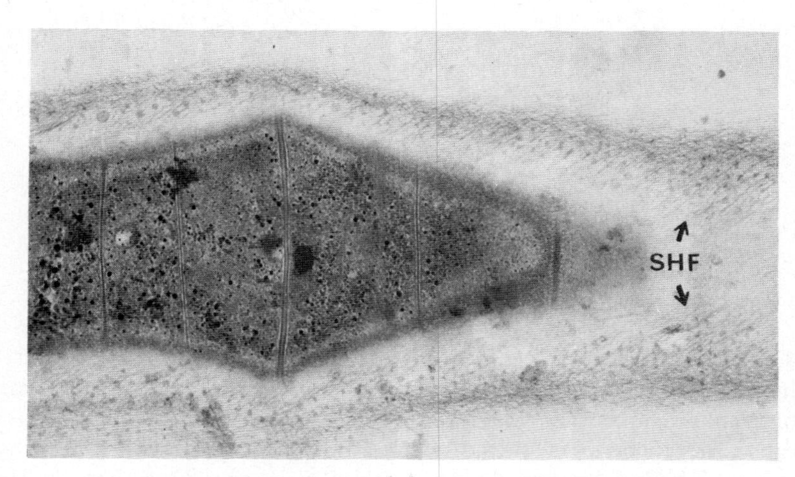

FIG. 3.30. Grazing longitudinal section of *Lyngbya* sp. displaying the orientation of sheath microfibrils (SHF). Electron micrograph (× 22,000). By courtesy of H. C. Lamont (1969) *J. Bacteriol.* **97**, 350, Fig. 6c.

of glucuronic acid. The polysaccharide is evidently a complex molecule. Attempts to determine its structure from partial hydrolysis products were unsuccessful, small quantities of all six sugars being released on subjecting it to conditions causing incomplete breakdown of the molecule. A parallel situation is again found with *Anabaena flos-aquae* strain A-37. Moore and Tischer (1965) determined the sugar content of extracellular polysaccharide which they thought might be sloughed off from the surface of the filaments. They demonstrated glucose, xylose, ribose and glucuronic acid in the molar ratios of 88 : 39 : 3 : 1, again suggesting a complex molecule or group of molecules.

Chapter 4

Cell biology

The protoplasm of a healthy blue-green algal cell appears, under the light microscope, as a highly viscous gel without protoplasmic streaming, Brownian movement or large sap vacuoles. It is not a rigid structure because, irrespective of the shapes of the cells from which they were prepared, isolated protoplasts are spherical in shape (Crespi *et al.*, 1962; Fulco *et al.*, 1967; Biggins, 1967a).

On average there is, on a dry weight basis, 50% protein, 30% carbohydrate, 5% lipid and 15% ash in blue-green algal protoplasm (Collyer and Fogg, 1955). Thus, the protein content is higher than in other algal groups (see Table 4.I), and the carbohydrates and lipids make up a correspondingly smaller proportion of the cellular contents. There is not an unusually high concentration of solids in the cell. Lund (1964) obtained the following values for dry wt mm^{-3} of living cell for various bloom-forming algae: *Aphanizomenon flos-aquae*, 0·35 mg; *Oscillatoria agardhii*, 0·41 mg; *O. bourellii*, 0·61 mg and *Agmenellum quadruplicatum*, 0·56 mg. These are in the middle of the range of values observed for other algal groups, which is 0·18–1·23 mg mm^{-3}. Fay (1969b) obtained a value of 1·12 mg mm^{-3} for the vegetative cells of *Anabaena cylindrica*.

TABLE 4.I Percentage protein content on a dry weight basis of algae representing various groups (from Collyer and Fogg, 1955)

Chlorophyceae	*Chlorella vulgaris*	12–28
	Scotiella sp.	11–21
Euglenophyceae	*Euglena gracilis*	32–47
Xanthophyceae	*Monodus subterraneus*	9–23
	Tribonema aequale	6–19
Bacillariophyceae	*Navicula pelliculosa*	9–31
Rhodophyceae	*Porphyridium cruentum*	19–40
Cyanophyceae	*Anabaena cylindrica*	43–56
	Microcoleus vaginatus	41–47

Compared with other algal cells the protoplasm of blue-green algae is resistant to treatment with concentrated urea solutions which break hydrogen bonds, but it is susceptible to thioglycollate treatment which breaks the S—S valence bond (Fogg, 1969a). This suggests that the polypeptide chains of the cellular proteins are linked by main valency bonds rather than by hydrogen or other homopolar cohesive bonds. Amino acid analyses of the bulked algal proteins are no different from those of other algal groups (Fowden, 1962). Strongly hypertonic urea or glucose solution may cause dissolution of the cell wall (Stadelmann, 1962) and hypertonic glucose solutions lyse *Anabaena flos-aquae* (Walsby and Buckland, 1969) and *A. cylindrica* (Fay and Lang, 1971).

Permeability

Permeability studies using inorganic salts such as potassium nitrate, sodium chloride, calcium chloride and aluminium nitrate (1–3 M) show that these rapidly penetrate the cells. However the standard method of determining cell permeability using non-electrolytes and observing plasmolysis and the rate of deplasmolysis is of doubtful value when applied to blue-green algae because with the absence of large sap vacuoles and the close association of the protoplast and cell wall plasmolysis does not occur readily. Permeability constants of various low molecular weight organic substances obtained by Elo (1937) and Stadelmann (1962) for two species of *Oscillatoria* using the plasmolytic method are higher than the average for most other algae. Furthermore, there is a fairly good inverse correlation between permeability constant and molecular weight of the solute (Fig. 4.1). This contrasts with eukaryotic algae in which permeability constants are more closely related to lipoid solubility than to molecular weight. It indicates perhaps that the penetration of the solutes through pores on the sieve principal is more important in determining permeability than lipoid solubility. Davson and Danielli (1943) reached similar conclusions from studies on the apochlorotic prokaryote *Beggiatoa mirabilis*.

There are problems in such a simple interpretation, however, because different dyes penetrate in different ways. In *Oscillatoria bornetii*, for example, dyes such as methylene blue and toluidine blue penetrate only through the apices and transverse septa (Drawert, 1949), perhaps via plasmodesmata, whereas other dyes penetrate through the longitudinal walls. Variable staining occurs when neutral red dye is used, with two adjacent cells scarcely ever showing the same degree of staining (Drawert, 1949).

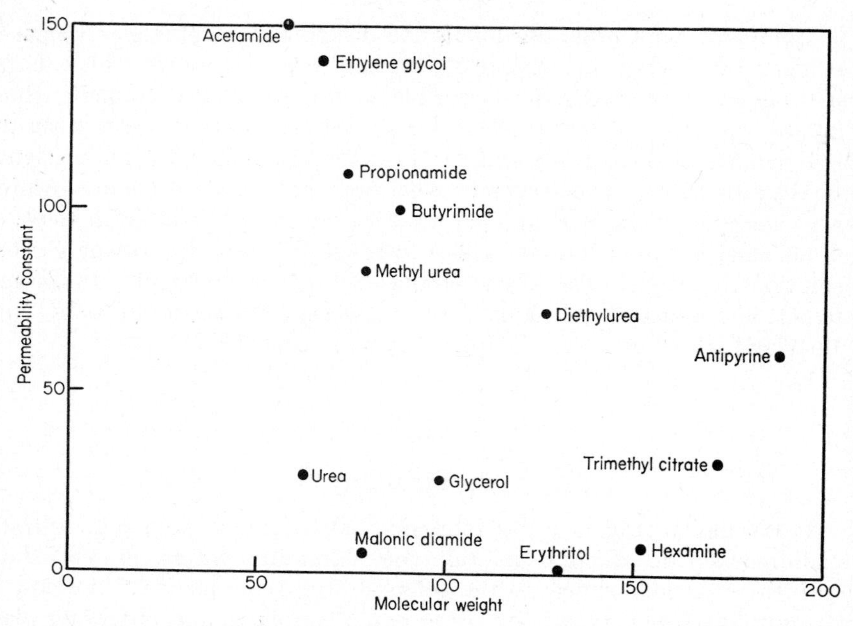

FIG. 4.1. Permeability of *Oscillatoria princeps* to various organic substances in relation to their molecular weight; data of Elo (1937) from H. Davson and J. F. Danielli (1943) "The Permeability of Natural Membranes". Cambridge University Press.

Osmotic relationships

The osmotic relationships of blue-green algal cells differ from those of higher plant cells in several respects. In the latter the vacuolar membrane is usually regarded as the principle semipermeable structure and the solutes in the vacuolar sap as the principle determinants of the osmotic pressure. As noted above, large sap vacuoles are lacking in the blue-green algae. Also, inbibition may play an important part in the osmotic relations of the latter organisms (Drawert, 1949), that is, τ, the matric pressure in the expression:

$$\Psi = P - \pi - \tau$$

has an appreciable value whereas it is zero in the vacuolar sap (Ψ = water potential, P = hydrostatic pressure and π = osmotic pressure). If blue-green algae do in fact have a rather higher concentration of protein in their cells than eukaryotic organisms, it may be that their matric pressure is correspondingly higher.

Experimental studies on osmotic relationships in blue-green algae have been handicapped by the indefinite nature of their plasmolysis. Some of the values reported may be overestimates (see p. 82 below). Nevertheless, the published results, reviewed by Guillard (1962), show that the osmotic pressures (which would include matric pressure) of the cells of blue-green algae are of the same order as those of other algae. Approximate values of 5 atm were obtained for freshwater blue-green algae and 11 atm excess over the local sea water for marine forms. Fulco *et al.* (1967) obtained similar results except for an unusually high value of 14·3 atm for *Nostoc muscorum*, a freshwater species, grown in culture. Hof and Frémy (1933) found that 4–5 molar sodium chloride solutions were required to produce 50% "plasmolysis" in the halophilic *Aphanocapsa litoralis*, but it would be unwise to deduce a value for osmotic pressure from this since, as in the halophilic bacteria, the cell membrane may be permeable to sodium chloride.

The studies of Walsby (1971; see also p. 100) provide a new approach to investigation of the osmotic relations of the blue-green algae. He showed that gas vesicles collapse at definite pressures and this provides a means of measuring directly the hydrostatic pressure inside a cell. If collapse curves at a series of externally applied pressures are determined for gas vesicles in turgid cells and in similar plasmolysed material, the displacement between the two curves is a measure of the turgor pressure in the former. When *Anabaena flos-aquae* is suspended in a series of sucrose solutions from 0–0·5 M and the pressure required to collapse 50% of the vesicles

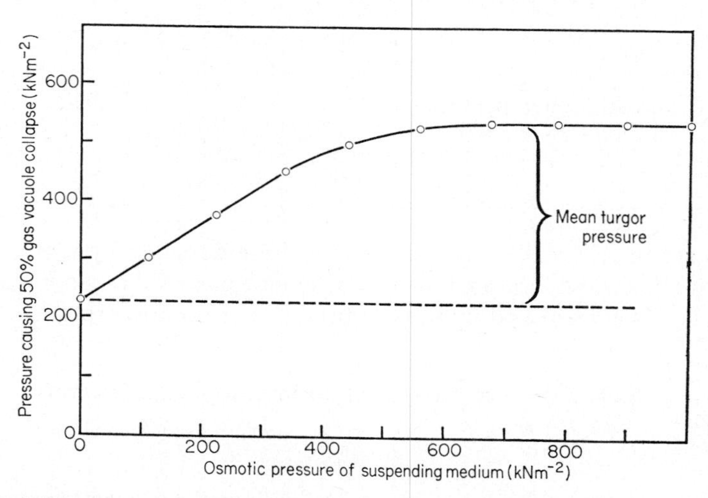

FIG. 4.2. The effect of removing the cell turgor of *Anabaena flos-aquae*, by suspending it in sucrose solutions of various osmotic pressures, on the pressure required to collapse gas vacuoles (redrawn from A. E. Walsby (1971) *Proc. R. Soc. Lond.* B **178**, 301–326).

determined, the results (Fig. 4.2) are consistent with the idea that, as the osmotic pressure of the suspending medium is increased, the turgor pressure acting on the vesicles in the cells decreases until it reaches zero in hypertonic medium. Figure 4.2 shows that the rise in externally applied pressure required to bring about collapse is not quite equivalent to the rise in external osmotic pressure to the isotonic point, suggesting that sucrose has penetrated the cells. This idea is supported by the observation that the longer the cells are left to equilibrate with the sucrose solutions, the greater does the discrepancy become. The penetration of solutes (plasmolytes) such as sucrose may be a cause of turgor pressures having been overestimated by the plasmolytic method. Another cause is that the plasma membrane does not separate easily from the adjacent wall layers, the cells shrinking, but not plasmolysing readily, in hypertonic solutions. Walsby (1971) considers that cell turgor pressures estimated by the plasmolytic method may exceed those measured by the gas-vacuole method, by a factor of 2. Turgor pressures determined by subtracting the pressure causing 50% collapse of gas vesicles in cells suspended in water from that required for cells suspended in hypertonic sucrose solution are given in Table 4.II. These

TABLE 4.II. Turgor pressures in cells of *Anabaena flos-aquae* grown under different conditions (after Walsby, 1971)

Growth conditions		*Turgor pressure*	
		($kN\ m^{-2}$)	*(atm)*
Bubbled with air in high light (*ca.* 7000 lux)		355	3·50
Without bubbling, in low light (300 lux)	day 5	255	2·50
	day 6	210	2·07

values are in reasonable agreement with those obtained by more conventional methods and indicate that a considerable part of the osmotic pressure in the cell is contributed by soluble products of photosynthesis.

Resistance to extremes of physical conditions

1. TEMPERATURE

Blue-green algae are well known for their ability to withstand and to grow at extremes of temperature. Together with green algae and diatoms they are prominent in terrestrial habitats in the Arctic and Antarctic and only bacteria surpass them in ability to survive in hot springs at temperatures approaching boiling point (see pp. 277, 339). Low temperatures exert at least two effects: physical damage produced by ice crystals, and a slowing down of the chemical reactions. At high temperatures an imbalance of metabolism results from the differential acceleration of its component reactions as well as from a denaturation of the proteins. Growth under these adverse conditions obviously calls for more adaptations than does mere survival.

Much information about the effects of exposure to low temperatures has come from studies on the freeze-drying of algal cultures. The most important findings are that most blue-green algae survive freeze-drying better than do green algae (see Fogg, 1969a) and that the results obtained may vary according to the conditions used. For example slow freezing to −30°C at a rate of about 1°C per min results in better survival of freeze dried algae than does rapid freezing (within about 25 sec) to the same temperature (Hwang and Horneland, 1965). There is also species to species variation, as Whitton (1962) found. Out of 18 strains of growing algae which he tested, 11 survived freezing at −15°C for 2 months. There is variation depending on the growth stage of the culture. For example, exponentially growing *Anabaena cylindrica* cells are quickly and completely killed by freezing while spores survive. Different results may be obtained according to the source from which the algae were isolated. For example Holm-Hansen (1967) obtained different results using algae isolated from Antarctica and from the U.S.A. The former isolates survived both quick and slow freezing, as well as repeated cycles of freezing and thawing, while isolates from the United States were unable to survive even one freeze-thaw cycle. The general viability of *Nostoc muscorum* samples freeze dried at −25°C was as good as those kept at room temperature (Holm-Hansen, 1963a, 1967). Whitton (1962) found that with *Anacystis nidulans* (*Synechococcus sp.*), which did not survive prolonged freezing, about 1 in 10^6 cells remained viable after 4 h freezing, suggesting that there are distinct "initial" and "storage" effects of freezing as in bacteria. Freeze-dried *Anacystis nidulans* (*Synechococcus sp.*), although evidently non-viable, is nevertheless capable of photosynthesis, with reduction of carbon dioxide and evolution of oxygen continuing for as long as a day on resuspension in phosphate buffer (Papageorgiou and Govindjee, 1967).

Some blue-green algae grow at appreciable rates at temperatures approaching 0°C under ice in lakes in temperate regions (p. 258) or in frigid lakes in the Antarctic (p. 273). Nitrogen fixation occurs in samples of *Nostoc commune* growing at near freezing point in the Antarctic (Fogg and Stewart, 1968, see p. 329).

There is some disagreement as to the exact limit at which blue-green algae survive exposure to high temperatures. Brock (1967b) put the upper limit for growth of blue-green algae at 75°C although he found bacteria, including "flexibacteria" which have been considered to be apochlorotic blue-green algae, in the same locality at 91°C. These environmental aspects are considered later (p. 277).

Although an ability to withstand these high temperatures is shown by a few species only, the blue-green algae as a group are able to tolerate higher temperatures than most other kinds of algae. Species such as *Anabaena variabilis*, *A. cylindrica* and *Nostoc muscorum*, isolated from temperate habitats, have temperature optima for growth in the region 32·5–35°C although they fail to grow at 41°C (Kratz and Myers, 1955a). For most strains of *Chlorella*, isolated from similar habitats, the optimum is around 25°C, and 30°C is lethal. Desiccated cells are able to withstand higher temperatures better than are hydrated ones. Thus Glade (1914) found that cells of *Cylindrospermum majus* survived temperatures of 95–100°C when dry. Spores, when produced, are probably more resistant to high temperatures than are vegetative cells but there is little published evidence supporting this idea. Glade (1914) stated that spores of *C. majus* survive high temperatures better than do vegetative cells and Wieringa (1968) found the same for other blue-green algae.

Enzymes isolated from thermal algae are more stable at higher temperatures than those from other organisms. For example, $NADPH_2$-cytochrome *c* reductase extracted from the thermal alga *Aphanocapsa thermalis* showed unimpaired activity after heating to 85°C for 5 min whereas that from *Anabaena cylindrica* was completely inactivated by similar treatment. The thermostability of the *Aphanocapsa* enzyme is not due to a dialyzable substance with unspecific protective action, since *Aphanocapsa* extracts do not confer heat stability on the enzyme from *Anabaena*. It is also known that cytochrome *c* reductase and catalase from *Aphanocapsa* are not so readily inactivated by solutions of acetamide or urea, reagents which dissociate hydrogen bonds, as are the corresponding enzymes from *Anabaena* (Marré, 1962). Kubín (1959) studied the thermal inactivation of catalase extracted from *Oscillatoria* sp. and found in Arrhenius plots that there was no difference between the characteristics of the inactivation and those reported for catalases extracted from beef liver and from various seaweeds. However, this *Oscillatoria*, although from a hot spring, was stated to have an optimum temperature for growth of 35°C, so it was probably not particularly temperature resistant. The ability of thermal Cyanophyceae to carry out photosynthesis at high temperatures does not appear to depend on any special properties of phycocyanin. Moyse and Guyon (1963) found that in *Aphanocapsa thermalis* exposed to temperatures of 50°C the phycocyanin-dependent photochemical activity decreased before that dependent on chlorophylls (see p. 145).

2. DROUGHT RESISTANCE

An ability to withstand desiccation is a characteristic feature of blue-green algae which accounts, in part at least for their dominance, together with unicellular green algae, in the microbial flora of tropical deserts, rocks and buildings. Blackish growths of *Scytonema*, for example, are characteristic of buildings in the tropics (Fig. 4.3). Viable blue-green algae have been recovered from soil samples stored in an air dry condition for 20 years or longer and *Nostoc commune* has been revived from a 107-year old herbarium specimen (Cameron and Blank, 1966).

The drier the habitat from which the algae are obtained, the greater is their resistance to dryness (Hess, 1962; Fig. 4.4). The rate of drying is of little consequence but algae desiccated in the dark appear to survive better than those desiccated in the light (Hess, 1962). In general the vegetative cells of species of the Oscillatoriaceae, which have no perennating cells, survive desiccation better than do the vegetative cells of *Nostoc*, *Calothrix* and *Scytonema* species. The spores and hormocysts of these latter genera show a high resistance to desiccation. Although some species showed great resistance to desiccation, growth of these does not occur at relative humidities of less than 80%. There is a general correlation between drought resistance and temperature resistance, but as Castenholz (1967a) has pointed out thermal algae do not in general survive desiccation.

The means whereby blue-green algae survive extremes of physical conditions have been insufficiently studied. There are indications that the mechanisms of resistance to temperature, desiccation and other extremes depend on the same basic feature and on a complex of secondary interacting factors (Fogg, 1969a). An essential feature of the cells must be the stability of the proteins. This may be conferred by a higher proportion of main-valency bonds, as opposed to weaker bonds sensitive to heat and mechanical forces, linking the polypeptide chains. As we have seen (p. 79) there is a little evidence for this. If it is correct that resistance towards extreme environmental conditions is at least partly dependent on the structural features of their proteins, this would be expected to be of a genotypic rather than phenotypic character. This is in accord with the general experience that these organisms cannot be trained to withstand higher temperatures.

Effects of invisible radiations and mutagenic agents

Few comparative studies of the effects of ionizing and ultraviolet radiations have been made with blue-green algae. Such little information as is available indicates that they are about as susceptible to damage by high

FIG. 4.3. Blue-green algae being cleaned from the cement rendering of a wall, Bombay, January 1960. Photo G. E. Fogg.

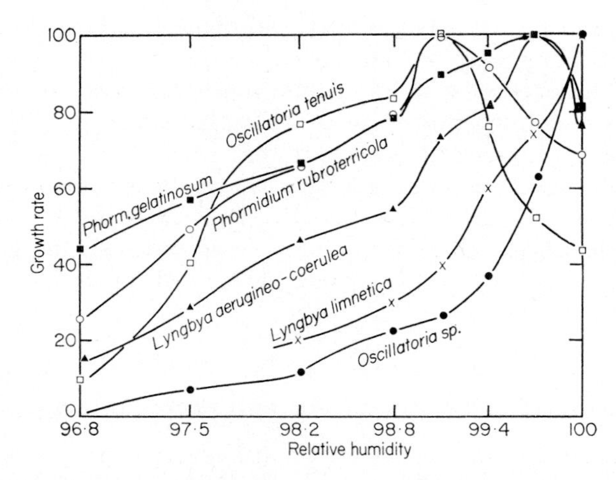

FIG. 4.4. Growth rates of six Oscillatoriaceae in relation to relative humidity (redrawn after U. Hess (1962) *Arch. Mikrobiol* **44**, 189–218).

energy rays and X-rays as are algae of other groups (Godward, 1962). They appear to be more resistant than bacteria to ultraviolet irradiation and this has been used as a method of obtaining axenic algal cultures (Gerloff *et al.*, 1950a), but it has a disadvantage in purification studies in that it may have a mutagenic effect.

There is evidence that, as with other micro-organisms, the sensitivity to radiation is affected by the environmental conditions before and after irradiation. Singh (1968) has isolated ultraviolet-sensitive and ultraviolet-resistant strains of *Anacystis nidulans* (*Synechococcus* sp.). Acriflavine given before irradiation increases resistance of the parent and resistant strains, but given after irradiation decreases their survival. Acriflavine has much less effect on the survival of the sensitive strain. This substance acts by inhibiting the DNA repair mechanism present in blue-green algae and its failure to affect the sensitivity of the sensitive strain suggests that this strain lacks the repair system present in the other two.

The above mechanisms are dark repair mechanisms, but there is also evidence that light stimulates a repair mechanism in the blue-green algae. When ultraviolet-irradiated cells are placed in the light they show less response to ultraviolet light than do cells kept in the dark. Werbin and Rupert (1968) reported on the presence of a photoreactivation system in *Plectonema boryanum* and Van Baalen (1968) studied the process in detail in *Agmenellum quadruplicatum*. He found that both the lethal effect and the decay in photosynthesis after ultraviolet (254 nm) treatment decreased markedly on illuminating the cells with light of a wavelength of approximately 430 nm and interpreted this as a photoreactivation of photosynthesis. However, more recently Stevens and Van Baalen (1969), in studying a repair mechanism in *A. quadruplicatum* after treatment with the mutagenic chemical NTG (N-methyl-N'-nitro-N-nitrosoguanidine), found that cells returned to growth conditions after NTG treatment showed increased survival on increasing the light intensity from darkness up to light saturating conditions for growth. They conclude that this repair mechanism in dim light is not in fact a photoreactivation effect and that the response to light may be no more than a requirement for photosynthesis. They suggest that the repair system in *Agmenellum* is similar to the dark repair system in bacteria and yeast.

Effects of growth inhibitors and antibiotics

Table 4.III lists values for the minimum concentrations of various antibiotics, antimetabolites and inhibitors which prevent the growth of blue-green algae in mineral media. These are approximate concentrations only, because the sensitivity of an alga towards a given agent may vary

TABLE 4.III. Minimum concentrations in mg l^{-1} (parts per million) of antibiotics, mutagens and inhibitors preventing growth of blue-green algae in inorganic culture media

SUBSTANCE	SPECIES	CONCENTRATION	REFERENCE
Albamycin	*Microcystis aeruginosa*	1000	9
Bacitracin	*Anabaena variabilis*	100	1
	Microcystis aeruginosa	100	9
Caffeine	*Fischerella muscicola*	1940	2
Candicidin	*Microcystis aeruginosa*	1000	9
Chloromycetin	*Anabaena variabilis*	10	1
Colchicine	*Fischerella muscicola*	>4000	2
Coumarine	*Fischerella muscicola*	>200	2
Dehydrostreptomycin	*Microcystis aeruginosa*	1	9
Dichloronaphthoquinone	Blue-green algae generally	0·03–0·055	3
Gliotoxin	*Nostoc* sp.	125–250	1
Gramicidin	*Anabaena variabilis*	1000	1
Isoniazid	*Anacystis nidulans*	>10	4
Lanthanum acetate	*Fischerella muscicola*	475	2
Maleic hydrazide	*Fischerella muscicola*	448	2
Manganous chloride	*Fischerella muscicola*	1500	2
Mitomycin	*Anacystis nidulans*	5	5
Neomycin	*Nostoc* sp.	4	1
	Phormidium sp.	4	1
	Anacystis nidulans	>100	5
	Microcystis aeruginosa	1	9
Patulin	*Microcystis aeruginosa*	10	9
Penicillin	*Anabaena variabilis*	0·1	1
	Microcystis aeruginosa	2	1
	Microcystis aeruginosa	1	9
	Anacystis nidulans	0·2	4
	Phormidium mucicola	0·1	6
Polymyxin A	*Nostoc* sp.	10–20	1
	Phormidium sp.	20–40	1
Polymyxin B	*Anabaena variabilis*	5	1
	Cylindrospermum licheniforme	2	1
	Microcystis aeruginosa	2	1
	Microcystis aeruginosa	10	9
	Anacystis nidulans	16	7
	Chlorogloea fritschii	80	7
Proflavin	*Anacystis nidulans*	>4	8
Ristocetin	*Microcystis aeruginosa*	100	9
Streptomycin	*Anabaena variabilis*	0·1	1
	Cylindrospermum licheniforme	2	1
	Microcystis aeruginosa	2	1
	Microcystis aeruginosa	1	9
	Nostoc sp.	2	1
	Phormidium sp.	2	1
	Anacystis nidulans	0·05	4

Table 4.III—*continued*

SUBSTANCE	SPECIES	CONCENTRATION	REFERENCE
Terramycin	*Anabaena variabilis*	10	1
	Microcystis aeruginosa	2	1
	Nostoc sp.	233	1
	Phormidium sp.	58–117	1
Thioglycollic acid	*Fischerella muscicola*	>92	2
Tyrothricin	*Microcystis aeruginosa*	10	9
Uranyl nitrate	*Fischerella muscicola*	500	2
Usnic acid	*Microcystis aeruginosa*	1000	9

References

1. Krauss (1962)
2. Singh and Subbaramaiah (1970)
3. Fitzgerald and Skoog (1954)
4. Kumar (1964)
5. Kumar (1968)
6. Srivastava (1969)
7. Whitton (1967b)
8. Kumar *et al.*(1967)
9. Vance (1966)

enormously according to the strain, its physiological history and the conditions under which the test is carried out. Nevertheless, there are some striking trends. For example, compared with *Chlorella pyrenoidosa*, *Anabaena variabilis* is equally sensitive to gramicidin and polymyxin, is one hundred times more sensitive to streptomycin and terramycin and is ten thousand times more sensitive to penicillin (Krauss, 1962). In fact blue-green algae are more sensitive to such substances than are most other plant groups including bacteria, a situation which appears paradoxical when one considers the wide range of environmental extremes which they withstand in the field.

In general there is no consistent resemblance in antibiotic sensitivity either to gram positive or gram negative bacteria, but the blue-green algae withstand actidione (cycloheximide), a potent inhibitor of fungi, Chlorophyceae, Xanthophyceae and Bacillariophyceae and which, as a result, is most useful in preparing unialgal cultures of blue-green algae (Zehnder and Hughes, 1958). Unfortunately it has little effect on most bacteria. Other fungicides such as candicidin, mycostatin and usnic acid also have little effect on blue-green algae (Vance, 1966). 2,3-dichloronapthoquinone has been used as a selective algicide for blue-green algae (Fitzgerald and Skoog, 1954) but because it is more expensive than copper sulphate has not found widespread application in the field (Whitton, 1965).

The effects of sub-lethal concentrations of antibiotics or inhibitors on blue-green algae are various. They include an increase in the length of the lag phase, an increase in cell length, the formation of filaments from unicells, and a temporary reduction in chlorophyll, carotenoid and phycocyanin formation (Kumar, 1964, 1968). In *Fischerellopsis muscicola*, Singh and Subbaramaiah (1970) found that mutagens, antimetabolites and chemicals having specific effects on proteins caused cell enlargement, vacuolation, and gas vacuole formation.

The way in which sensitivity to an antibiotic may vary in relation to other factors is illustrated in a study by Whitton (1967a) with polymyxin B. Above a critical population density the minimum lethal concentration towards *Anacystis nidulans* (*Synechococcus* sp.) or *Chlorogloea fritschii* was linearly related to population density, about 6×10^7 molecules being required to kill one cell of *A. nidulans*. Below this critical density it was almost independent of population density. Old cells of *A. nidulans* were slightly less sensitive than young cells and cells in the light were much more sensitive than those in the dark. Contrary to what has been found with bacteria, no interference with polymyxin toxicity by cations could be detected. The non-dialysable residues of extracellular material from *Anabaena cylindrica*, *Anacystis nidulans* and *Chlorogloea fritschii* are all effective in reducing the toxicity of polymyxin B towards each of these algae. The results suggest that the material offers alternative sites to those in the cell surface, with which the toxin forms complexes. The toxicity of 2,3-dichloronaphthoquinone towards *Anacystis nidulans* is also decreased by these extracellular products (Whitton and MacArthur, 1967).

Genetics

Sexual reproduction as found in eukaryotic organisms, with syngamy and meiosis, does not occur in the Cyanophyceae but, as in bacteria, genetic recombination takes place by parasexual processes.

The demonstration of recombination in a blue-green alga was first made possible by the production of strains having greater resistance to antibiotics than the wild-type. By successive subculture in gradually increasing concentrations of antibiotics, resistant strains of *Anacystis nidulans* (*Synechococcus* sp.) have been produced. They may have hundred-fold increased tolerance towards penicillin or several thousand-fold increased tolerance towards streptomycin. Such strains are stable and appear to originate by selection of mutants (Kumar, 1964). In a culture in which such penicillin- and streptomycin-resistant strains were grown together, Kumar (1962) demonstrated the appearance of a new strain having resistance to both antibiotics. This first demonstration of genetic recombination in a blue-green alga was not completely satisfactory because, at that time, techniques for growing *Anacystis* from single cells were not available, and thus clonal cultures of the recombinant strain could not be obtained. Pikalek (1967) criticized Kumar's experiment on the grounds that the medium in which the apparently recombinant cells were growing had lost penicillin through decomposition, but this objection cannot be sustained since the recombinant strain was found to be resistant to penicillin in separate tests (Kumar, personal communication). In any case, the occurrence of recombination has been confirmed using polymyxin B- and streptomycin-resistant strains with clonal isolation of a double resistant recombinant which remains stable during subculture over several months (Bazin, 1968). Genetic recombination has been reported also for *Cylindrospermum majus* (Singh and Sinha, 1965).

The frequency of recombination is low, varying from 4·5–491 per 10^9 cells in Bazin's experiments. It is probably sometimes brought about by conjugation between donor and recipient cells as in bacteria but this has not been demonstrated unequivocally although reports of cell fusions in blue-green algae have been reported (e.g. Lazaroff and Vishniac, 1962). Another mechanism for recombination is transformation by incorporation of exogenous DNA. This has been demonstrated in *Anacystis nidulans* by recovery of new erthyromycin-resistant strains from clones of non-resistant wild-type or strains with various genetic markers treated with purified DNA from an erythromycin-resistant strain (Shestakov and Khyen, 1970). Again the frequency of recombination is low but it is of interest to note that it is enhanced in the light as is cyanophage multiplication in blue-green algae. Gene recombination may also be brought about by transduction, a process in which a virus acts as a vector of certain genes, but this has not yet been demonstrated.

Another area of interest is in studies relating to nucleic acid hybridization. DNA–DNA hybridization has been investigated by Craig *et al.* (1969) using *Anabaena variabilis*, *Anacystis nidulans*, *Nostoc muscorum* and various bacteria. They found that *Anacystis* and *Nostoc* DNA bound to that of *Anabaena* with, respectively, 11% and 29% the efficiency of *Anabaena* DNA and that in general the binding of bacterial DNA to *Anabaena* DNA was even less efficient. These findings thus agree with the generally accepted view from morphological and physiological studies that *Anabaena* is more akin to *Nostoc* than to *Anacystis*.

A related aspect, which is of importance in connexion with the view that chloroplasts may have evolved from blue-green algae, is the relationship between the nucleic acids of blue-green algae and those of other plant chloroplasts. Pigott and Carr (1972) have reported on preliminary studies in which they examined the relationship of rRNAs from various blue-green algae, with *Euglena* chloroplast DNA, and various bacterial DNAs. They found hybridization with *Euglena* chloroplast DNA at 11–47% (depending on the algal species) of the efficiency obtained using *Euglena* rRNA. Lower degrees of hybridization were observed with bacteria and blue-green algae. These results are of interest, but before extrapolating them too far it will be necessary to compare possible homologies between nucleic acids of organisms which are generally considered not to be at all related.

Chapter 5

Gas vacuoles

Gas vacuoles have been known since the early light microscope studies on blue-green algae. They are clearly visible in the cells, appearing as blackish objects under low power and reddish under high power. Under phase-contrast microscopes they appear very bright, perhaps slightly pink, against the dark background of the cytoplasm (Fig. 5.1a, b). They sometimes occupy most of the cell or are situated at particular positions in it, as at the periphery in *Trichodesmium erythraeum* (van Baalen and Brown, 1969) or at the cross walls in *Oscillatoria redekei* (Whitton and Peat, 1969). In some algae they form only under certain physiological conditions, for example, in a strain of *Nostoc muscorum* they form on transfer to high light intensity (Waaland and Waaland, 1970) or on dilution of the culture medium (Waaland and Branton, 1969). In some more complex morphological forms they are restricted to particular structures such as the hormogonia in *Tolypothrix tenuis* and *Gloeotrichia ghosei* (Singh and Tiwari, 1970).

Gas vacuoles cannot always be distinguished with certainty from some of the other refractile granules present in blue-green algal cells by conventional light microscope observation alone, but there is a simple test which identifies them at once without ambiguity: they disappear completely and irreversibly under pressure. This was first demonstrated towards the end of the last century by three German microbiologists (Ahlborn, 1895; Klebahn, 1895; Strodtmann, 1895) using the now classical "hammer, cork and bottle" experiment. If a stout bottle is filled to the brim with algal suspension, a tight fitting bung inserted and hit with a hammer, the pressure generated destroys the gas vacuoles (Fig. 5.1c, d). The disappearance of the microscopic structures results in a dramatic macroscopic change in the appearance of the algal suspension from light, milky green to dark, translucent green (Fig. 5.2a) and two other events may follow. Firstly, if the alga is floating before pressure is applied, it loses its buoyancy and sinks (Fig. 5.2b). Secondly, if a very concentrated algal suspension is used, bubbles of gas may collect at the surface. It seemed clear to Klebahn

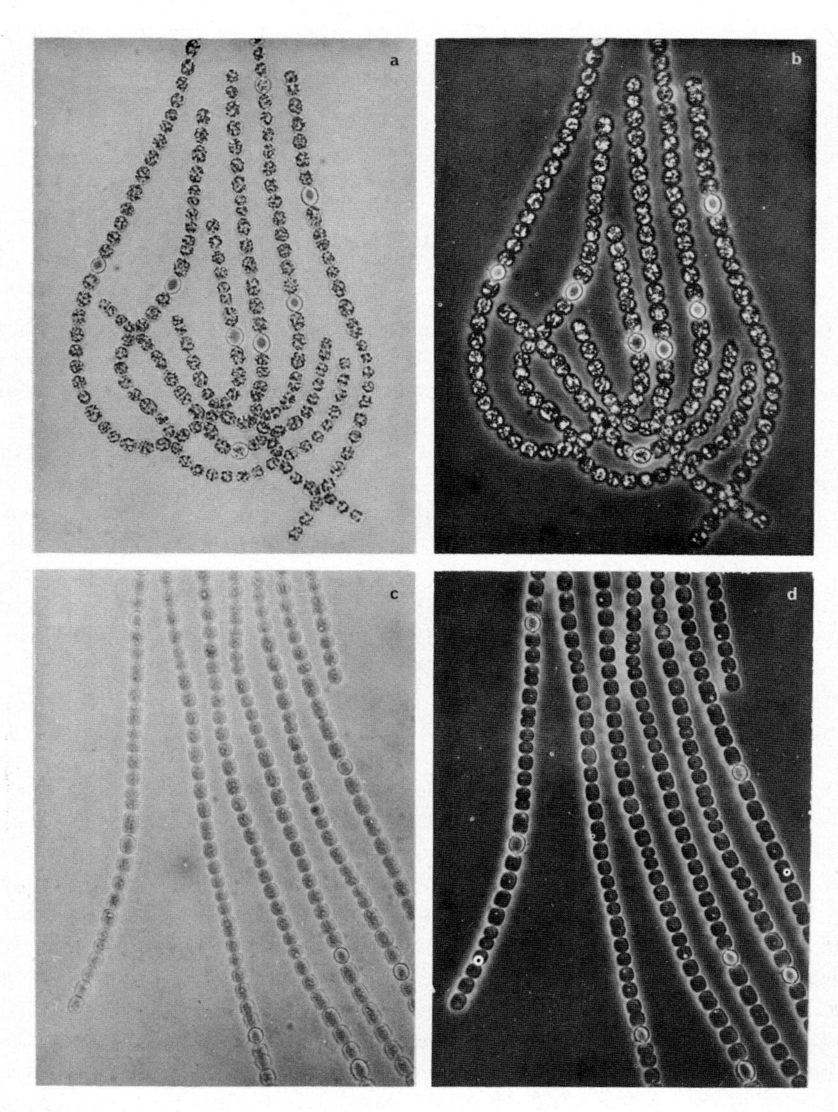

FIG. 5.1. Filaments of *Anabaena flos-aquae*; (a, b) with gas vacuoles; (c, d) filaments from the same sample after collapse of gas vacuoles under pressure. The light-microscope images of gas vacuoles are explained by Fuhs (1969). Light micrographs by A. E. Walsby; a, c bright field, b, d phase contrast illumination ($\times 600$).

(1895, 1922) that these microscopic structures contained gas, and he therefore coined the term *gas vacuole*, replacing *pseudovacuole*, the other name given to these structures. Apart from the fact that it was possible to obtain more gas from a gas-vacuolate suspension than from an equivalent suspension in which the gas vacuoles had been removed, Klebahn argued that, considering the relative changes in density and volume which accompanied

FIG. 5.2. The "hammer, cork and bottle experiment", showing two effects of collapsing the gas vacuoles in a suspension of the blue-green alga *Microcystis aeruginosa*. (a) The decrease in turbidity after striking the cork. (b) The loss of buoyancy after allowing the bottles to stand for 2h. From A. E. Walsby (1972) *Bact. Rev.*, **36**, 1, Fig. 1.

the destruction of gas vacuoles, the space which disappeared could only have been occupied by gas. The compressibility of the system also indicated gas, rather than solid or liquid. Analyses by Klebahn (1922) on the gas collected from gas vacuolate suspensions suggested that it was largely nitrogen, although the contamination of his samples with dissolved gases gave rise to uncertainty as to its exact composition.

Gas vacuole structure

Klebahn put forward several reasons for believing that the gas must be surrounded by a special membrane. Firstly, gas vacuoles persisted indefinitely under a vacuum and this suggested to him that the gas must be prevented from diffusing away by an impermeable layer. Secondly, the gas appeared to be at a much lower pressure than that expected in a bubble of comparable size, and suggested an enclosing rigid membrane which could withstand the pressures generated on it by surface tension and cell turgor.

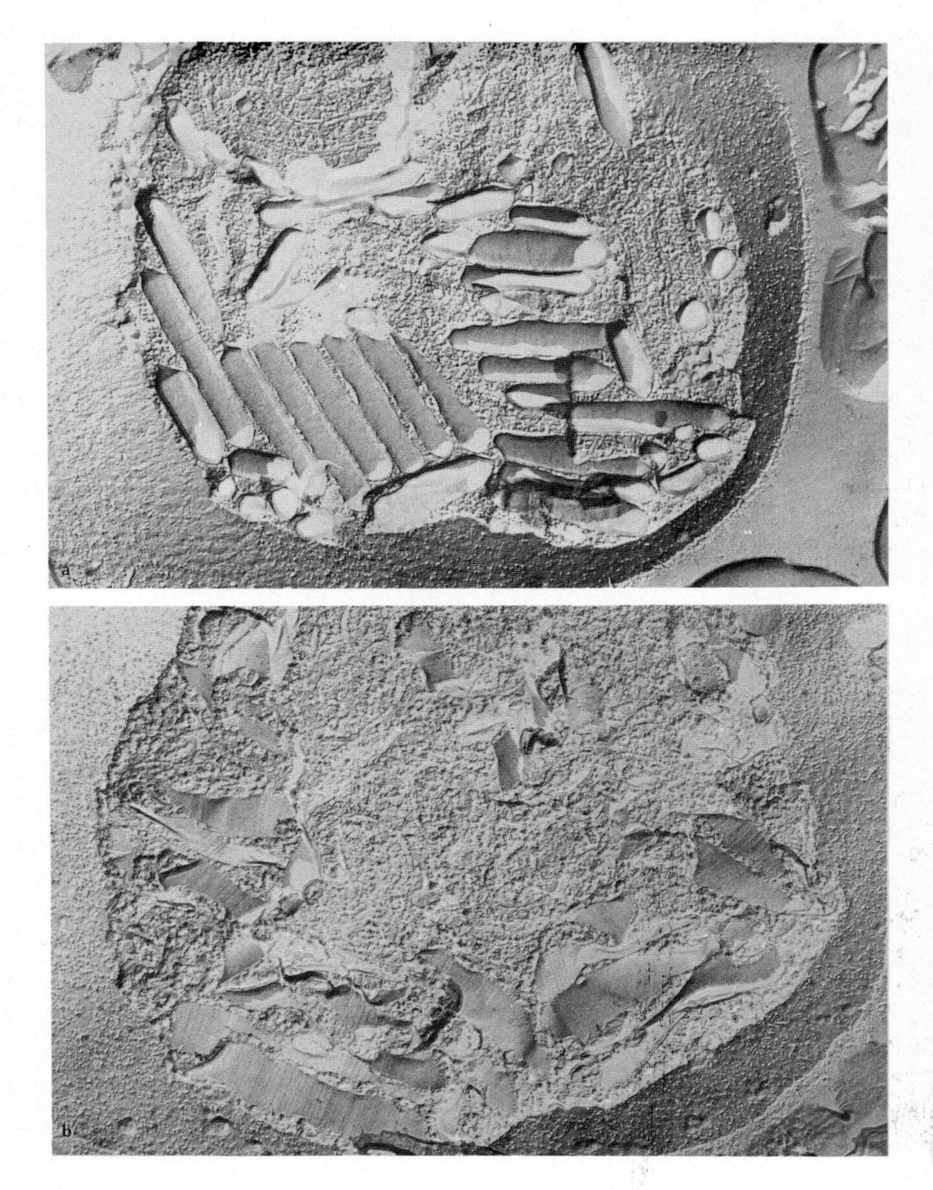

FIG. 5.3. Freeze-fractured cells of *Nostoc muscorum* (a) cell showing intact gas vesicles, (b) cell showing gas vesicles collapsed by exposure to hydrostatic pressure. Electron micrographs ×48,000, by courtesy of Dr. J. R. Waaland.

Klebahn's suspicions were not confirmed until electron microscope studies were carried out more than forty years later. Bowen and Jensen (1965) showed gas vacuoles of blue-green algae to be made up of cylindrical, hollow membranes, stacked closely together like the cells of a honeycomb. They termed these *gas vesicles* and demonstrated that the gas vesicles collapsed flat on the application of pressure (Fig. 5.3) and that this accounted for the disappearance of the gas vacuoles which they comprise. An identical sub-structure occurs in all gas-vacuolate algae so far investigated, and homologous structures occur in various bacteria (see Cohen-Bazire *et al.*, 1969). Walsby (1972b) lists a dozen genera from seven families of true bacteria which contain these structures. Several synonyms have been used to describe these sub-light-microscopic structures (e.g. gas vacuoles, gas vacuole sub-units, gas cylinders), but the term gas vesicle is universally applicable and has priority. Another term, *Hohlspindeln* (hollow spindles), was used by Jost (1965) to describe the same structures before it was realized that they were filled with gas, rather than water, and were components of gas vacuoles (Jost and Zehnder, 1966).

Gas vesicles vary in size and shape in different bacterial groups but in blue-green algae they always appear to have the same form: cylinders about 70 nm in diameter and of variable length. The mean length of those in *Microcystis aeruginosa* and *Anabaena flos-aquae* is 360 nm (Jost and Jones, 1970) and 327 nm (Walsby, 1971), respectively. The longest recorded are are up to 2 μm in *Nostoc muscorum* (Waaland, 1969). The ends of the cylinders are closed by conical caps, so that when the structures are very short they may appear spindle-shaped.

The thickness of the gas vesicle membrane is about 2 nm when fixed in osmium, and about 3 nm when fixed in permanganate, which gives rather poor preservation of the structure (Smith and Peat, 1967a). The freeze-etching studies of Jost (1965) showed the membrane to consist of ribs running orthogonal to the long axis (Fig. 5.3) at intervals of about 4 nm and seemed to indicate that the ribs were made up of particles, like beads on a string. Jost and Jones (1970) suggest, on the basis of measurements made on freeze-etched and metal-shadowed material, that the membranes are made up of a single layer of particles and postulate that the ribs represent turns of a helix rather than individual hoops; their suggestion is based on the observation that structures which appear to be uncoiled ribs, split from broken vesicles, sometimes exceed in length the circumference of a vesicle.

X-ray diffraction studies of partially orientated gas vesicles, and analysis of electron microscope images of gas vesicles by optical diffraction techniques have produced more exact measurements on the substructure of the membrane, giving 4·5 nm as the rib period, and 1·9 nm as the mean thickness of the membrane, and reveal a crystalline substructure in the ribs (Blaurock *et al.*, in preparation). Thus the gas vesicle membrane is fundamentally different from typical "unit" membranes (usually regarded as a bi-layer of lipid sandwiched between protein) a conclusion which is supported by chemical analysis of isolated gas vesicles.

Isolation of gas vesicles

Gas vacuole membranes may be isolated in a collapsed state by density gradient centrifugation as for other subcellular particles of narrow density range (Stoeckenius and Kunau, 1968) although as Jones and Jost (1970) point out, it is difficult to purify them completely in this way. Walsby and Buckland (1969) demonstrated that highly purified preparations can be obtained by isolating gas vesicles in an intact state (Fig. 5.4) provided the sample is not exposed to a pressure sufficient to cause their collapse. This is 200–700 kN m^{-2} (2–7 atm) for the vesicles of *Anabaena flos-aquae* (Fig. 5.7). Breaking the cells, osmotic shock, mechanical agitation, blending, ultrasonication, and disruption in a French press all cause partial or total gas vesicle loss. The use of lysogenic agents such as lysozyme is preferable (Jones and Jost, 1970) and with filamentous planktonic algae of the *Anabaena-Nostoc* type, shrinking the cells in hypertonic sucrose (0·7 M) provides a particularly rapid and effective method of inducing lysis without generating pressure (Walsby and Buckland, 1969).

Although they are very small, intact gas vesicles, being of extremely low density (about 0·01 g cm^{-3}, see Walsby, 1972b), float up much more rapidly than other cell components. This process may be speeded up by centrifugation so long as the pressure, generated by the acceleration acting on the column of the suspension, does not exceed the pressure causing the collapse of the vesicles. In practice, good recovery of intact vesicles can be obtained by spinning 60 mm layers of lysate at up to 350 *g* for about 4 h. The speed may be increased as the gas vesicles float to the surface (Buckland and Walsby, 1971). Resuspending the vesicles in water and repeating the procedure three times gives a fairly pure preparation which can be further improved by membrane filtration to remove remaining unlysed cells and soluble contaminants (Walsby and Buckland, 1969). Macromolecular sieving and liquid polymer partitioning may also remove certain classes of contaminants (Jones and Jost, 1970).

Chemical analysis of the gas vesicle

The intact gas vesicles purified from *Anabaena flos-aquae* by Walsby and Buckland (1969) and from *Microcystis aeruginosa* by Jones and Jost (1970)

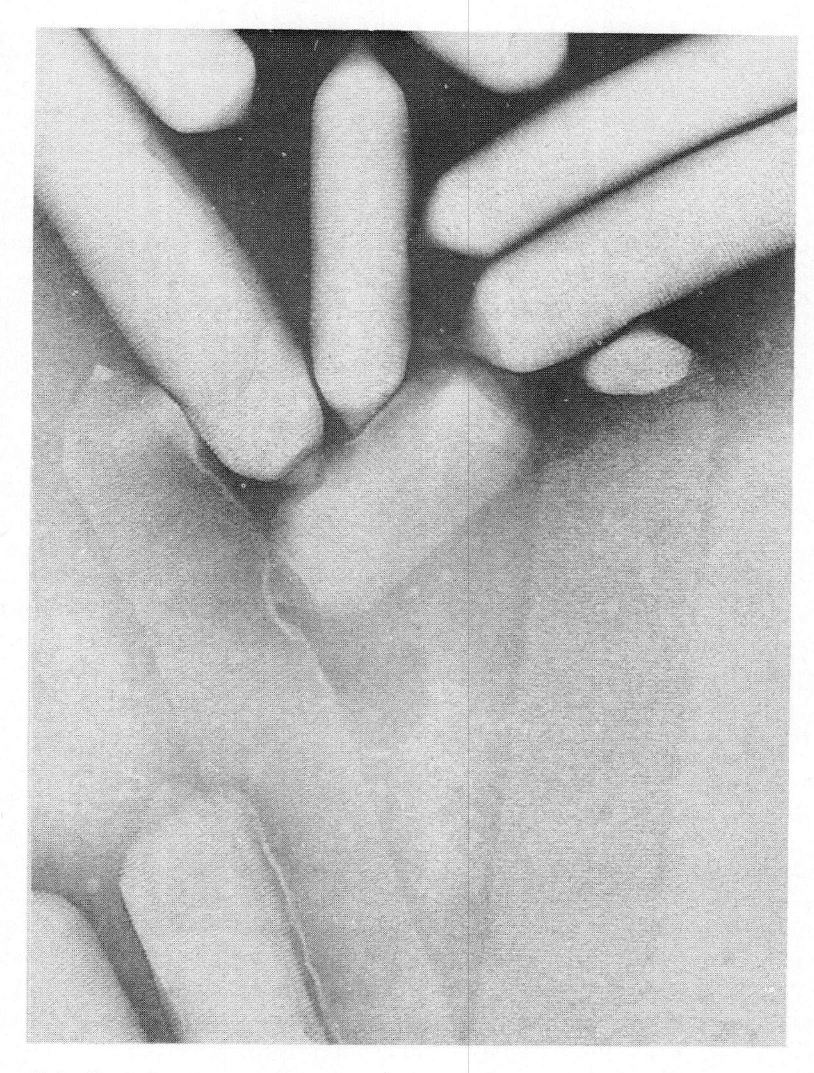

FIG. 5.4. Gas vesicles isolated from *Anabaena flos-aquae*, negatively stained with phosphotungstate. The two vesicles in the lower half of the picture have been collapsed flat by surface tension of the drying stain. Electron micrograph (× 150,000 From A. E. Walsby (1972) *Symp. Soc. exp. Biol.* **26**, 233–250.

form milky-white suspensions which become colourless and transparent when pressure is applied. Thus no pigments are present, contrary to what the earlier work of Jost and Matile (1966) had suggested.

Jones and Jost (1970) showed that protein accounted for over 95% of their preparations, which in view of the presence of a small quantity of contaminants, suggested that protein is the only component of the membrane. Of the 14 amino acids present in greater than trace amounts, alanine, valine, isoleucine, leucine, glutamate and serine accounted for about 70% of the total. The sulphur amino acids (methionine, cysteine and cystine) were absent. The amino acid composition of the *Anabaena* vesicle protein is very similar (see Walsby, 1972b).

The gas vesicle protein is difficult to solubilize completely but electrophoretic studies on preparations partially solubilized in acidified phenolic mixtures reveal only one protein of low molecular weight, perhaps in the region of 14,000 (Jones and Jost, 1971). The amino acid ratios obtained from the analysis of the *Anabaena* vesicles also support the idea of a single species of protein in this range of molecular weights (see Walsby, 1972b). X-ray diffraction techniques and studies in progress indicate that part of the protein is in a two-layered cross-β conformation (Blaurock *et al.*, in preparation) which may be important in providing the structure with rigidity.

Smith *et al.* (1969) have pointed out that gas vesicles are rather like certain proteinaceous viral structures and have made the interesting suggestion that they might have originated from viral infections. It may be worthwhile to make comparisons with the proteins from other species, to obtain information on the phylogeny of the gas vesicle.

Physical properties of gas vacuole membranes

1. PERMEABILITY TO GAS

Questions which have often been posed are—what is the gas inside gas vacuoles and how does it get there? The suggestion, first made by Klebahn (1922), that the gas vacuole membrane is impermeable to gases, still persists in the literature (see Bowen and Jensen, 1965; Jost and Matile, 1966; Pringsheim, 1966; Larsen *et al.*, 1967) but it is inaccurate. Walsby (1969) showed that the membranes are freely permeable to gases. Mass spectrometric analyses showed first that no more gas can be obtained from a gas vacuolate sample of alga after thorough evacuation to remove dissolved gases than from an otherwise identical sample in which the gas vacuoles had previously been destroyed. Secondly when cell suspensions containing a known volume of gas vacuoles are subjected to ultrasonic

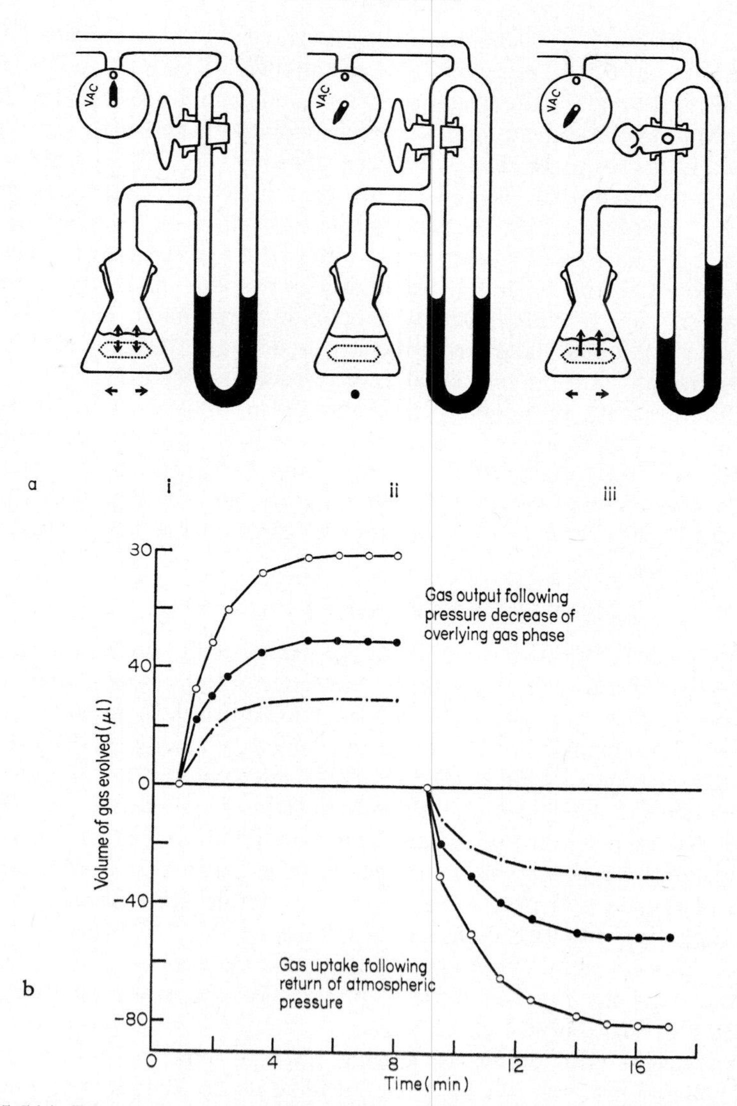

FIG. 5.5(a) Diagram summarizing the use of a modified Warburg apparatus to demonstrate the permeability of gas vesicle membranes to gases, and to determine the volume occupied by vacuole gas in the sample. (i) Gas-vacuolate suspension equilibrated at atmospheric pressure, with Warburg apparatus shaking. (ii) Apparatus brought to rest, pressure of gas phase reduced rapidly to minimize the loss of gas from the suspension by diffusion. (iii) Manometer tap closed. Apparatus set shaking. Gas evolved from the suspension measured by manometry.

(b) Gas exchanged by (○) gas-vaculoate suspension of *Anabaena flos-aquae*; by (●) a similar suspension in which the gas vacuoles have been collapsed; · — · difference between the two curves, using the apparatus described above (a). From Walsby (1972) *Bact. Rev.*, 36, 1–36, Fig. 7.

pulses to destroy the gas vacuoles, the pressure of the released vacuole gas is found to equal the pressure of the atmosphere overlying the suspension, and if this is decreased, the pressure in the vacuoles falls to the same level on equilibration. Modification of a Warburg apparatus has made it possible to follow gas exchange by the gas vacuoles directly (Fig. 5.5a, b) and it has been demonstrated that oxygen, argon and nitrogen equilibrate freely across the membrane. By measuring the resistance to collapse of gas vacuoles at the surface of a suspension under rising gas pressure it has also been shown that the vacuoles are highly permeable to oxygen, argon, nitrogen, carbon dioxide, carbon monoxide, hydrogen and methane (Walsby, 1971). Thus, the gases in the gas vacuole are the same as those in the surrounding solution and are at similar partial pressures.

The finding that gas vacuole membranes are permeable to gases necessitates a reconsideration of three aspects of gas vacuole physiology. Firstly, it is clear that pressure alone cannot support the structure. Secondly, gas vacuoles cannot be inflated by gas, and thirdly, gas vacuoles cannot store gas. These aspects will now be examined in more detail.

2. RIGIDITY AND PRESSURE RELATIONSHIPS

Several pieces of evidence show that gas vacuoles do not change volume significantly in response to a pressure difference between their outside and inside, so long as they do not collapse. Klebahn (1922) noted that gas vacuoles did not change noticeably in size before collapsing. Walsby (1969) found that creating a vacuum over an algal suspension did not increase the rate at which the alga floated up, as would have been expected if expandable bubbles had been present in the alga. He also observed that the degree of light scattered by a suspension of gas vesicles (which is related to the volume of the vesicles present) showed a reversible decrease of less than 1% on applying subcritical pressures (Walsby, 1971). Further evidence that gas vacuole membranes are rigid structures stems from Klebahn's realization that the pressure acting on a gas vesicle in an algal cell can exceed the pressure of the vacuole gas.

As with rigid engineering structures, such as pipes and boilers, the critical collapse pressure of a gas vesicle will be determined by the strength of the wall material, its thickness, and the size and shape of the structure. The ratio of wall thickness to cylinder diameter is of particular importance (Bienzeno and Koch, 1938) but this appears to be constant in blue-green algal vesicles. The critical pressure of the vesicles is not correlated with the length of the structure, which suggests that the variation in critical pressure is determined by variation in the strength of the membrane material itself (Walsby, 1971). Buckland and Walsby (1971) found that certain physical and chemical factors which affect the interactions of the component protein molecules (Jones *et al.*, 1969) also affect the strength of the vesicles. They are strongest in the pH range 7–9 and are weakened at temperatures above 40°C and in the presence of substances which interfere with hydrophobic interactions and hydrogen bonding. The inherent variation in the strength of gas vesicles from a given sample is probably a consequence of irregularities in the membrane, or of its having been exposed to proteolytic enzymes in the cells (Buckland and Walsby, 1971).

3. SURFACE PROPERTIES

The discovery that gas vacuole membranes are permeable to gas raises the question of how water is prevented from entering the structure. Although there is no direct evidence for or against, there is no reason to believe that water vapour is excluded unless it is discriminated against on account of its polar properties. The vacuole membranes are not any less permeable to another polar gas, carbon monoxide, than they are to non-polar gases. Carbon monoxide has, however, only one tenth of water's dipole moment (Walsby, 1971, and unpublished).

The vacuole membrane need not be completely impermeable to water molecules to prevent liquid water from accumulating inside the gas vesicles. If there are pores in the membrane and if these pores are lined with hydrophobic material or open out onto a hydrophobic surface, then a pressure would have to be applied to force the water in against surface tension. The pressure required would be $2(\gamma/r)\cos\theta$ (where γ is the interfacial tension, θ the contact angle, and r the radius of the pore at point of contact). If these conditions obtain, the pressure required would be enormous and far greater than the critical pressures of gas vesicles (Walsby, 1969). Again, a hydrophobic inner surface to the gas vesicle would militate against the condensation of water inside it, as the droplets which form would have a higher vapour pressure than the highly concave water surface in contact with the outside of the structure.

The outer surface of the vesicle is probably hydrophilic because hydrophobic surfaces would give rise to a large interfacial tension which would generate pressures tending to collapse the structure. For example an interfacial tension of 0·075 N m^{-1} (as exists at an air–water interface) would generate a pressure of over 2 MN m^{-2} (20 atm) on the structure, well in excess of the additional pressures (over the surface tension generated pressure) required to collapse gas vesicles (see Walsby, 1971). None of the surfactants investigated by Buckland and Walsby (1971) increased the apparent critical pressure of isolated gas vesicles, which suggests that the pressures generated at the interface are very small. Moreover, although

Jones and Jost (1970) have postulated a hydrophobic outer surface on the basis of the observation that gas vesicles adhere to the supposedly hydrophobic surface of Formvar, attempts to partition the vesicles between oil and water (Walsby, 1971) have shown that intact vesicles strongly prefer aqueous as opposed to oily phases. This again indicates a hydrophilic outer surface.

The contrasting properties of the inner and outer surfaces postulated here could be accounted for by the distribution of the polar and non-polar side groups of the amino acids which form the gas vacuole protein (Jost and Jones, 1970), and may be important not only in stabilizing the structure but also in its assembly.

Thus, the gas vesicle is a rigid structure from which liquid water is likely to be excluded and not at all like a balloon or a bubble. A more apt analogy is that of a rigid porous pot with a hydrophobic lining, through whose walls gases may freely diffuse, but from which water is excluded by surface tension.

Formation of gas vesicles

Walsby (1969) proposed that vesicles form by the aggregation of their component particles in such a way that, as a cluster of them increases in size, an empty space forms into which gas passes by diffusion. The occurrence of very small vesicles testified to this idea. Waaland and Branton (1969) also postulated, independently, the *de novo* synthesis of gas vesicles on observing that small gas vesicles were the first to form on inducing gas-vacuolation in *Nostoc commune*. They suggested that the cylindrical vesicles might grow by intussusception in the central region seeing that the juvenile forms which first arose appeared as very short cylinders with conical ends, and the older forms had similar ends but had longer, cylindrical middle regions. They also noticed in their pictures of freeze-etched vesicles that a rib near the centre of the cylinder often stands out from the others and speculated that this might represent the growing point. Walsby and Buckland (1969) found that isolated vesicles often split in two at a similar point, on drying, which also suggests that the vesicles are made in two halves. On the other hand Jost and Jones (1970) reported a variable number of prominent ribs on their freeze-etched vesicles and proposed that they represent either stacking faults or preparation artefacts.

Gas vacuoles reappear after several hours in algal cells which have been exposed to pressure and this suggested to Bowen and Jensen (1965) that the collapsed membranes might be re-inflated with gas. However, the same consequences of the membrane's permeability to gas apply equally for their re-erection as for their initial formation (Walsby, 1969). The sequence of small gas vesicles forming first, followed by vesicles of increasing length, has also been found in cells with recovering gas vesicles (Lehmann and Jost, 1971) and this shows that the same process of *de novo* synthesis occurs (Fig. 5.6). This is not to say, of course, that particles from the old, collapsed

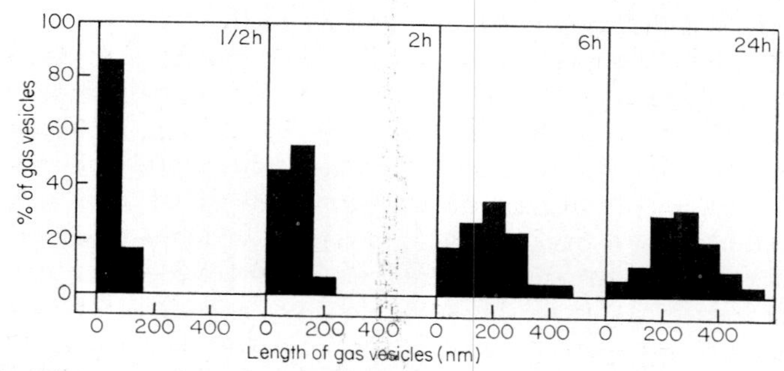

FIG. 5.6. Histograms showing the change in distribution of gas vesicle length with time, after collapsing gas vesicles in cells of *Microcystis aeruginosa* by ultrasonication. (Redrawn from Lehmann and Jost, 1971, *Arch. Mikrobiol.* **79**, 59–68).

membranes might not be used in the synthesis of new, intact ones, although as yet there is no information on this possibility.

Functions of gas vacuoles

The three events which accompany the destruction of gas vacuoles in algae subjected to pressure (the release of gas bubbles, the increase in the alga's density, and the change in its macroscopic appearance) have given rise to the idea that gas vacuoles might serve three functions: (1) storing gas, (2) providing buoyancy, and (3) light shielding. Because of their permeability to gas it seems clear that gas storage is not a role that gas-vacuoles could fulfil, and so old concepts of these structures storing nitrogen or gaseous fermentation products (Kolkwitz, 1928; Canabaeus, 1929) must be abandoned.

1. BUOYANCY

The change in density which occurs when gas vacuoles collapse shows without doubt that they are involved in the buoyancy of blue-green algal cells and studies by Walsby (1969) have demonstrated that the loss of the gas from the vacuole can account fully for the density changes. Klebahn (1922) found that the amount of gas-filled space required to make *Gloeotrichia echinulata* buoyant was in the region of 0·7% of the total cell volume, while the gas vacuole spaces actually occupied 0·8% (see Fogg, 1941). In

Anabaena flos-aquae, where the required volume is 2·1%, the gas vacuole void volume can be as high as 9·8% (Walsby, 1970). Smith and Peat (1967a) reported values up to 22% in the same two algae, and up to 39% in *Oscillatoria agardhii*, on the basis of the cross-sectional area they occupied in electron micrographs, but this figure is inclusive of the volume occupied by the membranes and the interstices between them.

Blue-green algae possessing gas vacuoles float towards the surface of natural waters where they become independent of the turbulence which other algae require to keep them in the photic zone (Fogg, 1969b). In fact turbulence tends to disperse the alga in the water column and its distribution will depend then on the degree of turbulence and speed with which the alga rises. According to Stokes' law, the rate at which a spherical body floats up varies as the square of its radius. Thus large bodies rise more quickly than smaller ones of the same density: colonies of *Gloeotrichia echinulata* achieve rates of up to 0·23 mm s^{-1}, whereas the highest rate observed with *Anabaena circinalis*, which forms much smaller colonies, is only 0·03 mm s^{-1}. In this context, it is interesting to find that, whereas many other planktonic algae exist as unicells or colonies of microscopic dimensions, so limiting the rate at which they sink, many of the planktonic blue-green algae aggregate into macroscopic colonies. Thus, the filaments of *Anabaena* become tangled into clumps, those of *Aphanizomenon* form large flakes, *Gloeotrichia* forms stellate clusters, and the unicells of *Coelosphaerum*, *Gomphosphaeria* and *Microcystis* are held together in large colonies by an extensive mucilaginous matrix (see Fig. 2.2). Reynolds (1967) has compared these different algae and finds that those which have the highest flotation potential are those which form the largest colonies. They are also those which are first to form surface blooms in certain Shropshire meres (see also p. 267).

Superficially, it might seem that the water surface, where both the light intensity and rate of arrival of carbon dioxide may be at their highest, would provide ideal conditions for algal growth (see Walsby, 1970). However, there are indications that the conditions which exist some way below the surface may be more suitable for blue-green algae and gas vacuoles may have an important role in maintaining them in such positions (see p. 263).

The primary mechanism proposed is that changes in the gas vacuole: cell volume ratio determines whether the alga floats or sinks. The ratio might be altered in two ways. Firstly, gas vacuoles might be effectively diluted out by growth as suggested by the results of Smith and Peat (1967b) who found that cells in the exponential phase of growth were less gas-vacuolate than those in the stationary phase. In line with this, it is found that both *Anabaena flos-aquae* (Walsby, 1969) and *Oscillatoria redekei* (Whitton and Peat, 1969) form gas vacuoles more abundantly under low light intensity when the growth rate is slower. Secondly, the existing gas vesicles may collapse under the turgor pressure of the cells containing them. While the critical collapse pressure of the vesicles varied from 200–700 kN m^{-2} in *Anabaena flos-aquae* (Fig. 5.7) the turgor pressure of the cells

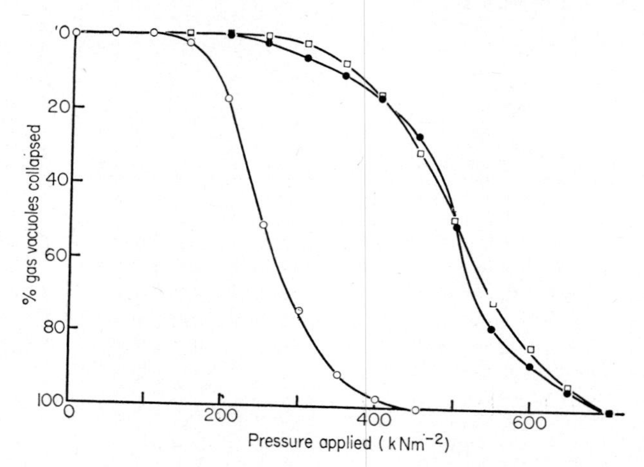

FIG. 5.7. Collapse of gas vesicles with pressure in the blue-green alga *Anabaena flos-aquae*. ○, Alga suspended in water; □, alga suspended in hypertonic sucrose solution; ●, gas vesicles isolated from the alga lysed in 0·7 M sucrose solution. The gas vesicles in the alga suspended in water collapse under a lower applied pressure because they are already subjected to the cell turgor pressure, equal to the mean separation between the two curves (□ minus ○). From Walsby (1971), *Proc. R. Soc.* B, **179**, 301, Fig. 6.

fluctuated over the range of 170–450 kN m^{-2} (1.7–4.5 atm) so that a proportion of the gas vesicles could be collapsed by a rise of turgor pressure alone. When the alga is transferred from low to high light intensity, in the presence of carbon dioxide-containing air, the turgor pressure rises by 100 kN m^{-2} or more. This serves to bring about collapse of sufficient gas vacuoles to make the cells lose their buoyancy (Dinsdale and Walsby, 1972). Responses of this sort can take place within an hour and allow the alga to respond relatively rapidly to changing conditions (Fig. 5.8). The possible ecological significance of this is considered in Chapter 12.

2. LIGHT SHIELDING

Gas vacuoles may be important in providing shielding from light as well as in providing buoyancy. This was suggested first by Lemmermann (1910) who noticed that vacuoles were invariably present in the surface

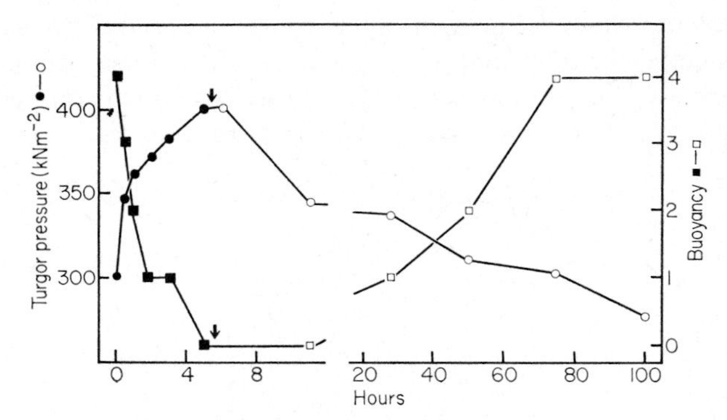

FIG. 5.8. Changes in the cell turgor pressure and buoyancy of *Anabaena flos-aquae* on transfer from low (50 lux) to high (10,000 lux) light intensity, and back to low light intensity again (arrowed). The rise in turgor pressure is brought about by the increased rate at which photosynthetic products are accumulated, and causes collapse of the weaker gas vacuoles in the cells. This in turn increases the density of the cells. Data from Dinsdale and Walsby (1972) *J. exp. Bot.* **23**, 561–570.

blooms of blue-green algae which could be exposed to high, and perhaps damaging, light intensities.

Despite the absence of pigmented components, the reddish appearance of gas vacuoles under the light microscope demonstrates that they interact with light and this might be a way which provides appreciable shielding. Coloration which is independent of differentially absorbing chromatophores, occurs widely in nature and is known as "structural" coloration. It may have three causes: refraction, interference, and scattering (Wright, 1964).

Petter (1932) first attributed the red colour of gas vacuoles to refraction on the grounds that the blue light would be bent further from the field of vision than the red, on passing from a liquid phase to a gaseous one with low refractive index. Whether the degree of differential refraction given by such a system would be sufficient to account for the colour is doubtful unless passage of light through a stack of gas vesicles in a gas vacuole might amplify this effect.

Interference is responsible for the rainbow colours of soap bubbles and oil films and occurs when light is internally reflected at more or less planar, parallel surfaces. Periodic arrangement of a number of such surfaces enhances the interference effect. Van Baalan and Brown (1969) have suggested that the light shielding and optical properties of gas vacuoles in *Trichodesmium erythraeum* might partly be a consequence of destructive interference. Although gas vesicle stacks would fulfil the requirement for parallel, repeating surfaces, the fact that the surfaces are not planar may mean that the interference effect is small. On the other hand, there may be something in their suggestion that internal reflection would divert light away from the centre of the cell.

The third possible cause of the red coloration is scattering. Light rays interact with very small particles in such a way that they emerge at an angle to the incident light path, without loss of energy. The intensity and angle of the scattered light depends on the difference in refractive index of the particle and the suspending medium, on the shape and size of the particle, and on a function of the wave-number (reciprocal wavelength). When the particle is very much smaller than the wavelength, the intensity varies as the fourth power of the wave-number. The relationship is more complex when the size of the particle approaches the wavelength, and the refractive index difference is very large (Stacey, 1956) as is the case with gas vesicles. Their dimensions of 0·07 μm by 0·1–2·0 μm compare with 0·4 μm for blue light and 0·65 μm for red; and the refractive index of their contents, air, should be close to 1·0. Fuhs (1968) gives the refractive index of whole gas vacuoles (which includes the gas vesicle membranes and interstitial cytoplasm as well as the air spaces) as 1·073, against 1·372 for the surrounding cytoplasm.

Gas vacuoles are strongly scattering objects. Simple nephelometry shows that they may scatter up to 6 times the amount of light scattered by the cells which possess them. Comparison of the spectra of purified, isolated gas vesicles in intact and collapsed states shows that the scattering is principally due to the air spaces, and is strongly wave-number dependent, so that blue light is scattered much more strongly than red.

It still has to be decided whether scattering, which only redirects the path of light, can produce significant shielding in the cell. It has been argued that in a surface water bloom, gas vacuoles result in increased back-scattering of light so that the amount penetrating the bloom is decreased (Walsby, 1970). Of course, the amount of light penetrating below the bloom and lost for purposes of photosynthesis will be decreased. In certain algae such as *Trichodesmium erythraeum* (van Baalen and Brown, 1969), *Nostoc muscorum* (Waaland *et al.*, 1971) and *Anabaena flos-aquae* (Shear and Walsby, in preparation) the gas vacuoles are located at the periphery of the cell, where they surround most of the photosynthetic apparatus. In *Nostoc* and *Anabaena* this arrangement is encountered only when the cells are grown under high light intensity suggesting that it may provide a significant light shield. Waaland *et al.* distinguished the apparent absorption (due to light scattering) of the gas vacuoles by the fact that it disappeared when the vacuoles were collapsed under pressure, and found that when the gas vacuoles were located in the algal cells (rather than isolated from them) their "absorption" was reduced in regions of the spectrum coincident with the absorption peaks of the photosynthetic pigments. This has been interpreted as evidence that the gas vacuoles were shading the light sensitive pigments, but Shear and Walsby (in preparation) have pointed out that the reduction is due in part to peak flattening, an artifact which arises when light absorbing (or scattering) material is not distributed homogeneously in solution but is restricted to packets (the cells) in suspension. The theory behind this idea has been considered by Duysens (1956). With *Anabaena flos-aquae*, Shear and Walsby (unpublished) found that cells with and without gas vacuoles showed no difference in photosynthetic activity at limiting intensities of white light, or in their resistance to ultraviolet irradiation. In blue light of limiting intensity photosynthesis was only about 21% less in algae with gas vacuoles than in corresponding material without gas vacuoles. Similar experiments should be carried out using *Nostoc muscorum*, but in the meantime the possible importance of gas vacuoles in providing light shielding remains unresolved.

Chapter 6

Movements

Many blue-green algae, and their colourless relatives the flexibacteria, show a curious sort of movement known as gliding, in which the cell or filament exhibits a slow, smooth progression without there being a visible means of propulsion. Gliding is seen most readily in the filamentous forms. Branched filaments do not glide, but the hormogonia which differentiate from their ends (as in *Tolypothrix* and *Stigonema*) are motile once released. Where the filament is sheathed the trichome moves within the sheath.

Compared with the movements of flagellates or ciliates, the gliding movements of filamentous blue-green algae are slow (about 1–10 $\mu m\ s^{-1}$). The movements of unicellular blue-green algae, which are reported to be intermittent and rather jerky (Geitler, 1936), are even slower (2–5 $\mu m\ min^{-1}$). They are best detected by time-lapse ciné photography or by logging the changing positions of individual cells. All filamentous forms can glide in both directions parallel to the long axis of the filament and show spontaneous reversal.

In addition to simple gliding, blue-green algae may show other movements: rotation, oscillation, bending, swaying, jerking and flicking movements. Very likely, these are either the direct consequence of gliding movement or of the activities of the gliding mechanism.

Gliding movement

Gliding movement occurs only in cells which are in contact with a surface. Solid surfaces, such as glass microscope slides or agar of a stiff consistency are particularly good. Filaments also show active movements over one another in liquid medium, although this results only in the relative displacement of the filaments involved as they are not capable of free swimming motion through liquids. It has been reported that gliding may be exhibited by filaments in contact with the meniscus at an air–water interface (Niklitschek, 1934) but it is difficult to distinguish true gliding from movements due to changes in surface tension in such situations.

A central problem is that gliding occurs without any obvious change in shape of the alga or movement of any organelle, as seen under the best light microscopes. Electron microscopy has revealed structures which may be involved in the gliding process but one cannot be certain that these structures, detected in fixed material, have a dynamic role in the process. Such problems give gliding movements an apparent insolubility which has challenged theoreticians and experimentalists alike. Theories on the mechanism of gliding movements have been reviewed in detail by Burkholder (1934), Fritsch (1945), Weibull (1960), and by Doetsch and Hageage (1968). The latter workers consider these under the four headings of (1) osmotic forces (2) surface tension phenomena (3) slime secretion and (4) contractile waves.

Theories invoking osmotic and surface tension phenomena have fallen from favour through lack of evidence and a failure to account for the observed facts of gliding. Briefly, Hansgirg (1883) proposed that filaments moved in response to a turgor pressure gradient in the cells, though it is difficult to see how motility resulted. Ulrich (1926) made use of turgor pressure theories to account for peristaltic waves of contraction passing along gliding filaments of *Oscillatoria* which he thought he had demonstrated by an ingenious stereoscopic technique. Schulz (1955) analysed the system which Ulrich described and concluded that the proposed contractions were optical artifacts.

The surface tension theory suggested that surface active material released by filaments provides the force required to move them (Coupin, 1923). A familiar example of such a mechanism is the camphor-driven toy boat: a piece of camphor placed at the rear of the boat slowly spreads on the water surface and the boat is driven forward before the spreading film. Such a mechanism might just operate under water also, but it would be necessary for the substances released to be immiscible with water and to be liberated from one end of the submerged object. Blue-green algae are known to liberate various chemical substances but these are usually of a hydrophilic nature. Moreover, within the scope of this theory, it is difficult to account for the observation that individual cells have independent gliding mechanisms, capable of operating in opposite directions in different parts of the same filament (Walsby, 1968b).

The theories of movement by mucilage secretion (Siebold, 1849; Correns, 1897; Niklitschek, 1934; Schulz, 1955) or by the production of contractile waves in one of the surface layers of the wall (Jarosch, 1959; Costerton *et al.*, 1961) have much more to offer. Current opinion appears to favour the latter (Halfen and Castenholz, 1970) but in certain respects the two theories are interdependent.

It has long been recognized that, during gliding, blue-green algae produce mucilages, probably of highly hydrated muco-polysaccharides. Bishop *et al.* (1954) isolated a long-chain complex polysaccharide from aged culture filaments of *Anabaena cylindrica* (see p. 75) which may be a component of the mucilage. It is sometimes possible to detect the mucilage, as it is produced during gliding, by staining with ruthenium red. It can also be detected by adding suspensions of particulate material, such as carmine or Indian ink (Niklitschek, 1934; Schulz, 1955). The mucilage then shows up as a light layer or halo surrounding the filament against the dark ground of the suspended particles (see Fig. 2.1) or else the particles become trapped in the mucilage which after some minutes appears as a darkly stained mass (Fig. 6.1). Walsby (1968b) concluded that the initial avoidance of the mucilage by ink particles is due to a like (negative) charge on both substances and found that the addition of calcium chloride at low concentration (2×10^{-3}M) resulted in the ink flocculating readily on the mucilage.

Trails of mucilage left behind by blue-green algae were probably seen first by Siebold (1849) and his observations have been confirmed many times since. Niklitschek (1934) published some fine light micrographs of slime trails and of mucilage rings which may accumulate around filaments. The continuous mucilage trails left by gliding filaments (Fig. 6.1c) demonstrate that mucilage is continuously produced, at least under some conditions, and it has been suggested repeatedly (see Siebold, 1849; Correns, 1897; Fechner, 1915; Prell, 1921; Schulz, 1955; and many others) that its secretion might be the cause of gliding movement. In summary, the theory is that mucilage is secreted, perhaps over the whole surface of the filament in a directional manner towards one end of the filament or the other. If the mucilage makes contact with a solid object it sticks to it and the further secretion of mucilage behind it results in a relative displacement of the filament and the object.

A useful analogy is that of a punt being pushed along by a pole, in which the filament is represented by the punt, the substrate on which the filament is gliding by the river bottom, and the mucilage trail by the pole, albeit an extremely long one which is continuously extruded! The analogy has several useful extensions. The method of movement is not very effective except in shallow water where the boat is close to the river bottom, and it becomes completely ineffective if contact with the bottom is lost. Nevertheless relative movement between adjacent boats can be obtained by one pushing off the other. Also the method affords only a rather slow means of progression. It is emphasized that the sort of movement intended is not one of jet or rocket propulsion (with mucilage being shot out at one end and the filament being displaced in the opposite direction with equal momentum) as suggested by Holton and Freeman (1965). This is

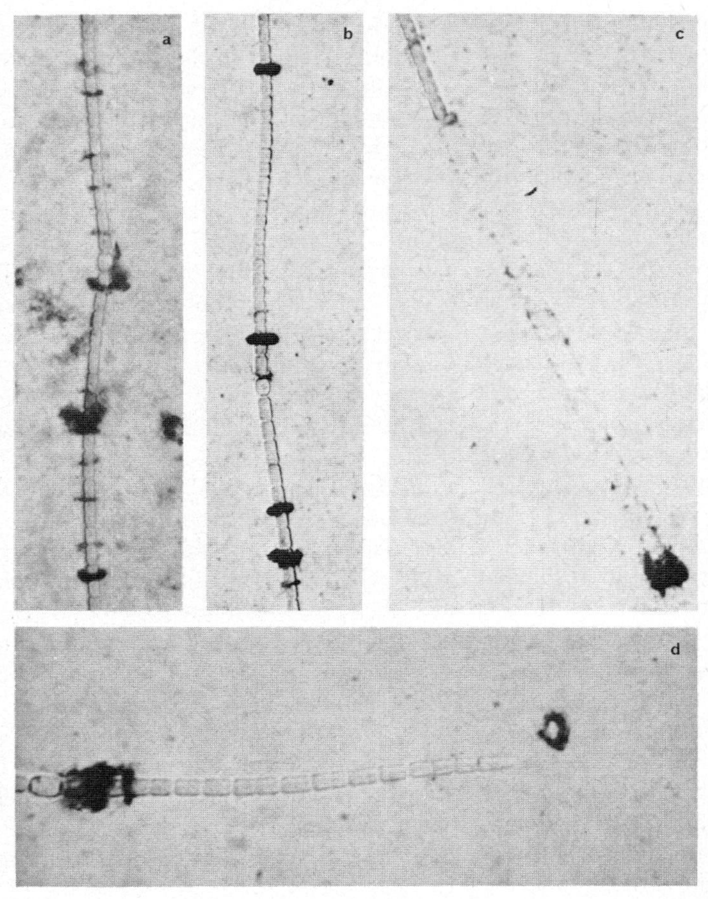

FIG. 6.1. Filaments of *Anabaena cylindrica* showing mucilage rings visualized by trapped Indian ink particles. (a) Freely suspended filament, about 2 min after mixing with ink, showing rings forming at alternate cross walls. (b) Another filament, after about 5 min; the rings have been moved along the filament and coalesced into larger ones. (c) Filament which has sunk into contact with the glass slide and started gliding, leaving a continuous trail of mucilage, and rings on the trail. (d) Ink-stained ring which has been transported to the end of the filament, and floated freely off. Light micrographs by A. E. Walsby; a, b, and c ×400 and d ×480.

unlikely to occur because freely suspended filaments show no movement relative to the surrounding water mass, even though they produce mucilage.

Careful observations of small particles being transported smoothly over the surface at speeds similar to that of gliding have provided further details of the gliding process (see Niklitschek, 1934; Schulz, 1955; Walsby, 1968b). Firstly, particles on different parts of the same filament may move in different directions. Secondly, rings of mucilage may form at cross-wall regions (Fig. 6.1a) of filaments of *Oscillatoria* (Niklitschek, 1934) and *Anabaena* (Walsby, 1968b). This may be where it is being secreted (Fig. 6.2a) or it

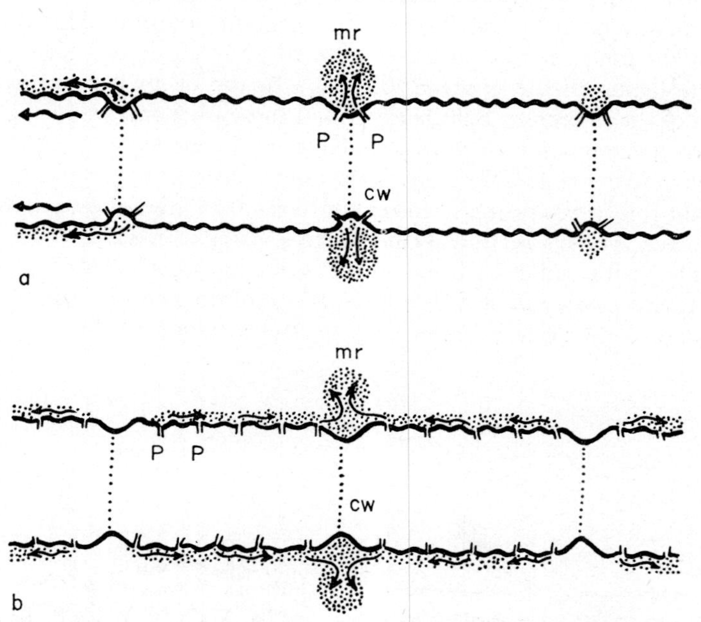

FIG. 6.2. Diagram summarizing two possible explanations for the formation of mucilage rings at the cross-wall region. (a) Mucilage accumulating at the site of its production. (b) Mucilage being piled up between two cells which are producing gliding movements in opposite directions. N.B. As yet there is no direct evidence that mucilage is secreted through distinct pores. (cw) cross wall; (mr) mucilage ring; (p) pore. From A. E. Walsby (1968) *Protoplasma* **65**, 223, Fig. 12.

may be that, under certain conditions, adjacent cells produce mucilage in opposing directions (Fig. 6.2b) and it piles up where the mucilages meet. Rings usually occur at alternate cross-wall regions rather than at every septum along the filament and this supports the latter idea. Thirdly, rings of mucilage may be transported along the filament, again in opposing directions and when they collide move off in one direction. Thus the gliding mechanism can show instant reversal.

The mucilage rings remain transverse to the filament axis as they pass along filaments. Walsby concluded from this that the gliding mechanism operates all round the filaments, rather than in a particular line or band.

Another feature is that on filaments which rotate as they glide, mucilage and adhering particles describe a helical path indicating that the gliding mechanism is not parallel to the long axis of the filament but is inclined at an angle to it. Schulz (1955) found that carmine particles adhering to filaments of *Oscillatoria sancta*, which exhibits a rotational component to its translational movement, remained immobile in the field of vision while the filament glided freely forward in constant rotation. However, the moment the filament was obstructed the carmine particles began to move in a helical line in the opposite direction to the original movement and rotation of the filament. Schulz explained these phenomena by suggesting the active participation of the extruded slime in movement. Others have not been convinced and although few would deny that the mucilage has an important role in the process, this is not to say that this role is active rather than passive. The distinction between active systems, in which the mucilage is extruded with sufficient force and in sufficient quantities to propel the filament, and passive ones where it is not, is summarized in Fig. 6.3. With light microscopy alone it is not easy to decide which of the two systems operates.

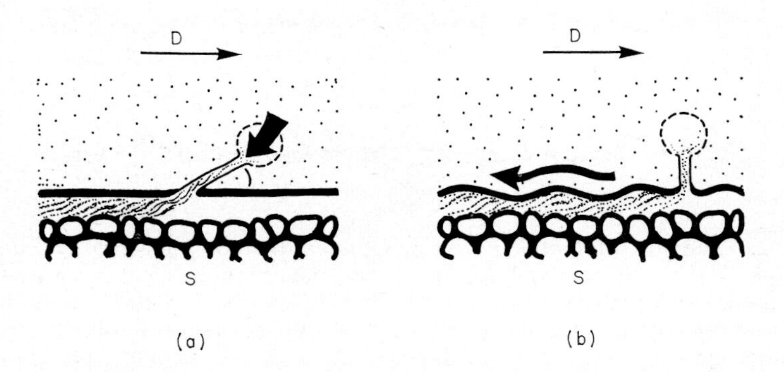

FIG. 6.3. Diagram illustrating two hypothetical systems of movement involving mucilage secretion. (a) Active involvement; the mucilage is extruded at an acute angle to the direction of movement with sufficient force to propel the filament. (b) Passive involvement; the filament glides by some other mechanism. Mucilage, which is continuously secreted, is left behind as a trail, and serves to increase the degree of contact between the filament and the substratum. Redrawn after Walsby (1968) *Protoplasma* **65**, 223.

Electron microscopy has revealed the presence of two groups of structures which may be associated with the gliding process: pores, which are possibly involved in mucilage secretion, and wall layers, which may possess contractile elements.

Pores and the mucilage-generating system

Early reports of pores in the side walls of blue-green algal filaments were unconvincing, according to Fritsch (1945) who considered that the mucilage secreted was derived by modification of the sheath or outer wall layers. However, two sorts of pore have been revealed by electron microscopy. One type is minute, about 2 nm in diameter, and numerous, with several hundred forming a girdle round the filament on either side of the cross walls (see p. 73, Fig. 3.28). According to Frank *et al.* (1962) the pores penetrate to the L_{II} layer and several authors have suggested that these may be the channels of mucilage excretion (e.g. Pankratz and Bowen, 1963). Walsby (1968b) pointed out that they were in just the position that the mucilage rings initially occupy when filaments are freely suspended (Fig. 6.1a). However, although these pores occur in filaments of *Oscillatoria* (with squarely abutting cells) they are not found in *Nostoc* and *Anabaena* filaments (with round-ended cells in chains). The second type of pore which is found only in the *Oscillatoria* type of filament is much larger and occurs in the most superficial layers of the wall or sheath (see p. 73).

Motile organelles

Under the electron microscope surface structures which might propel the filament are seen. Many species have an undulating surface wall layer (L_{IV}) in thin section, the pitch of the undulation being too small (about 1–2 nm) to be resolved at the light microscope level (see Echlin and Morris, 1965; Costerton, 1960; Lang, 1968). This undulating layer may sometimes be an artifact of shrinkage during preparation for electron microscopy and Jost (1965) found that the wall layers of *Oscillatoria rubescens*, although convoluted in embedded material, are smooth in freeze-etched preparations.

Wildon (1965) observed fine fibrous elements, 15 nm in diameter, on the outer surfaces of the vegetative cells of *Anabaena cylindrica* and postulated that they might be involved in gliding. Halfen and Castenholz (1971a) described smaller fibres (5–8 nm wide) from the L_{III} layer of *Oscillatoria princeps*, that is, just inside the region of the velocity gradient between the gliding trichome and the sheath within which it moves.

Jarosch (1962) described clearly how propagation of a wave set up in a surface fibre could cause the movement of the filament (in the opposite direction) or the shifting of secreted slime (in the same direction) as indicated in Fig. 6.3b. Costerton (1960) proposed similarly that the undulating outer layer which he found in *Vitreoscilla* and *Beggiatoa* participated in the gliding movement of these organisms. Evidence that such structures are dynamically involved in the gliding might be obtained by relating their form to particular aspects of movement as seen under the light microscope. In this way Halfen and Castenholz (1970, 1971a) contend that the superficial fibres they observed are involved in motility. They found that *Oscillatoria princeps* rotates as it glides so that a point on the surface of the filament describes a helix with a pitch of 30 degrees to the long axis of the cylindrical filament. Under the light microscope similar helical orientation could not be seen and under the electron microscope the only structures so arranged were parallel arrays of 5–8 nm wide fibrils, also at 30 degrees to the long axis. Halfen and Castenholz suggest that these fibrils are too fine to act as screw threads that guide the rotation of the filament as it glides forward (in the way that a spiral-ratchet screwdriver rotates as it is pumped) and conclude that waves propagated in the fibrils must propel the trichome. These could also displace attached particles or ensheathing mucilage in a similar helical path if denied contact with a solid substratum. This is perhaps the most coherent explanation yet produced for gliding motility, and accommodates most, if not all of the observations made with gliding trichomes. At the very least, it should stimulate fresh lines of investigation. If the fibril theory is correct, it poses further questions. Are the fibrils able to propagate waves in both directions, or is it necessary to postulate two sets, laid down in an antiparallel manner, to explain filament reversal? Also, if the fibrils provide the driving force, what is the function of the mucilage which is produced by gliding filaments?

Suggestions have been made by those who are unconvinced of the active role of mucilage in the gliding process that it may still be important or even indispensable in providing lubrication or adhesion. For instance Costerton (1960) observed that filaments of *Vitreoscilla* move with increased speed when they encounter the mucilage tracks left by previous filaments. He postulated that mucilage secretion is not causal of gliding movement but that the substance effectively increases the contact between the outer scalloped cell wall layer and the substratum. This could be particularly important in view of the very small pitch of the undulations thought to be involved in transmitting movement. Walsby (1968b) considered that if the motile organelle or mechanism was distributed over the whole cell surface, as his observations suggest, then a layer of mucilage which completely ensheaths the trichome adjacent to it, would also provide indirect contact with the motility mechanism on the other side of the trichrome. Thus the whole surface could participate and increase the efficiency of the process.

With species such as *Anabaena* and *Nostoc*, the mucilage dissipates in the surrounding water, although it may accumulate locally in aged growths to produce jelly-like masses. Old shake cultures of *Anabaena* will sometimes set to a gel after standing for about an hour and become liquid again when shaken for a short time, indicating that the mucilage is a thixotrophic gel. One wonders whether this property is important to movement.

Other algae, such as species of *Lyngbya*, produce sheaths through which the trichomes glide. Costerton (1960) observed that an *Oscillatoria* trichome will often glide halfway out of its sheath, pause, and then reverse back into it. These sheaths may be formed of the same mucilage that is associated with movement. They are usually absent from young trichomes and hormogonia.

Lamont (1969a) described a microfibrillar system in the sheaths of both *Oscillatoria chalybea* and a species of *Lyngbya*. Like the fibrils described by Halfen and Castenholz (1970) these 3–10 nm wide fibrils are oriented parallel to the helical path taken by the gliding filaments. Being present in the sheath, it seems unlikely that these fibrils are actively engaged in the gliding process, and Lamont proposed instead that they are orientated by shear at the velocity gradient between the surface of the gliding trichome and mucilaginous sheath.

Other movements

Cursory microscopic observation suggests that filamentous blue-green algae show other sorts of movement apart from gliding. Trichomes wave, jerk, and exhibit sudden bending, flexing or straightening movements, particularly when tangled. Certain workers have regarded such movements as independent of gliding (Burkholder, 1934) but the simple hypothesis that there is only the one basic type of movement can account for most, if not all, of these observations.

Filament rotation, when it occurs, is always accompanied by translational movement and must be dependent on the gliding process (see above). The swaying movements, to which the genus *Oscillatoria* owes its name, are the most distinctive and occur when one end or both ends of the attached filament are free. As the filament glides forward rotating, its tip follows a straight line only if the axis of the filament is straight. If it is bent, however slightly, the tip will describe a conical path as shown in Fig. 6.4. Such filaments seen from one side, under the microscope, will thus appear to wave from side to side, and indeed in the restricted depth between coverslip and slide, the conical path may be distorted into a simple side-to-side oscillation. When the filament moves forward some distance it can be seen that the oscillation coincides with the frequency of rotation (Fig. 6.4). However, at the edge of a clump of oscillatorian filaments the distance moved backwards or forwards may be small, owing to frequent

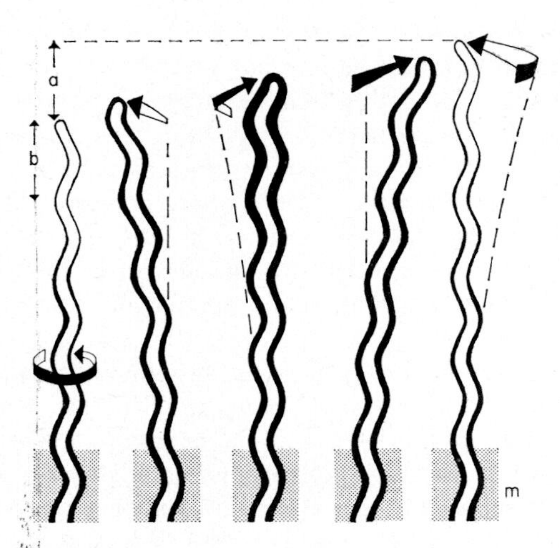

FIG. 6.4. Diagram showing five stages in the progress of a bent filament rotating in the direction of the arrow, as it glides through the mucilage (m). The end of the filament prescribes a conical path, as indicated by the thickness of the lines (thin lines into the page, thick lines out of the page). The wavelength (a) of the oscillations is the same as the distance travelled in one rotation and corresponds to the wavelength of the morphological spiral, where this occurs (b).

reversal. Also, the amplitude of the oscillation may actually well exceed the amount of translational movement due to the filament being markedly curved. Under such conditions with filaments gliding against the uneven moving surfaces of one another, the degree of lateral movement observed can be so great that it appears to take place in the absence of gliding.

It is conceivable, of course, that surface elements concerned with gliding may, by their contraction, cause distortion of the filament shape. Halfen and Castenholz (1971a) reported that the cylindrical wall casings isolated from *Oscillatoria princeps* by ultrasonication become coiled and turned inside out, indicating tension between adjacent wall layers. However there is no convincing evidence from observation of isolated filaments that a filament may autonomously bend itself.

Certain members of the Oscillatoriaceae such as species of *Spirulina* and *Arthrospira* have coiled filaments. The rotation of the filaments takes place in the same direction and period as that of the coils. Extending the observations of Halfen and Castenholz (1970), it is to be expected that the orientation of the gliding mechanism follows the basic helical morphology. The coils of *Arthrospira* are fairly uniform and of such a pitch that one trichome may glide up the groove formed by the coil of another forming a compact double helix.

Flexing movements in other algae do not always have such an obvious relationship with gliding. Trichomes of *Anabaena cylindrica* often show marked twitching movements particularly when tightly aggregated together. This appears to be due to filaments gliding free of mucilage (Fig. 6.8d, see p. 129).

The significance of gliding and other movements

Gliding, because of its slow speed, is unlikely to be important in dispersing the algae over large distances. Its limitation to the surfaces of various substrata demonstrates that it cannot be effective in moving suspended algae through water. Nevertheless, the widespread occurrence of gliding movement in blue-green algae testifies to its possible importance in maintaining cells or filaments under optimal light intensity and other environmental conditions, and in maintaining steep diffusion gradients.

RESPONSES TO LIGHT

Gliding filaments of blue-green algae exhibit *phototopotaxis* (orientation with respect to the direction of incident light) and *photophobotaxis* (responses to changes of light intensity in time independent of light direction—Haupt, 1965). Hansgirg (1883) demonstrated that filaments moved more rapidly in the light than in the dark, and Harder (1918) found that the speed of the young hormogones of *Nostoc* increased with light intensity up to 100 lx although Burkholder (1934) questioned whether temperature, which also has a marked effect on gliding rate, was kept constant in these experiments. Halfen and Castenholz (1971a) record that trichomes of *Oscillatoria princeps* glide in the dark.

Nienberg (1916) and Harder (1920) found that gliding filaments reversed when the light intensity was decreased suddenly, but that no reversal was shown on an increase in light. Apparently, by the same mechanism, a filament gliding from light into shade also reverses. This enables a filament to remain under optimal illumination in an unevenly lit habitat. Burkholder (1934) observed that under a given light intensity *O. splendida* showed negative phototaxis while eight other species of *Oscillatoria* behaved positively.

Although gliding filaments can move only backwards or forwards and are unable to steer a new course, it is often observed that a colony on an agar plate will come to occupy a region of optimal light intensity. Drews (1959) considered this as a form of phototopotaxis. He observed that if a filament lies along a light gradient it will glide to and fro but that the reversal occurs sooner when moving back towards low light (or inhibitory high light), so that the overall result is net movement towards the optimal light regime (see Fig. 6.5). Filaments that are initially orientated across

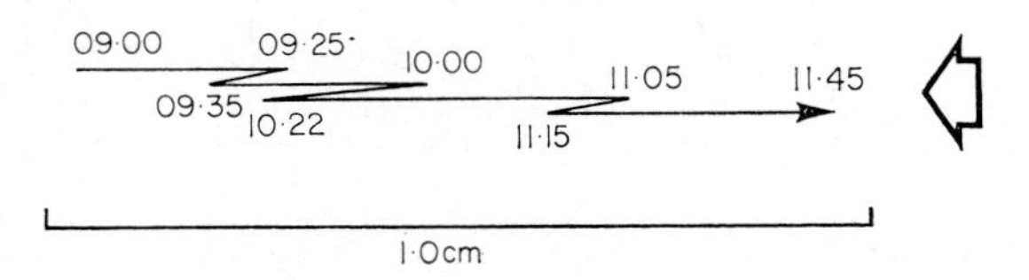

FIG. 6.5. Diagram representing the phototopotactic movement of *Phormidium uncinatum*. The filament reverses its direction of movement periodically (at the times shown) but glides for a longer time towards the light than away from it. Redrawn from Drews (1959) *Arch. Protistenk.* **104**, 389–430.

the axis of the gradients are unable to move to either end of it, but as Drews points out, filaments rarely move in perfectly straight lines and their course may be redirected by collisions with obstacles or by the passage over them of other gliding trichomes. Eventually, they also come to occupy approximately the same orientation by the same process.

The action spectra of the different responses have been investigated in detail by Nultsch (1961, 1962a, b) using *Phormidium uncinatum* and *P. autumnale*. The photophobotactic response showed a similar action spectrum to photosynthesis, which may indicate that the rate of gliding is stimulated directly by photosynthetic products. Phototopotaxis, on the other hand, appears to be insensitive to red light though responsive to light of shorter wavelengths; this suggests the involvement of carotenoids. Nultsch (1962a) also distinguished a third response to light, *photokinesis*, (an aggregation in response to light). The extent of this response, which was measured as the quotient of the distance covered in light and dark, varied with the wavelength of light in a way which suggested the participation of chlorophyll, rather than photosynthesis *per se*. The action spectra of the three responses are compared with the absorption spectra of the principal pigments from these species of *Phormidium* in Fig. 6.6a, b.

In the dense mats and webs that blue-green algae form on rocks, lake- and sea-shores, in streams and on waterlogged soils, the slow gliding movements could be important in placing an individual filament in just the right light intensity. The distance it must move to do this need not be very great, perhaps just far enough to move from the shade of overlying filaments, or between sand grains that may be sedimenting over it.

SIGNIFICANCE IN MAINTAINING DIFFUSION GRADIENTS

If an algal cell is suspended motionless in water, dissolved nutrients will diffuse to the surface of the cell and be taken up so that a concentration

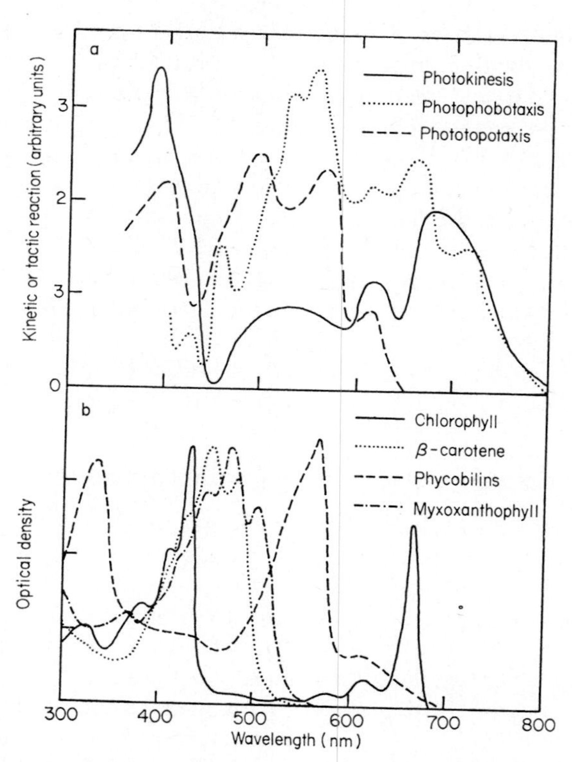

FIG. 6.6. The photokinetic, phototopotactic and photophobotactic reaction spectra of *Phormidium autumnale* (a), and absorption spectra of the major pigments extracted from the same alga (b). Redrawn from Nultsch (1961, 1962a, b).

gradient results. The slope of the diffusion gradient can be steepened by relative movements between cell and water. Such movements can be achieved actively or passively, as when cells rise or sink through the water (see p. 266). With blue-green algae, movements generated actively by the gliding process may also be important. The oscillating and swaying movements are presumably more effective than simple linear translation, as they result in much greater relative movement between the filament and the bathing solution. This may, in fact, be the main biological significance of these movements.

Gliding provides a means of movement while at the same time permitting the cell or filament to remain attached to a particular surface, a function which could not be duplicated by other mechanisms such as swimming. It is important in fast-flowing streams, on wave-beaten shores, rain-washed surfaces and similar habitats. It is also important in certain colony-forming blue-green algae, such as *Aphanizomenon* and *Trichodesmium*, where the filaments form floating rafts or flakes of considerable size (Fig. 14.7). These colonies change shape constantly as the filaments glide over one another. The movements here presumably result in irrigation of the colony with nutrients, and in each of its members sharing in the light falling on it. At the same time the gliding permits the filaments to stay together, which may be important to the floating habit (see p. 106).

One further function gliding might fulfil is to help free filaments of attached bacteria, as was suggested by Dodd (1960). Bacteria, which attach to the layer of slime, are left behind as the alga glides. This can be of use in separating algae from contaminating bacteria, but it is advisable to sever the filament from its trailing end, for otherwise there is a danger of transferring the invisible discarded sheath with the isolated filament.

Energy consumed by gliding movement

Having considered the possible advantages of gliding it is now worth assessing the costs at which these advantages are bought. Halfen and Castenholz (1971b) measured the gliding rate of filaments of *Oscillatoria princeps* (2·7 μm s^{-1}) at 25°C in a methyl cellulose-containing medium of known viscosity. They calculated the force of viscous drag that would act on a cylinder of similar dimensions to a filament (5.6 mm long and 35 μm wide) moving through such a medium at the gliding speed, as 0·0127 dynes (0·127 μN) and assumed that the inertia of such a small object would be negligible. On this basis they calculated that the total energy required by each cell of the gliding trichome would be provided by utilization of 7300 molecules of ATP per second. From measurements of the respiratory rate of the alga, this represents only about 0·05% of the ATP produced by oxidative phosphorylation, which Halfen and Castenholz (1970, 1971b) have shown to be the source of energy for gliding. Comparison with other motility systems suggests that the overall efficiency of the process might be as low as 1%, but even if this were the case, gliding would appear to consume only 5% of the respiratory energy of the cell, perhaps a small price to pay for the possible benefits.

Aggregation of dense suspensions and mats of filamentous algae

If dense suspensions of active filamentous blue-green algae are shaken and then left to settle, the filaments will contract into a discrete clump which may mimic the shape of the containing vessel. The contraction shown by *O. terebreviformis* (Castenholz, 1967b, 1968), a thermophilic blue-green alga with a thermal range of between 35 and 54°C, is particularly rapid, the edges of clumps drawing together at speeds up to 180 μm s^{-1}. *Anabaena cylindrica* contracts less quickly but produces clumps of an identical nature (Fig. 6.7). The facts (Walsby, 1967b, 1968b) that the rate of

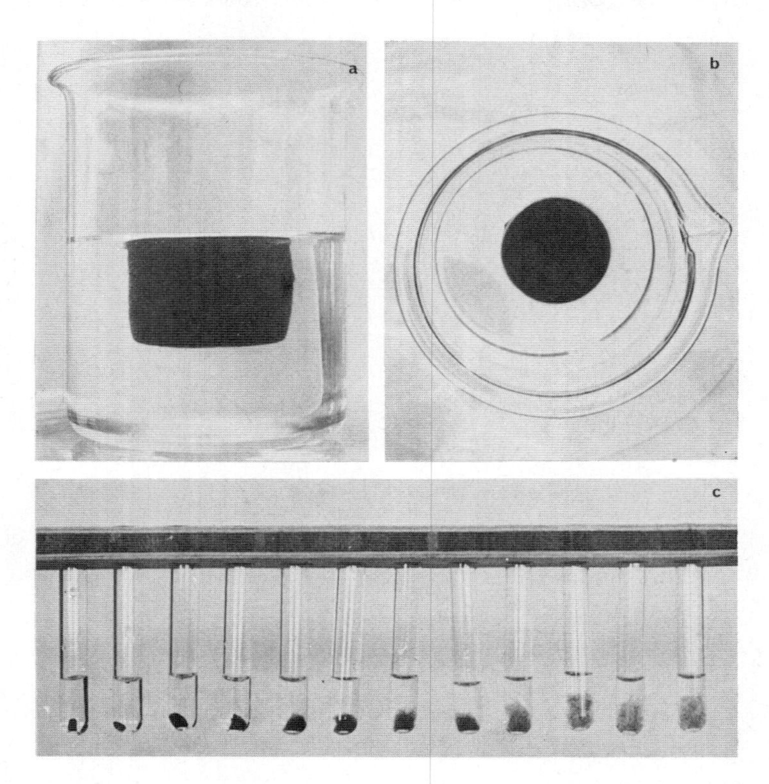

FIG. 6.7(a, b) A clump of *Anabaena cylindrica* filaments which formed on leaving an evenly-suspended culture to stand for 1h. (c) Clumping rate was used as an assay for factors affecting the gliding rate, or for toxic chemicals (in this case copper). Duplicates were exposed to a series of concentrations from zero (extreme left) to 1·0 ppm (extreme right). a and b from A. E. Walsby (1968), *Protoplasma* **65**, 223, Figs 9 and 10.

clump contraction greatly exceeds the observed gliding rates of the organisms, and that the filaments are randomly orientated in the clumps, make it clear that in each case the contraction cannot result from a topotactic response.

Walsby (1968b) concluded that on standing, mucilage which encircles the filaments sticks them together when they touch (Fig. 6.8a) and they continue to produce mucilage as they glide (Fig. 6.8b). Several filaments

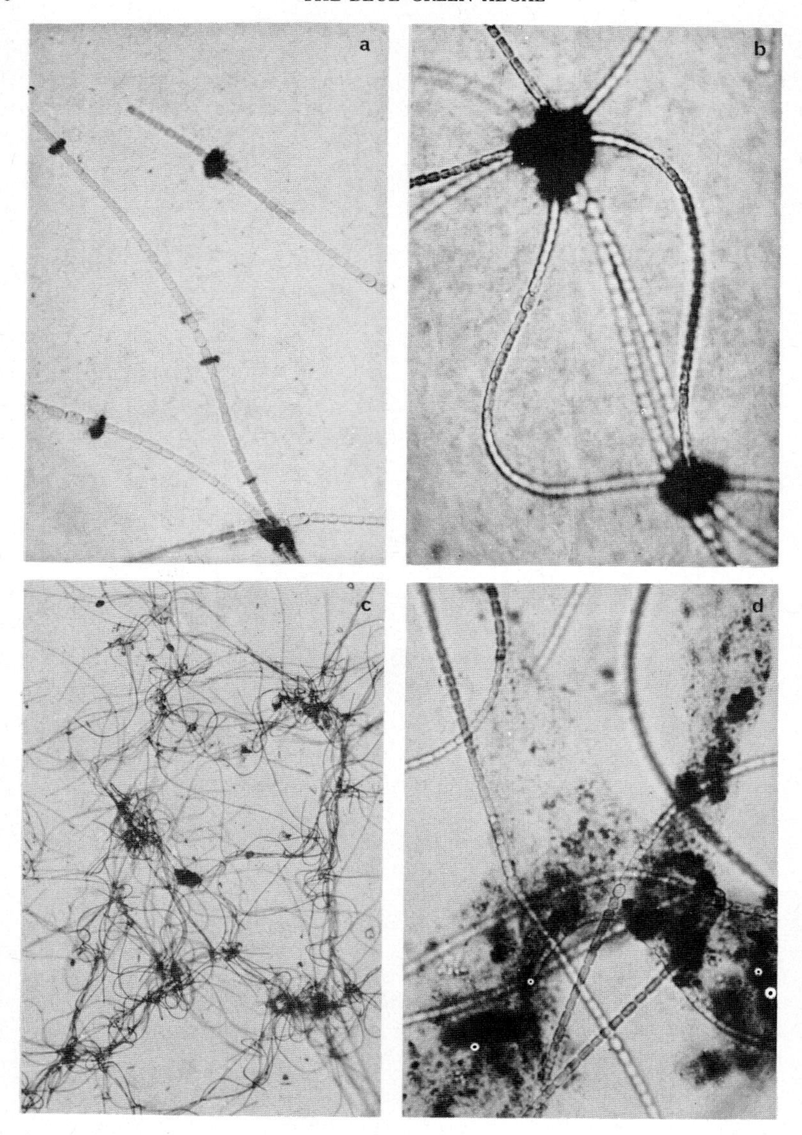

FIG. 6.8. Stages in clump formation by *Anabaena cylindrica* (a) Filaments drift into mutual contact and are held together by (ink-stained) rings of mucilage. (b) Filaments start gliding and contribute to the mucilage masses holding them together. (c) Long filaments involved in several aggregates, which are brought together by the gliding activity. (d) Skeins of mucilage (ink-stained) which accumulate in mature cultures. (Light micrographs; a, b and d $\times 350$, c $\times 20$) b and d from A. E. Walsby (1968), *Protoplasma* **65**, 223, Figs 15 and 16.

aggregate in clumps (Fig. 6.8 a-d), and the filaments continue to glide through the aggregates. Movement may occur in opposite directions in different parts of a single filament so that two aggregates may be drawn together at a rate twice that of the basic gliding rate if they are connected by a single filament. A series of such events could result in contraction of aggregates at many times the gliding rate (Fig. 6.9). Many of the

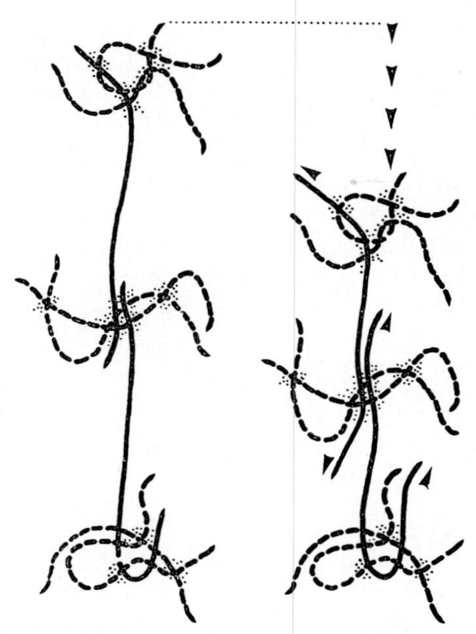

FIG. 6.9. Diagram showing how aggregates of filaments can be brought together at a rate which exceeds the gliding rate. The ends of the filaments (solid lines) bridging the aggregates are gliding in the directions denoted by the arrows. Other filaments are shown as broken lines, and the mucilage binding them together as dots.

filaments which bridge the aggregates are pulled out straight, giving the impression that they are under tension and responsible for the contraction. Sometimes the filaments break under the strain, or are suddenly pulled free and this may result in sudden jerks in the clump, and in the filaments flicking free. Castenholz (1967b) thought that these flexional movements might be instrumental in the contraction process but they seem rather to be a consequence of it.

This phenomenon can also occur in natural ecosystems. Castenholz (1968) has demonstrated that such rapid contractions not only occur with *Oscillatoria terebreviformis* in its natural habitat, but also that they are important and perhaps essential, to its survival there. The mats formed by this organism in hot springs contract rapidly, at up to 100 $\mu m\ s^{-1}$, when suddenly exposed to direct sunlight, when shadows shorten during the morning. The degree of self-shading which must occur in such highly contracted clumps evidently provides protection from excessive light. Castenholz also showed that clump contraction would occur in response to rises in temperature of the bathing water, above the maximum that the alga would tolerate. *O. terebreviformis* will grow at its maximal rate from 45–53·5°C but prolonged exposure to 54°C and over results in death. In hot streams showing temperature gradients rising towards their sources, the alga is found to occur right up to the 54°C isotherm. Castenholz showed that by diverting away hot water in the stream, the alga would increase its territory apparently by gliding, but that the algal mat would contract rapidly back to its former position when the hot water was allowed to resume its former course. Such contractions might well determine the survival of the alga in such hazardous situations.

Chapter 7

Culture, nutrition and growth

Advance in our knowledge of blue-green algae, whether of their biochemistry, genetics or role in the natural environment, depends largely on the use of cultures. The notable part which certain species have played in the general study of such basic processes as photosynthesis and nitrogen fixation has only been possible because these particular forms can be grown easily under laboratory conditions.

Micro-organisms used in research are grown generally in artificial isolation from other species because it is difficult otherwise to attribute a particular feature or response to the organism under investigation. Cultures containing only one kind of organism are referred to generally as *pure*, *bacteria-free*, or *axenic*. Apart from being axenic, laboratory cultures of algae are artificial in that the medium is synthetic and the physical conditions such as light and temperature to which they are exposed are controlled. Such conditions are selected normally to give optimum growth and provide the investigator with maximum yield of the algal material. This is important because often large quantities of material are required in order to extract sufficient amounts of specific substances present in minute amounts in the cells for biochemical analysis. The physiological activity of the algal material and the intensity of particular biochemical activities may vary considerably during the course of growth in culture and for most experimental purposes, material in the active phase of growth is used. Hence purity, yield and activity are the main criteria used in developing culture techniques.

It is important to stress at this point that the conditions under which laboratory cultures are grown usually bear little resemblance to those in the natural environment. Axenic populations rarely, if ever, occur in nature. There, a mixture of species interact with each other in a complex way and there is no ample and steady supply of nutrients or optimum and constant physical conditions. It is, therefore, essential to bear in mind, as Fogg (1965) has emphasized, that information obtained from study of laboratory cultures should be applied only with extreme caution to natural conditions.

Isolation and purification of blue-green algae

Methods for the isolation and maintenance of algae in pure culture have been based on the classical methods used in bacteriology. Patience, as Pringsheim (1946) stressed, is essential to the investigator setting out to obtain pure cultures of blue-green algae. In spite of many attempts to simplify and improve old techniques and develop more effective, less laborious and more rapid methods, the situation has not changed much—the isolation of blue-green algae is usually a tedious, lengthy, and in its outcome, an unpredictable exercise. This is especially true where the filamentous and mucilage-forming species of the group are concerned. The principal techniques which have been used are reviewed by Bold (1942), described in a monograph by Pringsheim (1946) and more recently summarized and up-dated by Lewin (1959), Droop (1967, 1969), Venkataraman (1969) and Stewart (1970a). More specific accounts of the cultivation of blue-green algae have been produced by Allen (1952) and Carr (1969). Stanier *et al.* (1971) have considered in detail the cultivation of non-filamentous blue-green algae.

In general there are three main stages in the isolation procedure.

1. SELECTION

The original material, if in soil, mud or rock scrapings is first flooded with water. Such preparatory cultures and natural water samples may be enriched with phosphate and nitrate to encourage algal growth and incubated in diffuse light of moderate intensity (1000 lux) at room temperature. Abundant growths of various algae then occur usually within a few weeks. To encourage selectively the growth of blue-green algae, various types of enrichment culture may be used. These include adjusting the pH to the alkaline side, incubating the cultures at 35° (Allen and Stanier, 1968b), adding 0·02–0·04% (w/v) sodium sulphide (Allen, 1952) which discourages the growth of green algae and many bacteria, but which promotes the growth of blue-green algae (Pringsheim, 1914; Knobloch, 1966; Stewart and Pearson, 1970), and in the case of nitrogen-fixing algae, using medium free of combined nitrogen. Such selective cultures usually develop mixed growths of different species but colonies or single filaments of a particular species may grow apart from others and can then be picked out easily from culture and transferred into sterile medium. This selection procedure may have to be repeated several times but in most cases will result in unialgal cultures.

2. GENOTYPE ISOLATION

Genotypic uniformity is desirable in cultures used for physiological and biochemical studies and may be achieved by the isolation of a single cell, filament or colony and by using this as the source of a clonal population.

With unicellular forms genotype isolation may be achieved by standard plating and/or dilution techniques (van Baalen, 1962) if the single isolated cells can be induced to grow. Sometimes this is not achieved easily, as workers such as Kumar (1963) found with *Anacystis nidulans* (*Synechococcus* sp.). Van Baalen (1965a) and Marler and van Baalen (1965) have evidence that the failure to multiply is due to hydrogen peroxide toxicity, the hydrogen peroxide being produced by a reaction in the medium between Mn^{2+} and citrate ions. Hydrogen peroxide levels may be reduced by lowering the concentration of manganese or omitting it altogether. Allen (1968a) has overcome the difficulty in getting unicells to multiply by using low agar concentrations (not more than 1·5%, w/v) and sterilizing it and the mineral media separately. This is thought to prevent toxic decomposition products which result when the agar and mineral media are heated together.

Single filaments can be isolated using capillary pipettes, and hormogonia and akinetes, isolated in the same way, are often free from bacteria (Bunt, 1961b; Wieringa, 1968).

3. PURIFICATION PROCEDURES

The extreme difficulty often experienced in obtaining axenic cultures of blue-green algae is usually due to the presence of epiphytic bacteria which become embedded in the mucilaginous algal sheaths. These can be eliminated by conventional methods of plating and dilution only by chance and although many methods of getting rid of them have been tried, the success of a particular procedure depends largely on the algae under investigation and, in particular, on the types of contaminating bacteria which are present. Successful techniques are based on the differential sensitivity of bacteria and blue-green algae to the agent used, and include washing with dilute chlorine water (Fogg, 1942), treatment with detergent and phenol (McDaniel *et al.*, 1962) and with antibiotics (Pintner and Provasoli, 1958). Exposure to ultraviolet light (Allison and Morris, 1930; Bortels, 1940; Gerloff *et al.*, 1950a) or to gamma radiation (Kraus, 1966) have also been used. The most successful is usually the ultraviolet irradiation treatment, which tends to kill off contaminating organisms, before it kills off the blue-green algae. A possible disadvantage of this treatment is its mutagenic effect, but often it is the only treatment which is successful in purifying an alga. Less drastic techniques include making use of the phototactic response of filamentous algae (Allen, 1952; Stanier *et al.*, 1971); differential temperature treatments (Wieringa, 1968; Allen and Stanier, 1968b), the isolation of akinetes and hormogonia by means of a micro-manipulator (Bowyer and Skerman, 1968) or the conventional method of large scale dilution and replication in accordance with the principle of "positive operator bias" (Tischer, 1965)—in other words making full use of one's skill and intelligence.

Nutrition

Most of the culture media in general use for the growth of blue-green algae were developed empirically by trial and error and by modification of previously used standard media (Allen, 1952). Only a few systematic and quantitative studies have been reported concerning the mineral requirements of blue-green algae (Kratz and Myers, 1955a; Allen and Arnon, 1955a; van Baalen, 1962; Volk and Phinney, 1968) and these have been limited in scope. The relative ignorance of the nutritional requirements of blue-green algae may be one of the reasons for the frequent failure to obtain a good yield in culture. It is also true that the cultural requirements are not a group character but rather a species-specific property, since the needs of individual species and their response to changes in composition of the medium may vary tremendously. It is thus extremely difficult, if not impossible, to design a general medium satisfactory for the growth of all blue-green algae.

On the other hand, many blue-green algae can tolerate considerable changes in the composition of the medium without any appreciable ill-effects (Gerloff *et al.*, 1950b; Allen, 1952; Kratz and Myers, 1955a) (see Table 7.I). Doubling or halving the total salt concentration

TABLE 7.I: Effect of variation in the concentration of mineral nutrients on the growth of *Coccochloris peniocystis*. After Gerloff *et al.* (1950b)

Concentration of element varied as compared to that in the basic solution	Growth as per cent of that in the basic solution						
	NO_3^-	PO_4^{3-}	SO_4^{2-}	K^+	Mg^{2+}	Ca^{2+}	Fe^{3+}
3x	123	103	94	92	98	93	—
2x	113	102	104	99	108	97	92
basic	100	100	100	100	100	100	100
1/2	42	99	102	94	93	106	—
1/4	24	101	101	88	96	105	95
1/10	10	79	75	61	94	107	100
1/20	7	53	54	43	91	101	94
1/40	7	35	37	39	70	112	87
None	—	12	20	25	17	110	60

of the medium have no effect on the yield of *Microcystis aeruginosa* (McLachlan and Gorham, 1961); growth and cell-nitrogen content of *Chlorogloea fritschii* is not greatly affected by up to fifty-fold increases of sodium chloride concentration in the standard medium (Fay and Fogg, 1962) and sodium: potassium ratios varying from 1:20 to 160:1 have no appreciable effect on the growth of several blue-green algae (Kratz and Myers, 1955a).

The major elements required for the growth of blue-green algae (N, P, S, K, Na, Mg and Ca, in addition to C, H and O), are no different from the requirements of other plant groups (Gerloff *et al.*, 1950b, 1952; Allen and Arnon, 1955a,b; Kratz and Myers, 1955a). The main source of carbon is carbon dioxide from the atmosphere taken up by photosynthesis and the growth rate can be increased in culture by passing air enriched with up to 0·5–5% carbon dioxide over the surface of the medium or bubbling it through the culture suspension. Care is required as higher carbon dioxide levels frequently result in a drastic decrease in pH even in well buffered medium (Allison *et al.*, 1937; Kratz and Myers, 1955a). Fogg and Than-Tun (1960) found 5% carbon dioxide to be toxic for cultures of *Anabaena cylindrica* although Allen and Arnon (1955a) found that this concentration was satisfactory for very dense cultures of this species at saturating light intensity. Inorganic carbon can also be provided in the form of bicarbonate (0·1 g l^{-1} of sodium bicarbonate is satisfactory) and certain species utilize organic carbon sources for growth in the light and in the dark (see p. 155).

Nitrogen assimilation is considered later (p. 180). It suffices here to say that nitrate is the most suitable nitrogen source for the growth of most blue-green algae in culture and can be supplied at high concentrations (about 0·02 M, Allen and Arnon, 1955a). Nitrate assimilation may cause a slight upward drift in pH of the medium but this does not affect growth seriously as most blue-green algae grow well in the pH range 7·0–8·5 (see p. 136). Ammonium-nitrogen which is assimilated in preference to nitrate–nitrogen is toxic in high concentrations, and even at low concentrations causes a pH decrease which may become inhibitory unless the medium is adequately buffered (Kratz and Myers, 1955a). Nitrogen sources for algal growth are considered in more detail later (p. 181).

Phosphorus is supplied in most algal culture media as dibasic phosphate at concentrations of up to 0·05%; this also provides a buffering capacity at around pH 7·5. Higher concentrations of phosphate (0·1–0·5%) are inhibitory, particularly to cells growing on elemental nitrogen (Allison *et al.*, 1937). Often, if phosphate is added to the culture medium prior to autoclaving, the phosphate precipitates out during autoclaving. It should therefore be autoclaved apart from the rest of the medium, or alternatively the precipitate can be dissolved out by bubbling the autoclaved medium with carbon dioxide.

Blue-green algae characteristically assimilate phosphorus in excess of their requirements, both in the light and in the dark, and store it as polyphosphate which can be used up under conditions of phosphorus deficiency (Batterton and van Baalen, 1968; Volk and Phinney, 1968; Stewart and Alexander, 1971) (see also p. 338). According to Whitton (1967a), *Nostoc verrucosum* accumulates large amounts of inorganic phosphate in the mucilage surrounding the colony.

The requirement for sodium and potassium first shown by Emerson and Lewis (1942) is now well established (Allen, 1952; Kratz and Myers, 1955a; Allen and Arnon, 1955b; Volk and Phinney, 1968) with the respective optimum levels being 5 ppm and 2·5 ppm. McLachlan and Gorham (1961) found that *Microcystis aeruginosa* has a very low requirement, if any, for sodium, and Batterton and van Baalen (1971) showed that the requirement for sodium is saturated at 1 ppm of sodium chloride in freshwater blue-green algae and at 100 ppm sodium chloride in marine blue-green algae. Potassium, lithium, rubidium or caesium cannot substitute for sodium, and sodium cannot replace potassium for growth of *Anabaena cylindrica*. Sodium deficiency decreases the rate of nitrogen fixation by *A. cylindrica*, whereas assimilation of combined nitrogen is enhanced under these conditions. Nitrate assimilation in sodium-deficient cultures causes toxicity and chlorosis, perhaps due to an imbalance between nitrate reduction and subsequent nitrite reduction (Brownell and Nicholas, 1967).

It has been stated that calcium is required for growth of *Nostoc muscorum* on elemental nitrogen but not on combined nitrogen. However in a critical study Allen and Arnon (1955a) found a low, but absolute, requirement for calcium on nitrate. In general blue-green algae require more calcium for growth on elemental nitrogen than on combined nitrogen (Fay, 1962) but there is no evidence that it is associated specifically with the process of nitrogen fixation.

Magnesium is a component of the chlorophyll molecule, and as such is essential for the growth of blue-green algae. It is usually supplied in culture medium at concentrations of approximately 20 ppm, although this is higher than necessary as optimum growth has been demonstrated at concentrations of 0·125 ppm or less (Volk and Phinney, 1968). High calcium concentrations alleviate magnesium toxicity in *Microcystis aeruginosa* (McLachlan and Gorham, 1961).

Iron, a constituent of components such as cytochromes, ferredoxin and nitrogenase, is generally supplied as a complex with chelating agents such as citrate (Kratz and Myers, 1955a), EDTA (ethylene-diamine tetra-acetic acid) (Allen and Arnon 1955a) (see Fig. 7.1), or EDDHA (ethylene diamine di-*o*-hydroxyphenylacetate) (Volk and Phinney, 1968). These

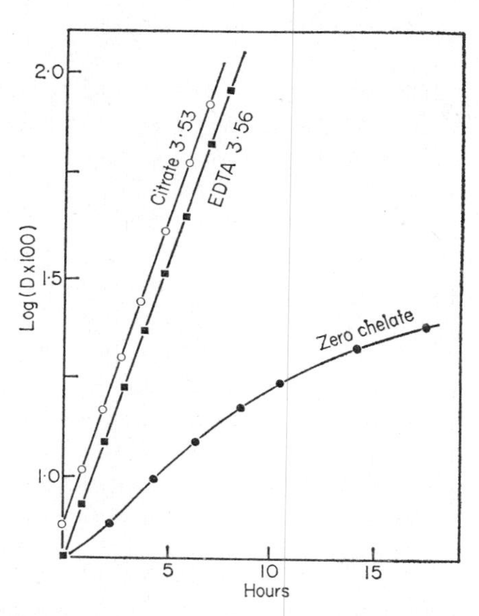

FIG. 7.1. Effect of chelating agents on growth of *Anacystis nidulans* (*Synechococcus* sp.) at 41°. Figures on the curves indicate specific growth rates ($\log_{10}$ units day^{-1}) ○—○, plus citrate; ■—■ plus EDTA; ●—● no chelating agent added. After Kratz and Myers (1955) *Am. J. Bot.* **42**, 282, Fig. 3.

compounds also maintain in solution other elements which may precipitate out in alkaline conditions.

Trace amounts of manganese, boron, molybdenum, copper, zinc and cobalt are also essential for the growth of blue-green algae. Eyster (1952) first demonstrated boron deficiency in *Nostoc muscorum*. Deficiency resulted in decreased growth rates and chlorosis and could be corrected by adding optimum boron levels of 0·1 ppm. Concentrations greater than 1 ppm were toxic. Bortels (1940) and Fogg (1949) have shown that molybdenum is essential for nitrogen fixation in blue-green algae. Molybdenum is a component of both nitrate reductase and nitrogenase and Wolfe (1954a, b) showed that *Anabaena cylindrica* required 0·2 ppm molybdenum for maximum growth on elemental nitrogen and 0·1 ppm for growth on nitrate–nitrogen. A cobalt requirement has been demonstrated by Holm-Hansen *et al.* (1954) for *Nostoc muscorum*, *Calothrix parietina*, *Coccochloris peniocystis* and *Diplocystis aeruginosa*. This requirement (0·4 $\mu g\ l^{-1}$) is irrespective of the nitrogen source supplied but is more pronounced in cultures growing on elemental nitrogen and can be met by the addition of Vitamin B_{12}, which contains cobalt. Johnson *et al.* (1966) found that cobalt was required for the symbiotic growth of the water fern *Azolla* when growing on elemental nitrogen fixed by the symbiotic *Anabaena*. There was no requirement for growth on nitrate-nitrogen. The discrepancy between the results of the above two groups of workers requires checking.

There is no evidence that freshwater blue-green algae require vitamins for growth but the discovery by Pintner and Provasoli (1958) that the marine species *Phormidium persicinum* required vitamin B_{12} and the subsequent report by van Baalen (1961) that eight out of fifteen marine species isolated from inshore waters require vitamin B_{12} has increased interest in the vitamin requirements of the group. There is no evidence so far of requirements for vitamins other than B_{12}. It is of interest that, because of its sensitivity to vitamin B_{12}, Burkholder (1963) used *Coccochloris elabens* as an assay organism for the estimation of vitamin B_{12} production by marine bacteria and algae.

The importance of growth substances, e.g. indole-3-acetic acid (IAA), for the growth and development of blue-green algae has not been investigated in detail. Bunt (1961b) found that hormogonium development in a *Nostoc* species from tropical soils depended on the presence of a bacterium (a *Caulobacter* species) in the culture and that hormogonia developed in axenic culture only when IAA was added to the medium. There was no direct evidence that the *Caulobacter* produced IAA, but sections of *Avena coleoptiles* increased in length in the presence of the bacteria which perhaps indicates the action of a growth substance. IAA has also been reported to stimulate the growth of blue-green algae at low concentrations (10^{-5}–10^{-9} M), but higher concentrations were inhibitory (Ahmad and Winter, 1968). More work on the effects of higher plant growth substances on algae is required and until these are carried out the available results should be treated with caution (Fogg, 1972).

Most blue-green algae grow best in the pH range 7·5–9·0 (Gerloff *et al.*, 1950b; Kratz and Myers, 1955a), but there are exceptions. *Microcystis aeruginosa*, for example, grows well at pH 6·5 on urea as nitrogen source (McLachlan and Gorham, 1962). In general no growth occurs below pH 5·7 (Allison *et al.*, 1937) and a great many failures to grow blue-green algae in culture arise from the unfavourable pH of the medium. It is thus important to use a well-buffered medium. Dibasic phosphate which buffers at about 7·5 is generally satisfactory. If air enriched with carbon dioxide is used, or if more alkaline conditions are desired, bicarbonate or carbonate (0·1–1·0%) may be included in the medium (Allison *et al.*, 1937; Fogg, 1951b). "Tris" (tris-hydroxymethyl aminomethane) at 10 mM concentration is satisfactory as a buffer for *Microcystis aeruginosa* cultures and increases the tolerance toward supra-optimal concentrations of monovalent cations (McLachlan and Gorham, 1961).

The composition of three widely used media for the growth of blue-green algae is given in Table 7.II.

TABLE 7.II: Comparison of the composition of three standard media used for the culture of blue-green algae. Values represent micromoles per litre

Compound	Kratz and Myers (1955a) medium C	Allen and Arnon (1955a) medium	Gorham *et al.* (1964) ASM–1 mediun
$NaNO_3$	—	—	2000
KNO_3	10,000	—	—
NaCl	—	4000	—
$MgCl_2$	—	—	200
$MgSO_4.7H_2O$	1000	1000	200
$CaCl_2$	—	500	200
$Ca(NO_3)_2.4H_2O$	105	—	—
K_2HPO_4	5750	1000	100
Na_2HPO_4	—	—	100
$FeCl_3$	—	—	4
$Fe_2(SO_4)_3.6H_2O$	15·4	—	—
Fe EDTA	—	72	—
Na_2 EDTA	—	—	20
Na citrate.$2H_2O$	1700	—	—
H_3BO_3	46	46	40
$MnCl_2.4H_2O$	11	—	7
$MnSO_4.4H_2O$	—	9·2	—
MoO_3	0·123	1	—
$ZnCl_2$	—	—	3·2
$ZnSO_4.7H_2O$	0·77	0·77	—
$CuSO_4.5H_2O$	0·32	0·32	—
$CuCl_2$	—	—	0·0008
$CoCl_2$	—	—	0·08
$Co(NO_3)_2.6H_2O$	—	0·17	—
NH_4VO_3	—	0·2	—
$NiSO_4.6H_2O$	—	0·1	—
$Cr_2(SO_4)_3.K_2SO_4.24H_2O$	—	0·1	—
$Na_2WO_4.2H_2O$	—	0·05	—
$TiO(C_2O_4)_x.yH_2O$	—	0·05	—

Growth

The relative growth rates and final yields of blue-green algae measured at saturating light intensity and optimum temperature are not appreciably different from those obtained under corresponding conditions with green algae of corresponding size (Kratz and Myers, 1955a) and failure to achieve high yields with blue-green algae is usually attributable to sub-optimal culture conditions. Under favourable conditions the yield of *Anabaena cylindrica* can be 7–8 g dry weight of algae per litre of medium (Allen and Arnon, 1955a), and the maximum relative growth rate (k) of *Anacystis nidulans* (*Synechococcus* sp.), at 3600 lux light intensity and 4.° temperature, is 3·55 $\log_{10}$ units day $^{-1}$ which corresponds to a doubling time of about 2 h (Kratz and Myers, 1955a) (Fig. 7.2). This is the shortest doubling time so far recorded for any alga.

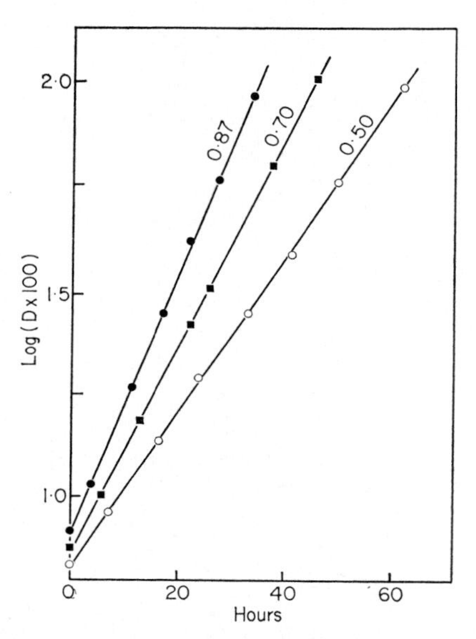

FIG. 7.2. Growth of *Anabaena variabilis* (■—■), *Nostoc muscorum* G (○—○), and *Anacystis nidulans* (*Synechococcus* sp.) (●—●) at 25°C at light saturating illumination and aerated with 0·5% CO_2 in air. Figures on the curves indicate specific growth rates ($\log_{10}$ units day^{-1}). After Kratz and Myers (1955a), *Am. J. Bot.* **42**, 282, Fig. 1.

Light intensity is a factor which appears to exert different effects in the field and in the laboratory. Some species of blue-green algae appear to exist in the field at light intensities which they are unable to tolerate in culture (Allison *et al.*, 1937). Cultures are thus generally maintained at low light intensity or with intermittent illumination but nevertheless given adequate supplies of carbon dioxide, nitrogen and mineral nutrients, growth increases with increasing light intensity up to 16,000 lux (Allen and Arnon, 1955a) (Fig. 7.3). Light intensities quoted generally refer to the light incident at the surface of the culture vessel. However, the mean intensity of light to which the cells are subjected is lower, since it falls off

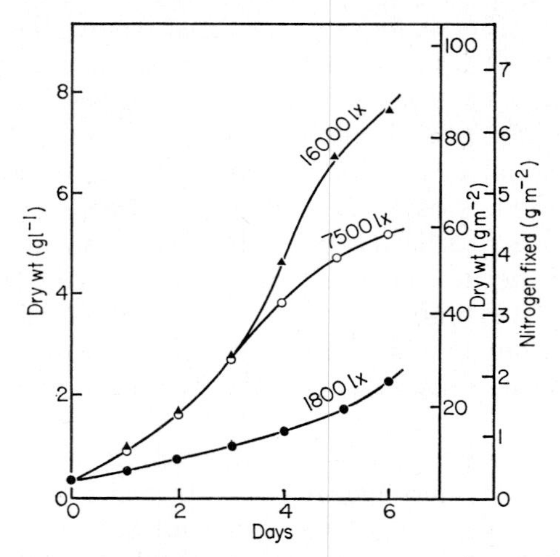

FIG. 7.3. Effect of light intensity on growth of *Anabaena cylindrica* under nitrogen-fixing conditions. After Allen and Arnon (1955) *Pl. Physiol.* **30**, 366, Fig. 5.

exponentially through successive layers of the cell suspension, and will continue to fall as the culture increases in density. For this reason most values have little meaning in absolute terms. Intermittent illumination (16 h light: 8 h dark) does not give better yields than continuous illumination and blue-green algae do not appear to require a diurnal alternation of light and dark periods. The spectral composition of the light does not appear to be critical for growth, cultures growing well with illumination from either tungsten or fluorescent lamps, but it may have striking effects on pigmentation (see p. 148).

Ranges of temperature suitable for the growth of blue-green algae are significantly wider than those for other algae and in general the optimum temperature for blue-green algae is near 35°C. Allen (1968a) has suggested that this fact can be made use of in purifying blue-green algae from other algal groups such as green algae which have lower temperature optima. Especially pronounced is the ability of certain thermophilic blue-green algae to tolerate upper extremes of temperature (see p. 277). These algae generally require a temperature of about 45° C for growth in the laboratory and usually will not grow below 30° C. Thermophilic *Synechococcus* species may not survive temperatures below 50° C (Peary and Castenholz, 1964), and can be grown in culture at temperatures up to 73°C, the maximum temperature at which they occur in natural situations (Castenholz, 1969). Many species are adapted to low temperatures (Whitton, 1962) (see p. 82). Others including some planktonic blue-green algae from temperate regions, such as *Anabaena flos-aquae* isolated from Lake Windermere, do not tolerate temperatures above 15–23° C in the laboratory.

Growth of the blue-green algae in batch or limited volume culture follows the usual pattern with more or less clearly defined lag, exponential, declining relative growth and stationary phases (Fogg, 1965). Heavy inocula are sometimes needed to establish cultures (Gerloff *et al.*, 1950a), and with some species the lag phase has been shown to vary inversely with the size of the inoculum (Eberley, 1967). Fogg (1944) found that *Anabaena cylindrica* exhibited two successive exponential phases, one while the filaments remained in free suspension and a second, of lower relative growth rate, following settlement of the filaments on the surface of the culture flask.

High growth rates and good yield depend on an adequate supply of carbon dioxide and the continuous removal of oxygen produced in photosynthesis. Growth on free nitrogen in a stagnant culture may be limited by the rate of diffusion of nitrogen to the cells (Allen and Arnon, 1955a). An efficient exchange of gases requires a large gas–liquid interface and this can be effected by various methods of aeration and agitation. Agitation also prevents sedimentation and clumping of the algae. This will maintain a uniform suspension and a sharp concentration gradient of nutrients at the interface between cells and medium which will promote nutrient uptake. Furthermore, agitation provides even illumination of the algal material. Although agitation in general is beneficial for the growth of blue-green algae in culture, the response to the intensity and nature of the agitation shows a wide variation depending on species. *Chlorogloea fritschii*, for example, tolerates only gentle agitation and will not grow in a vigorously shaken culture (Fay and Fogg, 1962). It grows best with agitation by a stream of air, or gas, reinforced by rotatory movement of the medium. *Anabaena cylindrica* grows well in cultures shaken at about 90 linear oscillations per minute but fails to grow when the rate of oscillation is increased to 140 per minute (Fogg and Than-Tun, 1960) (Table 7.III). This is probably due to the filaments being damaged by turbulent movement in the liquid culture.

TABLE 7.III: The effect of shaking on the growth of *Anabaena cylindrica*, as compared with unshaken cultures. After Fogg and Than-Tun (1960)

Oscillations per minute	0	65	80	90	100	120	140
Growth as %	100	143	258	288	197	130	0

Culture apparatus for the growth of blue-green algae is essentially similar to that used to grow other micro-algae, but may vary in detail and design according to species and experimental requirements. A culture vessel designed by Walsby (1967) is especially suitable for the growth of several unicellular and filamentous blue-green algae on a mechanical shaker. Using this type of vessel it is possible to maintain in uniform suspension filamentous species such as *A. cylindrica* which tend to form aggregates (see p. 124) or stick to the sides of the vessel.

In order to provide large quantities of the nitrogen-fixing species, *Tolypothrix tenuis*, Watanabe and his collaborators (1959) devised an open system of mass culture of 250 litres capacity and 5 m^2 surface yielding 6·4 g dry weight m^{-2} day^{-1}, but this has the disadvantage that axenic conditions cannot be maintained. Units designed for mass production of *Chlorella* and other green algae (Setlick *et al.*, 1970) may also be suitable for growth of blue-green algae on a larger scale.

Many studies, such as those on photosynthesis, require a highly uniform material in respect of rate of growth and metabolic activity. A turbidostat apparatus for the continuous culture of algae which can maintain steady state growth conditions and provide a continuous reproducible supply of experimental material was devised by Myers and Clark (1944) and has been successfully adapted for the culturing of blue-green algae (Myers and Kratz, 1955; Cobb and Myers, 1964). A simplified and partly improved version of this apparatus (Fay and Kulasooriya, 1973) is presented in Fig. 7.4. Bone (1971a) reported on the adaptation of the chemostat for the continuous culture of *Anabaena flos-aquae* strain A-37 under phosphate-limiting conditions and with dilution rates of 0·015–0·03 l h^{-1} (i.e. flow of 15–30 ml fresh medium per hour per litre of culture in the chemostat).

Studies which aim to examine physiological and biochemical changes during the life cycle of the alga depend on the successful synchronization of growth in culture. Synchronous growth of blue-green algae has not yet been achieved to the same degree as with other algae. Quasi-synchronous cultures of *Chlorogloea fritschii* have been obtained by a combination of light and temperature treatments (Fay *et al.*, 1964). However, the life cycle of this alga comprises several cell generations and is not directly comparable with the division cycle of a unicellular organism (see p. 251). The synchronization of the unicellular *Anacystis nidulans* (*Synechococcus* sp.) has been achieved by two different techniques. Lorenzen and Venkataraman (1969) used temperature shifts as a synchronizing factor in a cycle consisting of an 8 h period at 26° and 6 h period at 32°. This was coupled at the end of each cycle with periodic dilution of the algal suspension to a constant cell number. The culture was kept in continuous light. Herdman *et al.* (1970) induced synchrony in an exponentially growing

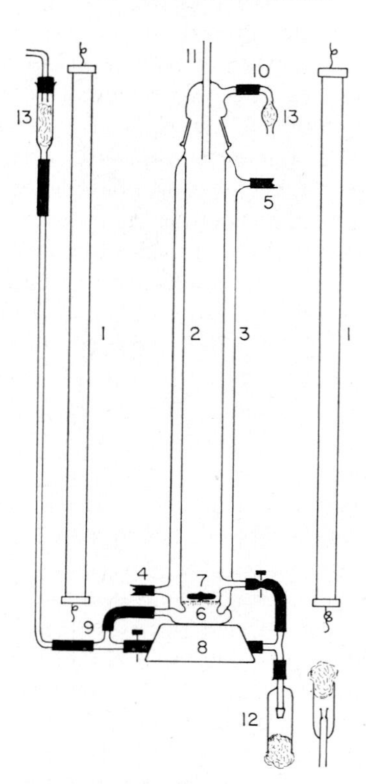

FIG. 7.4. A simple culture apparatus for continuous growth of blue-green algae, after Fay and Kulasooriya (1973). 1, fluorescent tube; 2, culture column; 3, constant temperature water jacket; 4, water inlet; 5, water outlet; 6, sintered glass disc; 7, magnetic follower; 8, magnetic stirrer; 9 gas inlet; 10, gas outlet; 11, medium inlet; 12, device for aseptic sampling; 13, cotton-wool filter. Fresh medium is admitted at a constant rate by a peristaltic pump.

culture by the termination of illumination and supply of carbon dioxide. Growth ceased after 30 min and the alga was kept in the dark for several hours until the RNA and DNA content remained constant. When light and carbon dioxide were restored, a synchronous growth followed and persisted for about three generations. In both techniques the degree of synchrony gradually decreased and was lost after several generations. Improved techniques for establishing synchronous cultures of blue-green algae and maintaining them over longer periods are greatly needed.

Chapter 8

Photosynthesis and chemosynthesis

Photosynthesis, the characteristic method of nutrition in the blue-green algae, is usually defined as the manufacture of organic substances from carbon dioxide and water by green plants in the light. This is not a single process but is a complex of reactions which do not necessarily have any fixed relationship to one another and it is difficult to devise a succinct definition which covers all the possible variants. The essence is that radiant energy is converted to the potential chemical energy of relatively stable compounds. The basic pattern of the processes whereby this is achieved is the same in all photosynthetic organisms and it need not be detailed again (general descriptions may be found in books by Fogg, 1968 and Rabinowitch and Govindjee, 1969). Here the outline will be sketched and variations in details characteristic of blue-green algae emphasized.

In photosynthesis in general, light of wavelengths from 400 to 700 nm is absorbed by a number of pigments in the photosynthetic apparatus. Once absorbed by a pigment molecule the quantum of energy may be transferred, by resonance or some other process not involving re-emission of radiation, to another pigment molecule of the same or a different sort. The crucial photochemical acts take place in special active centres (of which there is only one per 400 or so molecules of the principal photosynthetic pigment chlorophyll *a*) containing a chlorophyll *a* modification (P 700) as a principal component.

In the photochemical act a high-energy electron is transferred from the chlorophyll and yields up its energy as it falls back along a chain of electron carriers. Part of the released energy is incorporated as potential chemical energy in adenosine triphosphate (ATP), this process being known as photophosphorylation. There is good evidence that two distinct kinds of active centres, involved respectively in photosystems I and II (see p. 146) act in series. Photosystem I receives the electron transferred from photosystem II, after it has lost some of its potential energy, and raises it to a high enough energy level to reduce the electron carrier ferredoxin (Fig. 8.1). Reduced ferredoxin in turn reduces nicotinamide adenine dinucleotide

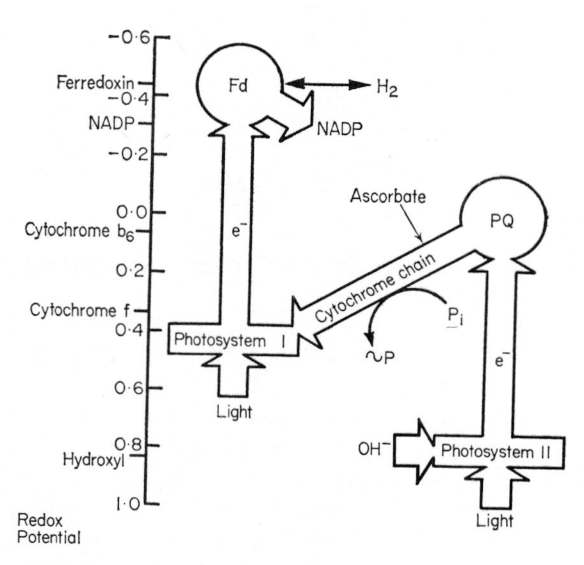

FIG. 8.1. Scheme illustrating electron transfer in photosynthesis in terms of redox potentials (redrawn after F. R. Whatley and M. Losada (1964) *In* "Photophysiology" Vol. **1** (ed. A. C. Giese). Academic Press, New York and London, pp. 111–154).

phosphate, NADP, to $NADPH_2$ and the hydrogen ion, H^+, necessary for these reductions is supplied by the dissociation of water. The electron is replaced in photosystem II from hydroxyl ion, OH^-, in green plants, or from an electron donor such as sulphide or thiosulphate in the photosynthetic bacteria. The problem of disposing of the residual oxidizing portion of the water molecule (OH) has been solved in photosynthetic eukaryotes and blue-green algae by the elimination of molecular oxygen. The net effect of the utilization of the hydrogen ion and hydroxyl ion is thus the splitting of the water molecule to yield hydrogen, in the form of the organic hydrogen donor $NADPH_2$, and oxygen. This entire sequence of reactions takes place in the thylakoids and ATP and $NADPH_2$ may be regarded as the end products of the photochemical reactions (Fig. 8.1).

The fixation of carbon dioxide is a separate process. Carbon dioxide is added on to a 5-carbon acceptor substance, ribulose diphosphate (RuDP), which thereupon splits to yield 2 molecules of a 3-carbon substance, phosphoglyceric acid (PGA). PGA is reduced to a triose sugar phosphate from which, by a series of sugar phosphate interconversions, the acceptor for carbon dioxide, RuDP, is regenerated. This cycle can be visualized as producing one glucose molecule for every 6 carbon dioxide molecules fed into it but in actual fact the fixed carbon can be taken out in a variety of forms. It is driven by the ATP and $NADPH_2$ supplied by the photochemical reactions (see Fig. 8.2).

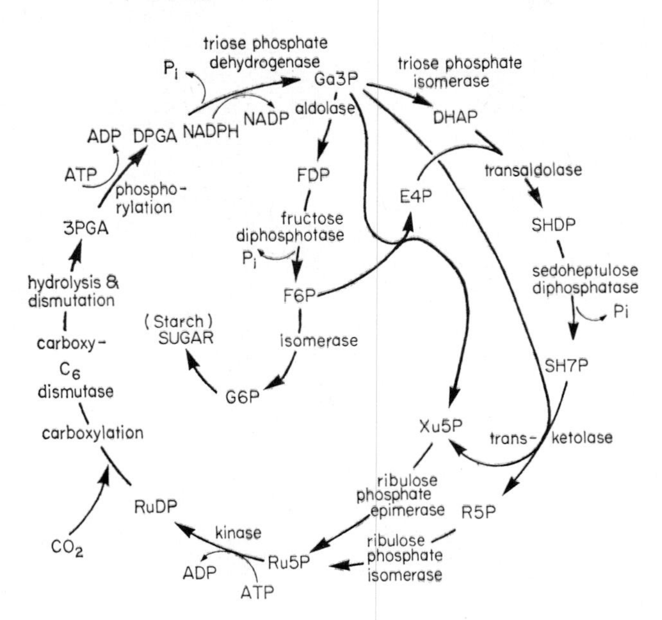

FIG. 8.2. The reductive pentose phosphate cycle or Calvin–Benson–Bassham cycle

RuDP: ribulose-1,5-diphosphate
3PGA: 3-phosphoglyceric acid
DPGA: 1,3 diphosphoglyceric acid
Ga3P: glyceraldehyde-3-phosphate
DHAP: dihydroxyacetone phosphate
FDP: fructose-1,6-diphosphate
F6P: fructose-6-phosphate
G6P: glucose-6-phosphate
E4P: erythrose-4-phosphate
SHDP: sedoheptulose-1,7-diphosphate
SH7P: sedoheptulose-7-phosphate
Xu5P: xylulose-5-phosphate
R5P: ribose-5-phosphate
Ru5P: ribulose-5-phosphate
C_6: C_6 compound
ATP: adenosine triphosphate
ADP: adenosine diphosphate
P_i: inorganic phosphate
NADPH/NADP: reduced and oxidized forms of nicotinamide adenine dinucleotide phosphate.

Light absorption

The photosynthetic pigments of blue-green algae, which include chlorophyll *a*, various phycobiliproteins (*c*-phycoerythrin, *c*-phycocyanin and allophycocyanin) and carotenoids (β-carotene, echinenone and myxoxanthophyll), have been described elsewhere (see p. 45 *et seq.*). Evidence that light absorbed by the phycobiliproteins, as well as that absorbed by chlorophyll, is effective in photosynthesis was obtained as long ago as 1883 by Engelmann. He used bacteria, of a kind which aggregate in regions of high oxygen concentration, as an index of photosynthesis by algal material, including the filamentous blue-green alga *Oscillatoria*, in different parts of the spectrum. His perspicacious conclusions were disputed by his contemporaries and had to wait nearly 60 years for confirmation. In 1942 Emerson and Lewis reported on the action spectrum of photosynthesis by *Chroococcus*, in which the relatively wide separation of the chlorophyll and phycocyanin absorption peaks renders investigation of the role of the individual pigments fairly straightforward. They found that the quantum yield was the same at 600 nm, where phycocyanin absorbed at least six times more light than chlorophyll, as in the region 660–680 nm where virtually all the light was absorbed by chlorophyll, and concluded that the photosynthetic efficiency of phycocyanin equals that of chlorophyll *a*. The quantum yield fell off in the region 420–550 nm, where absorption by the carotenoids is greatest, but best agreement between calculated and observed results was obtained when it was assumed that the photosynthetic efficiency of the carotenoids was a fifth of that of chlorophyll, rather than zero. Similar results have been obtained by other workers for other blue-green algae (Fig. 8.3a). Haxo and Blinks (1950), however, found that in *Anabaena* and *Oscillatoria* the photosynthetic activity of chlorophyll was less than that of phycocyanin (Fig. 8.3b).

More about the transfer of excitation energy between pigments has been learnt by investigating fluorescence spectra of *Oscillatoria* excited by equal quantum fluxes at 420 nm, strongly absorbed by chlorophyll, and at 578 nm, absorbed mainly by phycocyanin (Duysens, 1952). Light absorption by phycocyanin resulted in stronger chlorophyll fluorescence than that resulting from light absorption by chlorophyll alone. Examination of fluorescent spectra of various blue-green algae at low temperatures, which renders the peaks more distinct, shows that light energy absorbed by phycobilins is more readily transferred to some forms of chlorophyll than to others (Goedheer, 1968). These results demonstrate effective transfer of excitation energy from phycocyanin to chlorophyll but absence of transfer in the reverse direction. Although there seem to be no data for blue-green algae, experiments with red algae have also shown effective transfer from phycoerythrin to phycocyanin (Rabinowitch, 1956). In all this, the rule is obeyed that transfer only takes place to pigments having an absorption maxima at a longer wavelength than that of the pigment providing the energy.

These investigations were highly artificial in that the algal material was at any one time illuminated with a narrow wave-band only. As we now know, for complete photosynthesis, it is necessary that two distinct photosystems should operate and, since these have absorption maxima at different wavelengths, monochromatic radiation may excite one much more than the other with resulting loss of efficiency. Among the first definite evidence for the existence of two photosystems was the discovery by Emerson (1958) of the enhancement effect in *Chlorella*. A similar enhancement, that is production of a photosynthetic yield, by a combination of long and short wavelengths, which is greater than the sum of the yields from the two wavelengths given separately, was later shown with blue-green algae (Hoch *et al.*, 1963). Jones and Myers (1964) found that in *Anacystis nidulans* (*Synechococcus* sp.) the action spectrum of enhancement when the background illumination was monochromatic light of wavelength 690 nm corresponded approximately with the absorption spectrum of phycocyanin. With shorter wavelength, background illumination of 620 nm the action spectrum of enhancement was related to high chlorophyll *a* and carotenoid absorption (Fig. 8.3). The two photosystems thus

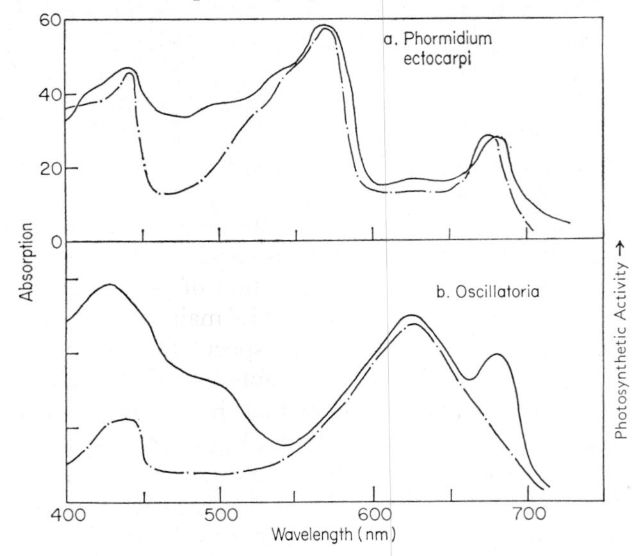

FIG. 8.3. Absorption (——) and photosynthetic action spectra (– · –) for two blue-green algae, *Phormidium ectocarpi* (redrawn after F. T. Haxo (1960) in "Comparative Biochemistry of Photoreactive Systems" (ed. M. B. Allen) Academic Press, New York and London, pp. 339–359) and *Oscillatoria* sp. (redrawn after L. N. M. Duysens (1952) Ph.D. Thesis, University of Utrecht, Holland).

characterized are those designated II and I respectively. Neither depends on a single pigment. A shoulder on the main peak in the action spectrum of photosystem II indicates the involvement of the chlorophyll *a* modification 670 but its general shape shows that phycocyanin is the main pigment concerned. Goedheer (1969) has shown that β-carotene, but not the xanthophylls, functions efficiently in light transfer in photosystem I in *Synechococcus*. The ultimate energy traps in both photosystems are thought to be chlorophyll modifications, P 690 in photosystem II and P 700 in photosystem I. The existence of these various chlorophyll modifications in the two photosystems explains the anomalies noted above, that light absorbed by phycocyanin is sometimes more efficient in photosynthesis and in exciting chlorophyll fluorescence than is light absorbed by chlorophyll itself. Thus in photosystem II energy is transferred to the chlorophyll modification P 690, which fluoresces strongly, but not to the more abundant P 670, which fluoresces less strongly.

Two separate kinds of pigment–protein complex, one of which contains components from photosystem I and the other from photosystem II (Boardman, 1970) can be separated from chloroplasts by treatment with detergents. As already mentioned (p. 51) Ogawa *et al.* (1968) studying *Anabaena variabilis* found that the heavier, photosystem I fraction obtained was not strongly fluorescent, whereas the lighter, photosystem II fraction was strongly fluorescent. The membrane system, containing photosystem I, isolated from *Anabaena variabilis* by treatment with the detergent Triton X–100 has been found to be rich in the chlorophyll *a* modification P 700 (Ogawa and Vernon, 1970). Action spectra show that phycocyanin, which solubilizes in the course of preparation, is present in both photosystems but is particularly important in photosystem II. Phycocyanin is not essential for the Hill reaction, which is carried out by photosystem II, or for cyclic photophosphorylation, a function of photosystem I (Petrack and Lipmann, 1961; Susor *et al.* 1964) and its main function appears to be that of light absorption. This leads to the speculation (Bishop, 1966) that mutants lacking phycocyanin may be able to photosynthesize, albeit inefficiently. Electron micrographs show that the biliproteins are contained in phycobilisomes (p. 52) but the relationship of these particles with the fractions that have been isolated remains to be established.

It is not yet clear whether photosystem I and photosystem II function independently at the level of excitation energy transfer—the "separate package" hypothesis—or whether excess quanta absorbed by photosystem II can be transferred to the other—the "spillover" hypothesis (see Rabinowitch and Govindjee, 1969). The fact that they are distinct physical entities does not necessarily mean that the former hypothesis is the correct one. In heterocysts, as we shall see (p. 234), photosystem I evidently occurs and functions alone.

Chromatic adaptation

Having demonstrated that light absorbed by the accessory pigments is used in photosynthesis, Engelmann (1883) suggested that the pigmentation of algae is usually such as to permit the most efficient use of the wavelengths available in the particular habitats in which they grow. The classic example of this is the zonation of algae on a rocky seashore. The bright green pigmentation of the Chlorophyceae, which generally grow near high water mark, is best adapted to absorb the orange-yellow wavelengths which predominate in unaltered sunlight. On the other hand, red and blue-green algae growing in deep clear water characteristically have pigmentation giving maximum absorption of the green wavelengths which are the major components of the sparse light which penetrates. The brown algae occupy an intermediate position. Examples of what seem to be chromatic adaptations of this sort have been noted for freshwater and for marine blue-green algae.

In examining the theory of chromatic adaptation more critically in relation to blue-green algae it must first of all be pointed out that their pigmentation may vary genotypically and phenotypically. The genetically determined pigmentation characteristic of the group as a whole has been supposed to be the result of chromatic adaptation to the quality of light prevailing at the time when the group underwent its first phase of evolution (Tilden, 1933). It seems likely that blue-green algae evolved under anoxic conditions (see p. 368) and that methane then present in the atmosphere may have given rise to orange-coloured hydrocarbons (Urey, 1952). Phycocyanin has maximum absorption for the wavelengths which would penetrate such an atmosphere so this is a plausible, if unprovable, hypothesis.

Blue-green algae are noted for phenotypic variation in pigmentation (Fritsch, 1945) and some of these pigment changes are not directly concerned in chromatic adaptation. For example, most blue-green algae assume a yellow-brown colour when nitrogen-deficient, as a result of the breakdown of chlorophyll and phycocyanin and retention of carotenoids. The original colour is rapidly regained if a suitable nitrogen source is supplied irrespective of whether the alga is in the dark or in the light (Boresch, 1913; Allen and Smith, 1968). Halldal (1958), who developed a technique for growing algae in crossed gradients of light intensity and temperature, found that the pigmentation of *Anacystis nidulans* (*Synechococcus* sp.) and an *Anabaena* species varied in a rather complex way in relation to these two factors and the age of the cells. As with other plants, the concentration of chlorophyll decreased with increasing light intensity whereas there was an increase in carotenoids, which appear to have some protective function against high light intensities. Phycocyanin concentration decreased with increasing light intensity except in *Anacystis* growing at 42–45°C in which, together with chlorophyll, it increased at the higher light intensities. Also, Eley (1971) has found that *Anacystis nidulans* growing in the presence of 1% carbon dioxide in air contains only about half the concentration of phycocyanin of cells growing in air alone.

Here, we are more concerned with pigment variation in relation to light quality. Many workers, such as Gaidukov (1902, 1923) and Boresch (1919) found that when cultures of certain *Oscillatoria* and *Phormidium* species were illuminated with a spectrum, the filaments assumed, in the course of a day or two, a colour complementary to that of the part of the spectrum in which they were growing. Such changes do not invariably occur in all species of blue-green algae (Demeter, 1956). When they do they may be dependent on intensity differences as much as on wavelength but more refined experiments have confirmed that the latter does have a remarkable influence. Hattori and Fujita (1959a) illuminated *Tolypothrix tenuis* with rather broad wave bands adjusted to equal incident energy and found that phycoerythrin synthesis was most strongly induced by blue and green light whereas synthesis of phycocyanin was best induced by red or orange light. These effects were largely independent of intensity and accompanying changes in chlorophyll and carotenoids were not found. A period of pre-illumination with a particular wavelength for as little as 3 min under appropriate conditions, had marked effects on the relative proportions of the two biliproteins formed in a subsequent dark period. If the light was green, around 500 nm, phycoerythrin, with an absorption peak at 576 nm, was mainly formed. On the other hand, if the light was red, 600–650 nm, then phycocyanin, with an absorption maximum at 620 nm, predominated (Hattori and Fujita, 1959b). There appears to be a common precursor which can be switched towards synthesis of either pigment by illumination with a wavelength close to its absorption maximum.

Pigment variations of a different type have been observed in *Anacystis nidulans* (*Synechococcus* sp.) by Jones and Myers (1965). When this alga is grown in continuous culture in red light of wavelength greater than 660 nm the chlorophyll content of the cells is only about a quarter of that of cells grown under normal tungsten light although phycocyanin and carotenoid contents show little change (Fig. 8·4). Jones and Myers regard this as an instance of the same sort of intensity control that results in "sun" leaves or algal cells having lower chlorophyll concentrations than their "shade" equivalents. In red light chlorophyll *a* absorbs more light than can be used by photosystem I, the activity of which is governed by that of photosystem II which is light limited. Thus the control operates to reduce the chlorophyll *a* concentration.

There can be little doubt, in view of these results, that there is some substance in the theory of chromatic adaptation. Pigment changes of the kinds described result in increased utilization during photosynthesis of the available wavelengths under limiting light conditions. The zonation of algae is, however, often other than what a simple theory of chromatic adaptation would predict (Crossett *et al.*, 1965). This generally happens in zones where high light intensities are available so that efficient use of the available energy is not critical. In places where light is always limiting,

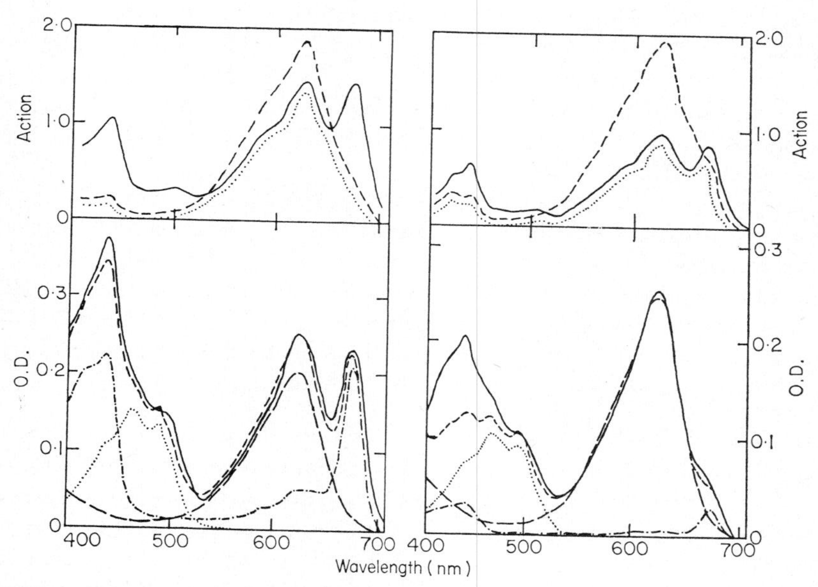

FIG. 8.4. The effects of growth in lights of different spectral composition on the pigment content and photosynthetic performance of *Anacystis nidulans* (*Synechococcus* sp.). The diagrams on the left show results with cells grown in white light, those on the right with cells grown in red light. The top two diagrams show action spectra for photosynthesis with complementary background illumination (– –, 690 nm background; ———, 620 nm background; · · ·, without background, i.e. unenhanced). The lower two diagrams show absorption spectra (———, total pigments *in vivo*; – – –, total extracted pigments; — —, phycocyanin; – · –, chlorophyll; . . . . . . , carotenoid). (Redrawn after L. W. Jones and J. Myers (1964) *Pl. Physiol.* **39**, 938–946).

algae with pigments which do not specifically absorb the wavelengths available may also be successful if they contain particularly high concentrations of poorly-absorbing pigments. In such habitats, the possession of pigments which ensure as complete absorption as possible consistent with balance between the two photosystems is a biological advantage. Red and violet pigmented blue-green algae found in deep water and shaded habitats (see pp. 274 and 308) are evidently adapted in this way.

The electron donor

Blue-green algae, like other algae and higher plants, normally use water as the ultimate hydrogen or electron donor in photosynthesis with the elimination of molecular oxygen as a means of disposal of the oxidizing fragment. Little is known of the mechanism by which this is accomplished except that, as in higher plants (Richter, 1961), manganese plays an essential part. In manganese-deficient *Anacystis*, oxygen evolution does not begin after restoration of manganese until the alga has been exposed to light for a short period (Gerhardt and Wiessner, 1967) and the amount of manganese required is about one atom per 100 chlorophyll molecules (Cheniae and Martin, 1969).

In common with certain other algae, however, blue-green algae are capable of utilizing other hydrogen donors in a process of photoreduction, which resembles bacterial photosynthesis. Frenkel *et al.* (1950) found that *Synechococcus elongatus* and *Chroococcus* sp. resembled the green alga *Scenedesmus* in possessing a hydrogenase which could be induced by incubation under anaerobic conditions. This enabled these algae to carry out photoreduction in accordance with the following equation:

$$CO_2 + 2H_2 \longrightarrow CH_2O + H_2O$$

De-adaptation by these algae was a more gradual process than in *Scenedesmus*, which reverts quickly to normal photosynthesis if exposed to moderately high light intensities. The short-term products of photoreduction are presumably similar to those of photosynthesis, as has been noted in green algae (Bishop, 1966). Frenkel *et al.* (1950) were not successful in adapting *Nostoc muscorum* or *Cylindrospermum* sp. to carry out photoreduction.

Blue-green algae occur frequently in habitats where hydrogen sulphide is produced. Nakamura (1938) grew an *Oscillatoria* in the presence of 1×10^{-3} M hydrogen sulphide and could detect no oxygen evolution (although this might have been because of direct reaction of evolved oxygen with sulphide) but observed the deposition of sulphur within the cells. He suggested that the form of photosynthesis represented in the following equation was being carried out:

$$CO_2 + 2H_2S \longrightarrow CH_2O + H_2O + 2S$$

This suggestion has still not been satisfactorily substantiated. Stewart and Pearson (1970) noted that hydrogen sulphide disappears during the growth of *Anabaena flos-aquae* under anaerobic conditions but did not observe the accumulation of large sulphur granules (Stewart and Pearson, unpublished). Investigating this further, they found that the alga ceased to reduce acetylene in the presence of hydrogen sulphide and DCMU (3(3,4-dichlorophenyl)1,1-dimethyl urea), a specific inhibitor for photosystem II.

Concentrations of another inhibitor of photosystem II, salicylaldoxime, which completely inhibited acetylene reduction under aerobic conditions were only partially inhibiting in the presence of sulphide. They took acetylene reduction as an indirect measure of photosynthesis and, assuming that if hydrogen sulphide could act as a source of electrons it would substitute for photosystem II, concluded that although this substance could not be utilized as sole electron donor in photosynthesis, it might act simultaneously with water in this capacity.

Electron transport, photophosphorylation and the production of reductant

Electrons are considered to be conveyed from photosystem II to P 700 in photosystem I by a series of carriers among which plastoquinone, plastocyanin and cytochrome C_{554} have been identified. This series appears to be the same in blue-green algae as in the other algal groups. Other electron carriers found in blue-green algae which may be implicated here are α-tocopherolquinone, vitamin K_1, a specific cytochrome C-reducing substance and pteridines (see Holm-Hansen, 1968). It may be noted that some of these factors e.g. cytochrome C_{554}, become solubilized in cell-free preparations with consequent loss of photochemical activity by washed lamellae. This circumstance makes blue-green algae particularly suitable objects for investigation of the biochemistry of photosynthesis (Biggins, 1967b). Rurianski *et al.* (1970) have used steady state relaxation spectroscopy to estimate rates of electron flow in photosynthetic systems. In isolated chloroplasts the data are consistent with a sequential electron flow from cytochrome to P 700 but this was not so in whole cells of *Anacystis nidulans* (*Synechococcus* sp.) and whole cells of *Porphyridium cruentum*, a red alga. These results are difficult to reconcile with the hypothesis of a series of electron carriers between the two photosystems.

ATP production is coupled with this electron transport. Cell-free preparations obtained from *Anacystis nidulans* are capable of carrying out both cyclic and non-cyclic photophosphorylation. In the latter process the ratio of molecules of ATP formed to two electrons transferred varied between 1 and 3 (Gerhardt and Santo, 1966). Photosynthesis in intact *Anacystis* cells is inhibited by 2-alkyl-benzimidazoles as a result of uncoupling of photophosphorylation and non-cyclic photophosphorylation is more sensitive than cyclic photophosphorylation in cell-free preparations (Büchel *et al.*, 1967). Cell-free preparations from *Anabaena variabilis* grown in the presence of diphenylamine, which inhibits carotenoid synthesis, have markedly higher cyclic photophosphorylation activities than similar preparations from normal cells (Neumann *et al.*, 1970).

The production of the reductant $NADPH_2$ in the photochemical reaction is mediated by electron transfers involving ferredoxin and phytoflavin. Ferredoxins have been isolated and characterized from *Anacystis nidulans* by Yamanaka *et al.* (1969) and from a *Nostoc* species by Mitsui and Arnon (1971). These are generally similar to chloroplast ferredoxin, which they also resemble in being 1—electron carriers rather than 2—electron carriers like the ferredoxins of *Clostridium* and *Chromatium* (Evans *et al.*, 1968). Another substance, phytoflavin, which may function in $NADPH_2$ production, has been identified in *Anacystis nidulans* under conditions of iron-starvation by Smillie (1965).

Carbon dioxide fixation

From a survey in which the products of photosynthesis in 27 different plants representing nine major groups were characterized, Norris *et al.* (1955) concluded that the path of carbon fixation is remarkably uniform throughout the plant kingdom. The blue-green algae investigated were *Phormidium* sp., *Nostoc muscorum*, *Nostoc* sp. and *Synechococcus cedrorum*. Among the few variations found was the production of an unknown compound which accounted for 7% and 21% of the total radiocarbon fixed by *Synechococcus* and *Nostoc muscorum* respectively during a 5 min period of photosynthesis. It was usually less than 1%, however, in other blue-green algae and in the plants belonging to other groups. Linko *et al.* (1957) identified this compound as citrulline. In some organisms citrulline plays an integral part in the conversion of ornithine to arginine which may then be hydrolyzed to urea and ornithine but there is no evidence for this in blue-green algae. Indeed, no labelled urea or arginine could be found by Hoare *et al.* (1967) in *Nostoc muscorum* supplied with ^{14}C-citrulline and it was concluded that a role of citrulline may be in the transfer of nitrogen from ammonia to amino acids.

The experiments of Norris *et al.* (1955) were done with the rather long exposure time to ^{14}C-carbon dioxide of 5 min. Experiments with only 5 s exposure were carried out by Kandler (1961) using *Anacystis*. These confirmed that phosphoglyceric acid was the first product although the high carbon dioxide concentration used would favour this product rather than others produced by carboxylases with lower carbon dioxide optima. Kindel and Gibbs (1963) showed that the labelling pattern of starch from blue-green algae on ^{14}C-carbon dioxide was similar to that from *Chlorella*. All this evidence is consistent with the idea that the carbon fixation cycle discovered by Calvin and his collaborators (Fig. 8.2) occurs in blue-green algae although further research may show that the situation is rather more complex than this.

Among the enzymes concerned in the Calvin cycle, aldolase is one which has excited most discussion in relation to the blue-green algae as it was at one time thought to be absent from them. Green algae and higher plants possess an aldolase Type I, inhibited by mercurials but unaffected by chelating agents, whereas various blue-green algae have been shown to contain a Type II aldolase, only slightly affected by mercurials but inhibited by chelating agents (Gibbs *et al.*, 1970). The enzyme glyceraldehyde-3-phosphate dehydrogenase which is responsible for producing a triose sugar from the first product of photosynthetic carbon dioxide fixation, phosphoglyceric acid, functions with either $NADH_2$ or $NADPH_2$ as hydrogen donor. In blue-green algae this is the only enzyme of its kind (Hood and Carr, 1967) but in eukaryotic plants there is a similar enzyme utilizing $NADH_2$ rather than $NADPH_2$ which is concerned in the glycolytic pathway in respiration. The existence of only the one enzyme, using both hydrogen donors, in blue-green algae may partly explain the special relationships between photosynthesis and respiration found in blue-green algae (p. 165).

The photoassimilation of organic substances

There is ample evidence, obtained with a great variety of photosynthetic bacteria and algae, that the uptake of certain exogenously supplied organic substrates is stimulated by light and may provide a considerable proportion of the total carbon requirement for growth (Wiessner, 1970). Many blue-green algae seem unable to grow on organic substrates in the dark but nevertheless they may utilize them in the light.

Photoassimilation of acetate, which has been demonstrated in *Nostoc muscorum* (Allison *et al.*, 1953), *Anabaena variabilis* (Pearce and Carr, 1967), *Anabaena flos-aquae*, *Anacystis nidulans* (*Synechococcus* sp.), *Chlorogloea fritschii* (Hoare *et al.*, 1967; Pearce and Carr, 1967) and *Gloeocapsa alpicola* (Smith *et al.*, 1967) have been investigated in most detail. It might be expected, from comparisons with other kinds of algae, that photo-assimilation of an organic substrate at limiting light intensity would increase the relative growth rate although, at light saturation, rates higher than those which can be achieved with carbon dioxide alone would not result. However this has not been observed with blue-green algae (Smith *et al.*, 1967; Hoare *et al.*, 1967) possibly because the experiments were not conducted under sufficiently light-limited conditions. Nevertheless, appreciable amounts of carbon may be assimilated from this source. Thus with *Anabaena variabilis* at relatively low light intensities, 18% of the newly assimilated carbon was found to come from acetate when this was supplied (Carr and Pearce, 1966). Also, van Baalen *et al.* (1971) found, with *Agmenellum quadruplicatum* and *Lyngbya lagerheimi*, that while glucose had no effect on growth in complete darkness or at 3500 lux it supported growth in dim light which barely permitted growth with carbon dioxide as sole carbon source. The biological advantage of photoassimilation in dimly lit situations may be considerable and its possible importance for freshwater plankton is discussed later (p. 264).

Hoare *et al.* (1967) found that photoassimilation of acetate was extremely sensitive to DCMU, a potent inhibitor of non-cyclic photophosphorylation, a fact that indicates that electron donors and ATP produced by the photochemical reactions are involved in its assimilation. A major pathway for assimilation of acetate leads to lipid (Allison *et al.*, 1953) but some acetate carbon may also enter into amino acids, such as glutamic acid, arginine, proline, leucine, and other organic acids. The pathways followed appear to be those found in other organisms and include parts of the interrupted tricarboxylic acid cycle (see p. 171) and, perhaps with the exception of *Anacystis nidulans* (Hoare *et al.*, 1967), the glyoxylate cycle (Pearce *et al.*, 1969). The finding that much of the carbon enters lipide may be because temporary inhibition of growth in the experimental suspensions diverted the flow of carbon towards storage materials. Although the conditions were similar to those used for growing algae, in the experiments of Allison *et al.* (1953) and Hoare *et al.* (1967), resuspension in fresh medium may have resulted in cessation of growth for several hours as Fogg and Than-Tun (1960) found with *Anabaena cylindrica*.

Sugars are photoassimilated by blue-green algae. These include glucose by *Tolypothrix tenuis* (Kiyohara *et al.*, 1962) and *Anabaena variabilis* (Pearce and Carr, 1969), sucrose by *Chlorogloea fritschii* (Fay, 1965) and fructose by *Anacystis nidulans*, *Coccochloris peniocystis* and *Gloeocapsa alpicola* (Smith *et al.*, 1967). Contrary to the findings with acetate, the photoassimilation of glucose accelerates growth of *Tolypothrix* under light limiting conditions (Kiyohara *et al.*, 1962) and Pearce and Carr (1969) found, using ^{14}C-glucose that up to 46% of the total dry weight of the organism could be derived from this source. Glucose assimilation is mainly via the photosynthetic pentose phosphate pathway in *Tolypothrix* (Cheung and Gibbs, 1966) and possibly also *Anabaena* (Pearce and Carr, 1969) and activities of various glucose-metabolizing enzymes are the same in *Anabaena* in the presence and absence of glucose (Table 9.V). This is the same situation as with acetate assimilation and confirms the view that this alga cannot adjust its metabolism as efficiently as, for example, bacteria or green algae. Control seems to be exerted at the level of enzyme activity rather than in enzyme synthesis.

There is little information about photoassimilation of other substances. Propionate, although somewhat toxic, can be utilized by *Anacystis nidulans* but this alga does not assimilate butyrate, citrate, glutamate or succinate to any appreciable extent (Hoare *et al.*, 1967). However, pyruvate, succinate, malate, glutamate, aspartate and leucine can all be utilized slightly by *Anacystis nidulans*, *Coccochloris peniocystis* and *Gloeocapsa alpicola* according to Smith *et al.* (1967). Photoassimilation of glycollate by *Anabaen flos-aquae* and *Oscillatoria* sp. has been demonstrated by Miller, *et al.* (1971).

Obligate phototrophy

It is remarkable that many blue-green algae seem to grow only in the light. It is, of course, impossible to prove a negative and it may be that under particular conditions and combination of substrates species which have hitherto defied attempts at culturing them heterotrophically will grow in the dark. Nevertheless even those species which grow in the dark on suitable substrates do so rather slowly, and supplying exogenous substrates usually has little effect on the respiration rate (see p. 163). Since, as we have just seen, blue-green algae assimilate organic substrates in the light, an inability of the substances to penetrate into the cells is not likely to be the reason for their obligate phototrophy.

The explanation probably lies in the fact that these algae do not possess a sufficient complement of enzymes to operate a full tricarboxylic acid cycle (see p. 171). This metabolic lesion and the absence of $NADH_2$-oxidase were thought by Smith *et al.* (1967) to prevent oxidative phosphorylation so that the organisms would be completely dependent on photophosphorylation to supply ATP for synthesis. However, it was later demonstrated that the obligate phototroph *Anabaena variabilis* has an $NADH_2$ oxidase and is able to carry out oxidative phosphorylation, albeit slowly (Leach and Carr, 1969). Carr *et al.* (1969) put forward the view that obligate phototrophy is due only partly to lack of any one or two specific enzymes and more to the relative inflexibility of the metabolic pattern which results from their inability to repress and de-repress enzyme synthesis. Thus when presented with an organic substrate the amount of enzyme system needed to metabolize it efficiently cannot be produced. The carbon dioxide fixation cycle, which is not repressed, is then still necessary to supply the variety of different metabolites required in growth.

Photosynthesis and the assimilation of nitrogen

The assimilation of nitrogen is discussed elsewhere (p. 181) but it may be noted here that the electrons and/or ATP required for the conversion of certain forms of combined nitrogen to ammonia, the compound in which nitrogen enters directly into general metabolism, may be supplied by photochemical reactions. There is evidence that the ATP required for nitrogen fixation in heterocysts is supplied mainly by photosystem I-dependent phosphorylation (p. 234). Nitrate, nitrite and hydroxylamine reduction are light stimulated in *Anabaena cylindrica* (Fig. 8.5). The

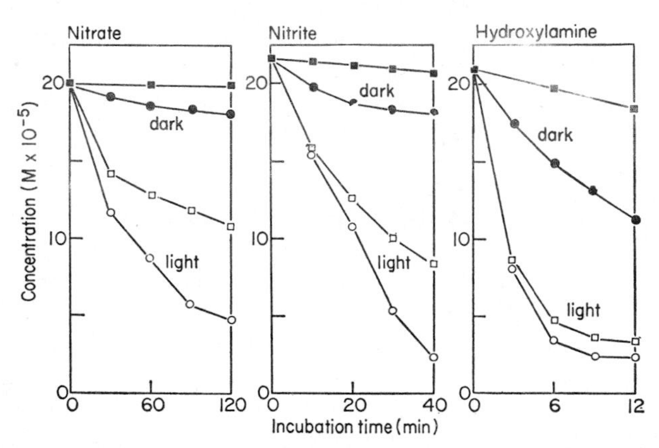

FIG. 8.5. Uptake of nitrate, nitrite and hydroxylamine by *Anabaena cylindrica* in the light and dark. ○, light aerobic; □, light anaerobic; ●, dark aerobic; ■, dark anaerobic (redrawn after A. Hattori (1962), *Pl. Cell Physiol.*, (Tokyo) **3**, 355–369).

reduction of these substances is saturated at a light intensity only about one tenth of that required to saturate photosynthesis and is partially inhibited by CMU, which specifically inhibits photosystem II, in the light but not in the dark (Hattori, 1962a). The nitrate and nitrite reductases can both accept electrons from ferredoxin, and so may be linked directly with photosystem I (Hattori and Myers, 1966, 1967).

General physiology

Generally speaking the physiological relationships and response to environmental factors by the photosynthesis of blue-green algae are similar to those of other algae but a few features call for special comment.

The relationships of rate of photosynthesis to light intensity and carbon dioxide concentration for *Anacystis nidulans* (*Synechococcus* sp.), *Anabaena variabilis* and *Nostoc muscorum* have been reported by Kratz and Myers (1955b; see p. 140). Fogg and Than-Tun (1960) found the optimum p_{CO_2} for *Anabaena cylindrica* growing at 5500 lux to be about 0·10% at 15°C and 0·25% at 25°C and above, that is, values of the same order as for other simple algae. Inhibition was observed at 0·5%. However this concentration, or even 5% has been routinely used in the culture of *A. cylindrica* and other blue-green algae with higher population densities and buffered media (see p. 135).

In their studies on the effect of oxygen on nitrogen fixation, Stewart and Pearson (1970) observed that photosynthesis, as measured by incorporation of ^{14}C from carbon dioxide, was higher at p_{O_2} values approaching zero than at 20% and almost completely inhibited at 100%. *Chlorella* and isolated chloroplast photosynthesis is affected in a similar way (Gibbs *et al.*, 1968) and it is likely that blue-green algae are not exceptional in this respect.

In *Anabaena cylindrica* photosynthesis is markedly dependent on the available nitrogen. When the p_{N_2} in nitrogen-fixing cultures grown for 48 h was reduced from the normal value, 78% to 2%, the rate of photosynthesis, expressed on a cell nitrogen basis, was reduced to less than a half, although the cell nitrogen content was scarcely affected (Fogg and Than-Tun, 1960). Reduction of photosynthesis to about half the normal rate has been observed in molybdenum deficient cultures of *Nostoc* grown in the absence of combined nitrogen (Arnon, 1958). Since molybdenum is essential for nitrogen fixation but not for photosynthesis (p. 137) this is evidently an effect similar to that described for *A. cylindrica*. The addition of molybdenum to deficient cultures 11 h prior to the experiment restored the full rate and caused a dramatic shift in the pattern of fixation of ^{14}C into assimilatory products, the percentage fixed in total amino acids rising from 6 to 26·5.

Certain blue-green algae are able to grow at temperatures up to 72–75° C and the photosynthetic rates are correspondingly among the highest known. *Synechococcus* sp. (*Anacystis nidulans*), which has been much used as an experimental organism in studies on photosynthesis, at its temperature optimum of 39°C has a generation time of about 2 h and a Q_{O_2} ($\mu l\ O_2\ h^{-1}$ mg dry wt^{-1}) of 309 at light saturation in Warburg buffer. This may be compared with the values of 7·7 h and Q_{O_2} 190 for *Chlorella pyrenoidosa* at 25°C (Kratz and Myers, 1955b) and 2·6 h and Q_{O_2} about 680 for *Chlorella pyrenoidosa* strain 7–11–05 at 39°C (Sorokin, 1959). The truly thermophilic blue-green algae, which also have minimum generation times of about 3 h, presumably have maximum rates of photosynthesis similar to that for *A. nidulans* in spite of their optimum temperatures being 30°C higher. Because thermal waters are unusually rich in carbon dioxide, and flowing, the conditions necessary for sustaining high rates of photosynthesis are met if the stream is unshaded. During the course of a summer day the rate of photosynthesis in a mat of thermophilic blue-green algae remains proportional to light intensity (Brock, 1967a). This is best explained by supposing that light is attenuated rapidly within the mat so that cells in the lower layers are always light limited. High light intensities are required to saturate the photosynthesis of thermal algae and light inhibition is rare unless the cells have been adapted to low light intensities

(Castenholz, 1969). Brock (1967b) has shown that the temperature optima for given samples of thermal algae are sharply defined and extraordinarily close to the actual temperatures in the habitat in which they were growing. He concludes that the upper temperature limit is set by some inherent characteristic of the photosynthetic apparatus since non-photosynthetic bacteria are found at higher temperatures.

Chemosynthesis

The ability to utilize energy released by the oxidation of inorganic substances to support growth, a process known as chemosynthesis or, more accurately, chemolithotrophy, is possessed by many prokaryotic organisms but not by eukaryotic forms. Some chemosynthetic organisms perhaps should be included in the blue-green algae. Thus, from consideration of their morphology and cytology Pringsheim (1949) concluded that *Beggiatoa*, *Thiothrix* and *Achromatium* spp. are colourless members of the Cyanophyceae. This conclusion has not been seriously contested although for pragmatic reasons authorities such as Soriano and Lewin (1965) have preferred to retain these and other colourless forms showing the gliding motion characteristic of blue-green algae, within the bacteria. The genera *Beggiatoa* and *Thiothrix* include chemosynthetic forms since some strains grow in the dark in the absence of organic matter if hydrogen sulphide is supplied. This is oxidized to sulphur, which appears as granules within the cells, and sulphur is further oxidized to sulphuric acid (Pringsheim, 1949). However, there have been no detailed studies on the biochemistry of these presumed chemosynthetic processes and many species of *Beggiatoa* are also capable of heterotrophic growth on substrates such as acetate (Pringsheim, 1964). *Achromatium*, a unicellular analogue of *Beggiatoa*, deposits sulphur within its cells and has been cultured in an inorganic medium containing hydrogen sulphide and so is evidently chemosynthetic (Pringsheim, 1963). *Thiospirillopsis* apparently also belongs to this chemosynthetic group. *Crenothrix*, an "iron-bacterium" of doubtful affinities which Pringsheim (1963) tentatively assigned to the Cyanophyceae, is possibly another chemosynthetic form, subsisting on the oxidation of ferrous or manganous ions.

Studies on chemosynthetic Eubacteriales, such as *Hydrogenomonas* and *Thiobacillus*, show that the oxidation of the inorganic substrate produces ATP and that carbon dioxide is fixed by a Calvin-type cycle (Peck, 1968). There thus appears to be no basic difference between the metabolism of chemosynthetic organisms and other forms of life and it may be assumed that this conclusion applies also to chemosynthetic species related to the blue-green algae.

Chapter 9

Heterotrophy and respiration

Early assumptions that blue-green algae utilized organic substances for growth in darkness or low light were based on observations that they are frequently abundant in habitats rich in dissolved organic matter (see p. 256), and on early reports (Bouilhac, 1897, 1901; Brunthaler, 1909) of their growth in culture at the expense of organic substrates. These early reports, however, were obtained using impure cultures and Pringsheim (1914) who first obtained pure cultures of blue-green algae, observed little or no stimulation by organic substrates and could not demonstrate growth in the dark. However, Harder (1917b), who isolated *Nostoc punctiforme* from a symbiotic association with *Gunnera* (see p. 350), showed that it grew in bacteria-free culture in the dark on various sugars and polysaccharides and Allison *et al.* (1937) showed that *Nostoc muscorum* (Allison strain) grew slowly on glucose or sucrose in the dark, both in the presence and absence of combined nitrogen. They were unable to demonstrate dark heterotrophic growth of *Synechococcus* sp. (*Anacystis nidulans*), *Anabaena variabilis* and *Nostoc muscorum* (Gerloff strain) using a variety of sugars, organic acids and proteins. Allen (1952) found that *Oscillatoria* sp., *Lyngbya* sp., *Phormidium foveolarum* and *Plectonema notatum*, but not several other species, could grow very slowly in the dark on medium containing glucose and yeast autolysate.

More recently heterotrophic growth of *Tolypothrix tenuis* has been reported by Kiyohara *et al.* (1960). Growth in the dark was supported by glucose, fructose and, to a lesser extent, by sucrose, in the presence of ammonium salts or casamino acids, both of which acted as sources of nitrogen for heterotrophic growth whereas nitrate did not. Analysis showed that of various amino acids present in casamino acids only arginine was utilized appreciably in the dark. When single amino acids were tested, arginine, phenylalanine, leucine, serine, alanine, glutamine, asparagine and glycine, in that order, but not glutamate and aspartate, supported some growth in the dark in the presence of glucose. Biotin, thiamine and vitamin B_{12} did not promote growth in the dark. In these studies, chlorophyll, carotenoids and phycocyanin, but not phycoerythrin, were synthesized in the dark.

Heterotrophic growth by *Chlorogloea fritschii* has been studied by Fay and Fogg (1962) and by Fay (1965). The latter worker found sucrose superior to glucose in supporting growth in the dark, especially under nitrogen-fixing conditions and some growth occurred on maltose, glycine and glutamine. The alga could be adapted to heterotrophic growth on certain substrates in time and continued to produce photosynthetic pigments in complete darkness. Sucrose assimilation was slower in the dark than in the light, in which uptake stimulated both the growth rate and the nitrogen fixation rate several-fold. In contrast to the photoassimilation of acetate in *Nostoc muscorum* (Allison *et al.*, 1953) (see p. 155), sucrose assimilation in the light increased in the absence of carbon dioxide. Nitrogen fixation, however, was greatest under conditions of simultaneous carbon dioxide fixation and sucrose assimilation in *Chlorogloea fritschii* (Table 9.I). Heterotrophic nitrogen fixation in *Anabaenopsis circularis*,

TABLE 9.I. ^{14}C-sucrose assimilation and $^{15}N_2$ fixation in *Chlorogloea fritschii* in the light and in the dark, with and without CO_2 in the gas phase. After Fay (1965)

Treatment	Radioactivity of cells (counts min^{-1})	Radioactivity of respiratory CO_2 (counts min^{-1})	Atom % ^{15}N excess
Light + CO_2	17,915	—	0·675
Dark + CO_2	10,390	—	0·093
Light − CO_2	30,240	95	0·378
Dark − CO_2	15,635	4640	0·106

reported by Watanabe and Yamamoto (1967), was best on glucose but also occurred on fructose, sucrose or maltose, although several other sugars and a number of organic acids were inactive in this respect. Khoja and Whitton (1971) also observed heterotrophic growth of certain other species and Hoare *et al.* (1971) observed heterotrophic growth, both under aerobic and anaerobic conditions by a *Nostoc* species isolated from root nodules of the cycad *Macrozamia lucida*. There were no significant differences in the ultrastructure of light-grown and dark-grown cells.

There is good evidence, therefore, that many blue-green algae grow heterotrophically in the dark, although usually at a much slower rate than in the light, and only on certain organic substrates which vary with the species. Stimulation of growth in the light by organic substances, which is quite distinct from heterotrophic assimilation, is discussed on p. 155.

Respiration

Under natural conditions photosynthesis is only possible during the day but many metabolic and biosynthetic processes may continue at night. These activities are then supported through the mobilization and utilization of polymeric storage material deposited in the form of cytoplasmic granules. This involves the breakdown of large molecules into small constituent units and further degradation of these in respiration. Aerobic respiration consists of the oxidation of organic substrates to carbon dioxide and water with oxygen acting as terminal electron acceptor, most of the energy liberated becoming available for metabolic use as ATP. In higher plants and animals respiration is accomplished through a complex series of reactions and interrelated metabolic pathways including the Embden-Meyerhof-Parnas pathway, the pentose phosphate pathway, the tricarboxylic acid cycle, the glyoxylate cycle, and the electron transport system associated with oxidative phosphorylation. There is much less known about the intermediary metabolism of blue-green algae but some progress has been made recently in demonstrating the existence of various respiratory pathways and the activity of their constituent enzymes. This is considered shortly (p. 171).

RESPIRATION RATES IN THE DARK

Dark endogenous respiration is usually measured as the rate of oxygen uptake ($Q_{O_2} = \mu l\ O_2$ consumed mg dry weight^{-1} h^{-1}) by an organism in the absence of an external substrate in the dark. The first quantitative measurements in a blue-green alga, an *Anabaena* species, were those of Webster and Frenkel (1953) who showed that the alga absorbed 4·2–4·8 $\mu l\ O_2$ mg dry weight^{-1} h^{-1} during the first 4 h in the darkness at 25° C. The respiratory quotient (RQ = volume of CO_2 evolved/volume of O_2 absorbed) was 0·9–0·95. Similar respiratory quotients (close to unity) have been obtained also for unicellular and filamentous blue-green algae (Kratz and Myers, 1955b; Biggins, 1969). These respiration rates are considerably lower than those for baker's yeast and *Chlorella pyrenoidosa* obtained under rather similar conditions (Gibbs, 1962) and for various bacteria, including flexibacteria (Dietrich and Biggins, 1971). The low respiratory rate in blue-green algae is related to their limited ability to utilize organic substrates for growth and for energy production. There is a limited response to organic substrates even after previous dark starvation (Webster and Frenkel, 1953; Kratz and Myers, 1955b; Biggins, 1969) and only a few sugars (glucose, fructose, galactose, sucrose) and organic acids (acetate, pyruvate, succinate, ascorbate) cause even a slight increase in the respiratory rate.

The rate of respiration in the dark depends on the physiological state of the alga, as well as on external conditions such as temperature, pH of the medium, etc. (Kratz and Myers, 1955b; Padan *et al.* 1971). The rate is higher when the alga is growing vigorously and declines rapidly to about 20% of the original rate when the alga is incubated in the dark (Webster and Frenkel, 1953; Kratz and Myers, 1955b). The type of variation which may occur, depending on growth rate and temperature, is shown in Table 9.II.

TABLE 9.II. Endogenous respiration of blue-green algae. (Measurements in 5×10^{-2} M phosphate at pH 7·8. Q_{O_2} : $\mu l\ O_2\ h^{-1}$ (mg dry wt of alga)$^{-1}$.) After Kratz and Myers (1955b)

Organisms	Preliminary conditions	Temperature °C	Growth rate, K	Q_{O_2}	R.Q.
Anabaena variabilis	growing	25	0·30	8·4	1·10
	dark starved			1·7	1·00
Anacystis nidulans	growing	25	0·55	1·6	1·00
	dark starved			0·3	0·93
	growing	39	1·00	4·7	1·10
	dark starved			1·9	1·10
	growing	39	2·50	7·5	1·05
	dark starved			2·9	—
Nostoc muscorum G	growing	25	—	4·4	—
	dark starved			1·1	—

The optimum pH for dark respiration is usually near 8 (Gibbs, 1962) but lower pH optima (6–7) have been recorded for algae such as the thermophilic *Schizothrix calcicola* (Everton and Lords, 1967) which is adapted to extreme conditions. Maximum rates of respiration usually occur at about 32·5° C but those of the thermophilic *Synechococcus* sp. (*Anacystis nidulans*) and *Oscillatoria subbrevis* occurred at temperatures near 40° C (Bünning and Herdtle, 1946). For *Schizothrix calcicola* the optimum temperature was 58° C (Everton and Lords, 1967).

The first experimental study on the effects of inhibitors on respiration in blue-green algae was carried out by Emerson (1927) who found that respiration of an *Oscillatoria* species was inhibited by 80% in the presence of 10^{-4} M cyanide, a concentration which scarcely affects respiration of green algae. Respiration of a species of *Anabaena* is inhibited by 20% in the presence of 10^{-5} M cyanide and by 80% in the presence of 10^{-1} M cyanide (Webster and Frenkel, 1953). Strong cyanide inhibition was observed also with *Synechococcus* sp. (*Anacystis nidulans*) and *Phormidium luridum* (Biggins, 1969). Other respiratory inhibitors include phenanthroline and azide (Webster and Frenkel, 1953; Everton and Lords, 1967). Carbon monoxide only inhibits respiration at very high ratios of $CO : O_2$. Thus a mixture of 97% carbon monoxide and 3% oxygen caused only 12% inhibition of respiration by *Oscillatoria* (Emerson, 1927). Endogenous respiration of *Anabaena* sp. is relatively insensitive to carbon monoxide while substrate-supported respiration is inhibited up to 58% at high $CO : O_2$ ratios (Webster and Frenkel, 1953). These findings suggest that the terminal oxidase mediating most of the respiratory electron transport in blue-green algae is probably a heavy metal-containing cytochrome system, but that this system may also contain a CO-insensitive flavoprotein component. The possible involvement of a cytochrome oxidase is indicated by the enhanced oxidation of ascorbate and hydroquinone in a reaction mixture containing fragmented *Anabaena* cells and cytochrome *c* (Webster and Frenkel, 1953). A stimulation of the rate of respiratory oxygen uptake by *Synechococcus* sp. (*Anacystis nidulans*) and *Phormidium luridum* by low concentrations of the uncouplers dinitrophenol and carbonyl cyanide *p*-trifluoromethoxy-phenylhydrazone (FCCP) and the subsequent reduction of the concentration of ATP (Biggins, 1969) suggest that oxidative phosphorylation is coupled with electron transport in blue-green algae as in other organisms.

RESPIRATION IN THE LIGHT

In estimating gross photosynthesis by blue-green algae and other plants it has often been customary to apply a correction for respiratory oxygen uptake occurring concomitantly with photosynthetic oxygen evolution using values obtained by separate measurements of oxygen uptake in the dark. The basic assumption here—that the rate of respiration is the same in the light and in the dark—is open to question.

Brown (1953) and Brown and Webster (1953) carried out a detailed study of respiration by blue-green algae in the light, using the oxygen isotope, ^{18}O, as a tracer and assaying it with a mass spectrometer. By this means it was possible to follow the concurrent processes of respiratory oxygen consumption and photosynthetic oxygen evolution. Brown and Webster established that although the rate of respiratory uptake in *Chlorella* and in the leaves of *Nicotiana tabacum* was usually similar in the dark and in the light, respiratory oxygen uptake in the light by the blue-green alga *Anabaena* was considerably more complex (see Fig. 9.1). They

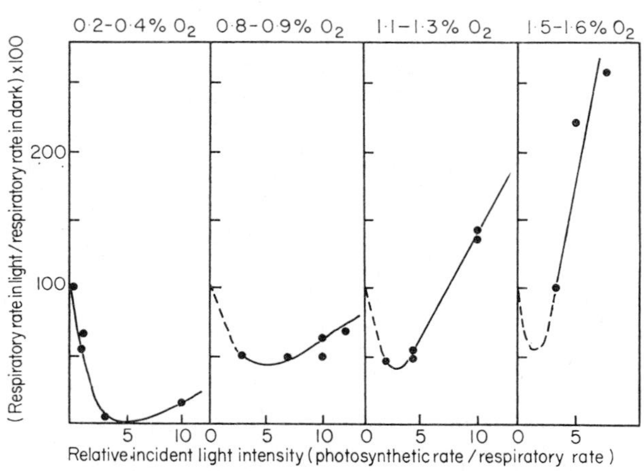

FIG. 9.1. The effect of light on oxygen consumption by *Anabaena* sp. shown as a function of oxygen tension and of relative light intensity. After A. H. Brown and G. C. Webster (1953) *Am. J. Bot.* **40**, 753, Fig. 4.

found an apparent photoinhibition of respiration at low oxygen tensions (below 0·5%) thus confirming an earlier observation on blue-green algae by Gaffron and Fager (1951). This effect was less pronounced at higher (up to 1·6%) oxygen concentrations. However, at high light intensities a distinct photostimulation of oxygen uptake took place (see Fig. 9.2). A similar effect was observed with *Synechococcus* sp. (*Anacystis nidulans*) by Hoch *et al.* (1963) who showed that the inhibition of oxygen uptake at low light intensity was sensitized by chlorophyll *a* and was not inhibited by DCMU whereas enhanced oxygen uptake at high light intensity was sensitive to DCMU. This enhanced oxygen uptake in the light is now referred to as "photorespiration", which has been defined in its broadest sense by Jackson and Volk (1970) as "all respiratory activity in the light regardless of the pathways by which carbon dioxide is released and oxygen consumed". Lex *et al.* (1972) studied enhanced oxygen uptake by *Anabaena cylindrica* in detail. They found that the photorespiration rate was up to twenty times the dark respiration rate and could approximate to that of true photosynthesis. This reaction was saturated at about 1500 lux and could be

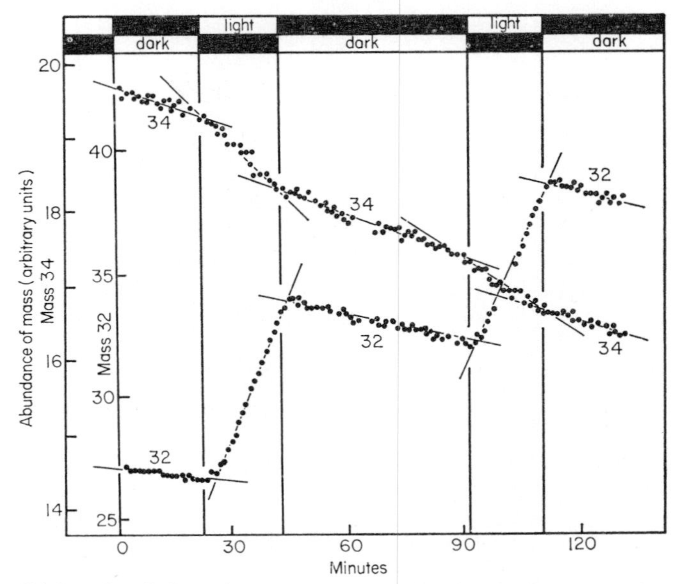

FIG. 9.2. Light stimulation of oxygen consumption by *Anabaena* sp. simultaneously with photosynthetic oxygen evolution, measured with a mass spectrometer. Mass 32 (^{16}O ^{16}O) was recorded to measure the rate of photosynthetic oxygen production, while mass 34 (^{18}O ^{16}O) indicates the rate of respiratory oxygen consumption in an atmosphere containing ^{18}O-labelled oxygen gas. After A. H. Brown and G. C. Webster (1953). *Am. J. Bot.* **40**, 753, Fig. 3.

distinguished from dark respiration in that it increased linearly with increasing p_{O_2} (up to 0·23 atm in their studies), while dark respiration was saturated at a p_{O_2} level near 0·05 atm (Fig. 9.3). The photorespiration rate was markedly affected by carbon dioxide concentrations, being stimulated by levels less than 0·0003 atm, and was inhibited by increasing bicarbonate concentration. Furthermore DCMU immediately inhibited photorespiration and revealed an oxygen-consuming process sensitive to KCN and approximating in rate to dark respiration. No direct evidence could be obtained that glycollate was the substrate for photorespiration in this alga, as it is in higher plants, and it was suggested that the uptake of oxygen in the light was due to the oxidation of a reduced product of photosystem I, possibly ferredoxin. There was also good evidence that photorespiration and nitrogenase activity competed for reductant (Lex *et al.* 1972).

The evidence for photorespiration was based on oxygen uptake data and the possible photoevolution of carbon dioxide characteristic of photorespiration in higher plants was not investigated. However, Döhler and

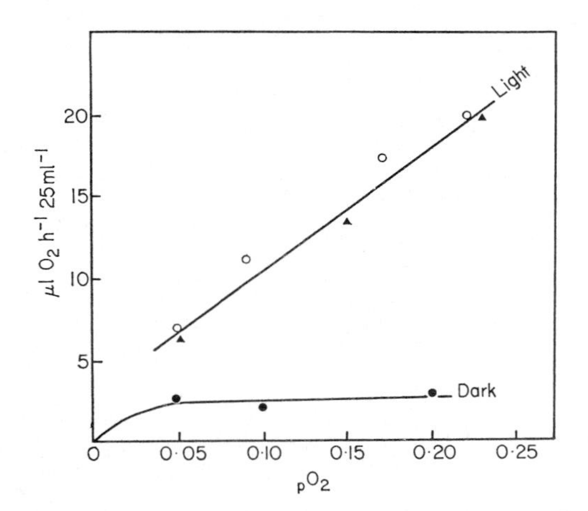

FIG. 9.3. Effect on partial pressure of oxygen on oxygen consumption by *Anabaena cylindrica* in the light and in the dark. After M. Lex, W. B. Sylvester, and W. D. P. Stewart (1972). *Proc. R. Soc.* B. **180**, 87, Fig. 3.

Braun (1971) have reported on a dark carbon dioxide outburst in *Synechococcus* (*Anacystis nidulans*) at the start of the dark incubation period which could be regarded as a photorespiratory release of carbon dioxide.

Interrelations between the processes of photosynthesis, photorespiration and dark respiration in blue-green algae are of obvious interest: the photosynthetic carbon fixation (Calvin) cycle includes reactions which are the reverse of the respiratory glycolytic reactions and the pentose phosphate pathway and there are several intermediates, enzymes and cofactors common to both respiration and photosynthesis. Furthermore, whereas photosynthesis, photorespiration and dark respiration are separated in eukaryotic cells within specialized membrane-bound organelles (chloroplasts, peroxisomes and mitochondria respectively) no such segregation exists within blue-green algae. Hence interactions between the various processes are not only possible but probably unavoidable.

The photoinhibition of respiration at low light intensities, which is usually referred to as the "Kok effect", may indicate a competition between photosynthesis and respiration for substances and sites involved in both reactions. Photosynthetic electron transport, for example, may inhibit or exclude respiratory electron transport. Some evidence for this was obtained in a study of the reduction of dichlorophenol-indophenol (DCPIP), a Hill oxidant, by Carr and Hallaway (1965) who showed that it is reduced by *Anabaena variabilis* in the light as well as in the dark. However, the two reactions differed in several respects: the dark reaction showed a linearity with time whereas the response increased exponentially with time in the light. Also the rate of dark reduction did not change with varying DCPIP concentration whereas the rate of light reduction increased proportionally with increasing DCPIP concentration. It was thus possible to distinguish between dark and light reactions and show that the dark reduction of DCPIP is suppressed at high light intensities. Another interaction between metabolic processes is that of glyceraldehyde-3-phosphate dehydrogenase which is operative in both glycolytic and pentose phosphate pathways, using $NADH_2$ as coenzyme in the former and $NADPH_2$ in the latter.

The finding that photosynthesis and respiration in blue-green algae may occur in the same subcellular fraction is further evidence that photosynthesis and respiration may share a common electron transport system and energy conversion mechanism (Petrack and Lippman, 1961; Susor *et al.*, 1964; Biggins, 1967b; Fujita and Myers, 1965a). Such fractions could also carry out respiratory reactions such as decarboxylation (Fay and Cox, 1966) and oxidation of pyridine nucleotides, ascorbate, hydroquinone, succinate and ferrocytochrome *c* (Horton, 1968; Biggins, 1969). The respiratory mechanism of flexibacteria is also associated with a particulate cell-free fraction (Webster and Hackett, 1966; Dietrich and Biggins, 1971).

Bisalputra *et al.* (1969) studied the site of respiratory activity in *Nostoc muscorum* by observing the reduction of potassium tellurite and of tetranitro-blue tetrazolium (TNBT) by electron microscopy. Tellurite, which is used to locate dehydrogenase activity, has been shown to accept electrons directly from the respiratory enzyme and to be reduced to an insoluble and highly electron scattering product which was deposited only over the photosynthetic lamellae. When the alga was incubated in the presence of TNBT, a fine insoluble formazan deposit was formed at the same sites (see Figs. 9.4 and 9.5). Both reactions occurred in the presence or absence of succinate as substrate in the light and in the dark.

RESPIRATORY PATHWAYS IN BLUE-GREEN ALGAE

Little is known about the presence or operation of the glycolytic pathway but the oxidative pentose phosphate pathway in blue-green algae has been studied in detail (Fewson *et al.*, 1962; Van Baalen, 1965b; Willard *et al.*, 1965; Hood and Carr, 1967; Pelroy and Bassham, 1972; Pelroy *et al.*, 1972). By following the metabolism of specifically-labelled ^{14}C-glucose and taking into consideration the pattern of $^{14}CO_2$ production, Wildon and Rees (1965) obtained good evidence for the participation of aldolase in the oxidation of glucose in *Anabaena cylindrica*.

Cheung and Gibbs (1966) found that in *Tolypothrix tenuis* about 60% of the glucose assimilated was converted to polysaccharide. Glucose

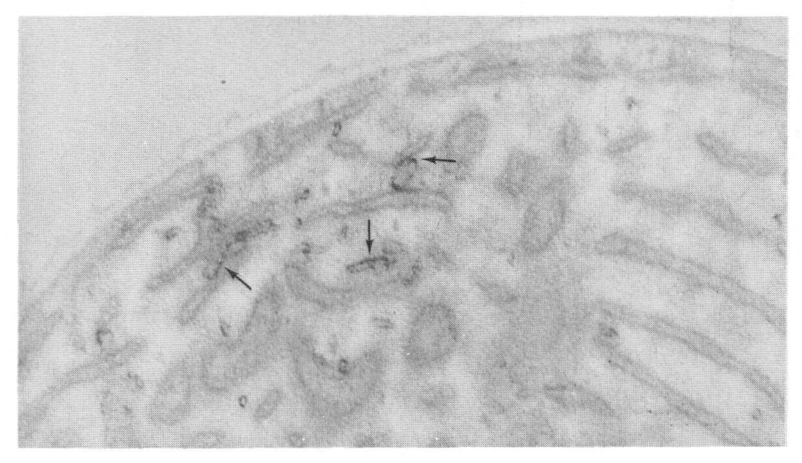

FIG. 9.4. Part of a section of *Nostoc sphaericum* incubated for 30 min in the presence of tetra-nitro-blue-tetrazolium (TNBT) previous to fixation, showing formazan deposition (arrows) over the photosynthetic lamellae. Electron micrograph (×80,000). By courtesy of T. Bisalputra, D. L. Brown and T. E. Weier (1969). *J. ultrastruct. Res.* **27**, 182, Fig. 3.

FIG. 9.5. Portion of a cell of *Nostoc sphaericum* treated with potassium tellurite, showing deposition of reduced tellurite (arrows) in regions co ntaining the photosynthetic thylakoids. Electron micrograph (×56,000). By co urtesy of T. Bisalputra, D. L. Brown and T. E. Weier (1969). *J. ultrastruct. Res.* **27**, 182, Fig. 10.

uptake and oxygen consumption were markedly resistant to high concentrations of arsenate (0·001 M) or iodoacetamide (0·005 M), the latter being known to inhibit specifically the triose phosphate dehydrogenase reaction. This suggests that the glycolytic pathway is of limited significance in the breakdown of glucose in this alga. The distribution of ^{14}C in polysaccharide, following assimilation of labelled glucose, also indicated the oxidative pentose phosphate cycle rather than the glycolytic pathway as the main route of carbohydrate metabolism. Pelroy *et al.* (1972) found that, in unicellular species in the dark, acetate conversion to cell material and carbon dioxide, presumably via the tricarboxylic acid cycle, was less than 1% of glucose utilization via the pentose phosphate pathway. However, in the light, the latter pathway may be switched off by ribulose 1,5-diphosphate generated by the Calvin cycle.

THE INTERRUPTED TRICARBOXYLIC ACID CYCLE

Studies on the Krebs' cycle (tricarboxylic acid cycle) in blue-green algae were first carried out by Allison *et al.* (1953) who investigated the pattern of ^{14}C incorporation from acetate in *Nostoc muscorum*. They found that the rate of acetate assimilation was greatest during photosynthesis, with little acetate incorporation in the absence of light and carbon dioxide. This ^{14}C uptake was not through secondary fixation of carbon dioxide produced by dissimilation of acetate, since only a negligible proportion of the photosynthetic products was derived from acetate. Acetate was incorporated primarily into lipids, glutamate and carboxylic acids, and was not incorporated into phosphorylated glycolytic intermediates (but see p. 156). The ^{14}C label in glutamate suggests that a tricarboxylic acid cycle operates in *Nostoc muscorum*. Light-stimulated assimilation of acetate and other organic compounds also occurs in obligate photo-autotrophic blue-green algae (p. 155). In these, more radioactivity is recovered in the cell material when 2-^{14}C-acetate rather than 1-^{14}C-acetate is supplied, whereas respiratory carbon dioxide is more highly labelled when 1-^{14}C-acetate is assimilated and ^{14}C-label is again found in glutamate (Pearce and Carr, 1967). These findings support the assumption that a tricarboxylic acid cycle operates in blue-green algae, although they do not rule out the possibility of an active glyoxylate shunt (see p. 173).

Hoare and Moore (1965) found in *Synechococcus* sp. (*Anacystis nidulans*) that about 60% of the label from acetate was incorporated into lipids and about 30% into proteins. Hydrolysis of the protein fraction showed that only four amino acids, i.e. glutamate, proline, arginine and leucine were labelled and, in particular, aspartate was not. Since the entry of acetate into glutamate involves the operation of the tricarboxylic acid cycle, the lack of ^{14}C label from acetate into aspartate and other amino acids of the aspartate family suggests that the flow of carbon from assimilated acetate proceeds to α-oxoglutarate and hence to glutamate, but further metabolism to succinate is blocked.

A similar pattern of acetate assimilation was found by Smith *et al.* (1967) in *Coccochloris peniocystis*, *Gloeothece linearis* and *Gloeocapsa alpicola* as well as in *Synechococcus* sp. (*Anacystis nidulans*). When ^{14}C-pyruvate was supplied, the subsequent distribution of radioactivity was restricted to valine, leucine, alanine and the glutamate family of amino acids (glutamate, proline, arginine). Again, no radioactivity was recovered in aspartate, threonine, serine and glycine, or in oxaloacetate. In extracts prepared from these algae, high concentrations of isocitrate dehydrogenase but only small concentrations of succinic and malic dehydrogenases were found and α-oxoglutarate dehydrogenase could not be detected. Smith and colleagues attribute this restricted distribution of labelling from ^{14}C-pyruvate to the lack of the α-oxoglutarate dehydrogenase required for the conversion of these substrates to oxalocacetate via α-oxoglutarate and suggested that this blockage in the tricarboxylic acid cycle was one of the principal reasons for the obligate autotrophy of blue-green algae (see p. 157). However, acetate and pyruvate assimilation showed a similar pattern in other blue-green algae (Van Baalen *et al.*, 1971), including several such as *Chlorogloea fritschii*, *Nostoc muscorum*, *Tolypothrix tenuis* and *Anabaenopsis* (Hoare *et al.*, 1969) when grown autotrophically. These algae can also grow heterotrophically in the dark. Hoare *et al.* (1967) demonstrated the presence in extracts of *Anacystis nidulans* of various enzymes involved in the incorporation of acetate into glutamate. These included acetic thiokinase, citrate synthetase, aconitate hydratase and isocitrate NADP dehydrogenase. They were unable to detect glutamate dehydrogenase but demonstrated glutamate formation by transamination from α-oxoglutarate and an appropriate L-amino acid in qualitative tests. The demonstration of transacetylase and ornithine transcarbamylase activity supported the concept of arginine synthesis from glutamate through the ornithine cycle which was earlier reported for *Nostoc muscorum* by Holm-Hansen and Brown (1963).

The presence of most of the tricarboxylic acid cycle enzymes in extracts of *Anabaena variabilis* has since been demonstrated by Pearce and Carr (1967) and Pearce *et al.* (1969). However, succinic dehydrogenase was present only in extremely low concentrations and α-oxoglutarate dehydrogenase as well as succinyl-CoA synthase could not be detected in *A. variabilis*, or in *Anacystis nidulans*. The lack of these enzymes prevents the cyclic flow of intermediates and is consistent with the incorporation of ^{14}C from acetate into a limited number of amino acids only. An alternative route for the synthesis of oxaloacetate and further to the aspartate group of amino acids is from pyruvate and phosphoenol-pyruvate.

Pearce and Carr (1967) have detected the presence of two key enzymes of the glyoxylate cycle, isocitrate lyase and malate synthetase in low concentrations in extracts of *Anacystis nidulans* and *Anabaena variabilis*. The glyoxylate shunt may permit a slow flow of carbon from isocitrate to succinate, and further towards the synthesis of tetrapyrrole compounds. The present concept of the interrupted tricarboxylic acid cycle in blue-green algae is presented in Fig. 9.6.

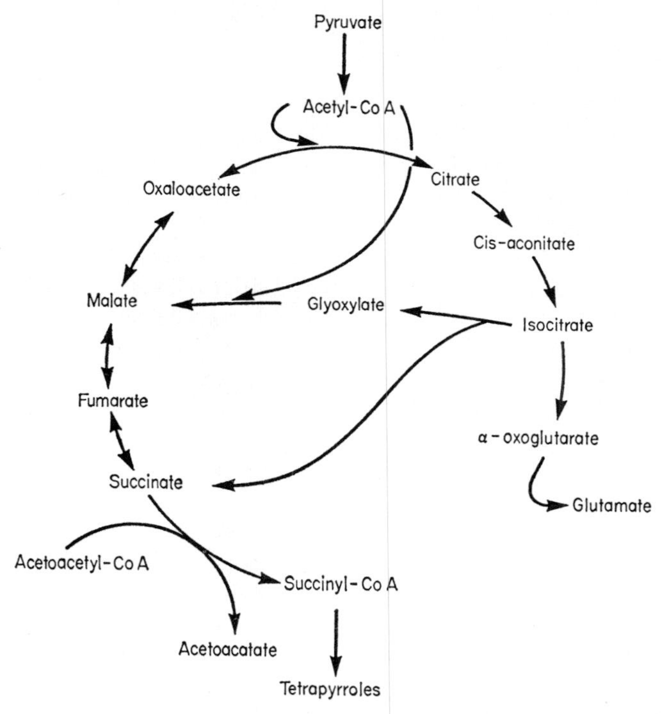

FIG. 9.6. The tricarboxylic acid cycle in blue-green algae. After A. W. Smith, J. London and R. Y. Stanier (1967). *Bacteriol.* **94**, 972, Fig. 4 and N. G. Carr, W. Hood and J. Pearce (1969) *Progress in Photosynth. Res. III* 1565, Fig. 2).

It is an interesting feature of blue-green algae that although certain species will not grow heterotrophically in the dark they will do so at very low light intensities which do not support purely autotrophic growth (Van Baalen *et al.*, 1971). This suggests perhaps that, in the dark, shortage of ATP and/or carbon limits heterotrophic growth. Furthermore, Pan (1972) has obtained results that may necessitate a reassessment of the importance of the interrupted tricarboxylic cycle in regulating obligate autotrophy in blue-green algae. He showed that *Plectonema boryanum*, which was regarded as an obligate autotroph, could grow heterotrophically in the dark on glucose if toxic products liberated by the algae during growth were removed. He thus considers that biochemical differences alone are not sufficient to prevent heterotrophic growth of *Plectonema* on glucose.

THE RESPIRATORY SYSTEM IN THE FLEXIBACTERIACEAE

Apochlorosis, the irreversible loss of photosynthetic pigments (Pringsheim, 1963) occurs in most of the eukaryotic algal groups and perhaps also in the Cyanophyceae. A group of aerobic filamentous prokaryotic organisms, comprising *inter alia* species of the genera *Vitreoscilla*, *Saprospira* and *Leucothrix*, have been considered to be apochlorotic blue-green algae (cf. Pringsheim, 1949, 1963). This assumption was based principally on the filamentous structure of these organisms and on their gliding movement, characteristic of the filamentous blue-green algae (p. 111), but is also supported by certain biochemical features such as the similarity of DNA base composition (Edelman *et al.*, 1967). Lewin and his colleagues, while recognizing these resemblances to blue-green algae, found it more convenient to incorporate these genera into the family of Flexibacteriaceae (Lewin, 1966; Soriano and Lewin, 1965; Lewin, 1969). Pierson and Castenholz (1971) have recently found two organisms which may represent a phylogenetic link between flexibacteria and blue-green algae. They are both filamentous gliding organisms forming mats in thermal streams. In culture they appear bright orange in consequence of their high carotenoid content, but the feature of particular interest is that they contain bacteriochlorophylls, in the one case *a* and in the other *c*. Under anaerobic conditions both of these organisms were shown to assimilate ^{14}C-labelled bicarbonate without evolving oxygen.

In a comparative study of the respiratory reactions of algae Webster and Hackett (1965, 1966) employed apochlorotic counterparts of pigmented algae to avoid the complication of interference by the presence of photosynthetic pigments, and in the course of these studies *Vitreoscilla* sp., *Saprospira grandis* and *Leucothrix mucor* were used to represent colourless Cyanophyta. Studies of difference spectra with whole cells and cell fractions indicated the presence of *b* and *c* type cytochromes and the absence of *a* type cytochromes in these organisms. $NADH_2$ oxidase activity could be demonstrated and was found to be associated with particulate fractions of the cells. The response of $NADH_2$ oxidase activity to inhibitors, such as cyanide, 2-heptyl-4-hydroxyl quinoline-N-oxidase (HOQNO), antimycin A and rotenone, differed from that observed in preparations from other algae and higher plants, and resembled the response shown by the $NADH_2$ oxidase obtained from certain bacteria.

Cytochrome oxidase activity was absent in these flexibacteria, and the spectral characteristics of a carbon monoxide binding pigment suggested that the terminal oxidase is an *o*-type cytochrome.

These results, on the whole, have been confirmed in a more detailed study on the respiratory mechanism of flexibacteria by Dietrich and Biggins (1968, 1971). However, in *Saprospira grandis* they detected an additional functional *a*-type cytochrome, similar to cytochromes a and a_3 found in mitochondria and in several bacteria. This cytochrome was found to be usually the most oxidized and may therefore function as a terminal oxidase.

THE ELECTRON TRANSPORT SYSTEM IN BLUE-GREEN ALGAE

The inhibition of endogenous respiration of *Anabaena*, by cyanide, azide, phenathroline and carbon monoxide (Webster and Frenkel, 1953) suggested the presence of an electron transport system incorporating heavy metal-containing enzymes (p. 165). Of the metal-containing oxidases associated with plant respiration, polyphenol oxidase was found to be absent. Evidence for ascorbic acid oxidase was equivocal but the presence of cytochrome oxidase was indicated by the effect of inhibitors and by the finding that the oxidation of hydroquinone (or of ascorbate) by fragmented cells of *Anabaena* was stimulated by cytochrome *c*.

Holton and Myers (1963, 1967a, b) isolated and characterized three water-soluble cytochromes from *Synechococcus* sp. (*Anacystis nidulans*). Cytochrome C_{554} resembled photosynthetic cytochrome *f* of higher plants. C_{552} resembled mammalian cytochrome *c* and C_{549} was similar to the *o*-type cytochrome present in *Vitreoscilla* (Webster and Hackett, 1964, 1965) and proposed as the terminal oxidase in flexibacteria.

Smith *et al.* (1967) did not detect $NADH_2$ oxidase in extracts of *Anacystis nidulans*, *Coccochloris peniocystis* and *Gloeocapsa alpicola* and considered this deficiency as the most important factor responsible for the inability to couple the oxidation of organic substrates to ATP generation, and thus the most characteristic feature of obligate autotrophs. They suggested that oxidative phosphorylation did not occur in blue-green algae although ATP might be generated by substrate-level phosphorylation in the glycolytic pathway. However, $NADH_2$ oxidase was demonstrated later in extracts from several blue-green algae (Horton, 1968; Leach and Carr, 1968) including *Anacystis nidulans* and *Anabaena variabilis* and the flexibacterium *Leucothrix* sp. (Table 9.III), although the rates of $NADH_2$ oxidase activity in blue-green algae were about 10 times lower than in *Leucothrix*, and cytochrome *c* oxidase, succinoxidase and $NADH_2$-cytochrome *c* reductase activities were not detected. Furthermore, the effect of

TABLE 9.III. $NADH_2$ oxidase activity in cell-free particles, of blue-green algae and of a flexibacterium, sedimented at 35,000 *g*. After Horton (1968)

Organism	NADH oxidase activity nmol (mg dry wt)$^{-1}$ min^{-1}
Anacystis nidulans	10
Anabaena variabilis	9
Leucothrix sp.	100

various respiratory inhibitors on $NADH_2$ oxidase activity differed from that on the respiratory enzymes of eukaryotes and resembled more the typical response of certain bacteria. Azide caused little inhibition, supporting the suggestion that the terminal oxidase in these organisms is different from that found in other plants. The complete inhibition of $NADH_2$ oxidase activity under anaerobic conditions provides good evidence that the enzyme system investigated was a true oxidase.

Leach and Carr (1968) further demonstrated, using *Anabaena variabilis*, *A. cylindrica*, *Nostoc muscorum*, *Mastigocladus laminosus* and *Fremyella diplosiphon*, that $NADPH_2$ oxidase was much more active than $NADH_2$ oxidase in these algae. They were able to couple $NADPH_2$ oxidase with oxidative phosphorylation in *Anabaena variabilis* (Leach and Carr, 1969, 1970). The initial rate of ATP formation was linear with time and corresponded to about 0·45 nmol of ATP formed min^{-1} mg cell protein^{-1}. This provides evidence that although they can do so only slowly compared with heterotrophic bacteria, blue-green algae can obtain ATP through the oxidation of organic substrates in the dark. The specific activity of $NADPH_2$ oxidase was about 4·1 nmol min^{-1} mg cell protein^{-1}. This indicates a coupling factor of 0·1 between $NADPH_2$ oxidation and ATP formation which is comparable to that found in some heterotrophic bacteria. $NADH_2$ was about 50% less effective than $NADPH_2$ in oxidative phosphorylation (Fig. 9.7). Anaerobic conditions almost completely suppressed ATP synthesis in the dark (Batterton and Van Baalen, 1968; Leach and Carr, 1970).

The low rate of oxidative phosphorylation in blue-green algae is apparently related to their low rate of respiration and cannot be significantly increased by the addition of exogenous substrates. Nevertheless, this low rate of oxidative phosphorylation may be of importance under natural conditions by providing "energy of maintenance" in the dark or at low light intensities (Leach and Carr, 1970).

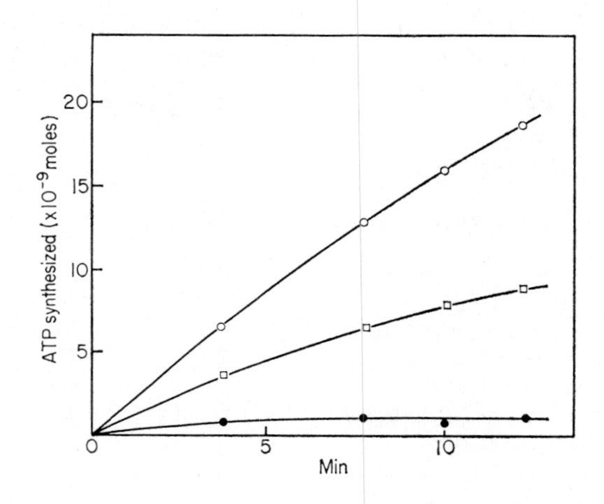

FIG. 9.7. Coupling of $NADH_2$ and $NADPH_2$ oxidation with oxidative phosphorylation in cell-free extracts of *Anabaena variabilis*. ○—○, plus $NADPH_2$; □—□, plus $NADH_2$; ●—● reduced pyridine nucleotides omitted. After C. K. Leach and N. G. Carr (1970). *J. gen. Microbiol.* **64**, 55, Fig. 7.

Metabolic control in blue-green algae

Research on blue-green algae has always been characterized by puzzling observations and unexpected findings. A recent surprise has been the discovery by Carr and his colleagues of an apparent lack of metabolic control by repression of certain enzymes.

In early studies of enzyme induction during the course of acetate metabolism, it was found (Allison *et al.*, 1953; Hoare and Moore, 1965) that, although acetate carbon was incorporated into cell material in *Synechococcus* sp. (*Anacystis nidulans*) and *Anabaena variabilis*, the presence of acetate in the medium had no effect on the activities of either the acetate activating enzymes or of the glyoxylate and tricarboxylic acid cycle enzymes (Pearce and Carr, 1967; Table 9.IV). Similarly there was no appreciable difference in the activities of enzymes of the glycolytic and pentose phosphate pathways when glucose was assimilated by *Anabaena variabilis* (Pearce and Carr, 1969; Table 9.V).

These observations suggest that blue-green algae, unlike most other prokaryotes, do not respond to the presence of certain substrates by the usual mechanism of enzymic induction and repression. That is, there is no control of enzyme synthesis by "repressor" molecules. These appear to block the synthesis of specific messenger-RNA molecules by combining with corresponding sites on the DNA molecule. The attachment of a

TABLE 9.IV. Enzyme activities in extracts of *Anabaena variabilis* grown in the presence or absence of acetate. Activities expressed as nmol mg^{-1} min^{-1}. After Carr *et al.* (1969)

Enzyme	Growth on	
	CO_2	CO_2 + acetate
Acetate kinase	4·1	4·5
Phosphotransacetylase	1·4	1·7
Isocitrate lyase	0·37	0·38
Malate synthetase	0·80	0·82
Isocitrate dehydrogenase	4·7	5·0
Citrate synthetase	5·7	6·1

TABLE 9.V. Enzyme activities in extracts of *Anabaena variabilis* grown in the presence or absence of glucose. Activities expressed as nmol mg^{-1} min^{-1}. After Carr *et al.* (1969)

Enzyme	Growth on	
	CO_2	CO_2 + glucose
Hexokinase	1·2	1·2
Phosphofructokinase	8·1	7·8
Fructose diphosphate aldolase	3·6	3·6
Glucose-6-phosphate dehydrogenase	8·7	7·8
6-phosphogluconate dehydrogenase	12·0	12·0

specific inducer to the repressor molecule is thought to inactivate the repressor and permit enzyme synthesis. The activity of acetolactate synthetase involved in the synthesis of valine from pyruvate in *Anabaena variabilis*, for example, remained the same whether the alga was grown autotrophically or in the presence of L-valine, L-leucine and L-isoleucine. This indicates the lack of enzyme repression and of feed-back inhibition. Similarly, the alga failed to repress threonine deaminase activity when grown in the presence of isoleucine or valine, leucine or isoleucine. However, the enzyme was competitively inhibited by its end product, isoleucine (Hood and Carr, 1968). Five of the enzymes concerned with arginine biosynthesis and two enzymes involved in the breakdown of arginine were neither repressed nor induced in *Anabaena variabilis*, following the addition of arginine to the culture medium (Hood and Carr,

1971). However N-acetylglutamate phosphokinase, which catalyzes the conversion of N-acetyl glutamate to N-acetyl ornithine, was inhibited "allosterically" in several blue-green algae in the presence of the end product, arginine (Hoare and Hoare, 1966). Such inhibition may be effected by the reversible combination of the end product with a site other than the active site of the enzyme, thereby transforming the enzyme and preventing it from combining with the substrate. A further example of an apparent end product inhibition is that of phosphofructokinase in *Anabaena variabilis* (Carr *et al.*, 1969). The enzyme showed maximum activity in the presence of 1 mM ATP and only 10% of this in the presence of 6 mM ATP. The inhibition was complete in the presence of AMP.

A special case of enzyme regulation occurs when a single enzyme is involved in two different pathways, as with glyceraldehyde-3-phosphate dehydrogenase (p. 169), which may function in both glycolysis and the pentose phosphate pathway in *A. variabilis* according to its association with the corresponding coenzyme, $NADH_2$ or $NADPH_2$. Hood and Carr (1967) have suggested that the existence of a common active site on the enzyme molecule may prevent the binding of both coenzymes simultaneously and thereby provide metabolic control.

3-deoxy-D-arabinoheptulosonic acid-7-phosphate-synthetase (DAHP) which catalyzes the first reaction in the multi-branched pathway of aromatic acid synthesis, has been shown by Weber and Böck (1968) to be controlled in *Synechococcus* sp. (*Anacystis nidulans*) and *Anabaena variabilis* through an allosteric end product inhibition of tyrosine and phenylalanine.

It appears that the failure of blue-green algae to increase their growth rate in the presence of substrates which can actually be assimilated is, at least partly, due to the limited response of their mechanisms for enzyme biosynthesis. This and the presence of an interrupted tricarboxylic acid cycle seem sufficient explanation for their limited ability to utilize organic substrates for growth. However, absence of mechanisms for enzyme repression and de-repression is not invariable, indeed induction of some enzymes, such as nitrate- and nitrite-reductase in *Anabaena cylindrica* has been demonstrated (Ohmori and Hattori, 1970).

Chapter 10

Nitrogen metabolism

The nitrogen content of healthy blue-green algae is usually within the range 4–9% on a dry weight basis. This varies depending on the stage of growth and while exponentially growing cells may have nitrogen contents around 8%, this may fall to as low as 3% as the algae age and accumulate lipid and polysaccharide. Nitrogen-limited cultures grown under a high light intensity usually show a marked decrease in nitrogen content with time and develop "nitrogen chlorosis" (see also p. 149). This is a yellowing of the culture as the nitrogen-containing pigments, *c*-phycocyanin (see Allen and Smith, 1969; Stewart and Lex, 1970; Neilson *et al.*, 1971; Van Gorkom and Donze, 1971; p. 181; Fig. 10.1) and chlorophyll *a*, are utilized as nitrogen sources and the non-nitrogen containing carotenoid pigments persist. On the other hand at low light intensities and with ample supplies of nitrogen, cultures have a high nitrogen content, are deep blue-green in colour and develop abundant structured granules (see p. 64) which serve as nitrogen resources.

The morphological form of the algae determines in part the percentage nitrogen content. Under optimum growth conditions it is high, for example, in algae such as *Oscillatoria* and *Synechococcus* which have thin sheaths and lower in algae such as *Gloeocapsa* or *Tolypothrix* which have thick sheaths. Similarly, forms without long terminal hairs usually have a higher gross nitrogen content than do forms with hairs, the protoplasmic contents of which tend to be moribund. The variation which may occur depending on the abundance of different cell types in the culture is exemplified by the studies of Fay (1969b) who showed, using three-week old *Anabaena cylindrica* cultures, that the percentage nitrogen content of vegetative cells, heterocysts and spores was, respectively, 7·1, 2·5 and 4·8 of the dry weight.

Sources of nitrogen for growth

Blue-green algae can readily utilize inorganic nitrogen compounds such as nitrate, nitrite and ammonium salts. Some, but not all species, assimilate

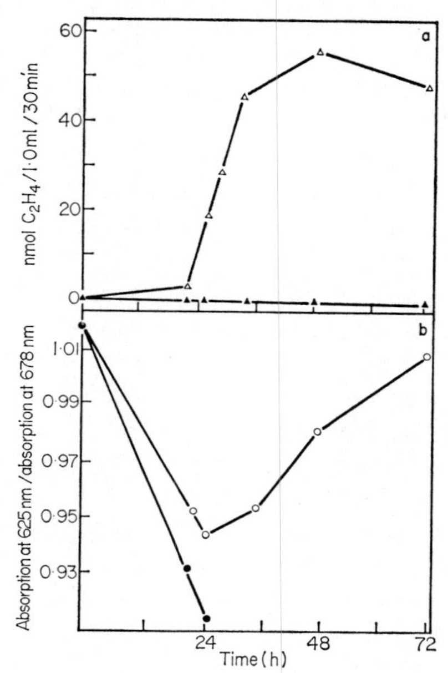

FIG. 10.1. (*a*) Acetylene reduction by $N_2/A/CO_2$-(△—△) and air-(▲—▲) bubbled cultures of *Plectonema boryanum* 594 in medium free of combined nitrogen; (b) the variation in the phycocyanin/chlorophyll *a* ratio in the $N_2/A/CO_2$-(○—○) and the air-(●—●) cultures. After W. D. P. Stewart and M. Lex (1970), *Arch. Mikrobiol.* **73**, 250–260.

elemental nitrogen (N_2) from the atmosphere, and certain species can use organic nitrogen compounds.

INORGANIC NITROGEN ASSIMILATION

Using carefully controlled cultures and optimal nitrogen levels it is possible to obtain equally good growth of blue-green algae on nitrite, ammonium-nitrogen, nitrate and also on elemental nitrogen if the algae fix nitrogen (Allen and Arnon, 1955a; Singh and Srivastava, 1968). These are the most usual findings but other workers such as Kratz and Myers (1955a) have reported lower relative growth rates when the alga depends on elemental nitrogen as its source of nitrogen than when nitrate is available. This is probably due to the conditions for growth on elemental nitrogen being sub-optimum.

Ammonium-nitrogen, although energetically the more favourable nitrogen source, often supports poorer growth than does nitrate supplied at comparable levels (Singh and Srivastava, 1968) and may cause cell lysis, as Pintner and Provasoli (1958) found with *Phormidium persicinum* and Stewart (1964b) found with *Calothrix scopulorum*. These effects are caused in part at least, by a rapid selective accumulation of ammonia within the cells and concomitant drop in the pH of the external environment and are most noticeable when high ammonium-nitrogen concentrations are supplied. For example, Singh and Srivastava (1968) noted, using *Anabaena doliolum*, that concentrations above 0·4 M were toxic, although good growth occurred at lower concentrations. Toxicity also varies with the environmental conditions, as Stewart (1964b) noted with *Calothrix scopulorum*. He found that ammonium-nitrogen concentrations which were toxic at pH 8·4 allowed growth at pH 7·4; this can be accounted for in terms of the greater concentration of undissociated ammonium hydroxide entering the cells at the higher pH.

Nitrate-nitrogen is the preferred nitrogen source in most culture media. It can be supplied at high concentration (100 mg nitrate-nitrogen l^{-1}) without detriment to the algae and there are no adverse downward pH drifts of the culture medium during growth which are characteristic when ammonium salts are used as a nitrogen source. Nitrate-nitrogen is assimilated rapidly, although in cells grown previously on elemental nitrogen or on ammonium nitrogen, there may be a short lag (Allen, 1956) before it can be used because the enzymes which reduce nitrate to ammonia are inducible and not constitutive (Hattori, 1962b; Carr, 1967). The similar distribution of the ^{15}N label in algae supplied with ^{15}N-labelled nitrate and ^{15}N-labelled ammonia (Magee and Burris, 1954) provides support for the idea that nitrate is reduced to ammonia before its entry into general nitrogen metabolism.

Nitrite, which is reduced quantitatively to ammonia in *Anabaena cylindrica*, is as good a source of nitrogen for the growth of blue-green algae as is nitrate. The optimum nitrite concentration for growth of *Anabaena doliolum* is 0·02 M with decreased growth occurring at 0·004 M.

Several workers have studied nitrogen uptake by blue-green algae when different sources are available. Kratz and Myers (1955a), who grew *Synechococcus* sp. (*Anacystis nidulans*) on ammonium nitrate, observed a drop in the pH of the medium during growth which indicated preferential uptake of ammonia, while Allen (1956) found that high levels of nitrate nitrogen had no effect on nitrogen fixation, although ammonium-nitrogen inhibited the process. Stewart (1964a), using ^{15}N as tracer, showed in short-term experiments that nitrogen fixation could still occur in the presence of 50 mg l^{-1} of ammonium-nitrogen and Dugdale and Dugdale (1965) and Goering *et al.* (1966) using ^{15}N and field populations found that nitrate, ammonia and elemental nitrogen were assimilated concurrently. Thus, nitrate-nitrogen, ammonium-nitrogen and elemental nitrogen can be assimilated concurrently but whether the algae do so or not depends on various factors such as the levels and types of combined nitrogen supplied, the length of the treatment with combined nitrogen etc. This aspect is considered again later (p. 196).

NITRATE REDUCTION

Hattori (1962a) demonstrated that, as in higher plants, the sequence of nitrate reduction in blue-green algae is as follows:

$$NO_3^- \longrightarrow NO_2^- \longrightarrow NH_2OH \longrightarrow NH_3$$

He also found that under aerobic conditions the amount of ammonia recovered was about one third less than predicted, suggesting that some of the ammonia entered organic combination. An elegant technique, which confirms that the reduction of nitrate to ammonia is via nitrite, has been used by Stevens and Van Baalen (1970). They obtained mutants of *Agmenellum quadruplicatum* and found that those which were deficient in nitrate reductase grew well only on nitrite and that nitrite reductase deficient mutants grew well only on ammonium–nitrogen. Such mutants should prove particularly useful in further studies on nitrogen reduction by algae.

Nitrate reductase, which reduces nitrate to nitrite, is a molybdenum-containing protein complex. The enzyme in *Anabaena* is induced by nitrate but not by elemental nitrogen (Hattori, 1962b). Hattori and Myers (1967) obtained it in a particulate fraction which could be centrifuged down at 10,000 *g* for 20 min after sonication and removal of unbroken cells and cell debris. It could also be obtained in particulate form by acetone treatment of the cells. Ferredoxin, reduced either by photosystem I or by $NADPH_2$, can act as a source of electrons for nitrate reductase obtained by sonication, but $NADPH_2$ cannot act directly, and $NADH_2$ acts only in the acetone-dried preparations. More recently Hattori (1970) has been able to solubilize over 90% of the *Anabaena* enzyme by taking acetone treated preparations and sonicating these in the presence of the detergent, Triton-X 100 (1%). The solubilized enzyme can then be purified further by treatments with acetone and calcium phosphate gel. The nitrate reductase preparations so obtained are not reduced by dithionite-reduced ferredoxin, but dithionite-reduced methyl viologen is active. FAD and FMN both reduce nitrate reductase when an $NADPH_2$-regenerating system is present.

Anabaena nitrite reductase is induced by nitrite but not directly by nitrate. Unlike nitrate reductase, it is solubilized readily when the algal cells are sonicated and remains in the supernatant after centrifugation at 104,000 *g* for 5 min (Hattori and Myers, 1966). It shows low rates of activity without additives but $NADH_2$ (1×10^{-3} M) and reduced methylene blue (4×10^{-4} M) increase activity by 50% and $NADPH_2$ (1×10^{-3} M) gives a 300% increase in activity. Ferredoxin mediates in the transfer of electrons from $NADPH_2$ to nitrite and artificial electron donors such as methylviologen, benzylviologen and diquat can also be used. Hattori and Uesugi (1968a, b) have partially purified nitrite reductase by acetone precipitation and chromatography on diethylaminoethyl (DEAE) cellulose. The enzyme has an optimum pH for activity near 7·4, it can be stored at −15°C for one month without loss of activity and has a molecular weight near 68,000. It can use photochemically reduced ferredoxin, or $NADPH_2$ in the presence of diaphorase, as electron donors and Hattori and Uesugi have proposed the following scheme:

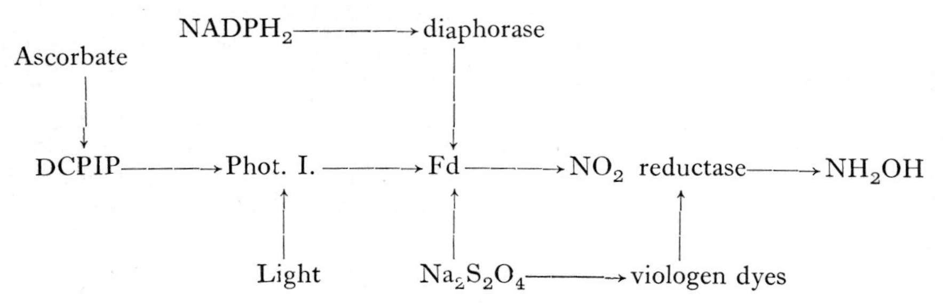

One mole of nitrite is reduced per mole of ammonia formed and the Michaelis constant for nitrite is 5×10^{-5} M.

Hydroxylamine reductase which reduces hydroxylamine to ammonia in the presence of Mn^{2+} and $NADH_2$ can be separated from nitrite reductase on DEAE cellulose (Hattori and Uesugi, 1968a). The ratio of hydroxylamine reduced to ammonia produced is 1 : 1 in intact cells (Hattori, 1962a).

The inter-relationships between nitrate reduction and cell metabolism in blue-green algae are not well understood. In higher plants and in eukaryotic algae there are distinct nitrate reductases in the chloroplasts and in the cytoplasm (see Grant, 1968) but whether there are different kinds of nitrate reductases in the cells of blue-green algae is not known. The significance of the fact that nitrate reductase is tightly bound to the photosynthetic lamellae, whereas nitrite reductase is not, is not known. If chloroplasts have been derived from unicellular blue-green algae, as has been suggested (see p. 371), it may be that cyanophycean nitrate reductase will closely resemble that found in the chloroplasts of eukaryotes.

Nitrate reduction in blue-green algae is stimulated in the light. The question is whether there is a direct photochemical reduction of nitrate *in vivo* and/or whether photosynthesis acts indirectly by supplying substrates which are then used up in respiration, which in turn supplies the necessary energy and reductant. Hattori (1962a) who studied light-induced reduction of nitrate, nitrite and hydroxylamine found that in *Anabaena cylindrica* in the light under anaerobic conditions extra oxygen was produced and that the stoichiometry of ammonia production was such as to suggest that direct photoreduction of these substrates may occur (Fig. 8.5). However, Grant (1968) has shown with green algae that such a stoichiometry is not constant and suggests that the evolution of extra oxygen in the light depends on the relative activities of the carbon dioxide fixing system and the photosynthetic electron transport system. Lex *et al.* (1972) have shown with *Anabaena cylindrica* that gross oxygen evolution may vary markedly depending on the rates of photorespiration (see p. 165).

Studies using inhibitors have provided only limited information on the sources of reductant for nitrate reductase but CMU, which specifically inhibits photosystem II (see p. 148) rapidly inhibits nitrate reduction, suggesting that photochemical reduction may occur. It is possible using cell-free extracts to reduce nitrate and nitrite photochemically when ferredoxin acts as a mediator in the system (Hattori and Uesugi, 1968a, b). As mentioned earlier, it has also been shown *in vitro* that a dark reduction of nitrite can occur when $NADPH_2$, $NADPH_2$-diaphorase and ferredoxin are supplied and that ferredoxin can be replaced in these studies by methylviologen, benzylviologen and diquat but FAD and FMN do not act as electron donors. Nitrate is also reduced in the dark with $NADPH_2$ and ferredoxin, and with $NADH_2$ (Hattori and Myers, 1967). It thus appears from *in vitro* studies that nitrate may be reduced to ammonia either by direct photoreduction or by reductant generated in the dark and although one must be cautious in applying *in vitro* findings to *in vivo* situations, it seems likely that both dark and light-generated reductant can supply reducing power in intact algae and that the system which they use at a particular time depends on the environmental conditions.

Nitrogen fixation

As we saw in the introductory chapter there was uncertainty for 40 years as to whether the suggestion, first made in 1889, that blue-green algae are able to assimilate elemental nitrogen was correct. Now there is unequivocal evidence that many species do fix nitrogen.

The early controversy arose partly because not all blue-green algae fix elemental nitrogen and because the methods used to demonstrate nitrogen fixation were open to error. It is now accepted that if an alga possesses the nitrogen-fixing enzyme, nitrogenase, in an active form, it should be possible to demonstrate that the alga can: (i) increase in total combined nitrogen when grown in medium free of combined nitrogen, (ii) incorporate elemental nitrogen enriched with the tracer ^{15}N into algal protein and (iii) reduce acetylene to ethylene.

Increases in total nitrogen are measured reliably by Kjeldahl analysis (Burris and Wilson, 1957; Bremner *et al.*, 1965) so long as adequate controls are included in the experiment. The method used by Fogg (1942) is usually followed in studies with blue-green algae. The alga is inoculated into culture medium free from combined nitrogen and bubbled with air from which traces of oxides of nitrogen and of ammonia have been removed by first passing the air through sodium hydroxide and then through acid. The combined nitrogen content of the cultures is measured at the start and at the end of the experiment and increases not found in uninoculated control flasks, or in control flasks inoculated with dead algae, are considered as due to fixation of elemental nitrogen by the algae.

^{15}N-labelled gaseous nitrogen was first used as a tracer with blue-green algae by Magee and Burris (1954) who showed its incorporation into cell material of pure cultures of *Nostoc muscorum* and an *Anabaena* sp. and of unialgal cultures of *Gloeotrichia*. The method, which is much more sensitive than the Kjeldahl method, depends on the fact that nitrogen-fixing plants do not discriminate between ^{14}N and ^{15}N. The technique is to grow the alga in medium free of combined nitrogen and then to expose it to a gas phase containing elemental nitrogen enriched with ^{15}N. After several hours, the cells are harvested and their nitrogen analysed for ^{15}N enrichment by mass spectrometry. An enrichment of 0·015 atom % over the ^{15}N content of unexposed controls, which usually contain background levels of 0·360–0·370 atom % ^{15}N, is considered as the minimum acceptable evidence of nitrogen fixation (see Stewart, 1966). In fact nitrogen-fixing algae often show considerably higher ^{15}N enrichments if they are exposed to nitrogen containing 95 atom % ^{15}N for more than a few minutes.

The acetylene reduction technique (Schöllhorn and Burris, 1966; Dilworth, 1966) was used first in studies on blue-green algae by Stewart *et al.* (1967, 1968), who demonstrated acetylene reduction by pure cultures of algae, by natural populations of freshwater plankton algae, soil algae and by cycad root nodules containing blue-green algae. The method depends on the fact that the enzyme complex, nitrogenase, not only reduces elemental nitrogen to ammonia, but also reduces a variety of other substrates as well and these are listed in Table 10.I. Substrates such as azide and cyanide affect cell metabolism generally and thus they are not satisfactory substrates for assays using whole cells. Acetylene on the other hand has little effect on general metabolism, at least in short term studies, and both it and its product ethylene are easily detected by gas chromatography. The technique, therefore, is to expose the test samples to a gas phase containing acetylene and after a short incubation period

TABLE 10.I. Substrates reduced by the nitrogen-fixing enzyme complex, nitrogenase

Nitrogen gas	N_2	$\rightarrow 2NH_3$
Nitrous oxide	N_2O	$\rightarrow N_2 + H_2O$
Azide	N_3^-	$\rightarrow N_2 + NH_3$
Acetylene	C_2H_2	$\rightarrow C_2H_4$
Cyanide	HCN	$\rightarrow CH_4 + NH_3 + CH_3NH_2$[a]
Isocyanide	CH_3NC	$\rightarrow CH_4 + C_2H_4$[a] $+ C_2H_6$[a] $+ CH_3NH_2$[a]

[a] Denotes traces only formed.

(30 min is usually sufficient) a sample of the gas phase is taken and analysed for ethylene production by gas chromatography. Compared with the ^{15}N method the technique has advantages of sensitivity, simplicity, cheapness and ease of manipulation. It is however an indirect method and it should be remembered that insufficient data are available yet to say categorically that the rate of acetylene reduction truly reflects the rate of nitrogen fixation under all conditions. From the few available data it appears that with intact filaments of blue-green algae (Stewart *et al.*, 1968) the rate of acetylene reduction is about 3 times greater than the rate of nitrogen reduction. This value agrees with data obtained by other workers with other *in vivo* nitrogen-fixing systems (Klucas, 1967; Schöllhorn and Burris, 1967; Hardy *et al.*, 1968; Bergersen, 1970). Details of the method as applied to blue-green algae are given by Stewart *et al.* (1967) and Stewart (1971a).

The method, although simple, is not foolproof and it is worthwhile mentioning some of the problems. One is to ensure that the ethylene produced has originated by a biological reduction of acetylene. We have found (Stewart, 1971a) that in some instances, particularly with field samples, the use of acids to terminate the experiments is not satisfactory because of the chemical production of ethylene. A particularly useful technique to overcome this has been employed by Schell and Alexander (1970) in which, at the end of the exposure period, a gas sample is sealed into a previously evacuated container and is stored there until it can be analysed. A second problem is that ethylene may dissolve in the rubber bungs, or serum stoppers, syringes etc. and may be released subsequently into the gas phase. The third problem is the question of establishing the ratio of nitrogen reduction: acetylene reduction and to ensure that in doing so facts such as the different solubilities of nitrogen, acetylene and ethylene are taken into account. In making these points we are not decrying the method. It is an excellent method which has revolutionized the measurement of nitrogenase activity but its simplicity does not allow us to dispense with caution, care and adequate controls in our experiments.

THE OCCURRENCE OF A CAPACITY TO ASSIMILATE ELEMENTAL NITROGEN

There are three main groups of nitrogen-fixing blue-green algae. First, there are heterocystous species. These fix nitrogen aerobically and have been investigated most extensively (see Fogg and Wolfe, 1954; LaPorte and Pourroit, 1967; Stewart, 1967c, 1969, 1970b). The second group are the aerobic nitrogen-fixing unicellular forms. To date, positive data are available only for two strains of *Gloeocapsa* (Wyatt and Silvey, 1969; Rippka *et al.*, 1971) and the capacity to fix nitrogen appears to be rare in unicellular forms. The third group comprises various non-heterocystous filamentous forms which do not fix nitrogen in air, but do so under micro-aerophilic conditions (Stewart and Lex, 1970; Stewart, 1971b, 1972b; Kenyon *et al.*, 1972). The discovery by Wyatt and Silvey (1969) of nitrogen fixation by *Gloeocapsa* was the first demonstration of nitrogen fixation in a non-heterocystous alga and this, together with the initial report by Stewart and Lex (1970) of nitrogen fixation by non-heterocystous algae under micro-aerophilic conditions only, opened up a new chapter in the study of nitrogen fixation by blue-green algae.

The species of blue-green algae which are known to fix nitrogen in pure cultures are listed in Table 10.II but these represent only a small proportion

TABLE 10.II. Blue-green algae for which there is evidence of nitrogenase activity in pure cultures incubated under aerobic and microaerophilic conditions

Type	Species or strain	Aerobic	Microaerophilic	References
Unicellular	*Gloeocapsa* 795	+	+	Wyatt and Silvey (1969; Stewart (1971a)
	Gloeocapsa 6501	+	+	Rippka *et al.* (1971)
Non-heterocystous filamentous	*Lyngbya* 6409	−	+	Kenyon *et al.* (1972)
	Oscillatoria 6407	−	+	Kenyon *et al.* (1972)
	Oscillatoria 6412	−	+	Kenyon *et al.* (1972)
	Oscillatoria 6506	−	+	Kenyon *et al.* (1972)
	Oscillatoria 6602	−	+	Kenyon *et al.* (1972)
	Plectonema boryanum 594	−	+	Stewart and Lex (1970)
	Plectonema 6306	−	+	Kenyon *et al.* (1972)
	Plectonema 6402	−	+	Kenyon *et al.* (1972)
Heterocystous filamentous	*Anabaena ambigua*	+	+	Singh (1942)
	A. azollae	+	+	Venkataraman (1962)
	A. cycadae	+	+	Winter (1935)
	A. cylindrica	+	+	Bortels (1940); Fogg (1942)
	A. fertilissima	+	+	Singh (1942)

TABLE 10.II.—*continued*

Type	Species or strain	Aero-bic	Micro-aero-philic	References
Heterocystous filamentous —*continued*	*A. flos-aquae*	+	+	Davis *et al.* (1966)
	A. gelatinosa	+	+	De (1939)
	A. humicola	+	+	Bortels (1940)
	A. levanderi	+	+	Cameron and Fuller (1960)
	A. naviculoides	+	+	De (1939)
	A. variabilis	+	+	Drewes (1928); Bortels (1940)
	A. sp.	+	+	Drewes (1928)
	Anabaenopsis circularis	+	+	Watanabe (1951, 1959)
	A. sp.	+	+	Watanabe (1951, 1959)
	Aulosira fertilissima	+	+	Singh (1942)
	Calothrix brevissima	+	+	Watanabe (1951, 1959)
	C. elenkinii	+	+	Taha (1963, 1964)
	C. parietina	+	+	Williams and Burris (1952)
	C. scopulorum	+	+	Stewart (1962)
	Chlorogloea fritschii	+	+	Fay and Fogg (1962)
	Cylindrospermum gorakhporense	+	+	Singh (1942)
	C. licheniforme	+	+	Bortels (1940)
	C. majus	+	+	Bortels (1940)
	C. sphaerica	+	+	Venkataraman (1961a)
	Fischerella major	+	+	Pankow (1964)
	F. muscicola	+	+	Pankow (1964); Mitra (1961)
	Hapalosiphon fontinalis	+	+	Taha (1963, 1964)
	Mastigocladus laminosus	+	+	Fogg (1951b)
	Nostoc calcicola	+	+	Bortels (1940)
	N. commune	+	+	Herisset (1952)
	N. cycadae	+	+	Watanabe and Kiyohara (1963)
	N. entophytum	+	+	Stewart (1962)
	N. muscorum	+	+	Burris *et al.* (1943)
	N. paludosum	+	+	Bortels (1940)
	N. punctiforme	+	+	Drewes (1928)
	N. sphaericum	+	+	Pankow and Martens (1964)
	Scytonema arcangelii	+	+	Cameron and Fuller (1960)
	S. hofmanni	+	+	Cameron and Fuller (1960)
	Stigonema dendroideum	+	+	Vankataraman (1961b)
	Tolypothrix tenuis	+	+	Watanabe (1951, 1959)
	Westiellopsis prolifica	+	+	Pattnaik (1966)

Not all species of heterocystous algae listed here have been tested for nitrogen fixation under micro-aerophilic conditions but there is no reason to suppose that they do not do so.

of the total species. In addition to the algae presented in this list there is evidence of nitrogen fixation by some other algae as well. The most notable of these are species of the genus *Trichodesmium*, a non-heterocystous planktonic form, natural populations of which fix elemental nitrogen as rapidly as they assimilate nitrogen from nitrate (see p. 341).

(a) *Heterocystous algae*

These nitrogen-fixing blue-green algae have been studied most extensively and are the most important contributors of combined nitrogen in aerobic environments (see Chapter 16). All species tested have been shown to fix nitrogen although some, as might be expected, have non-nitrogen-fixing strains (Drewes, 1928; Singh and Tiwari, 1969).

The direct correlation in aerobic tests between the presence of heterocysts in blue-green algae and their capacity to fix nitrogen was noted by Fogg (1949) and in 1968 Stewart, Fitzgerald and Burris published data using the newly discovered acetylene reduction technique which showed a close correlation during the growth of a culture of *Nostoc muscorum* between nitrogenase activity and heterocyst production (Fig. 10.2). These findings

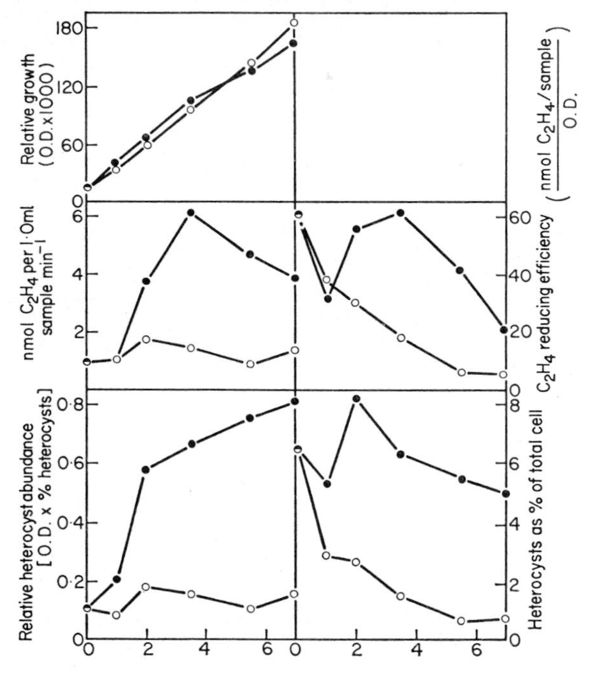

FIG. 10.2. Acetylene reduction and heterocyst production by *Nostoc muscorum* grown for seven days in batch culture on elemental nitrogen (●—●) and in the presence of 100 mg/l of nitrate-nitrogen (○–○). After Stewart *et al.* (1968) *Arch. Mikrobiol.* **62**, 336–348.

have been confirmed since (Ogawa and Carr, 1969; Neilson *et al.*, 1971; Thomas *et al.*, 1972) and Fay *et al.* (1968) on the basis of these data and various other results obtained by several laboratories, put forward in detail the hypothesis that heterocysts may be the sites of nitrogen fixation in blue-green algae. They pointed out, in particular, that various morphological features, such as the positioning of spores and the development of polarity, are readily explicable if other cells are assumed to be dependent on heterocysts for their supply of combined nitrogen (see p. 24). Furthermore, heterocysts show reducing activity (see p. 231) and the fixation of nitrogen is a reductive process with the nitrogen-fixing enzyme being inactivated by oxygen in cell-free extracts (see p. 199). This is consonant with heterocysts lacking the pigments necessary for photosystem II activity and oxygen evolution (Thomas, 1970), yet showing high respiratory activity (Fay and Walsby, 1966), and absorbing oxygen (Bradley and Carr, 1971). Thus they appear to be an ideal location for the efficient functioning of an oxygen-sensitive enzyme such as nitrogenase. They also carry out photophosphorylation which may supply the necessary energy for nitrogen fixation (Scott and Fay, 1971). Lastly, they have a complex contorted lamellar system (Lang and Fay, 1971) which may protect the nitrogenase from oxygen, just as the membranous system in *Azotobacter* may do (Oppenheim and Marcus, 1970).

The first direct evidence of nitrogen fixation by heterocysts was obtained by Stewart *et al.* (1969) who demonstrated acetylene reduction by isolated heterocysts supplied with an exogenous supply of ATP and reductant (sodium dithionite). Their findings of nitrogenase in heterocysts have since been confirmed by Wolk and Wojciuch (1971a, b) who were able to show light dependent nitrogenase activity by isolated heterocysts and that an ATP-generating system and dithionite could substitute for light. They concluded that heterocysts could account for up to 30% of the nitrogenase activity of intact filaments of aerobically grown *Anabaena cylindrica*. Another line of investigation by Stewart *et al.* (1969) which provided support for the idea of nitrogenase activity in heterocysts was to treat aerobically-grown cultures with a 0·1% (w/v) solution of triphenyltetrazolium chloride (TTC). They found that the TTC was reduced first in the heterocysts, where distinct formazan crystals were formed, and that a stage could be reached when almost all the heterocysts but few of the vegetative cells showed such deposits. When such TTC-treated filaments were exposed to acetylene or ^{14}C-labelled bicarbonate, the results in Table 10.III were obtained. That is, formazan crystals in the heterocysts inhibited acetylene reduction by over 95% but ^{14}C fixation continued almost uninhibited. The most logical explanation of these findings is that in these cultures the rapid deposition of formazan crystals in the heterocysts inhibits nitrogenase activity there and that there was little, if any, nitrogenase activity in the vegetative cells of their aerobically grown *Anabaena cylindrica*.

Despite the above findings the view that heterocysts fixed nitrogen was

TABLE 10.III. C_2H_2 reduction and $^{14}CO_2$ fixation by TTC-treated *Anabaena cylindrica* and *A. flos-aquae* (after Stewart *et al.*, 1969)

	A. cylindrica		*A. flos-aquae*	
	TTC-treated	control	TTC-treated	control
Percentage of heterocysts with formazan crystals	96	0	96	0
Percentage of vegetative cells with formazan crystals	17	0	15	0
nmol C_2H_4/sample/min	0·022	0·088	0·042	0·190
^{14}C uptake from $NaH^{14}CO_3$ (c.p.s.)	1770	1880	513	547
Percentage inhibition of C_2H_2 reduction in TTC-treated filaments	75		78	
Percentage inhibition of ^{14}C-fixation in TTC-treated filaments	2		6	

not accepted readily at first (Wolk, 1970; Smith and Evans, 1970). The latter group, who obtained a soluble nitrogenase in preparations which they obtained by sonication and which they considered were obtained only from ruptured vegetative cells and not heterocysts, concluded on this basis that the heterocysts did not fix nitrogen (Smith and Evans, 1971). However Fay and Lang (1971) showed subsequently, by electron microscopy, that the sonication treatment used by Smith and Evans damages heterocysts severely so that the soluble nitrogenase could equally have come from the heterocysts.

Another possible reason for the results of Smith and Evans was that their extracts were obtained from anaerobically grown cells. As the non-heterocystous alga *Plectonema boryanum* 594 (Stewart and Lex, 1970) and several other non-heterocystous algae (Stewart, 1971b; Rippka *et al.*, 1971) develop nitrogenase under micro-aerophilic conditions it may be, as Stewart (1971b, 1972b) suggested, that in aerobically grown cells there is a nitrogenase activity only or mainly in the heterocysts but that under micro-aerophilic conditions the vegetative cells also develop an active nitrogenase. Van Gorkom and Donze (1971) have obtained supporting evidence for this hypothesis using an elegant technique of assaying phycocyanin fluorescence as an indirect measure of nitrogenase activity. Phycocyanin which acts not only as a pigment in photosystem II but also as a nitrogen reserve becomes depleted rapidly under conditions of nitrogen deficiency and increases when there is ample nitrogen available. These workers showed that when nitrogen-starved cultures were incubated aerobically the re-appearance of phycocyanin occurred first in the vegetative cells adjacent to the heterocysts and that gradients of decreasing phycocyanin content occurred away from the heterocysts. Nitrogen-starved cells incubated anaerobically, on the other hand, showed no such gradients and phycocyanin developed equally in all vegetative cells. Van Gorkom and Donze (1971) also showed that under micro-aerophilic conditions the "extra" nitrogenase activity which developed was sensitive to oxygen concentrations which had no effect on nitrogenase activity by aerobic cultures—thus supporting the idea that a highly oxygen sensitive nitrogenase may be present in vegetative cells of *Anabaena*.

(b) *Filamentous non-heterocystous algae*

The ability of non-heterocystous filamentous algae to fix nitrogen was disputed until recently and the discovery that certain of them do so, but only under micro-aerophilic conditions, has considerable physiological and ecological implications. Early reports of fixation by species of *Lyngbya*, *Phormidium*, *Plectonema*, *Oscillatoria* and *Trichodesmium* were made (see Stewart, 1971b, 1973c, d for references) but were in doubt either because they were based on studies with impure cultures or else critical tests for nitrogenase activity had not been carried out. The data for the marine genus *Trichodesmium*, in particular, are intriguing although not conclusive. Dugdale *et al.* (1961) were the first to report on nitrogen fixation by natural populations which they collected from the south coast of Bermuda and subsequently fixation has been confirmed by the ^{15}N technique in samples from the Atlantic Ocean, the Indian Ocean and the Arabian Sea. The fixation rates obtained were variable but often high, and were much higher in the light than in the dark. More recently Bunt *et al.* (1970) reported on acetylene reduction by almost pure populations of *Trichodesmium* from off the coast of Florida. Evidence for fixation by pure cultures has also been put forward by Ramamurthy and Krishnamurthy (1968) but their results can be interpreted as showing inhibition of growth by high nitrate levels in their control cultures rather than nitrogen fixation in the experimental series.

The first unequivocal evidence of nitrogenase activity in a non-heterocystous filamentous alga was obtained by Stewart and Lex (1970) using *Plectonema boryanum* 594. This discovery led on from the findings by

Stewart and Pearson (1970) that in the light heterocystous blue-green algae reduced nitrogen best at p_{O_2} levels at or below 0·2 atm and that nitrogen-fixing cell-free extracts are particularly oxygen-sensitive (Fay and Cox, 1967; Haystead *et al.*, 1970; Smith and Evans, 1971) in spite of the fact that vegetative cells of blue-green algae evolve oxygen during photosynthesis. Stewart and Lex (1970) considered that perhaps all the cells of blue-green algae have a capacity to synthesize nitrogenase but that in vegetative cells, which evolve oxygen, there may be an inactivation of nitrogenase activity, or even of nitrogenase synthesis. They therefore chose a strain of *Plectonema*, an alga which has no heterocysts but which, unlike other members of the Oscillatoriaceae, shows false-branching, a characteristic of some heterocystous algae such as *Scytonema* and *Tolypothrix*. They incubated *Plectonema* under a gas phase of nitrogen and carbon dioxide (99·96/0·04 per cent, v/v) and found that under these micro-aerophilic conditions the algae reduced acetylene to ethylene (Fig. 10.1), assimilated $^{15}N_2$, and grew readily in medium free of combined nitrogen. They found further that *in vivo* the enzyme was particularly sensitive to oxygen but that the inhibition was reversible in short term studies (Stewart, 1971b). Since then they have obtained evidence of nitrogenase activity under micro-aerophilic conditions in pure cultures of other *Plectonema* strains and in some unialgal cultures of species of *Lyngbya*, *Phormidium* and *Oscillatoria* (Stewart, 1971b, and unpublished) and Kenyon *et al.* (1972) have since obtained similar results using strains of *Plectonema* *Oscillatoria* and *Lyngbya* (see Table 10.II).

(c) *Unicellular algae*

The presence of nitrogenase in unicellular blue-green algae was first demonstrated convincingly in 1969 by Wyatt and Silvey using a strain of *Gloeocapsa*. Earlier reports of nitrogen fixation by *Gloeocapsa minor* (Odintzova, 1941), *Chroococcus rufescens* and *Aphanocapsa grevillei* (Cameron and Fuller, 1960) had been discounted because the cultures used were not pure. In *Gloeocapsa* nitrogenase activity is light-dependent and under aerobic conditions the rates are, according to Wyatt and Silvey (1969), comparable with those obtained for heterocystous algae. The protective mechanism against oxygen inactivation in *Gloeocapsa* is unknown but Wyatt and Stewart have shown that under micro-aerophilic conditions *Gloeocapsa* may reduce acetylene at a higher rate than aerobically (see Stewart, 1971b). It is possible that the thick mucilaginous sheaths in which *Gloeocapsa* cells are embedded may provide protection against oxygen in some way. It has been suggested that SH-groups in the sheaths remove excess oxygen in *Aphanizomen flos-aquae* (Sirenko *et al.*, 1968) Perhaps there may be a similar mechanism in *Gloeocapsa*, although it is difficult to see how it could operate on a continuing basis. The respiratory rate in blue-green algae (see chapter 9) does not seem to be sufficiently high to scavenge oxygen from the nitrogen-fixing sites. Kulasooriya (1971) obtained little nitrogenase activity by this alga grown in air. More recently Rippka *et al.* (1971) demonstrated acetylene reduction in a strain, *Gloeocapsa* 6501, which may be identical to that tested by Wyatt and Silvey. Activity occurred over a short period of the growth cycle only and all tests were carried out under N_2/CO_2 (99·5/0·05 v/v). They found no difference in ultrastructure in cells grown on elemental nitrogen and those grown on combined nitrogen.

As emphasized by Stewart and Lex (1970) and Stewart (1971b, 1973c), it is possible that under the appropriate experimental conditions a variety of blue-green algae hitherto unsuspected of fixing nitrogen can do so. This means that it is now necessary to re-examine all the species reported not to fix nitrogen on the basis of aerobic tests and to check further species for fixation under both aerobic and micro-aerophilic conditions. It seems, moreover, that the contribution by blue-green algae in natural ecosystems may be higher than the previous data obtained on the basis of aerobic tests suggest.

PHYSICOCHEMICAL FACTORS AFFECTING WHOLE CELL NITROGENASE ACTIVITY

Blue-green algae grow more slowly than heterotrophic nitrogen-fixing bacteria such as *Azotobacter* and *Clostridium* or photosynthetic bacteria such as *Rhodospirillum*. This is due to an inherently slower metabolism rather than to inferior growth on elemental nitrogen compared with combined nitrogen, because growth rates are often as high on elemental nitrogen as on combined nitrogen (see p. 181). Rates of acetylene reduction by pure cultures of heterocystous algae are usually within the range of 1–10 nmol ethylene (mg algal protein)$^{-1}$ min^{-1} for aerobically grown cultures. However, higher values can be obtained for filamentous algae grown under microaerophilic conditions (Stewart and Lex, 1970) and for *Gloeocapsa* (Wyatt and Silvey, 1969). The physico-chemical factors which affect growth in general (see p. 132) affect nitrogen fixation (Fogg, 1956a), and here only those which most directly affect nitrogen fixation will be considered.

(a) *Inorganic nutrients*

The nutritional requirements for the growth of blue-green algae are simple. Heterocystous algae grow on inorganic media and it is only for certain marine non-heterocystous species, including *Lyngbya majuscula*, which is thought to fix nitrogen (Van Baalen, 1962), that an organic growth factor, vitamin B_{12}, is required.

Combined nitrogen is one of the most important nutrients affecting nitrogen fixation because, when supplied at high concentration, it may inhibit nitrogenase activity. In whole cell studies ammonium-nitrogen is more effective in this respect than is nitrate–nitrogen.

The inhibitory effect of combined nitrogen on nitrogen fixation may be threefold. First, it may inhibit the synthesis of the nitrogenase enzyme; secondly, it may inhibit the activity of the existing nitrogenase, and thirdly it may cause some breakdown of existing nitrogenase. Stewart *et al.* (1968; Fig. 10.2) found, using batch cultures, that when 100 mg l^{-1} of nitrate-nitrogen was added to a culture previously grown on elemental nitrogen there was a fairly steady decrease in nitrogenase activity which appeared to be due simply to a diluting out of the enzyme with growth rather than to an inhibition or breakdown of existing nitrogenase activity.

It is more difficult to show that end-product repression of nitrogenase activity occurs using whole cells, because the addition of large amounts of combined nitrogen also inhibits other metabolic processes. For example, ammonium–nitrogen inhibits photophosphorylation (McCarty and Coleman, 1969) and may cause cell lysis at high pH levels (Stewart, 1964b). However, in cell-free extract studies Dharmawardhene and Stewart (unpublished) have failed to demonstrate end-product repression of nitrogen fixation using up to 200 mg l^{-1} of ammonium-nitrogen. It is questionable therefore whether end-product repression of nitrogen fixation occurs in blue-green algae. There is no evidence that breakdown of the enzyme in the presence of ammonium nitrogen occurs in blue-green algae although Davis *et al.* (1972) found evidence of this in *Azotobacter*.

High concentrations of combined nitrogen also inhibit heterocyst formation (see p. 236) and a broad correlation exists between heterocyst formation and nitrogen fixation (Fig. 10.2; Stewart *et al.*, 1968; Ogawa and Carr, 1969; Jewell and Kulasooryia, 1970). This correlation is modified, of course, according to the metabolic state of the algae and strict correlation between nitrogenase activity and heterocyst frequency can only be made if the metabolic state of the culture is also taken into account. Kale and Talpasayi (1969) and Wilcox (1970) consider that ammonium-nitrogen acts by suppressing the development of proheterocysts into mature heterocysts (see p. 238). Neilson *et al.* (1971) have studied the kinetics of nitrogenase activity and heterocyst numbers in *A. cylindrica*. They showed that when nitrate-grown cells are transferred to nitrate-free medium there is a lag in the development of nitrogenase activity until about 70% of the phycocyanin disappeared, a finding noted previously with *Plectonema boryanum* by Stewart and Lex (1970). Heterocyst formation and nitrogenase activity then increased concomitantly, suggesting that nitrogenase activity and heterocyst formation are controlled by the same mechanism. They also observed that there were more heterocysts and higher nitrogenase activity under argon that under elemental nitrogen and suggested that elemental nitrogen is unlikely to act as an inducer of nitrogenase activity.

Apart from combined nitrogen, other inorganic nutrients which have a direct effect on nitrogen fixation are iron and molybdenum, both of which are essential components of the enzyme nitrogenase in all micro-organisms from which it has been extracted. It is likely that the two fractions obtained by Smith *et al.* (1971) from *Anabaena* were partially purified preparations of fraction I (an iron–molybdenum-containing protein) and fraction II, (an iron-containing protein) which fix nitrogen only when combined (see p. 201). The iron–molybdenum protein has been obtained from *Azotobacter* in crystalline form by Burns *et al.* (1970) as white needle-like crystals, the bi-molybdo crystallized form having a molecular weight of 270,000–300,000 and a molybdenum: iron ratio near 1 : 20. The iron protein of *Azotobacter* has a molecular weight of 30,000–40,000 and contains two atoms of non-haem iron per mole of protein (Bulen and LeComte, 1966). Although iron and molybdenum are the main metal components of the purified nitrogenase, it has also been suggested by Kajiyama *et al.*, 1969) that zinc is present in trace amounts. The necessity of molybdenum for nitrogen fixation was first demonstrated in the blue-green algae *Anabaena* and *Nostoc* by Bortels (1940), Wolfe (1954) showed that the optimum molybdenum concentration for nitrogen fixation by *Anabaena cylindrica* was 0·2 ppm and that only half this amount was required for optimum growth on nitrate–nitrogen. No molybdenum requirement could be detected in ammonium–nitrogen grown cultures. The necessity of iron for nitrogenase activity in blue-green algae is less easily demonstrated with intact cells because it is also required as a component of cytochromes and ferredoxin. Cobalt, although not part of the nitrogenase *per se*, is required specifically for growth on elemental nitrogen (Holm-Hansen *et al.*, 1954). For example, Johnson *et al.* (1966) found that in the water fern *Azolla*, which contains *Anabaena* as endophyte, cobalt is required for growth on elemental nitrogen but not for growth on nitrate–nitrogen. In the field it is possible to obtain a marked stimulation of nitrogen fixation by adding small amounts of these elements. De and Mandel (1956), for example, showed that the addition of 0·3 kg ha^{-1} of molybdenum increased the rice yield in paddy soils by 23% and that the addition of lime, superphosphate and molybdenum together resulted in an even greater stimulation. Cobalt, similarly, has been shown in Australia to increase nitrogen fixation by legumes when supplied in trace quantities (Powrie, 1964) but its effect on nitrogen fixation in the field by blue-green algae has apparently not been investigated.

(b) *Oxygen*

Blue-green algae are unique among nitrogen-fixing micro-organisms in that they evolve oxygen during photosynthesis. Nevertheless the nitrogen-fixing enzyme *in vitro* is highly sensitive to oxygen. This very feature, the extreme oxygen sensitivity of algal nitrogenase, has profound physiological, morphological and ecological implications. First, algae such as *Anabaena cylindrica* which fix nitrogen in air must have a particularly efficient protective mechanism against inactivation not only by air but also by the oxygen evolved during photosynthesis. This is probably achieved under aerobic conditions by active nitrogenase being restricted to non-oxygen evolving heterocysts (see p. 192). Oxygen sensitivity in non-heterocystous algae such as *Plectonema boryanum* 594 means that these algae can fix nitrogen only under microaerophilic conditions. Thus the optimum p_{O_2} for an alga will be that at which sufficient oxygen is available for processes such as respiration to proceed as rapidly as possible, but yet is insufficiently high to inhibit nitrogenase activity. In the light when photosynthesis is occurring this level is usually at or below 0·2 atm for heterocystous algae and, as Fig. 10.3 shows, levels above 0·2 atm may inhibit nitrogen fixation

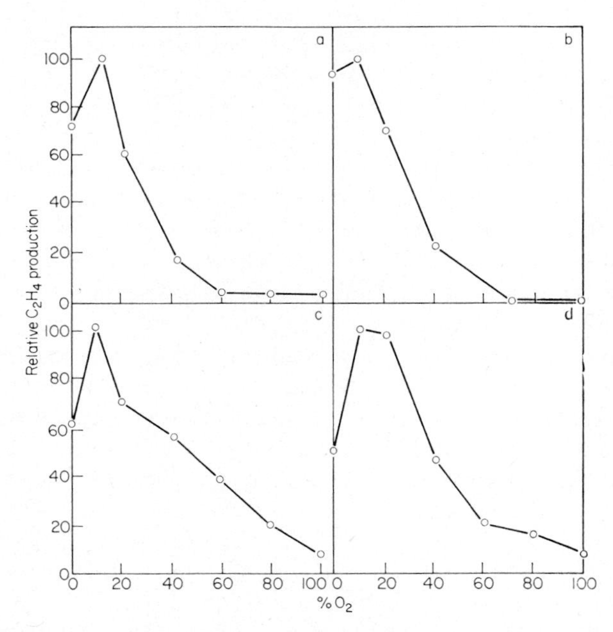

FIG 10.3. Effect of various p_{O_2} levels of C_2H_2 reduction in the light by (a) *Anabaena flos-aquae*; (b) *Gloeocapsa* 795; (c) natural populations of *Rivularia* sp. and (d) natural populations of *Gloeotrichia* sp. After Stewart (1971).

markedly. In the dark where no photosynthetic oxygen evolution is occurring the optimum p_{O_2} is usually about 20% (Davies and Stewart, unpublished). The oxygen inhibition of nitrogenase activity is irreversible in cell-free extracts but *in vivo* it is reversible. The rate of reversal is rapid, being complete within 2–6 h (Stewart and Pearson, 1970), but despite this Bone (1971b) obtained kinetic data which suggest that the recovery of nitrogenase after exposure to oxygen in whole cells is due to the synthesis of new nitrogenase and noted that inhibitors of protein synthesis stopped the re-appearance of nitrogenase.

GENERAL CHARACTERISTICS OF CELL-FREE EXTRACT NITROGENASE

Nitrogen-fixing cell-free extracts of blue-green algae were reported first by Schneider *et al.* (1960) who used *Mastigocladus laminosus*, but although extracts of low and sporadic activity were obtained subsequently (see Cox and Fay, 1967) and although active cell-free preparations of various bacteria were readily available (see Burris, 1969) it was not until 1970 that nitrogen-fixing cell-free extracts of high and consistent activity were obtained (Bothe, 1970; Haystead *et al.*, 1970; Smith and Evans, 1970). The successful preparation of these extracts was made possible through an appreciation of the fact that nitrogenase extracted from these oxygen-evolving organisms is highly oxygen sensitive (Fay and Cox, 1967) with activity of the extracts inhibited rapidly and irreversibly when exposed to air for as little as 5 min (Table 10.IV). In this respect algal

TABLE 10.IV. Oxygen inhibition of nitrogenase activity of *Anabaena cylindrica* and *Plectonema boryanum* 594 extracts. After Haystead *et al.* (1970)

Alga	Atm O_2 in gas phase	n moles C_2H_4(mg protein)$^{-1}$ min^{-1}	% inhibition
Plectonema boryanum 594	0·000	1·07	—
	0·001	0·60	44·0
	0·010	0·33	69·2
	0·050	0·15	86·0
	0·100	0·03	97·2
Anabaena cylindrica	0·000	1·32	—
	0·001	0·61	53·8
	0·010	0·26	80·3
	0·050	0·00	100·0
	0·100	0·00	100·0

Samples of 40,000 *g* × 15 min supernatant were exposed to O_2 in argon for a 5 min period, then regassed with argon and assayed for acetylene reduction over a 30 min period. Each value is the mean of triplicates.

nitrogenase resembles that of the strict anaerobic nitrogen-fixing bacterium *Clostridium* rather that of the aerobic *Azotobacter*.

The requirements for algal nitrogenase are similar to those of nitrogen-fixing bacteria in that a source of energy and a source of reducing power are required for its efficient functioning. Such requirements were demonstrated first by Stewart *et al.* (1969) who obtained nitrogenase activity in isolated heterocyst preparations only when an energy source (ATP) and reductant ($Na_2S_2O_4$) were supplied. When these additives are added to cell-free extracts together with Mg^{2+}, which is required for nitrogenase activity, there is vigorous acetylene reduction (Table 10.V) which in crude

TABLE 10.V. Requirements for acetylene reduction by 40,000 *g* for 15 min supernatant fractions of *Plectonema boryanum* 594 and of *Anabaena cylindrica*[a]

Reaction mixture	nmol C_2H_4 (mg protein)$^{-1}$ min^{-1}	
	Plectonema	*Anabaena*
Complete	3·82	0·83
−ATP generating system	0·00	0·00
−$Na_2S_2O_4$	0·00	0·15
−Mg^{2+}	0·C0	0·00

Each value is the mean of triplicates.
[a]After Haystead *et al.*, 1970.

extracts is about 10–25% of that at the whole-cell level and which continues at a linear rate for at least 30 min. The optimum ATP concentrations and $Na_2S_2O_4$ levels are near 2·0 mM (Fig. 10.4) and 4–10 μM (Fig. 10.5) respectively and the Mg^{2+} : ATP ratio at the optimum ATP concentration

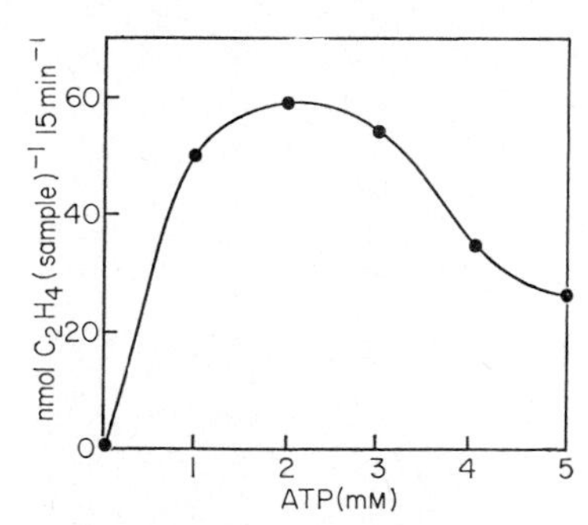

FIG. 10.4. Effect of ATP concentration on C_2H_2 reduction by a (40,000 *g* for 20 min) supernatant fraction of a cell-free extract of *Anabaena cylindrica*. $MgCl_2$ was supplied at a final concentration of 3 mM and ATP was supplied as indicated. After Haystead and Stewart (1972).

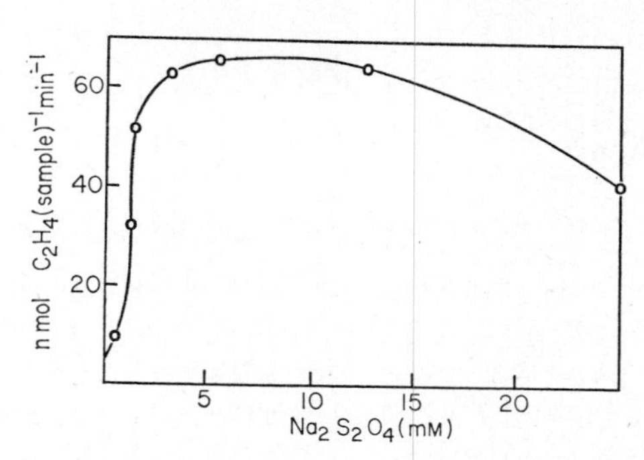

FIG. 10.5. Effect of $Na_2S_2O_4$ concentration on C_2H_2 reduction by a 40,000 *g* for 20 min supernatant fraction of a cell-free extract of *Anabaena cylindrica*. After Haystead and Stewart (1972).

is within the range 0·75–1·5 (Bothe, 1970; Haystead and Stewart, 1972). Mg^{2+} can be replaced by Co^{2+} and Mn^{2+} but not by other divalent cations, Ca^{2+}, Cu^{2+} or Zn^{2+} (Haystead and Stewart, 1972).

In general characteristics, algal nitrogenase from both heterocystous (Bothe, 1970; Smith and Evans, 1970; Haystead *et al.*, 1970) and non-heterocystous filamentous species (Haystead *et al.*, 1970) resembles that of other plant groups. These are: the Michaelis constant is within the range 0·002–0·006; the enzyme is cold labile, although not to the same extent as Dua and Burris (1963) found with *Clostridium pasteurianum*. That of *Gloeocapsa* LB795 is stable at 0°C (Gallon *et al.*, 1972). It will also reduce cyanide, it exhibits ATP-dependent hydrogen evolution, and it reduces cyanide, acetylene and elemental nitrogen. The optimum pH for activity is broad and within the range 7·0–7·5, the enzyme is a metalloprotein which contains iron and reduced thiol groups at its active site and the redox capacity of the enzyme involves a possible valency change in the iron (Haystead and Stewart, 1972).

Despite the basic similarities, algal nitrogenase has not been purified to the same extent as have those of bacteria. However, the partially purified enzyme from *Anabaena* and *Plectonema* which remains in solution after centrifugation at 40,000 *g* for 3 h (Haystead and Stewart, 1972) has been separated into two fractions by column chromatography (Smith *et al.*, 1971). Although the fractions obtained showed little activity alone, they exhibit nitrogenase activity (8 nmol C_2H_4 (mg protein)$^{-1}$ min^{-1}) when combined. Furthermore fraction 1 of *Anabaena* will cross with fraction 2 of the photosynthetic bacterium *Chloropseudomonas ethylicum* to give more active nitrogenase activity than that obtained using the two *Anabaena* fractions. The reciprocal cross was non-active. The highest activity so far reported is 50–100 nmol ethylene (mg protein)$^{-1}$ min^{-1} (Haystead and Stewart, 1972). Gallon *et al.* (1972) report that the nitrogenase of *Gloeocapsa* is sedimented down at 10,000 *g* .

NATURAL ELECTRON DONORS FOR NITROGENASE ACTIVITY IN ALGAE

One of the main aims of physiological and biochemical studies on the nitrogenase of blue-green algae is to obtain a better understanding of the *in vivo* sources of reducing power and energy for algal nitrogenase, and to see how this fits in with the various theories on the site of nitrogenase activity in these algae.

Reductant, in theory, may be supplied directly by photosynthetic electron transport processes, that is by a direct photoreduction of nitrogen, or it may be generated by processes which can occur in the dark, or it could be generated both by light and dark processes. Fogg and Than-Tun (1958) were the first to suggest that reducing power may be supplied directly by photoreduction. This suggestion was based on stoichiometric data which showed that "extra" oxygen evolved during nitrogen fixation by *Anabaena cylindrica* could be accounted for if water acted as the sole source of reductant for nitrogen fixation. Supporting results were obtained subsequently by Fay and Fogg (1962) and Cobb and Myers (1964) using whole cells.

In more recent studies evidence that light-generated reducing power can reduce *Anabaena* nitrogenase has been obtained using cell-free extracts. Bothe (1970) found that, in the light, ferredoxin from *Anabaena* or from spinach, phytoflavin from *Synechococcus* sp. (*Anacystis nidulans*) could provide reductant for nitrogenase. He thus concluded that ferredoxin was a cofactor in nitrogen fixation by blue-green algae but because he found that in intact cells nitrogenase activity was not inhibited by DCMU (see below) was not absolutely convinced that a direct photoreduction of nitrogen occurred *in vivo*. Smith and Evans (1971) and Smith *et al.* (1971) confirmed in cell-free extract studies that photosynthetic ferredoxin could act in the electron transport chain which transferred light-generated reductant to nitrogenase using 2,6-dichlorophenol-indophenol (DCPIP) and ascorbate, rather than water, as ultimate source of reductant. On the basis of these results they concluded that ferredoxin reduced by a photosynthetic electron transport chain provides the bulk of reductant for nitrogenase. However, it is unwise, as pointed out by Stewart (1973a, b) to conclude that because a direct photoreduction can be achieved *in vitro*, using various electron carriers, a similar system necessarily operates *in vivo*. Furthermore, there is a major objection to this theory if the heterocysts are the sole sites of nitrogen fixation under aerobic conditions, because heterocysts do not have a photosystem II activity, reductant must therefore come from the fixed carbon which is transferred from the vegetative cells into the heterocysts. Also, nitrogen fixation occurs in a variety of species in the dark, albeit slowly, and then direct photoreduction of nitrogen cannot be occurring.

Indeed, there is good evidence from studies on whole cells which suggests that water is not a source of reductant for nitrogen in aerobic cultures of *Anabaena cylindrica*. Cox (1966) found that the apparent stoichiometric relationship between nitrogen fixation and enhanced oxygen evolution, which suggested to early workers that direct photoreduction was occurring, could be varied by altering the environmental conditions, and Lex *et al.* (1972) observed similarly in studies on the relationship between nitrogen fixation and oxygen evolution. Secondly, there is the well-established finding (Cobb and Myers, 1962; Cox and Fay, 1969; Bothe, 1970) that DCMU, or the closely related CMU, does not inhibit nitrogen fixation immediately. This was shown particularly well by Cox (1966) who found, using nitrogen starved cells with an excess of fixed carbon, that she could eliminate the dependency of nitrogen fixation on photosynthesis (see also Fig. 10.6). She also showed that CMU(10^{-4} M) inhibited photosynthesis more than it inhibited nitrogenase activity and that cyanide and chlorpromazone, both inhibitors of respiration, inhibited nitrogen fixation more than photosynthesis. On the other hand, phenylurethane, which chiefly

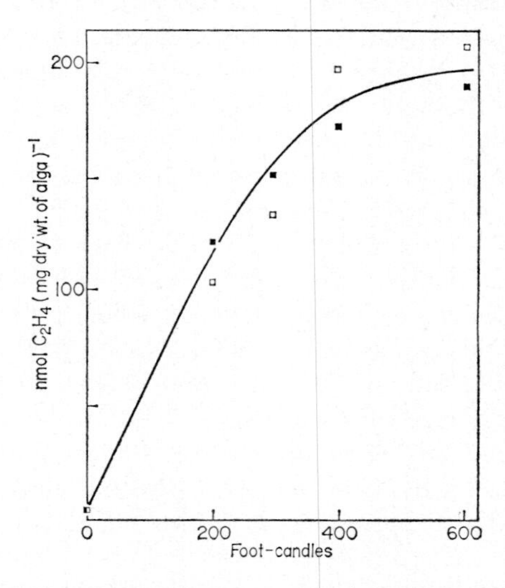

FIG. 10.6. Acetylene reduction by nitrogen-starved *Anabaena cylindrica* in relation to light intensity in the presence (■) and absence (□) of CMU(*p*-chlorophenyl-1,-1-dimethylurea.) After Cox and Fay (1969) *Proc. R. Soc.* B. **192**, 357–366.

inhibits the light reaction, inhibited photosynthesis to a greater extent than it inhibited nitrogen fixation. These findings suggest that in the light, respiration, rather than photosynthesis, supplies the reducing power.

Cell-free extract studies also support the idea of dark-generated sources of reductant for nitrogenase. Bothe (1970) obtained 50% of the rate obtained with light-generated sources of reductant using a system involving glucose-6-phosphate, glucose-6-phosphate-dehydrogenase, NADP, ferredoxin and ferredoxin-NADP reductase in the dark. Smith *et al.* (1971) also showed that pyruvate or isocitrate could provide electrons for nitrogenase via NADP and ferredoxin, but using this system, these workers obtained only 5% of the light-saturated rate of nitrogenase activity and considered that the stimulation observed by Cox and Fay (1967) was probably due to NADP in their extracts being reduced by ferredoxin. Such a view, however, ignores studies which show the lack of an inhibitory effect of DCMU on nitrogenase activity in short term studies. The inhibitory effect of DCMU in longer-term studies (2 h) can be attributed to an eventual shortage of carbon compounds (Stewart and Pearson, 1970).

Another possible source of dark-generated reductant is hydrogenase. Haystead and Stewart (1972) were able to show that in the presence of *Anabaena* ferredoxin, hydrogenase from *Clostridium kluyveri* could supply reductant to algal nitrogenase.

The source of dark-generated reductant *in vivo* in aerobic cultures is not known with certainty. However Fay and Cox (1966) and Cox and Fay (1969) have suggested that it may be pyruvate. Cox (1966) showed that pyruvate stimulates nitrogen fixation in whole filaments of *Anabaena cylindrica* both in the light and in the dark in the presence and absence of carbon dioxide. Pyruvate is also reported to stimulate nitrogen fixation in cell-free preparations (Cox and Fay, 1967) probably in a similar way to the mechanism that occurs in various nitrogen-fixing bacteria (Hardy and Burns, 1968), and there is a close correlation between pyruvate decarboxylation and nitrogen fixation. Cox and Fay (1969) obtained a stimulation of nitrogenase activity on adding pyruvate to whole cells and observed that the decarboxylation of pyruvate was coupled to nitrogen fixation. The observed ratio of 1 : 3 is the ratio which would be expected if pyruvate acted as the sole source of electrons for nitrogen fixation. Leach and Carr (1971), although they were unable to show a pyruvate dehydrogenase in a non-nitrogen-fixing strain of *Anabaena variabilis*, did show the presence of a pyruvate–ferredoxin oxidoreductase. When ferredoxin was added to the extract, pyruvate decarboxylation occurred and reduced ferredoxin and acetyl coenzyme A were produced. The presence of this enzyme in extracts of nitrogen-fixing *Anabaena* species has still to be confirmed but, if present, it could provide a pathway for reductant from pyruvate via ferredoxin.

It seems probable, therefore, from the available evidence, that dark processes are the main source of reductant to nitrogenase in aerobic heterocystous algae where heterocysts are the nitrogen-fixing sites. It is also possible, as suggested by Lex and Stewart (1973), that photosystem I which occurs in the heterocysts could also be involved in the transport of electrons from dark-generated sources to nitrogenase. It also seems likely that, if nitrogenase originated in heterotrophic prokaryotes, dark-generated reductant was probably the source to which the enzyme was originally adapted.

ENERGY SOURCES FOR NITROGENASE ACTIVITY

Cell-free extract studies have shown an absolute requirement of nitrogenase for energy supplied as ATP. In whole cells the bulk of the ATP probably becomes available from the processes of photophosphorylation, oxidative phosphorylation and perhaps from the breakdown of stored polyphosphate (see Kuhl, 1968).

There is good evidence that photophosphorylation can supply ATP and may in the light supply the bulk of that required. Evidence for this has been obtained primarily by Cox and Fay using the following lines of evidence. First, Cox and Fay (1969) showed that in the presence of CMU (10^{-4} M), which inhibits photosystem II activity, acetylene reduction, which is independent of carbon skeletons, increases on increasing the light intensity (Fig. 10.6). Secondly, Fay (1970) studied the action spectra of acetylene reduction and oxygen evolution by intact *Anabaena cylindrica* filaments and found that while maximum oxygen evolution occurred where phycocyanin absorbed most strongly, maximum rates of nitrogenase activity occurred at 675 nm where chlorophyll *a* absorbs. Activity in wavelengths absorbed mainly by phycocyanin (625 nm) was much less (Fig. 10.7). These results therefore suggest an involvement of photosystem I in nitrogenase activity and Fay (1970) concludes that this is via ATP being supplied by cyclic photophosphorylation.

When blue-green algae are fixing nitrogen heterotrophically, oxidative phosphorylation presumably is the main source of ATP, and the fact that dark respiration rates, and thus the generation of ATP, in blue-green algae are low may be a reason for the low dark rates of nitrogen fixation.

CARBON SKELETONS FOR NITROGENASE ACTIVITY

The nature and source of the carbon skeletons necessary to accept newly fixed ammonia in blue-green algae have not been studied extensively. This is probably because early studies on the aminating mechanism in

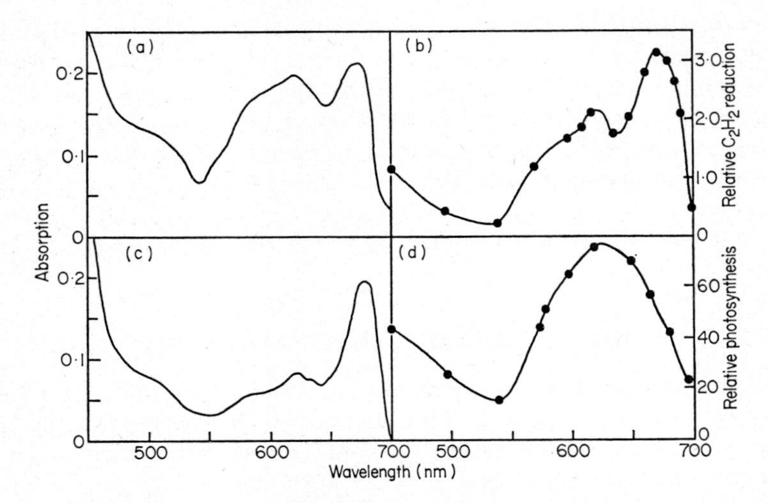

FIG. 10.7. Absorption spectra of (a) intact filaments and (c) isolated heterocysts of *Anabaena cylindrica.* (b) C_2H_2 reduction and (d) net oxygen evolution at different wavelengths by *Anabaena cylindrica.* After Fay (1970) *Biochim. biophys. Acta.* **216**, 353–356.

nitrogen-fixing micro-organisms suggested that, as in most plants assimilating ammonia, nitrate or elemental nitrogen, the usual route was via the keto acid α-oxoglutarate (see Wilson, 1958; Zelitch *et al.*, 1951). However, Hoare *et al.* (1967) failed to demonstrate the presence of glutamate dehydrogenase in various blue-green algae, although they obtained evidence of transmination with L amino acids of which leucine, isoleucine and valine were the best amino donors. Several other possible pathways have also been considered. For example, Linko *et al.* (1957) suggested that in *Nostoc muscorum* entry may occur via the ornithine cycle, a suggestion based on their finding that on exposing *Nostoc* to $^{14}CO_2$ for 5 min, over 20% of the labelling was found in citrulline which is a component of the ornithine cycle. Buchanan *et al.* (1964) suggested that another route of ammonia incorporation may be via pyruvate in the nitrogen-fixing bacterium *Chromatium*.

Recent evidence has confirmed the presence of the enzyme glutamate dehydrogenase in nitrogen-fixing blue-green algae (Pearce *et al.*, 1969; Scott and Fay, 1972; Neilsen and Doudoroff, 1973) as well as of the enzymes glutamine synthetase (Dharmawardene *et al.*, 1972, 1973) and alanine dehydrogenase (both NADP-specific (Scott and Fay, 1972) and NAD-specific (Neilsen and Doudoroff, 1973; Stewart, 1973)). All these enzymes may be involved in ammonia incorporation. In view of its low K_m for ammonia of 1·0 mM (Dharmawardene *et al.*, 1973), glutamine synthetase may be particularly beneficial to a nitrogen-fixing alga because it can scavenge low levels of ammonia within the cells. This may help to prevent the accumulation of ammonia which, in high concentration, would inhibit both nitrogenase synthesis and heterocyst production. The pathway of newly fixed nitrogen into organic combination is a subject which needs further investigation.

Organic nitrogen assimilation

Blue-green algae are particularly characteristic of waters rich in organic matter but in laboratory culture they appear in general to grow less well on organic nitrogen than on inorganic nitrogen.

Urea was reported by Allen (1952) to be an excellent nitrogen source for certain thermophilic unicellular blue-green algae, although others such as *Synechococcus cedrorum* cannot use it. Certain *Lyngbya* and *Nostoc* strains also grow on urea and Van Baalen (1962) observed good growth of marine blue-green algae on this nitrogen source. Gorham *et al.*, (1964) found that although *Anabaena flos-aquae* grew well on urea initially, the alga subsequently lysed. The toxic effect may be due to the accumulation of ammonia within the cells because Allison *et al.* (1954) concluded from studies using ^{14}C-labelled urea that it was decomposed to ammonia and carbon dioxide before it was utilized. There is obviously some variation in ability to utilize urea from one organism to another. Kratz and Myers (1955a) found, for example, that *Nostoc muscorum* and *Anabaena variabilis* both utilized urea while *Synechococcus* (*Anacystis nidulans*) did not.

Uric acid utilization and metabolism has been investigated extensively by Van Baalen and co-workers. They found (Van Baalen, 1962) that certain marine blue-green algae utilized uric acid even though they became unhealthy in appearance. Later Van Baalen and Marler (1963) demonstrated that both uric acid-utilizing (some strains of *Plectonema terebrans* and *Lyngbya lagerheimii*) and non-uric acid-utilizing species (some strains of *P. terebrans* and *Phormidium persicinum*) occurred. There appear to be two types of uric acid-utilizing algae, those which remain healthy in appearance and those such as *Agmenellum quadruplicatum* which become yellow, show characteristic symptoms of nitrogen deficiency, and accumulate polyglucoside when grown on uric acid. The former group may have a normal uricase and a uric acid $\rightarrow$ allantoin $\rightarrow$ urea system, whereas the latter group may break down uric acid by non-specific peroxidase activity. Van Baalen (1965) concluded more recently that *Synechococcus* sp. (*Anacystis nidulans*), which grows poorly on uric acid, could not use it as a nitrogen source until it was oxidized non-enzymatically to allantoin.

This agrees with an earlier observation of Birdsey and Lynch (1962), who were unable to detect uric acid or xanthine utilization by a variety of blue-green algae and found that *Synechococcus* sp. (*Anacystis nidulans*) decomposed uric acid to allantoin.

Complex compounds and mixtures of amino acids such as peptone, casein hydrolysate and serum albumin serve as excellent nitrogen sources for certain blue-green algae according to Allen (1952). Some organisms, such as *Synechococcus cedrorum*, for example, possess strong proteolytic activity. Van Baalen (1962), however, obtained evidence that the marine blue-green algae which he studied could utilize some, but not all, amino acids. Single amino acids were reported to be good nitrogen sources by Allen (1952) but others have found that certain amino acids are not used. Pintner and Provasoli (1958) found that *Phormidium persicinum* did not respond to lysine, histidine, glutamic acid and aspartic acid and that growth on glycine was poor. Van Baalen (1962) could not detect the utilization of glycine or lysine by his species of blue-green algae although *Chlorogloea fritschii* can grow slowly in the dark on glycine and glutamine (Fay, 1965). Wyatt *et al.* (1971) reported that unlike inorganic combined nitrogen, glycine and arginine do not inhibit nitrogenase activity. Hood *et al.* (1969) showed by ^{14}C-tracer studies that the carbon skeleton of arginine was rapidly incorporated by *Anabaena* and that this may account for up to 5% of the total cell biomass.

Some amides are excellent sources of nitrogen for the growth of blue-green algae. Pintner and Provasoli (1958) found that asparagine, supplied at concentrations of $0{\cdot}2$–$0{\cdot}8\ g\ l^{-1}$, maintained the healthy purple pigmentation of *Phormidium persicinum* and served as a slowly utilized reservoir of available nitrogen. There are no reports, so far as we know, of the utilization of succinamide and glutamine although, next to urea, they are utilized most readily after ammonia as an organic nitrogen source by most other groups of algae.

The extracellular nitrogen products of blue-green algae

Blue-green algae characteristically liberate substantial quantities of extracellular nitrogenous compounds into the medium irrespective of whether they are growing on elemental nitrogen or on combined nitrogen. The kinetics of this process were studied first by Fogg (1952) who found that the amount of extracellular combined nitrogen increased during the growth of *Anabaena cylindrica* cultures and that, as a percentage of the total nitrogen fixed, it was highest during the lag and stationary phases of growth and lowest during the exponential phase (Fig. 10.8). This pattern has been confirmed for *Nostoc entophytum* (Stewart, 1962), *Calothrix*

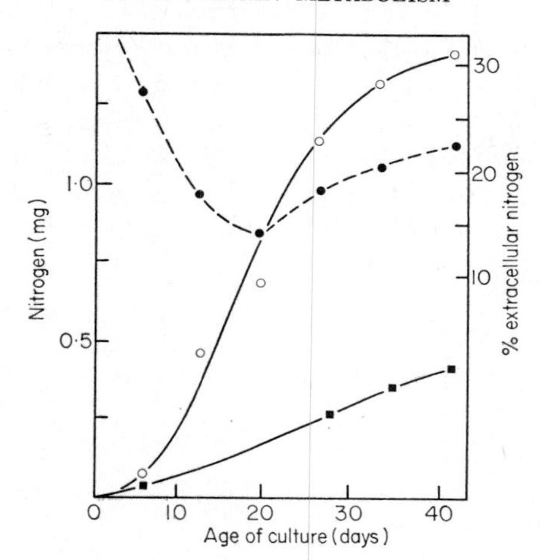

FIG. 10.8. The production of extracellular nitrogen during the growth of *Anabaena cylindrica* ○—○, intracellular nitrogen; □—□, extracellular nitrogen; ●—●, extracellular nitrogen as a percentage of total nitrogen. After Fogg (1952) *Proc. R. Soc.* B. **139**, 372, Fig. 1.

scopulorum (Stewart, 1962; Jones and Stewart, 1969a) and *Westiellopsis prolifica* (Pattnaik, 1966).

Besides depending on the stage of growth of the cultures, the relative amount of extracellular combined nitrogen also varies according to environmental conditions and perhaps also to the strain of alga used. Usually, between 20 and 40% of the total nitrogen assimilated appears in extracellular form, but Magee and Burris (1954) reported that less than 5% of the nitrogen fixed was liberated extracellularly by *Nostoc muscorum*. On the other hand, as mentioned below (p. 210), algae growing in symbiotic associations liberate a very high proportion of the nitrogen which they fix. Certain strains of *Chlorogloea fritschii* produced by treatment with mutagens, although fixing less nitrogen than the wild strain, liberate much more extracellular nitrogen (Fay *et al.*, 1964).

The origin of the extracellular products of blue-green algae is not clear. Fogg (1952) and others showed that autolysis of dead cells contributes no appreciable proportion of the extracellular material in healthy cultures. Hood *et al.* (1969) consider that liberation of extracellular products occurs because end product repression does not occur in blue-green algae so that excess amino acids accumulate. This suggestion could, according to Hitch and Stewart (1973) help to explain why symbiotic blue-green algae liberate such high quantities of extracellular nitrogen. This was noted first by Bond and Scott (1955) in studies on the fixation of $^{15}N_2$ by *Nostoc* colonies in symbiotic association with thalli of the liverwort *Blasia*. They concluded that a considerable proportion of the nitrogen fixed by the alga becomes available to the partner and similar conclusions were arrived at in studies on the lichen *Peltigera praetextata* (Scott, 1956) and *Peltigera apthosa* (Millbank and Kershaw, 1969). Labelling of *Nostoc* in the cephalodia of this lichen increases up to 4 days after exposure to $^{15}N_2$ and then remains steady, whereas the labelling of the remainder of the thallus increases steadily with time (Fig. 10.9).

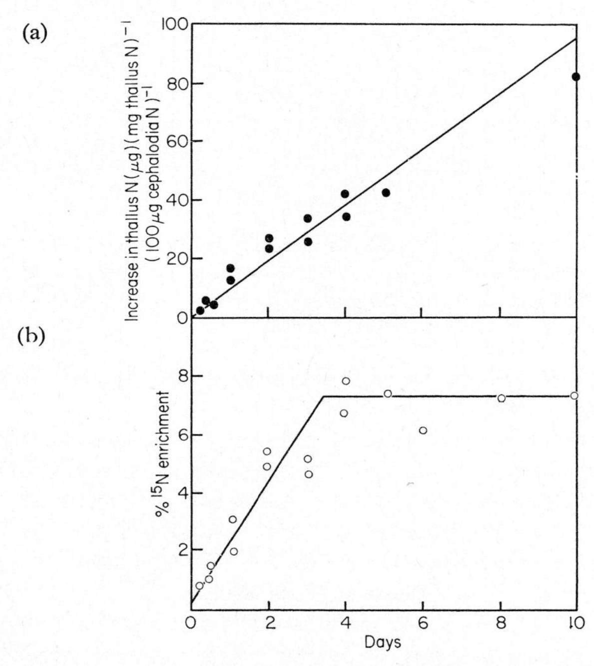

FIG. 10.9. Nitrogen uptake and transfer in the lichen *Peltigera apthosa*. There is a steady increase in the total mycobiont nitrogen with time (a), although there is no increase in the ^{15}N label of the phycobiont after 4 days. (b). This indicates that the fixed ^{15}N which the phycobiont obtains from the atmosphere is being transferred at a steady rate to the fungus. After J. W. Millbank and K. A. Kershaw (1969) *New Phytol.* **68**, 721–729.

A large scale transfer of fixed nitrogen from the alga also occurs in algal symbioses with higher plants. Bergersen *et al.* (1965) showed that *Macrozamia* root nodules fix nitrogen rapidly and that within 48 h this can be detected readily in the roots, leaf bases and stems of the gymnosperm. Similarly in the angiosperm *Gunnera* which has *Nostoc*-containing leaf glands, Silvester and Smith (1969) found within 1·5 h of exposure to $^{15}N_2$ that nitrogen had been transferred from the alga to the leaves, internodes and roots of the plants (Table 17.II). Thus the alga can supply all the nitrogen required for the growth of the plant. In all these symbiotic associations the algae are growing in the dark or at very low light intensities. Thus, if enzyme repression does not occur in blue-green algae and carbon skeletons to accept the fixed nitrogen are in short supply, it is not surprising that such a high proportion of the nitrogen fixed is liberated extracellularly in symbiotic systems.

Even if the bulk of the extracellular nitrogenous products are liberated as a result of a failure to repress enzyme synthesis, it is unlikely that this is the sole source of these products. Goryunova and Rzhanova (1964) consider that the extracellular nitrogen production of *Lyngbya aestuarii* results mainly from the decay of separation discs which are formed when hormogonia are released (see p. 26). This supposition is supported by the electron microscope investigations of Lamont (1969b) who has shown clearly that lysis of the separation discs does occur. However, as Fogg (1966) pointed out, not all filamentous algae form hormogonia or separation discs in culture. It is conceivable that a proportion of the extracellular nitrogen may be derived from the mucilaginous sheath (Jones and Stewart, 1969b). This is produced only in small quantity during exponential growth and is usually more abundant during the stationary phase. It is known that in many species this sheath material is water soluble and becomes dispersed in the medium.

Just as the extracellular nitrogen becomes available in symbiotic systems there is also evidence that it can be assimilated by a variety of non-nitrogen-fixing plants and may constitute an important source of nitrogen for growth of these plants in natural ecosystems. This is indicated by the many studies which show that higher plant crops such as rice increase in nitrogen in the presence of blue-green algae (see Singh, 1961) but in addition several studies have shown, using ^{15}N-labelled elemental nitrogen, that there is transfer of extracellular nitrogen from the alga to associated plants. Mayland *et al.* (1966) demonstrated this in Arizona soils and found that it occurred at a constant rate over a 520-day period, and Stewart (1967b) showed transfer of fixed nitrogen from *Nostoc* to the moss *Bryum pendulum* and the angiosperms *Glaux maritima*, *Suaeda maritima* and *Agrostis stolonifera* in a sand-dune slack soil. Rather similar findings have also been reported from Russia where Pankratova and Vakhrushev (1969) showed uptake of ^{15}N-labelled extracellular products and autolytic products of *Nostoc muscorum* and *Anabaena cylindrica* by barley plants. Jones and Stewart (1969a, b) showed that some of the nitrogen liberated by the marine blue-green alga *Calothrix scopulorum* was assimilated by various marine macroalgae, and that in addition a considerable proportion could be absorbed onto these algae, where presumably it would be broken down by bacterial action and subsequently assimilated.

Undoubtedly much of the confusion and controversy which surrounds the origins and functions of extracellular substances has arisen because although many studies have been made on the classes of compounds (e.g. combined nitrogen and carbon, amino acids, organic acids, and sugars; peptides and carbohydrates; mucilages and pigments) only in a few cases have the exact identities of the chemical compounds released been established.

Fogg (1952) reported the occurrence of traces of free amino acids in the culture filtrates of *Anabaena cylindrica* but these were not of quantitative significance. He also identified the amino acids released by acid hydrolysis of concentrated filtrates (serine and threonine being the most abundant) which established that the extracellular substances did not simply result from autolysis of the cells. Whitton (1965) found that the release of combined nitrogen by this alga was correlated with the accumulation of a brown, extracellular pigment which he suggested might be a complex polypeptide. Preliminary studies by Walsby (1965, 1970) indicate that a substantial proportion of the nitrogenous material is released initially in the form of small molecules (molecular weight of less than 1000) containing a small number of serine and threonine residues, and a nitrogenous, ultraviolet fluorescent, moiety of unknown composition. Polymerization of these molecules occurs slowly on exposure to light, apparently through the fluorescent moiety, giving rise to a family of pigmented molecules of high molecular weight (up to 5,000–10,000). These serine–threonine containing compounds were not produced when the alga was grown in the presence of the ammonium ion, suggesting involvement of either nitrogen fixation or heterocysts in their production. Their biological role is unclear as yet; they absorb strongly in the ultraviolet and they are capable of forming stable complexes with iron (up to 4% by weight) but the significance of the latter is unclear. Whitton (1965) found that crude, non-dialyzable concentrates of the extracellular substances of *A. cylindrica* were capable of reducing the toxicity of polymyxin to this and other blue-green algae, but this activity does not appear to reside in the serine–threonine pigment complex (Walsby, 1970).

The fluorescent moiety in the precursor of the pigmented molecule has several features in common with pteridines known to be produced by *A. cylindrica* and other blue-green algae. Pteridines, so called because of their occurrence in butterfly wings (Pfleiderer, 1964) are derivatives of a bicyclic ring system, pyrimido [4, 5–b] pyrazine (designated "pteridine") large numbers of which have been isolated and characterized from diverse plants and animals. Their presence in blue-green algae was first reported by Forrest *et al.* (1957) who isolated derivatives of 2-amino-4-hydroxy-pteridine

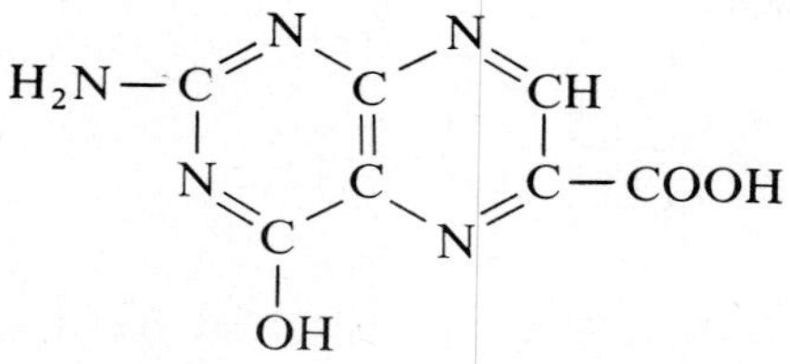

```
            N       N
         //   \   /   \\
H2N—C         C         CH
       |          ||         |
       N          C         C—COOH
         \\   /   \   //
            C         N
            |
           OH
```

from *Anacystis nidulans.* This compound was released when the alga was suspended in water at 4°; parallel decrease in photosynthetic activity suggested that it might be involved in photosynthesis. A number of other pteridines were subsequently obtained from *Anacystis* and other blue-green algae, several of them apparently being isolation artifacts formed from glycosides of biopterin (Forrest *et al.*, 1958), with various sugar residues, including galactose, glucose 6-deoxy-d-glucose, xylose, xyluronic acid and ribose, attached (Hatfield *et al.*, 1961).

Chapter 11

Differentiation, reproduction and life cycles

Blue-green algae, like bacteria, have a relatively simple morphology and show little cellular differentiation. Nevertheless most multicellular forms produce at certain times in their life cycle specialized structures which may promote survival and reproduction or have specific metabolic functions. These structures are produced from vegetative cells by differentiation. In theory, any vegetative cell should be genetically capable of differentiation, but, in order for it to do so, certain other conditions may have to be satisfied; for example the cell may have to occupy a special position within the multicellular organization, or there may be some particular environmental factor which determines the site and nature of differentiation. These conditions may be met at a certain stage in growth so that a definite pattern of life cycle results.

Cell division

Propagation and differentiation in blue-green algae are generally associated with cell division, the simplest and most important method of multiplication.

Cell division usually occurs as a consequence of cell expansion and there is a distinct relationship between size and cell division which is often characteristic of the species. However, cell division is not necessarily dependent on an increase in cell size. For example, in many Chroococcales several cell divisions may occur before cell expansion and endospore formation results from multiple divisions which are not accompanied by cell enlargement (p. 27). In many filamentous blue-green algae a steady state of cell division, in which expansion and division proceed simultaneously, is characteristic of active growth. A similar state of affairs occurs in the unicellular forms, *Anacystis nidulans* (*Synechococcus* sp.) and *Gloeocapsa alpicola* (Allen, 1968c). This is shown in Fig. 11.1 where the initial layers of the next transverse septum are seen to be deposited before the daughter cells separate completely.

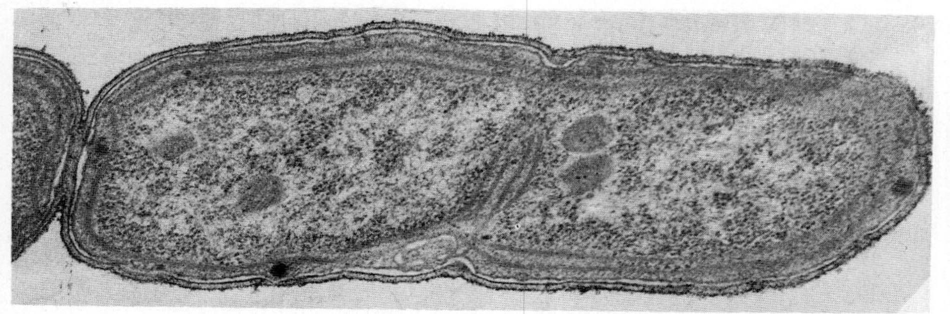

FIG. 11.1. Longitudinal section of a cell of *Anacystis nidulans* (*Synechococcus* sp.) in the process of cell division. A second division has begun before the daughter cells of the previous division have separated completely. Electron micrograph (× 42,000). By courtesy of M. M. Allen (1968) *J. Bacteriol.* **96**, 842, Fig. 3.

Cell divisions and extension in filamentous forms may be more or less evenly distributed along the trichomes, or they may be restricted to a meristematic region as in the Rivulariaceae. Here the meristem is usually presnet in a subterminal position adjacent to the terminal heterocyst but another meristematic zone may develop in some species in an intercalary position beneath the colourless hair and give rise to a hormogonium (Fritsch, 1945; Fig. 11.2). Similarly, in the Scytonemataceae the meristem may be situated

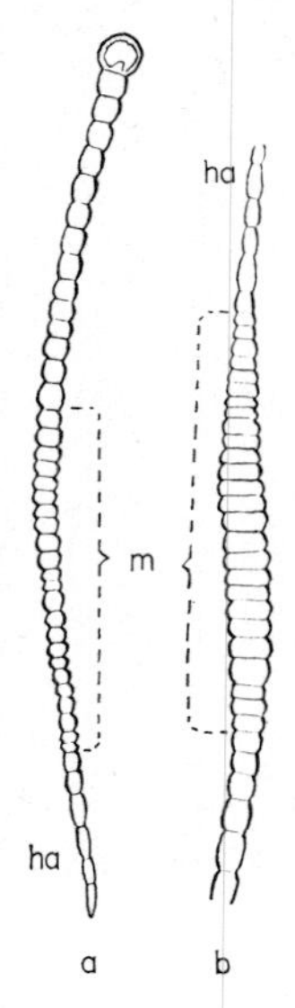

FIG. 11.2. Trichomes of *Gloeotrichia pisum* (a) and of *Biasolettiana polyotis*, (b) with meristematic zone (m) beneath terminal hair (ha). Drawings after Schwenderer, from Fritsch (1945).

in a terminal or intercalary position. In the latter case localized cell divisions may lead to loop formation and then to false branching (Fig. 11.3). As a

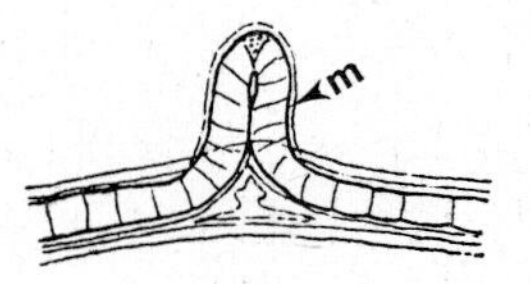

FIG. 11.3. Portion of trichome of *Scytonema pseudoguyanense* showing geminate branching with meristematic activity (m) resulting in loop formation. After Y. Bharadwaja, from Fritsch (1945).

result of the frequent division, cells within the meristematic zone appear short and disc-like and the cell contents appear more homogeneous (Geitler, 1936).

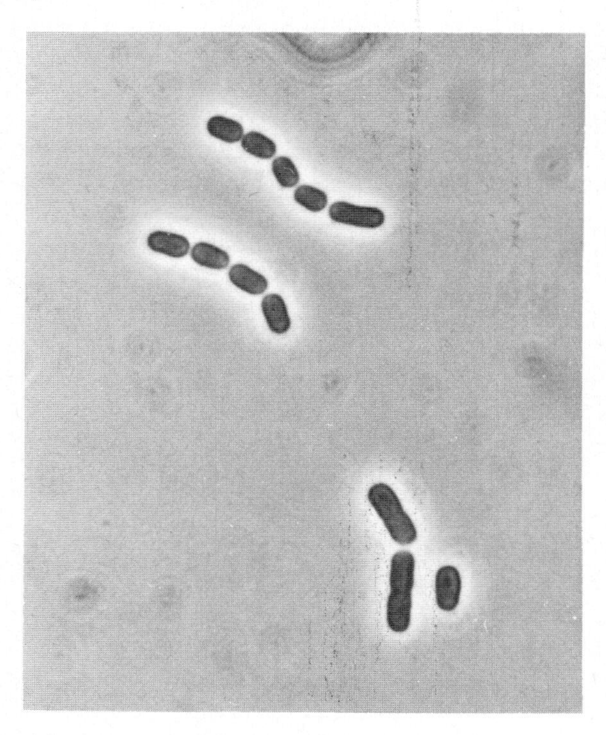

FIG. 11.4. *Synechococcus* sp., strain 7002, forming short chains of cells by successive cell division along a single plane. Light micrograph ($\times$2,000). By courtesy of R. Y. Stanier, R. Kunisawa, M. Mandel and G. Cohen-Bazire (1971) *Bacteriol. Rev.* **35**, 171, Fig. 1.

There may also be a periodicity in the timing of cell division. Rhythmical phases of cell divisions with a resting period between have been reported in *Merismopedia*, *Chroococcus* and *Gloeocapsa* (Geitler, 1925).

Geitler (1951) suggested that the plane of cell division, which is as a rule perpendicular to the longest axis of the cell, might be determined by mechanical factors. He viewed the cell as an ellipsoid with three axes. Cell division occurs in a single plane perpendicular to the long axis of rod-shaped cells such as *Gloeothece*, *Aphanothece*, *Anacystis* or *Coccochloris* (Fritsch, 1945; Allen and Stanier, 1968a) and elongation consistently occurs in a direction perpendicular to the plane of division. That is, multiplication is by transverse binary fission as in bacteria (Fig. 11.4). Consequently the cells may appear in short, mostly four-celled chains, owing to weak transitory attachments between daughter cells, especially when grown on solid media (Allen and Stanier, 1968a). In *Merismopedia* (Fig. 11.5) the

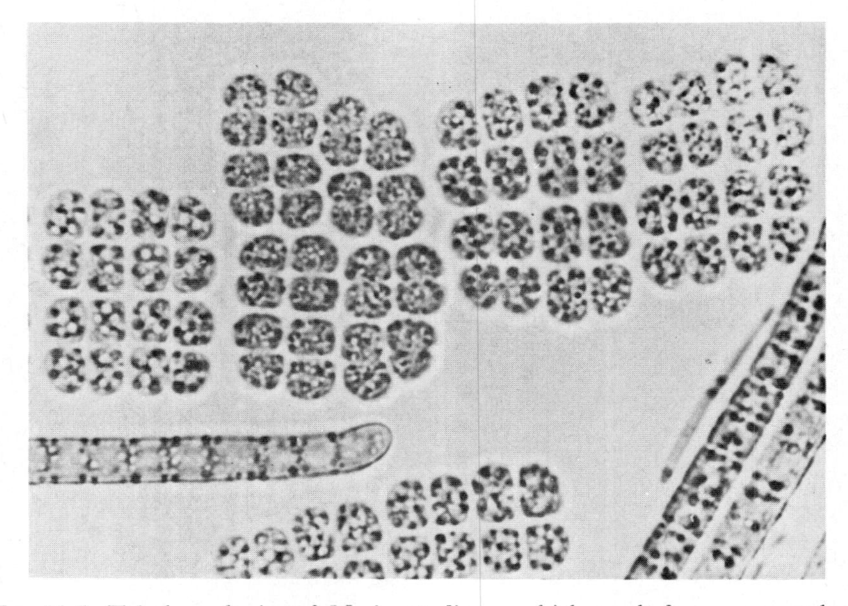

FIG. 11.5. Tabular colonies of *Merismopedia* sp. which result from more or less synchronous alternate cell divisions along two perpendicular planes (×1440). By courtesy of M. Lund.

daughter cells expand in a direction perpendicular to the long axis of the original mother cell and parallel to the plane of the completed cell division. Thus cell division occurs alternately along two perpendicular planes and this leads to the formation of tabular colonies. In *Eucapsis* (Fig. 2.2h) the

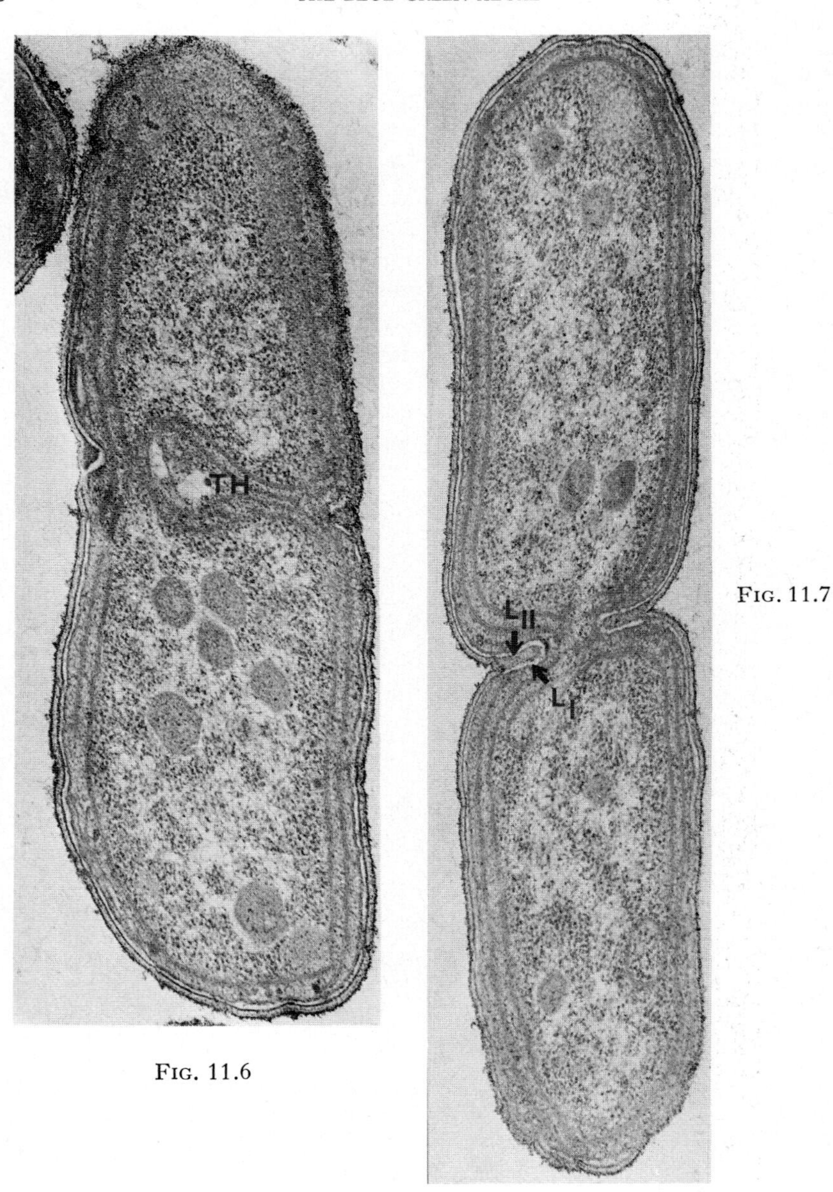

FIG. 11.6

FIG. 11.7

FIG. 11.6. Longitudinal section of *Anacystis nidulans* (*Synechococcus* sp.) in an early stage of cell division showing asymmetric invagination of thylakoid lamellae (TH). Electron micrograph (× 48,000). By courtesy of M. M. Allen (1968) *J. Bacteriol.* **96**, 842, Fig. 2.

FIG. 11.7. Longitudinal section of *Anacystis nidulans* (*Synechococcus* sp.) in a late stage of cell division showing the two inner cell wall layers (L_I, L_{II}) nearly separating the daughter cells. Electron micrograph (× 48,000). By courtesy of M. M. Allen (1968). *J. Bacteriol.* **96**, 842, Fig. 4.

situation is elaborated further in that the daughter cells elongate successively in three directions at right angles. This results in cell division along three perpendicular planes and in the production of cubical colonies. In the spherical forms, such as *Gloeocapsa* and *Chroococcus*, it is not always easy to decide whether division occurs along two or three planes, as post-divisional changes in cellular orientation may obscure the relations between successive planes of division (Allen and Stanier, 1968a).

The mechanism of cell division appears to be basically similar in all blue-green algae. The cell becomes constricted in the median plane and the two inner wall layers (L_I and L_{II}) appear to grow inwards in the form of an annulus, pushing the photosynthetic lamellae towards the centre and finally cutting them across as the septum is completed like the closing iris diaphragm of a camera (Hall and Claus, 1963, 1965; Figs 11.6, 7 and 8).

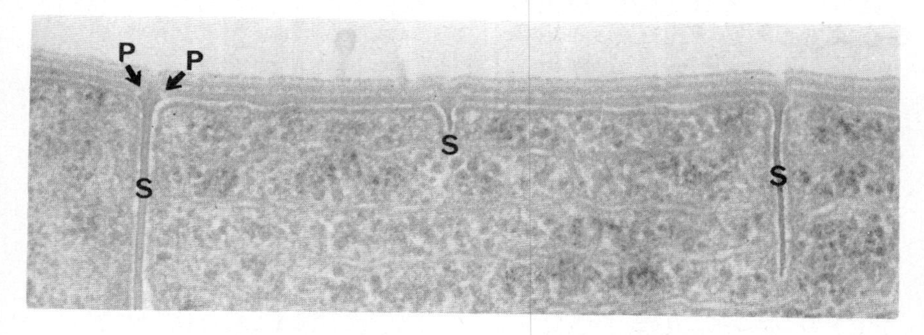

FIG. 11.8. Portion of longitudinal section of *Lyngbya* sp. illustrating the continuous process of cell division. Cross walls (S) are seen in various stages of formation and junctional pores (P) at both sides of the cross wall. Electron micrograph (× 62,000). By courtesy of H. C. Lamont (1969), *J. Bacteriol.* **97**, 350, Fig. 4.

Electron microscope observations (Echlin, 1964a; Allen, 1968c) show that cell division is initiated by the invagination of the photosynthetic lamellae and the plasma membrane before any constriction of the wall occurs (Fig. 11.6). As a result of invagination, the thylakoids become arranged in peripheral concentric layers on both sides of the developing cross wall and plasmalemma. The daughter cells are separated by the completed septum but stay within the enclosing outer wall layers as a single structural unit. This completes the process of cell division in the filamentous forms but in the unicellular forms the daughter cells become isolated from one another by constriction of the outer wall layers and by gradual splitting of the newly formed septum.

In rapidly growing filamentous blue-green algae each cell of a filament may possess up to seven partially completed cross walls. This process is essentially different from that found in eukaryotic algae where cross wall formation accompanies and completes the mitotic cycle. The continuous cross wall formation in blue-green algae suggests a kind of "steady state of cytokinesis" (Pankratz and Bowen, 1963). Cross wall initiates appear in a regular sequence and frequently with geometrical precision as new septa are initiated midway between completed or developing septa. The mechanism is otherwise similar to that described for Chroococcales, with only two layers, L_I and L_{II}, contributing to cross wall formation (Jost, 1965). The inward growing cross wall seems to "push" the thylakoid lamellae towards the centre causing their subdivision (Pankratz and Bowen, 1963). The cut ends of the thylakoid membranes apparently fuse with each other, thereby closing the lumen of the fragmented thylakoids.

Paradoxically, the mode of cell division in the complex, truly branching, members of the Stigonematales is more akin to that of the unicellular forms, if the situation described in *Fischerella ambigua* is typical of this group. Thurston and Ingram (1971) have demonstrated by electron microscopy that in *Fischerella* the invagination of the plasma membrane and the L_I wall layer which initially separates the two daughter cells is followed by invagination of the middle and outer wall layers, and of the sheath itself. Thus, only the sheath remains uninterrupted along the length of the filament (Fig. 11.9).

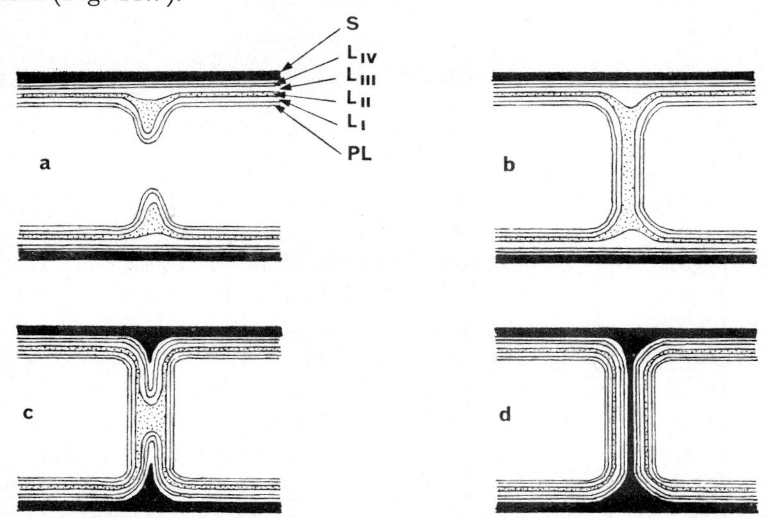

FIG. 11.9. Sequence of cell division in *Fischerella ambigua*. The invagination of plasmalemma (PL) and septum formation by the inner cell wall layers (L_I and L_{II}) are followed by the invagination of the outer wall layers (L_{III} and L_{IV}) and the sheath (S). The sheath is continuous throughout the length of the filament. After E. L. Thurston and L. O. Ingram (1971), *J. Phycol.* **7**, 203, Fig. 12.

We may presume that DNA replication and RNA synthesis will show a regular periodicity related to the division cycle. Using synchronous cultures of *Anacystis nidulans* (*Synechococcus* sp.) Herdman *et al.* (1970) demonstrated that DNA replication takes place before cell division and extends over a period of 60 min in a well synchronized culture under optimum conditions. This is followed by a lag of 30 min during which cell division is completed. Venkataraman and Lorenzen (1970) showed that the maximum amount of DNA per cell in synchronously grown *Anacystis nidulans* coincided with the onset of cell division. The rate of RNA synthesis was highest in the period between two successive cell divisions and declined before cell division, perhaps because DNA engaged in replication was not available as a template for RNA synthesis.

Heterocysts

In bacteria the only differentiated cellular structure is the spore. In blue-green algae there is a variety of reproductive structures and in addition there is the unique, and for a long time enigmatic, structure, the heterocyst, which appears to be mainly vegetative rather than reproductive in function. Heterocysts are produced from vegetative cells, particularly under nitrogen-limiting conditions during active growth of the alga, and are an example of division of labour in a relatively primitive multicellular organism.

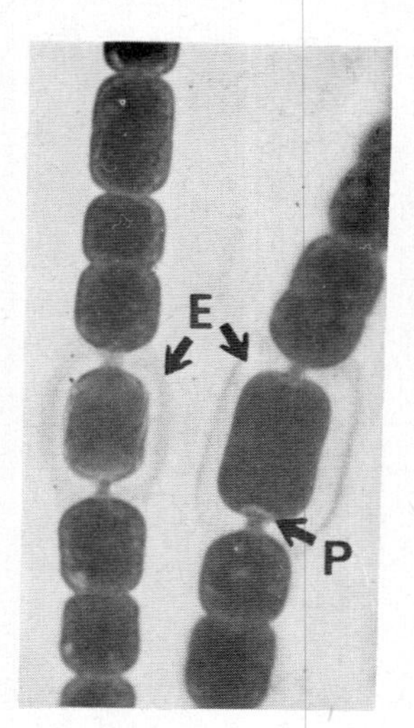

FIG. 11.10. Parts of filaments of *Anabaena cylindrica* with heterocysts; illustrating the relatively homogeneous cytoplasm, the polar thickenings (P) and the massive envelope (E). Ultraviolet micrograph (× 3,000). From G. E. Fogg (1951), *Ann Bot.* **15**, 23, Fig. 7.

1. HETEROCYST STRUCTURE AND CHEMICAL COMPOSITION

Heterocysts can generally be distinguished under the light microscope without difficulty. Their main features are (a) the partial separation by a constriction from the adjacent vegetative cells, (b) the presence of a massive cellular envelope, (c) reduced pigmentation, as compared with vegetative cells, (d) relatively homogeneous cell contents, and (e) the presence of refractive structures, called "polar nodules", near their attachment to the vegetative cells (Fig. 11.10). However, these features are not always evident, as, for example, in *Chlorogloea fritschii* (Fay *et al.*, 1964), which forms heterocysts that are smaller than the vegetative cells and are of varied form and structure (Whitton and Peat, 1967). This was a feature which resulted in the alga being placed originally in the Chroococcales (Mitra, 1950; Fay and Fogg, 1962). Although it has chroococcalean affinities (Stanier *et al.*, 1971) the presence of heterocysts in its multicellular stage

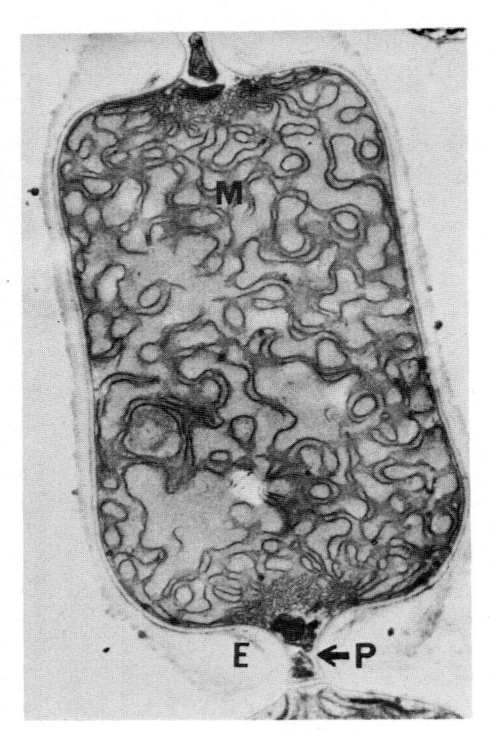

FIG. 11.11. Median longitudinal section of heterocyst of *Anabaena cylindrica* displaying the elaborate membrane system (M), pore channels (P) and the multi-layered envelope (E). Electron micrograph (× 14,000). From P. Fay and N. J. Lang (1971), *Proc. R. Soc.* **B**, 178, 185, Fig. 1.

make it clear that it is not properly classified in the genus *Chlorogloea*. Schwabe and El Ayouty (1966) have renamed it *Nostoc fritschii*.

Heterocysts may sometimes resemble spores superficially under the light microscope but there is no difficulty in recognizing heterocysts under the electron microscope. The changes occurring in the process of differentiation result in a characteristic ultrastructure (Wildon and Mercer, 1963b; Grilli, 1964; Lang, 1965; Fig. 11.11) and, now that a biochemical function has been attributed to these cells interest, in this ultrastructure has increased.

A prominent feature of the heterocyst is the thick envelope which is laid down outside the cell wall and which extends over the heterocyst except at the polar region (Wildon and Mercer, 1963b; Lang, 1965; Fig. 11.12).

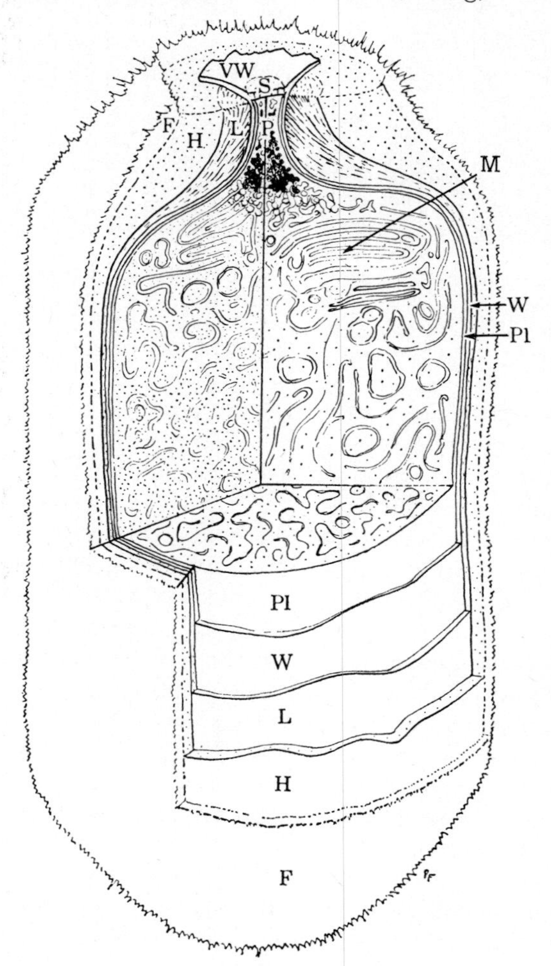

FIG. 11.12. Three dimensional view of heterocyst sectioned in three planes and illustrating the construction of surface layers. Pl, plasmalemma; W, cell wall; L, laminated, H, homogeneous and F, fibrous layers of the envelope; M, membrane system, P, pore-channel; S, septum; VW, wall of vegetative cells; from N. J. Lang and P. Fay (1971), *Proc. R. Soc.* B. **178**, 193, Fig. 18.

This envelope has three distinct layers: (a) an outermost loose fibrous layer of irregular thickness, (b) a broad and homogenous middle layer which is particularly well developed about the pore channel, and (c) a laminated innermost layer which is thick around the pore channel and which thins out towards the median region (Lang and Fay, 1971; see Fig. 11.13). A clear distinction between the cell wall, present before

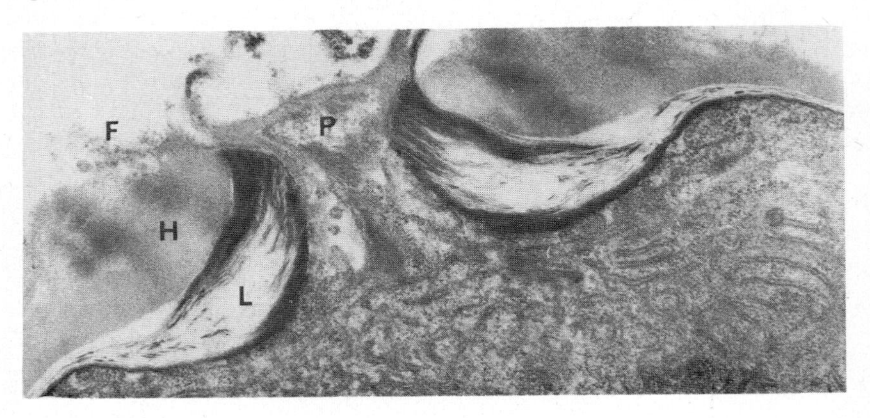

FIG. 11.13 Polar region of heterocyst of *Anabaena cylindrica* showing the three envelope layers (F, fibrous, H, homogeneous, L, laminated) surrounding the pore channel (P). Electron micrograph (×48,000). From N. J. Lang and P. Fay (1971), *Proc. R. Soc.* B, **178**, 193, Fig. 1.

differentiation into a heterocyst, and the additional envelope produced in the course of differentiation (see p. 239) is essential in order to avoid confusion about the construction, chemistry and function of these surface structures (Lang and Fay, 1971).

The intercellular connection between the vegetative cell and heterocyst is restricted to a small area of septum across the narrow pore channel between the two cells. The septum is similar to that between two vegetative cells with two inner cell wall layers, L_I and L_{II}. Fine connections across the septum (*microplasmodesmata*) were noted by Wildon and Mercer (1963b) and Lang and Fay (1971; Fig. 11.14). These connect the plasma membranes of the heterocyst and vegetative cell and may be a pathway for the exchange of metabolic products which may take place. (Fogg, 1951a; Wolk, 1968; Fay *et al.*, 1968; Stewart *et al.*, 1969).

Early biochemical studies on heterocysts were carried out using the conventional cytochemical methods developed for higher plant and animal cytochemistry. A more recent, and in some ways more satisfactory, approach is the chemical analysis of isolated heterocysts which can be prepared by differential disruption to break the vegetative cells and

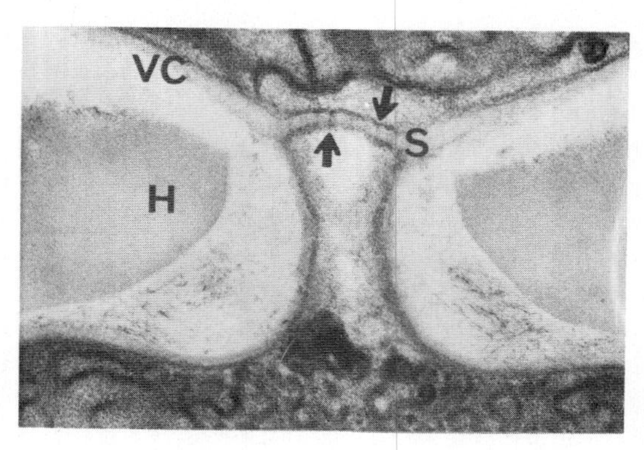

FIG. 11.14. Section through the septum (S) between heterocyst (H) and vegetative cell (VC) of *Anabaena cylindrica* displaying microplasmodesmata (arrows). Electron micrograph (× 46,000) from N. J. Lang and P. Fay (1971), *Proc. R. Soc.* B, **178**, 193, Fig. 8.

leave the more robust heterocysts apparently intact (Fay and Walsby, 1966). Heterocysts isolated by mechanical disruption of the filaments using a pressure cell, or sonication, or osmotic shock treatment may appear undamaged when viewed with the light microscope, but when examined in the electron microscope the internal structure of heterocysts treated in these different ways showed varying degrees of damage including the disruption of the wall and plasma membrane (Fay and Lang, 1971). This may result in a loss of cytoplasmic contents. A technique in which heterocysts were freed from vegetative cells by selective destruction of the cell wall with the enzyme lysozyme (Fay and Lang, 1971) yields heterocysts which appear intact in the electron microscope and compare well with untreated controls. These may be particularly useful in providing material for biochemical investigations on the heterocyst.

Chemical analysis of isolated heterocysts showed that compared to intact filaments heterocysts have a greater proportion of carbohydrate, less protein, and an unchanged total lipid content (Fay, unpublished). The increased carbohydrate content is probably due to the massive envelope. Chemical analysis of a so-called "wall" fraction (probably mostly composed of envelope material) obtained from isolated heterocysts of *Anabaena cylindrica* by Dunn and Wolk (1970) showed that on a dry weight basis 62% was carbohydrate and that the envelope accounted for about 52% of the total heterocyst dry weight. Lipids contributed 15% while amino compounds constituted 4%. In contrast, the vegetative cell wall showed a high amino content (65%), contained less carbohydrate (18%) and little (3%) lipid. The composition of the heterocyst wall is probably similar to that of the vegetative cell wall since its structure shows no apparent change (Lang and Fay, 1971; Kulasooriya *et al.*, 1972). The composition of the mucilage sheath isolated from whole filaments is similar (with 66% carbohydrate, 5% amino compounds) to that of the heterocyst envelope except for the absence of lipids (Dunn and Wolk, 1970). The electron opacity of the laminated envelope layer suggests that this probably has a higher lipid content than the fibrous and homogeneous envelope layers (Lang and Fay, 1971; Winkenbach *et al.*, 1972). The latter resemble the filament sheath, though they possess a more compact structure with tightly intertwined fibrils (Leak, 1967b), and probably have a strong mechanical construction.

Early suggestions that cellulose is present in the heterocyst wall and envelope were made by Hegler (1901) and Klein (1915) on the basis of cytochemical evidence. Brand (1903) noticed that the cellulose reaction in heterocysts is not consistent (see also Fritsch, 1945) but in spite of this the cellulose content of the heterocyst "wall" has often been assumed (Frey-Wyssling and Stecher, 1954; Kale and Talpasayi, 1969). Recent investigations by Dunn and Wolk (1970) do not support this idea; the heterocyst envelope-wall fraction is neither soluble in Schweitzer's reagent nor in 85% phosphoric acid, and is not digested by cellulase. A preliminary X-ray diffraction analysis of a heterocyst envelope fraction of *Anabaena cylindrica* (R. D. Preston, personal communication) yielded no evidence for cellulose.

Another erroneous belief about heterocysts is that they are empty, non-pigmented, structures. However, Fogg (1951a), in a detailed cytochemical investigation, demonstrated that heterocysts contain several types of material which varied in proportion as the cells matured. He showed the presence of Feulgen positive and arginine-containing substances and observed their mobilization, migration and ultimate disappearance during heterocyst development. A definite zonation of the protoplast occurred in an intermediate stage and this was followed by the concentration of substances at the poles. He suggested that materials eventually pass through into the adjoining vegetative cells and noticed that much less ash residue is left behind by heterocysts than by vegetative cells after micro-incineration. This agrees well with the disappearance from heterocysts of polyphosphate-containing particles (Tischer, 1957; Talpasayi, 1963).

Electron micrographs show that there is a definite fine structure in heterocysts and that they lack polyphosphate bodies, structured granules and polyhedral bodies (Fig. 11.15), but polyglucan granules and lipid droplets may be present in young heterocysts. Ribosomes, although apparently fewer in numbers than in vegetative cells, are present at all stages of heterocyst development. The nucleoplasmic region is less conspicuous than in the vegetative cells (Wildon and Mercer, 1963b; Lang, 1965), but

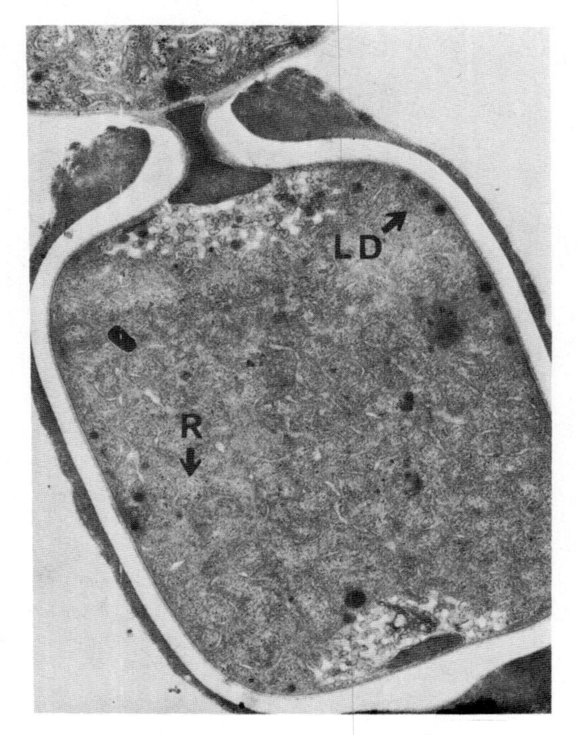

FIG. 11.15. Longitudinal section of heterocyst of *Anabaena cylindrica* after osmium fixation showing ribosomes (R) and lipid droplets (LD). Electron micrograph (× 20,500) from P. Fay and N. J. Lang (1971), *Proc. R. Soc.* B, **178**, 185, Fig. 2.

nevertheless the presence of Feulgen positive material has been confirmed using an electron microscope autoradiography technique in which tritium-labelled thymidine was incorporated in the nucleoplasmic region of the heterocyst (Leak, 1965).

Changes in the nature and distribution of membraneous structures occur during heterocyst maturation. The typical photosynthetic lamellae of the vegetative cells with their generally concentric and peripheral arrangement are replaced by a reticulate lamellar system which is more evenly distributed within the protoplast (Wildon and Mercer, 1963b; Grilli, 1964). Later, an increasing degree of contortion is observed with simultaneous concentration of lamellae towards the polar regions. Here a honeycomb or lattice-like tubular configuration develops which resembles in some respects the "prolamellar bodies" present in proplastids of higher plants (Lang, 1965; Fig. 11.16). An extensive membrane elaboration has been observed often in heterocysts of *Anabaena cylindrica* (Lang and

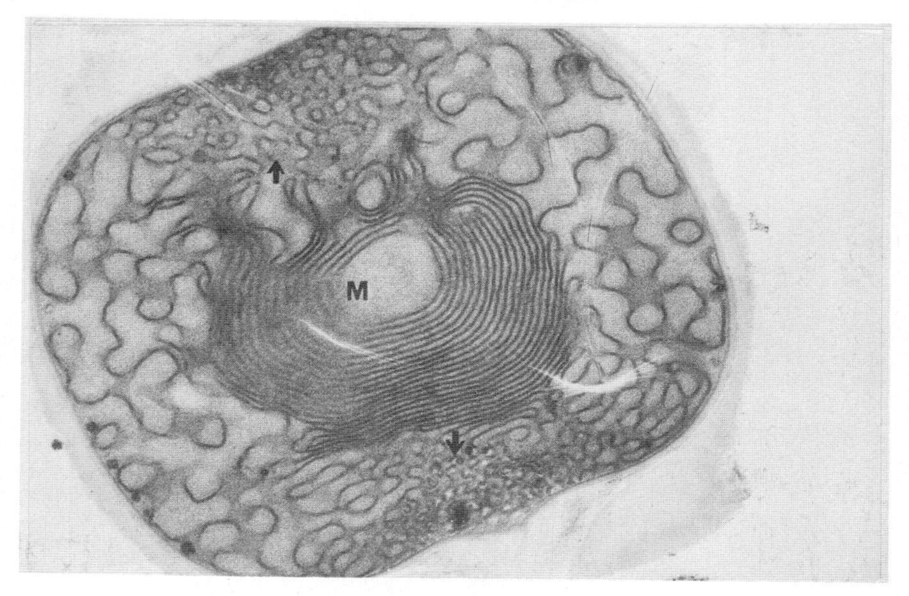

FIG. 11.16. Nearly median section through a heterocyst of *Anabaena cylindrica* showing the elaborate membrane system (M) and the honey comb-like configuration (arrows) in the polar region. Electron micrograph (× 22,500) by courtesy of N. J. Lang and P. Fay (1971), *Proc. R. Soc.* B, **178**, 193, Fig. 9.

Fay, 1971). The membranes are more densely packed than the photosynthetic thylakoids and have a greatly reduced inter-lamellar space; they are without polyglucan granules or phycobilisomes. They resemble bacterial mesosomes or the stacked thylakoids present in photosynthetic bacteria such as *Rhodopseudomonas viridis* (Drews and Giesbrecht, 1965). They develop mostly near the poles and are connected with the contorted lamellae and the honeycomb-like structure at the entry of the pore channel. It seems likely that new membranes are synthesized during differentiation. Eventually the contorted membranes of the honeycomb-like structure degenerate and breakdown is accompanied by an accumulation of electron dense material, the polar "plug", in the pore channel (Lang and Fay, 1971; Fig. 11.17).

The re-organization of the intracellular membrane system and the disappearance of typical thylakoids are reflected in changes in the pigment content of heterocysts. Fritsch (1951) emphasized that heterocysts remain pigmented after differentiation although some change in colour occurs. *In vivo* absorption spectra obtained with suspensions of isolated heterocysts of *Anabaena cylindrica* and pigment analysis performed on extracts from isolated heterocysts by Fay (1969b, 1970) and Wolk and Simon (1969)

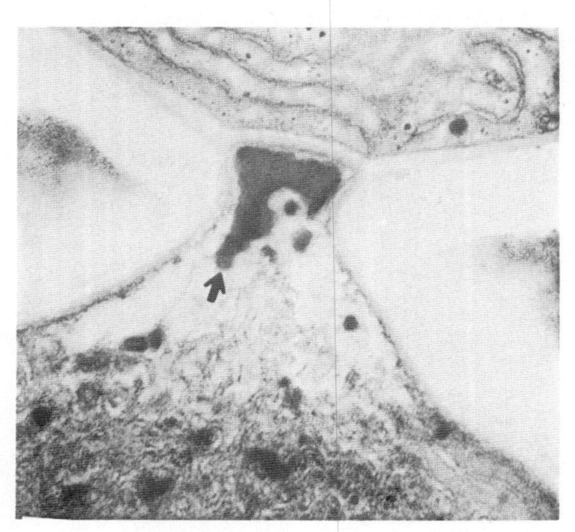

FIG. 11.17. Section through the polar region of heterocyst of *Anabaena cylindrica* showing electron opaque material (arrow) accumulated in pore channel. Electron micrograph (× 43,200) from N. J. Lang and P. Fay (1971) *Proc. R. Soc.* B, **178**, 193, Fig. 13.

showed that heterocysts contain significant though decreased amounts of chlorophyll *a* and carotenoid pigments but little if any phycocyanin (Fig. 10.7). *In vivo* micro-spectrophotometric determinations performed on individual heterocysts in intact filaments of *Anabaena* by Stewart *et al.* (1969) and Thomas (1970) agree well with the above analyses and have also indicated the absence from heterocysts of two other phycobiliproteins, allo-phycocyanin and phycoerythrin, both of which were present in small quantities in the vegetative cells (Thomas, 1970).

The presence of chlorophyll *a* and a higher ratio of β-carotene to xanthophyll (Fay, 1969b) suggests that heterocysts possess an active photosystem I. This is supported by evidence from isolated heterocysts of *Anabaena cylindrica* which were found to contain a higher proportion of the chlorophyll *a* modification P 700 (see p. 148) than the intact filaments of the same alga (Donze *et al.*, 1972). There was a ratio of 1 mol P 700 to 90 mol chlorophyll *a*, as compared with a ratio of 1 mol P 700 to 170 mol chlorophyll *a* in intact filaments. This ratio is similar to that estimated by Ogawa and Vernon (1969) for photosystem I particles isolated from *A. variabilis*. The absence of phycocyanin seems to indicate the lack of a carbon dioxide-fixing and oxygen-evolving photosynthetic mechanism in the heterocysts (see p. 233).

Walsby and Nichols (1969) studied changes in the membraneous and lipid-containing structures during heterocyst development (Fig. 11.18).

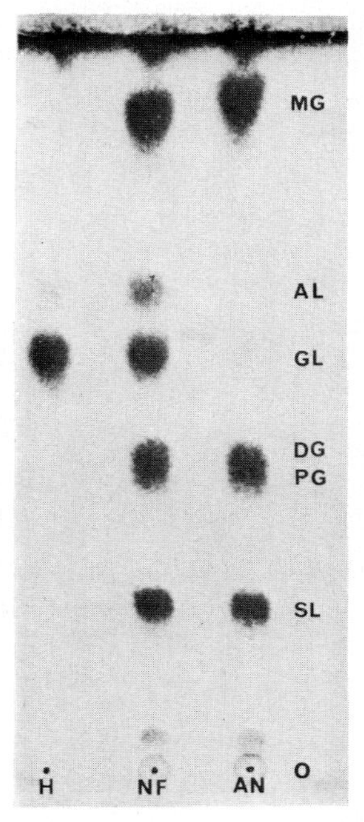

FIG. 11.18. Thin-layer chromatograph of lipid extract from cell preparations of *Anabaena cylindrica*. H, heterocyst preparation; NF, alga grown in nitrogen-fixing conditions; AN, alga grown with ammonium-nitrogen; MG, monogalactosyl diglyceride; AL, unidentified acyl lipid; GL, unidentified glycoside; DG, digalactosyl diglyceride; PG, phosphatidyl glycerol; SL, sulpholipid; O, origin. From A. E. Walsby and B. W. Nichols (1969) *Nature* **221**, 673, Fig. 1.

Four acyl lipids (monogalactosyl diglyceride, digalactosyl diglyceride, phosphatidyl glycerol and a sulpholipid) which all contain the polyunsaturated fatty acid α-linolenic acid, and which are associated with the oxygen-evolving photosystem II in higher plant chloroplasts, are absent in heterocysts although present in vegetative cells. In place of these lipids two other novel kinds of lipids were found. The major components are glycolipids, mono-glucoside and mono-galactoside derivatives of long-chain, unsaponifiable polyhydroxy alcohols, mainly 1,3,25-trihydroxy-hexacosane, accompanied by small quantities of another 26-C alcohol, probably 1,3,25,27-tetrahydroxyoctacosane (Bryce *et al.*, 1972). The structure of the minor component, an acyl lipid yielding a fatty acid on hydrolysis, is not yet known.

It has been suggested that these novel lipids which occur in several heterocystous algae investigated (Nichols and Wood, 1968) may replace the polyunsaturated lipids as components of the heterocyst photosynthetic lamellae (Walsby and Nichols, 1969). This might be responsible for the changed conformation of these membranes (see p. 240). A survey of the distribution of fatty acids and lipids in prokaryotic photosynthetic organisms has shown that digalactosyl diglycerides are confined to, and apparently universal among those which possess photosystem II (Kenyon and Stanier, 1970). If these lipids are actively involved in this oxygen-evolving process, their elimination from heterocysts may be a prerequisite to establishing anaerobic conditions inside these differentiated cells.

2. METABOLIC ACTIVITY IN HETEROCYSTS

The first experimental evidence of metabolic activity in heterocysts was obtained using cytochemical methods. Drews (1955) observed that blue-green algae are able to reduce tetrazolium salts and that the formazan precipitate produced in the reaction is deposited primarily at the periphery of the cell. Tetrazolium salts have a low redox potential and are used extensively as artificial hydrogen acceptors to indicate general reducing activity. Drawert and Tischer (1956) tested several heterocystous algae (*Anabaena*, *Cylindrospermum*, *Nostoc*, *Calothrix*, *Scytonema*) and noticed that heterocysts reduce triphenyl tetrazolium chloride (TTC) more quickly than vegetative cells (within 10 min in contrast to 30–35 min in vegetative cells; Fig. 11.19) even without an added substrate (succinate). They also tested the algae for the Nadi reaction (an index of cytochrome oxidase activity) and found that the reagent (dimethyl *p*-phenylenediamine) was oxidized by heterocysts (in the presence of α-naphthol) to indophenol at much faster rates than in vegetative cells. Both these cytochemical tests indicate a stronger reducing ability or a higher respiratory activity of heterocysts as compared with vegetative cells. Tischer (1957) employed several other redox dyes (Nile blue A, Janus Green B, methylene violet and neutral red) and found that the dyes with lower redox potential stained the heterocysts more quickly while no such differences were noticed in the vegetative cells. She concluded that heterocysts possess a high metabolic activity and should be regarded as cells with an active function. The TTC reaction has been confirmed by other workers (Stewart *et al.*, 1969; Fay and Kulasooriya, 1972) and in addition high reducing activity of heterocysts was demonstrated by their ability, unlike vegetative cells, to reduce silver salts in solution (Talpasayi, 1967) and in a photosensitive emulsion (Stewart *et al.*, 1969; Fig. 11.20).

A more direct measurement of respiratory activity in isolated heterocysts of *Anabaena cylindrica* was performed by Fay and Walsby (1966)

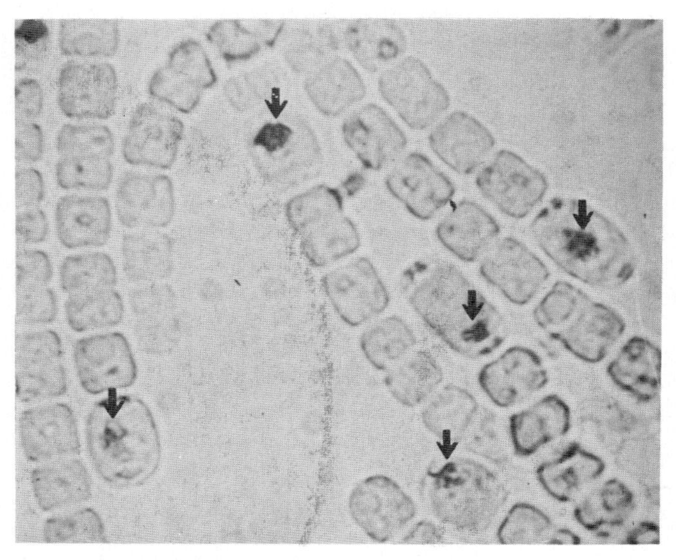

FIG. 11.19. Filaments of *Anabaena cylindrica* showing formazan crystals in the heterocysts (arrows) after a 20 min exposure to triphenyl tetrazolium chloride. Light micrograph (×2200), from P. Fay and S. A. Kulasooriya (1972) *Arch. Mikrobiol.* **87**, 341–352, Fig. 1.

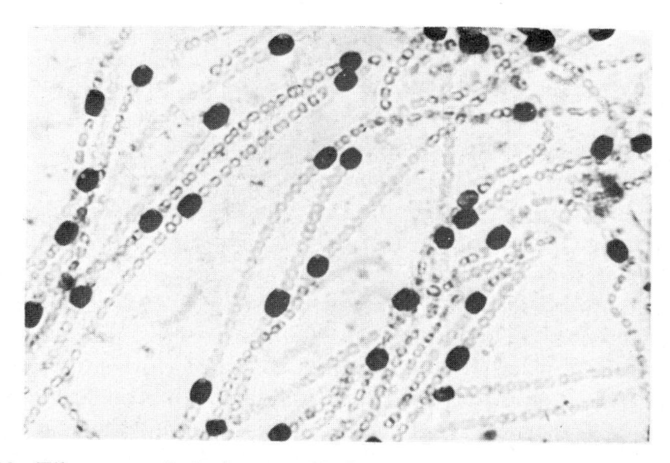

FIG. 11.20. Filaments of *Anabaena cylindrica* covered in a nuclear emulsion and showing the deposition of silver grains around the heterocysts. Photograph (×400) W. D. P. Stewart, A. Haystead and H. Pearson (1969) *Nature, Lond.* **224**, 226–228.

using a manometric technique. They found that on a cell nitrogen basis the rate of oxygen uptake was over 40% higher in the heterocysts than in intact filaments. It may be that this enhanced activity was due to damage (see p. 192).

Fay and Walsby (1966) also examined the photosynthetic activity of isolated heterocysts using the radiocarbon technique and found no carbon dioxide fixation. This is as to be expected from the absence of phycobilins which are the principal light absorbers for photosystem II in blue-green algae (see p. 147). Furthermore, isolated heterocysts show no Hill reaction activity. The absence of a functional photosystem II is also indicated by the low yield of chlorophyll fluorescence, the high concentration of P 700, and the absence of photoreduction of P 700 in isolated heterocysts (Donze *et al.*, 1972). Neither do heterocysts isolated by treatment with lysozyme evolve oxygen when illuminated (Bradley and Carr, 1971). To summarize, heterocysts in contrast to vegetative cells (a) lack an oxygen-evolving photosystem and (b) show higher reducing activity. Together these should produce a low intracellular oxygen tension in heterocysts.

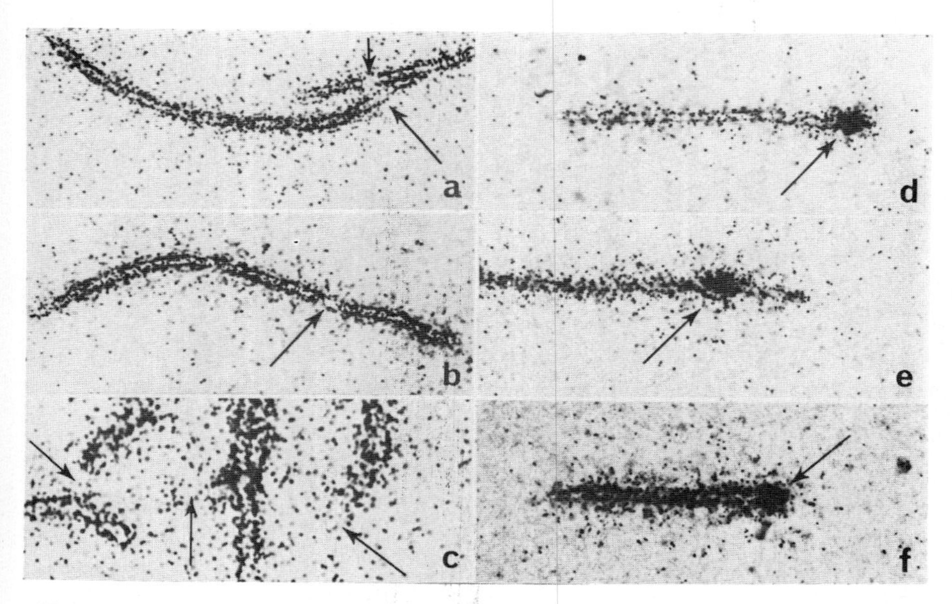

FIG. 11.21. Autoradiograph of filaments of *Anabaena cylindrica* following a short exposure to ^{14}C labelled carbonate. a, b, c: after 3–4 min exposure to light, very little radioactivity in the heterocysts (arrows) as compared with the vegetative cells: d, e, f; after 3–4 additional hours in the dark heterocysts (arrows) show equal or more radioactivity than vegetative cells (×330). By courtesy of C. P. Wolk (1968) *J. Bacteriol*, **96**, 2138, Fig. 2.

Wolk (1968), in an elegant experiment employing autoradiographic techniques, confirmed the lack of carbon dioxide fixation in heterocysts (Fig. 11.21). He also demonstrated that the radioactive carbon assimilated by vegetative cells in the light passed into the heterocyst during subsequent dark incubation. Stewart *et al.* (1969) were able to show a gradation in ^{14}C-labelling in heterocysts, it being absent from old heterocysts and present in developing heterocysts. It thus appears that vegetative cells provide heterocysts with fixed organic carbon which may be used as a source of reductant and energy in their metabolism.

There is experimental evidence that photosystem I, the presence of which is indicated by the component pigments, is active in heterocysts. The evidence is based partly on light-induced cytochrome oxidation (Donze *et al.*, 1972) and partly on photophosphorylation detected in isolated heterocysts (Scott and Fay, 1972; Fig. 11.22).

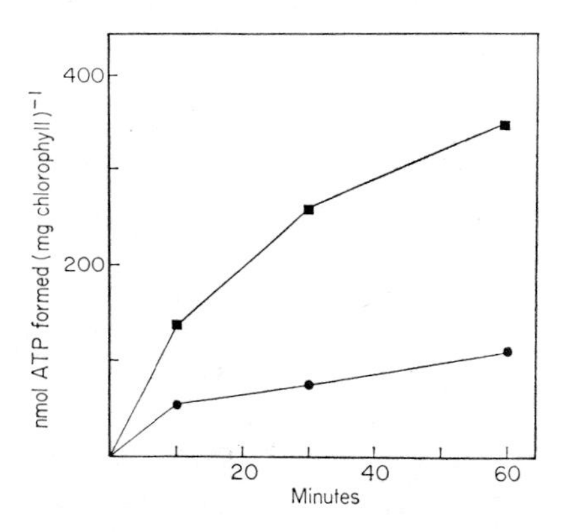

FIG. 11.22. Time course of photophosphorylation in isolated heterocysts of *Anabaena variabilis* in the presence (●) and absence (■) of free oxygen. After W. E. Scott and P. Fay (1972).

When isolated heterocysts of *Anabaena cylindrica* were exposed to a gas phase containing $^{15}N_2$ no uptake of the tracer was observed (Fay and Walsby, 1966). Stewart *et al.* (1969), taking into account the high oxygen sensitivity of nitrogenase in blue-green algae (Fay and Cox, 1967; Stewart, 1969; Stewart and Pearson, 1970) and the necessity of providing the heterocysts when isolated from vegetative cells with a reductant and ATP, were able to demonstrate acetylene reduction. Wolk (1970) using a similar reaction mixture confirmed the results of Stewart *et al.* (1969) and recovered in isolated heterocysts 11–12% of the original nitrogenase activity present in the intact filaments of *A. cylindrica*. Further, Wolk and Wojciuch

(1971a) reported on light-dependent nitrogenase activity in isolated heterocysts in the absence of any cofactor (i.e. artificial reductant or ATP generating system) which accounted for about 30% of the activity measured in intact filaments. This implies that the nitrogenase activity of heterocysts is several times higher than that of vegetative cells and when it is considered that the recovery of nitrogenase activity in extracts from various other nitrogen-fixing organisms is in general not higher than 20%, suggests that heterocysts are the only site of nitrogen fixation. Nevertheless, the possibility of the vegetative cells also possessing nitrogenase cannot be excluded. This possibility must be considered in view of the evidence that nitrogen-fixing activity is present in some non-heterocystous blue-green algae (see p. 188).

3. THE DIFFERENTIATION OF HETEROCYSTS

Fritsch (1951) in his presidential address to the Linnean Society of London, gave, under the title "The Heterocyst: A Botanical Enigma" an excellent account of heterocyst differentiation in a *Nodularia* species. He observed that a heterocyst invariably arises by the transformation of a vegetative cell in an actively growing filament. The first sign of differentiation is a slight enlargement of a cell with less pigment and fewer granules. Such an "incipient" heterocyst then rounds off and separates slightly from adjacent cells, as a hyaline envelope develops, so that only a narrow strand of protoplasm connects the heterocyst with the vegetative cells. At this stage the developing heterocyst is still granulated although both it and the adjoining vegetative cells have fewer granules than usual. The next stage is a marked thickening of the inner investment round the polar region. Fritsch believed that these conspicuous structures are essentially surface modifications of the envelope and did not subscribe to the view of Geitler (1936) that they result from internal deposition of substances and the formation of a cyanophycin-containing "plug" in the pore region. They resisted treatment with 35% chromic acid, which dissolved away all other cell contents, including cyanophycin granules, in the vegetative cells. They are thus a modified part of the envelope which eventually extends from the polar region all over the original inner investment. The cell contents of a fully differentiated heterocyst are clear and homogeneous. Fritsch thought that change of pigmentation need not imply a cessation of functions other than of photosynthesis. He concluded that heterocysts are highly specialized cells possessing distinct and specialized biochemical activity and maintaining cytoplasmic continuity with adjacent vegetative cells. He attached importance to the facts that at an early stage germinating akinetes form a terminal heterocyst at one or both ends of the new filament, and that as this increases in length by division new intercalary heterocysts are produced. Fogg (1951a) in a cytological study of heterocyst development in *Anabaena cylindrica* observed an essentially similar sequence. Using ultraviolet photomicrography and cytochemical techniques he demonstrated various changes in the protoplast and envelope during differentiation and maturation. The general pattern of these changes may now be related to those in fine structure but the significance of, for example, the redistribution of DNA which he observed in the developing heterocyst remains unexplained.

Fogg (1942) presented evidence showing that nitrogen fixation in *A. cylindrica* does not take place in the presence of high levels of available combined nitrogen such as ammonium- or nitrate-nitrogen. Subsequently he found that heterocyst frequency is inversely related to cell nitrogen content (Fogg, 1944) and that substances which provide a readily available source of combined nitrogen tend to suppress heterocyst formation whereas a supply of assimilable organic carbon promotes their production (Fogg, 1949). Nitrate, glycine and asparagine caused a transient inhibition of heterocyst formation but inhibition by ammonium lasted as long as an appreciable concentration remained in the medium (Fig. 11.23). The close

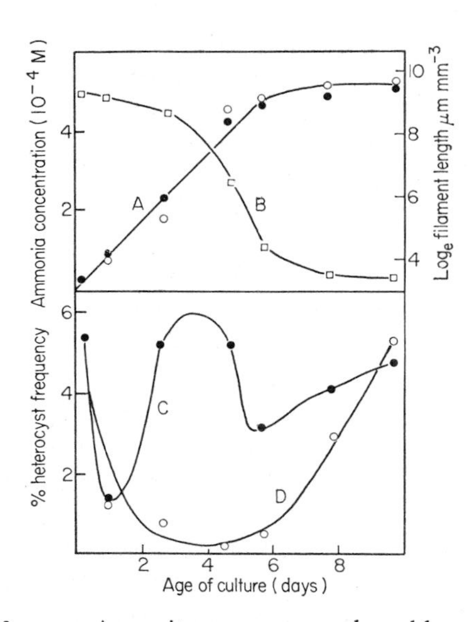

FIG. 11.23. Effect of ammonium-nitrogen on growth and heterocyst production in *Anabaena cylindrica*. A, growth as log$_e$ filament length in μm mm^{-3}; B, ammonia concentration in medium (10^{-4} M); C, heterocyst frequency under nitrogen-fixing conditions; D, heterocyst frequency in the presence of ammonia. After G. E. Fogg (1949) *Ann. Bot.* **13**, 241, Fig. 1.

correlation between the presence of heterocysts and the ability to fix nitrogen, and the finding that combined nitrogen, particularly ammonium nitrogen, suppresses both heterocyst formation and nitrogen fixation gave rise to speculations that heterocysts might be involved in the process of nitrogen fixation which, as we have seen (p. 190), have proved to be correct.

The inhibition of heterocyst formation by combined nitrogen was observed in many other species such as *Chlorogloea fritschii* (Fay *et al.*, 1964), *Anabaena flos-aquae* (Mickelson *et al.*, 1967), *Nostoc muscorum* (Stewart *et al.*, 1968), *A. variabilis*, *A. inaequalis*, *Aphanizomenon flos-aquae*, *Tolypothrix distorta* and *Gloeotrichia echinulata* (Ogawa and Carr, 1969), belonging to various groups of filamentous blue-green algae. Heterocyst numbers were also found to vary inversely with the concentration of nitrate nitrogen (Ogawa and Carr, 1969).

Fogg (1949) supposed "the formation of a heterocyst from a normal cell to occur when the concentration within it of a specific nitrogenous inhibiting substance, probably ammonia or some simple derivative of ammonia, falls below a critical level". He pointed out that in the filament, where growth is diffuse, periodic concentration gradients would be set up and heterocysts would form at the points of lowest concentration. Now that it is known that heterocysts are sites of nitrogen fixation it is likely that these structures produce the inhibitory substance. At a point midway between two existing heterocysts its concentration will be least and it is here that the formation of a new heterocyst will be induced. Exogenously supplied compounds containing nitrogen will suppress heterocyst formation to an extent depending on the readiness with which they are incorporated into cells and converted to the inhibitory substance. This hypothesis provides an attractive explanation for the even distribution of heterocysts along the filaments of Nostocaceae.

A similar hypothesis in slightly more general terms was put forward by Fritsch (1951) who suggested that the substance secreted by the heterocyst stimulates growth and cell division in vegetative cells. Again ammonia could be this substance. Wolk (1967) noticed that the fragmentation of the filament results in the stimulation of heterocyst formation and considered this as evidence that the spacing of heterocysts during normal growth is caused by the heterocysts themselves inhibiting transformation of nearby vegetative cells into heterocysts.

Talpasayi and Kale (1967) and Talpasayi and Bahal (1967) noticed that polyphosphate granules decreased in number during the early stage of heterocyst development, and used this criterion to identify "incipient" heterocysts, which are otherwise similar in shape, size and pigmentation to normal vegetative cells. They also observed such cells in cultures supplemented with ammonium chloride and considered these to be incipient heterocysts which fail to complete their transformation. They concluded that the loss of polyphosphate in the cell is the first of the changes leading to heterocyst formation and that the initial stage of heterocyst development is not under the control of ammonium ions. They further suggested (Kale and Talpasayi, 1969) that only certain cells termed "potential cells" which occupy a particular position in the filament possess the genetically determined ability to develop into "incipient" or *pro-heterocysts* and further, in the absence of combined nitrogen, into mature heterocysts. This theory received support from a study by Wilcox (1970), who observed by phase contrast microscopy a regular pattern of pro-heterocysts in the filaments of *Anabaena cylindrica* grown in the presence of ammonium chloride. On transfer into a medium free of combined nitrogen these cells were invariably transformed to mature heterocysts. This observation seems to confirm that ammonia itself cannot be the morphogen which determines that a cell develops the potential to become a heterocyst, but it still provides for the possibility that endogenously produced ammonia emanating from heterocysts can inhibit the differentiation of neighbouring cells which have become so determined.

Sequential observations were made in studies on *Anabaena cylindrica* by Kulasooriya *et al.* (1972) in which changes in microscopic appearance,

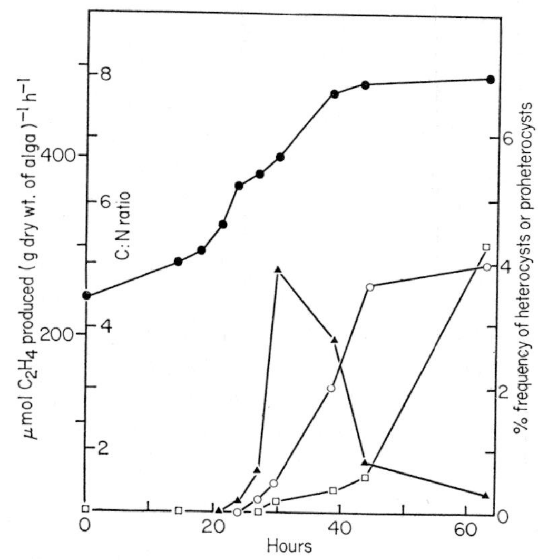

FIG. 11.24. Course of pro-heterocyst (▲—▲) and heterocyst (○—○) formation and of nitrogenase activity (□—□) of *Anabaena cylindrica* in relation to changes of cellular C:N ratio (●—●) during aerobic incubation in the light. After S. A. Kulasooriya, N. J. Lang and P. Fay (1972) *Proc. R. Soc.* B, **181**, 199–209, Fig. 1.

ultrastructure, chemical composition and metabolic activity during differentiation were followed using comparable material. The transformation was followed from an initial state of undifferentiated filaments, grown in the presence of ammonium phosphate, to the full development of heterocysts after transfer into nitrogen-free medium (Fig. 11.24). In contrast to the work of Wilcox described above, no pro-heterocysts were detected in the initial material either by phase-contrast microscopy (Fig. 11.25) or electron microscopy. Nitrogenase activity and heterocyst

FIG. 11.25. Undifferentiated filament of *Anabaena cylindrica* grown in the presence of ammonium-nitrogen. Phase contrast light micrograph (× 400) from S. A. Kulasooriya, N. J. Lang and P. Fay (1972) *Proc. R. Soc.* B, **181**, 199–209, Fig. 4.

differentiation were completely suppressed by an ammonium ion concentration sufficient to maintain a cellular C : N ratio of about 4.5 : 1. When the alga was deprived of ammonium-nitrogen under photosynthetic conditions, the cellular C : N ratio gradually increased. An initial stage of heterocyst development was detected, but only by electron microscopy, when the C : N ratio increased to about 5 : 1. Here there was a gradual dissolution of large cytoplasmic inclusions (structured granules, polyphosphate granules and polyhedral bodies), a slight decrease in the amount of polyglucan granules, and deposition of loose fibrous material external to the cell wall (Fig. 11.26). Thylakoids were generally intact at this stage and no nitrogenase activity was detectable. With phase contrast microscopy pro-heterocysts were first recognized when the C : N ratio had risen to a value of 6 : 1 (Fig. 11.27). At this stage the deposition of the fibrous envelope layer was complete and the middle homogeneous layer was being formed. Constrictions between developing heterocysts and adjacent vegetative cells were developing; the larger cytoplasmic inclusions had almost disappeared; a progressive breakdown of the thylakoid system was apparent; and the polyglucan granules originally present in the interthylakoidal space showed a distinct concentration at the periphery of the differentiating cell, possibly indicating a role in the deposition of envelope material (Fig. 11.28). There was no nitrogenase activity at this stage. As the cellular C : N ratio increased to 7 : 1 pro-heterocysts appeared more distinct by phase contrast microscopy. Polar nodules first appeared when

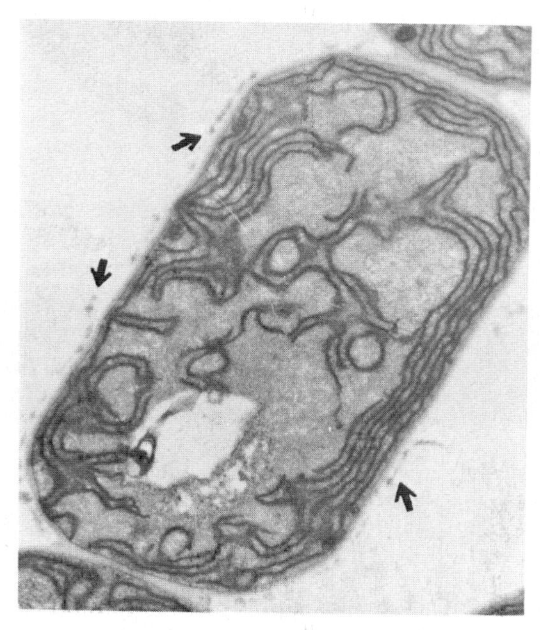

FIG. 11.26. Longitudinal section of *Anabaena cylindrica* showing a cell beginning differentiation into a heterocyst. Arrows indicate thin deposition of loose fibrous material external to the cell wall. Electron micrograph (× 18,000) from S. A. Kulasooriya, N. J. Lang and P. Fay (1972) *Proc. R. Soc.* B, **181**, 199–209, Fig. 7.

the C : N ratio was about 8 : 1 and their appearance coincided with the deposition of the third and innermost envelope layer. The formation of this layer coincided with the appearance of nitrogenase activity. At the same time, important changes were observed inside the differentiating protoplast and before the original thylakoid system had completely disintegrated, new membrane development started in various places by the internal budding of small vesicles, growing into small tubules which then fused and expanded into extensive membrane systems characteristic of the mature heterocyst (Fig. 11.29).

The main conclusions from these investigations are: firstly, the carbon: nitrogen balance is an important factor regulating heterocyst differentiation; secondly, the primary function of heterocysts is nitrogen fixation, and thirdly, this function is performed only when heterocysts are fully developed, furnished with all three envelope layers and equipped with a specific type of intracellular membrane system.

The significance of the cellular C : N balance for heterocyst formation is borne out by earlier studies. Thus, a distinct correlation between carbon

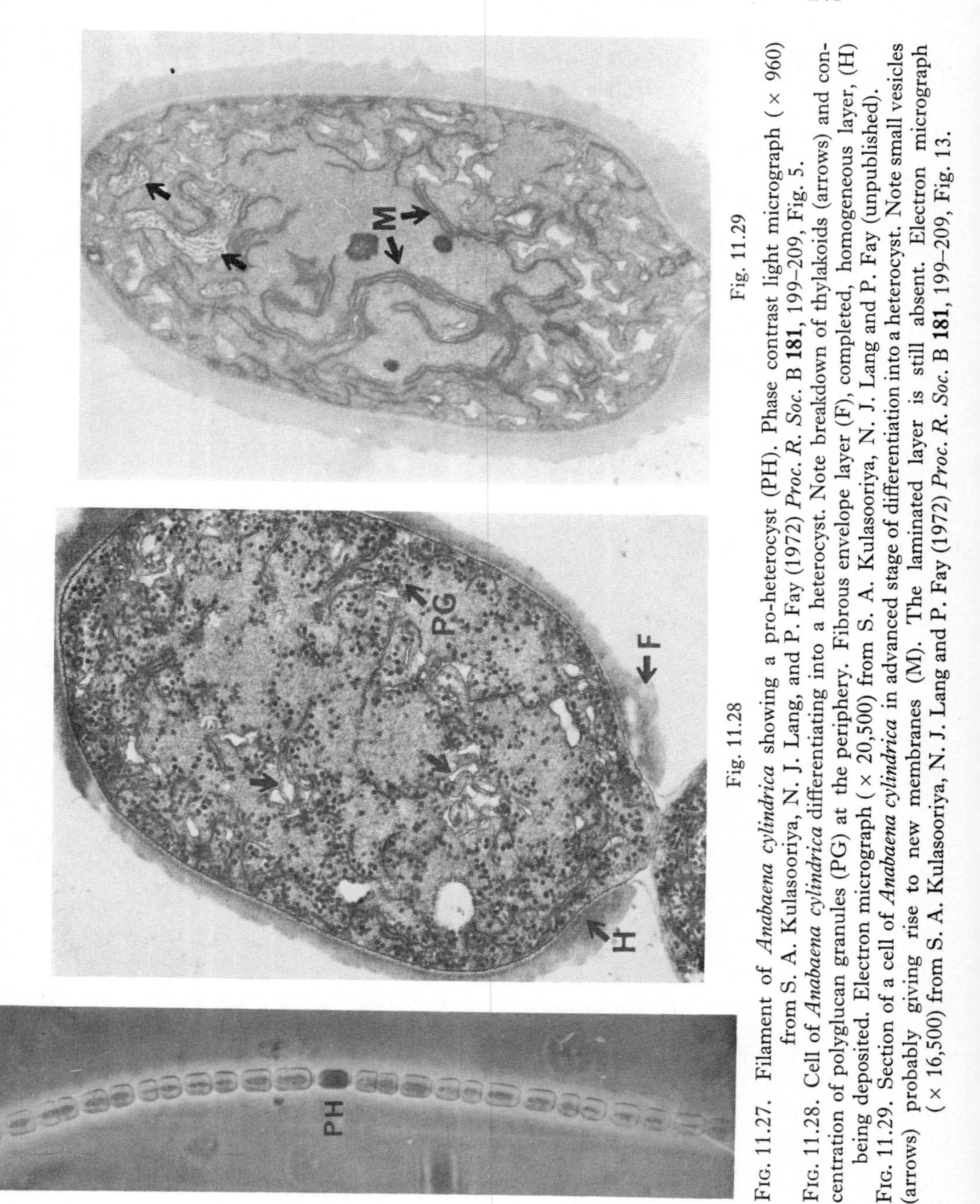

FIG. 11.27. Filament of *Anabaena cylindrica* showing a pro-heterocyst (PH). Phase contrast light micrograph (× 960) from S. A. Kulasooriya, N. J. Lang, and P. Fay (1972) *Proc. R. Soc.* B **181**, 199–209, Fig. 5.

FIG. 11.28. Cell of *Anabaena cylindrica* differentiating into a heterocyst. Note breakdown of thylakoids (arrows) and concentration of polyglucan granules (PG) at the periphery. Fibrous envelope layer (F), completed, homogeneous layer, (H) being deposited. Electron micrograph (× 20,500) from S. A. Kulasooriya, N. J. Lang and P. Fay (unpublished).

FIG. 11.29. Section of a cell of *Anabaena cylindrica* in advanced stage of differentiation into a heterocyst. Note small vesicles (arrows) probably giving rise to new membranes (M). The laminated layer is still absent. Electron micrograph (× 16,500) from S. A. Kulasooriya, N. J. Lang and P. Fay (1972) *Proc. R. Soc.* B **181**, 199–209, Fig. 13.

and nitrogen assimilation in *Anabaena cylindrica* was demonstrated by Fogg and Than-Tun (1960). In a study on the relations between photosynthesis and nitrogen fixation Cobb and Myers (1964) established a close relationship between changes in cellular C : N ratio and changes in C : N assimilation ratio (Fig. 11.30). From this they concluded that the

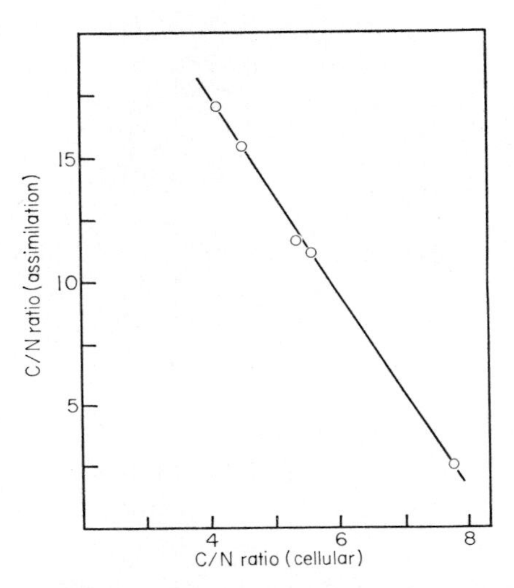

FIG. 11.30. Correlation between cellular C:N ratio and C:N assimilation ratio in *Anabaena cylindrica*. After H. D. Cobb and J. Myers (1964) *Pl. Physiol.* **51**, 753, Fig. 9.

alga possesses a remarkably exact mechanism controlling rates of carbon assimilation and nitrogen fixation. This control may be exerted largely via the differentiation of heterocysts, that is, there is a mechanism whereby ammonia and other sources of combined nitrogen suppress heterocyst formation and glucose and other readily available sources of organic carbon increase heterocyst production (Fogg, 1949; Kale and Talpasayi, 1969).

4. FACTORS AFFECTING HETEROCYST FORMATION

The effects of various substances and different conditions on the differentiation of heterocysts have been examined by several workers. Canabaeus (1929), for example, reported that in the presence of 0·2 % sodium chloride heterocyst frequency in *Anabaena variabilis* was three times greater than in cultures lacking the salt. However, in batch culture heterocyst frequency is dependent on the stage of growth and it is likely that effects observed by

Canabaeus and other early workers may have been more dependent on altered growth than direct effects on heterocyst differentiation. Fogg (1949) used growth analysis techniques to distinguish between these possibilities and found that molybdenum deficiency increased heterocyst production. Fay (unpublished) demonstrated that increased heterocyst production was accompanied by decreased nitrogenase activity under molybdenum deficient conditions (Fig. 11.31). This suggests that when nitrogenase

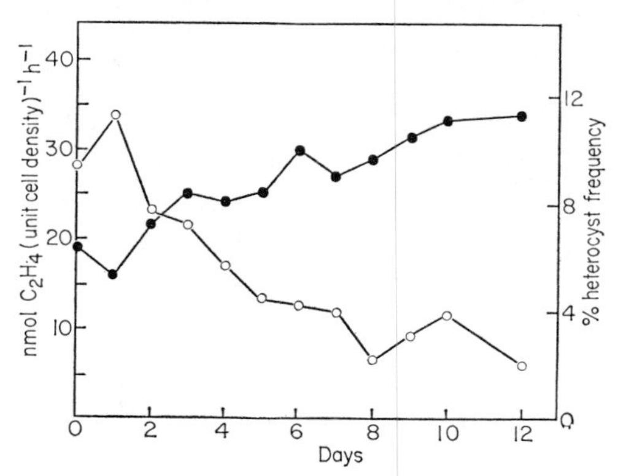

FIG. 11.31. Effect of molybdenum deficiency on the course of heterocyst formation (●—●) and nitrogenase activity (○—○) of *Anabaena cylindrica*. After P. Fay (unpublished).

synthesis is inhibited by the absence of the component metal and the decreased rate of nitrogen fixation cannot maintain the cellular C : N balance, the alga will respond with the differentiation of new heterocysts.

Heterocyst frequency, as well as nitrogenase activity, increased when *A. cylindrica* was incubated under microaerophilic conditions (in the light, aerated with nitrogen plus carbon dioxide) as shown by Kulasooriya *et al.* (1972) (Fig. 11.32). Further increase of heterocyst formation occurred under conditions of nitrogen starvation.

A light requirement for heterocyst formation and a failure to observe differentiation of heterocysts in the dark were reported by Kale and Talpasayi (1969) but it is not clear how far the requirement of light is independent of photosynthesis. Certainly, *Nostoc muscorum*, *Chlorogloea fritschii* and *Anabaenopsis circularis*, which grow and fix nitrogen in the dark (Allison *et al.*, 1937; Fay, 1965; Watanabe and Yamamoto, 1967) in the presence of suitable organic substrates, continue to produce heterocysts (Wildon and Mercer, 1963b; Fay, 1973). Further experiments are necessary to establish the role of light and how far the reported stimulation by red light and inhibition in the blue and green spectral region (Kale and

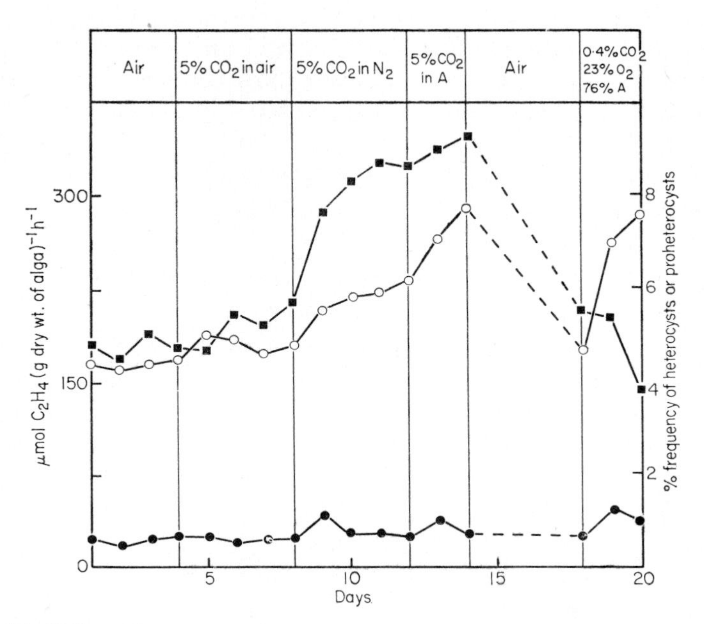

FIG. 11.32. Effect of oxygen and of nitrogen starvation on proheterocyst (●—●) and heterocyst (○—○) production and nitrogenase activity (■—■) of *Anabaena cylindrica*. After S. A. Kulasooriya, J. N. Lang and P. Fay (1972) *Proc. R. Soc.* B, **181**, 199–209, Fig. 3.

Talpasayi, 1969) are specific for the differentiation of heterocysts. Similarly, inhibition by magnesium ions and phosphate deficiency (Ogawa and Carr, 1961) require further study.

The effects of metabolic inhibitors, such as chloramphenicol, thiol compounds, azide, DNP (dinitrophenol) or CMU (*p*-chlorophenyl dimethylurea) have been examined (Talpasayi and Kale, 1967; Bahal, 1969; Bahal and Talpasayi, 1970 a,b; Tyagi 1973). Most of the above compounds inhibit heterocyst formation but their specificity remains to be established.

The genetic constitution of the alga may also govern heterocyst differentiation but nothing is known about the genetics of nitrogen fixation in blue-green algae and there is no explanation for the different heterocyst patterns in Nostocaceae, Scytonemataceae or Rivulariaceae. Fogg's (1949) hypothesis regarding heterocyst spacing is not directly applicable to the development of heterocyst pattern in algae other than members of the Nostocaceae.

The implication that heterocysts are produced in a "pre-formed pattern" (Talpasayi and Kale, 1967; Kale and Talpasayi, 1969; Wilcox, 1970), which incidentally was not confirmed by Kulasooriya *et al.* (1972), is not clear. Presumably all cells of a heterocystous alga carry the genetic information necessary for heterocyst differentiation and this differentiation is induced by the presence or absence of some specific substance whether or not this be ammonium ion.

Another question is that of the reversibility of heterocyst differentiation, the so called "de-differentiation". Kale and Talpasayi (1969) refer to heterocyst germination under this heading but obviously germination is not a simple reversal of heterocyst development. The reversibility of heterocyst development was tested by Fay and Kulasooriya (in preparation). They transferred *Anabaena cylindrica* showing the initial stages of heterocyst development from nitrogen-free medium into medium containing ammonium nitrogen and found that development could be arrested in young heterocysts but that there was no reversion of pro-heterocysts or heterocysts to a previous stage of development. This suggests that the transformation of a vegetative cell into a heterocyst is a one-directional, irreversible process, that is, a true differentiation.

Heterocysts in general are unable to divide and consequently their life span is limited (Tischer, 1957). Degenerative changes are manifest in the breakdown and depletion of cytoplasmic content followed by vacuolation. In this final stage the degenerate heterocyst may well appear as an empty cell, as the resistant envelope can survive for a longer period. The depleted heterocyst may disintegrate *in situ* or may become detached from the filament. Both result in fragmentation of the filament.

Spores

The occurrence of endospores in certain blue-green algae has already been considered in Chapter 2. The endospores of *Pleurocapsa* are formed by successive division in three planes of the enlarged mother cell. They possess an ultrastructure similar to that found in vegetative cells but with a more evident nucleoplasmic region and with dense granulation between the photosynthetic lamellae (Beck, 1963). There appears to be no other information concerning the differentiation of either endospores or exospores.

Akinetes are produced in heterocystous members of the Hormogonales and particularly in the Nostocaceae and Rivulariaceae. They are particularly resistant to adverse conditions and remain viable for long periods. They are distinguished readily under the light microscope by their large size, characteristic shape, modified pigmentation and the presence of many large cytoplasmic granules (Fig. 11.33). In contrast with that of the

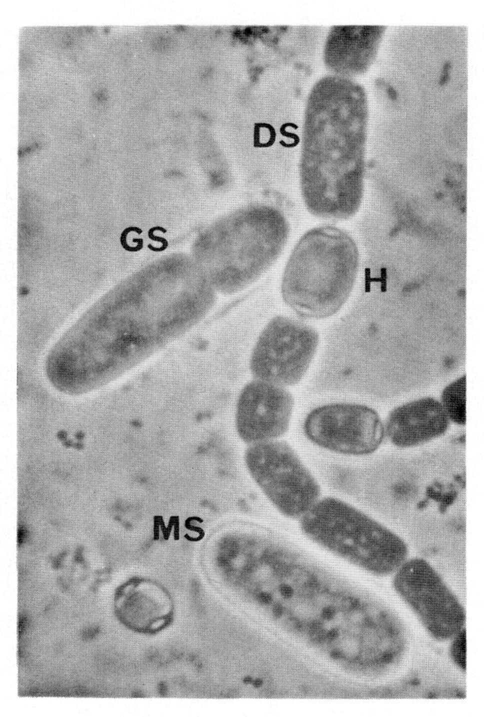

FIG. 11.33. Sporulation of *Anabaena cylindrica*. Developing spore (DS) adjacent to heterocyst (H), detached mature spore (MS) and germinating spore (GS). Phase contrast light micrograph (× 2,500), P. Fay (unpublished).

heterocyst the akinete envelope completely surrounds the cell and thus separates it from the adjacent cells (Fritsch, 1945). Under the electron microscope (Wildon and Mercer, 1963b; Leak and Wilson, 1965; Miller and Lang, 1968) sheath condensation, deposition of several envelope layers external to the original cell wall and an accumulation of large cyanophycin granules are also noted (Fig. 11.34). The mature spore retains photosynthetic thylakoids, a nucleoplasmic region and polyhedral bodies (Miller and Lang, 1968). As the vegetative cell enlarges and condensation of the mucilaginous sheath occurs there is a localized deposition within it of dense amorphous and fibrillar material. The latter gradually spreads and develops into a continuous thick sculptured layer surrounding the whole cell which in *Cylindrospermum* is protruded outward into rays as the akinete matures (Clark and Jensen, 1969). This layer presumably has a protective function. The sheath remains outside the dense fibrillar layer, inside of which further layers are deposited, with the innermost layer being formed last.

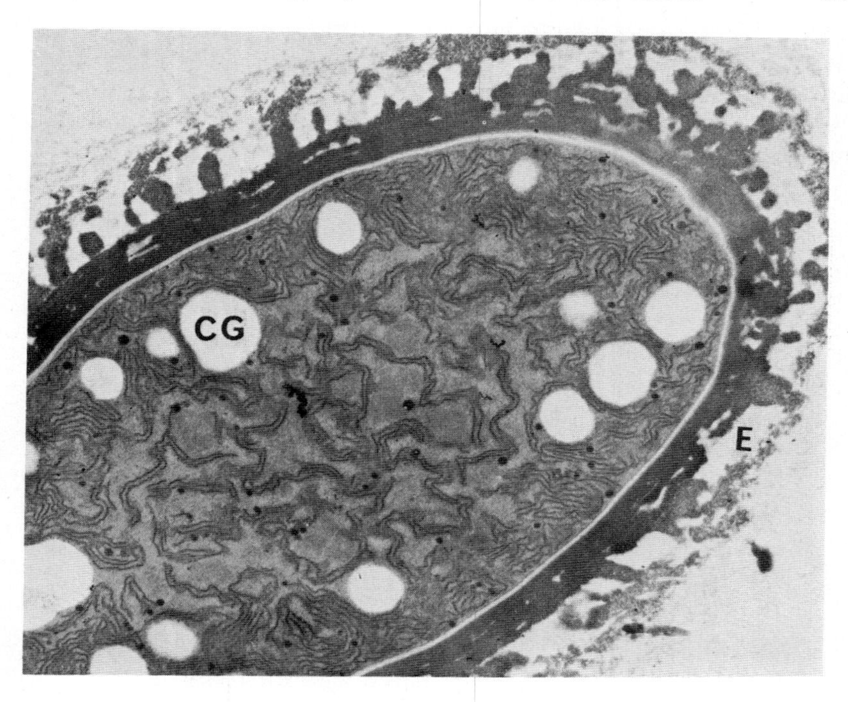

FIG. 11.34. Akinete of *Cylindrospermum* sp. with dense fibrillar envelope (E) surrounding the cell. Electron transparent spaces indicate cyanophycin granules (CG). Electron micrograph (× 9,500) by courtesy of M. M. Miller and N. J. Lang (1968) *Arch. Mikrobiol.* **60**, 303, Fig. 5.

The composition of the cell wall resembles that of vegetative cells (see p. 74) except for the thickening of the intermediate L_{III} layer (Jensen and Clark, 1969). The envelope stains with ruthenium red as does the original sheath. Cytochemical tests, while indicating the presence of polyglucoside, are negative for cellulose and for chitin (Miller and Lang, 1968). Chemical analysis of the spore envelope fraction of *Anabaena cylindrica* by Dunn and Wolk (1970) showed a composition of 41% carbohydrate, 24% amino compounds and 11% lipid on a dry weight basis.

Concomitantly with the formation of the envelope, cyanophycin granules accumulate at the periphery of the cell but polyphosphate granules disappear (Talpasayi, 1963). Polyhedral bodies and ribosomes are retained in the nucleoplasmic region. The photosynthetic lamellae become contorted and distributed throughout the protoplast. Polyglucan granules are present but their numbers gradually decrease in the resting period (Miller and Lang, 1968).

Isolated spores have a lower nitrogen content (4·8%) compared with the whole filament (7·7%). Pigment analysis suggests that in the mature akinetes phycocyanin is largely absent, chlorophyll is partly replaced by phaeophytin, the β-carotene content is reduced, and the xanthophyll content is increased (Fay, 1969b).

In batch culture the onset of sporulation coincides with the termination of the exponential phase of growth (Fay, 1969a; Fig. 11.35). Under steady

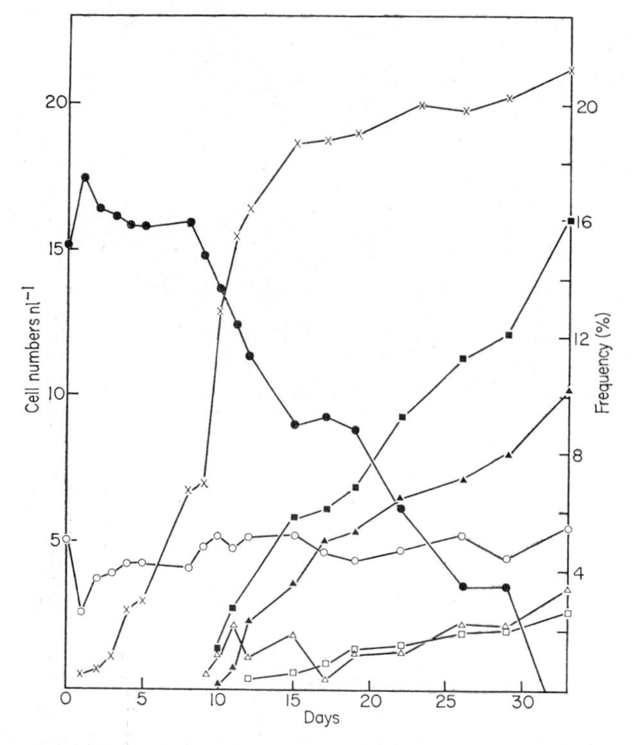

FIG. 11.35. Course of cell differentiation during growth of *Anabaena cylindrica* in culture. ×—× growth; ●—● vegetative cells; ○—○ heterocysts; △—△, developing spores; ▲—▲ mature spores; □—□ germinating spores, ■—■ total spores. After P. Fay (1969) *J. exp. Bot.* **20**, 100, Fig. 2.

state growth conditions no sporulation occurs. Spore production increases during the post-exponential phase.

The most significant single factor controlling sporulation is phosphate level (Wolk, 1965b; Gentile and Maloney, 1969). Further, the presence of sodium acetate, relatively high concentrations of calcium ion, the use of an efficient buffering agent in the medium, optimum light intensity and a variety of other factors stimulate sporulation in *Anabaena cylindrica* (Wolk, 1965b).

Although cell division ceases in the developing spore, photosynthesis continues at a gradually decreasing rate, so that a several-fold increase of dry weight results. The absence of photosynthesis in mature spores corresponds with the observed changes in the pigmentation. Nitrogenase activity is also absent but the rate of respiration per unit dry weight in isolated akinetes is higher than that in whole filaments. This high rate of respiration presumably occurs in developing and germinating spores rather than in mature ones (Fay, 1969a,b).

Germination of spores has been observed frequently in cultures (Fritsch, 1945). About 16% of akinetes were found to be germinating during the post-exponential phase of growth in a culture of *Anabaena cylindrica* (Fay, 1969a; Fig. 11.36). The first sign of germination is a decrease in the

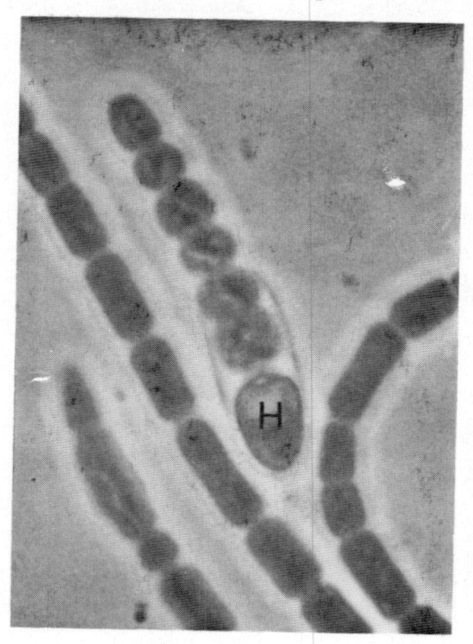

FIG. 11.36. Germinating akinete of *Anabaena cylindrica*. Germling with a terminal heterocyst (H). Phase contrast light micrograph (× 2,500) from P. Fay (unpublished).

number of cyanophycin granules, which coincides with the recovery of pigmentation and with the appearance of numerous small thylakoid vesicles (Lang and Fisher, 1969; Fig. 11.37). Cell division usually begins while the germling is still within the envelope but sometimes starts only after the germling has emerged (Fig. 11.37).

Little is known about the factors which induce germination. According to an early observation by Harder (1917), akinete germination in *Nostoc*, *Anabaena* and *Cylindrospermum* species takes place only in the light except

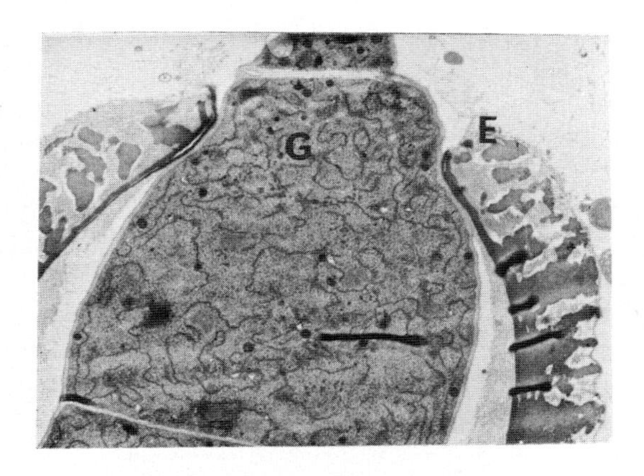

FIG. 11.37. Portion of a germinating akinete of *Cylindrospermum* sp. showing germling (G) emerging from spore envelope (E). Electron micrograph (× 6,500) by courtesy of M. M. Miller and N. J. Lang (1968) *Arch. Mikrobiol.* **60**, 303, Fig. 7.

where sucrose is supplied, when it can occur in the dark. Kaushik and Kumar (1970) have suggested that light stimulation of spore germination of *Anabaena doliolum* is not associated with photosynthesis.

Development and life history

In batch culture reproductive bodies are usually produced at some definite stage of growth of blue-green algae. This indicates that a distinct life cycle may be possessed by some species although its precise nature varies considerably according to conditions. The first suggestion that a distinct alternation of morphological stages did, in fact, occur was made by Sauvageau (1897) but even now only a limited amount of information is available and that has been obtained mainly with three species: *Nostoc muscorum* (Lazaroff and Vishniac, 1961, 1962), *Chlorogloea fritschii* (Fay *et al.*, 1964) and *Nostoc commune* (Robinson and Miller, 1970).

The life history of *Chlorogloea fritschii* is simpler and may be considered first. Fay *et al.* (1964), using quasi-synchronous cultures, found changes in morphological characters related to variation in chemical composition of the alga. Four stages in the developmental cycle were distinguished (Fig. 11.38).

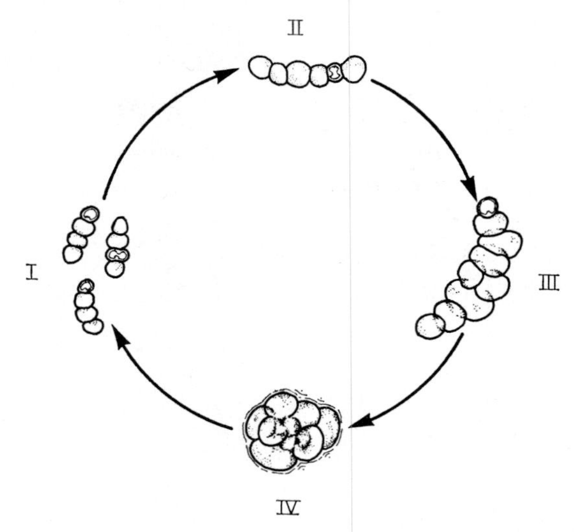

FIG. 11.38. Diagram showing the four stages in the life history of *Chlorogloea fritschii*. After P. Fay (unpublished).

Stage I. Endospores, which remain adhering to each other, form short motile filaments, usually of 4–6 cells, resembling a hormogonium. These filaments are liberated from the parent envelope. The cells are small and pale blue-green in colour. A single heterocyst is usually present in a terminal or intercalary position.

Stage II. The hormogonium develops into a longer filament by growth and transverse divisions of the cells, which become deep blue-green in colour.

Stage III. Cell division also takes place in the longitudinal plane so that the filamentous character gradually disappears. The cells increase in size, become granulated and rather polygonal in shape.

Stage IV. The cells become larger, more granulated, yellowish green in colour and divide to form endospores.

Cells in stage II, which predominate during active growth, show the highest pigment content and are the most active in nitrogen fixation. Under extreme conditions of high temperature and low light intensity stage I type short filaments are formed almost exclusively, while at intermediate conditions of light and temperature, stage IV cells are prevalent (Findlay *et al.*, 1968). The morphological features observed at different stages are related to changes in ultrastructure (Peat and Whitton, 1967;

Findlay *et al.*, 1970). In young filaments of stages I and II the thylakoids run parallel to the cell surface while at the endospore stage IV the thylakoids are scattered throughout the protoplast.

Lazaroff and Vishniac (1961, 1962) observed a more complex state of affairs in *Nostoc muscorum* which has several stages within two main cycles, the so-called heterocystous and sporogenous cycles (Fig. 11.39).

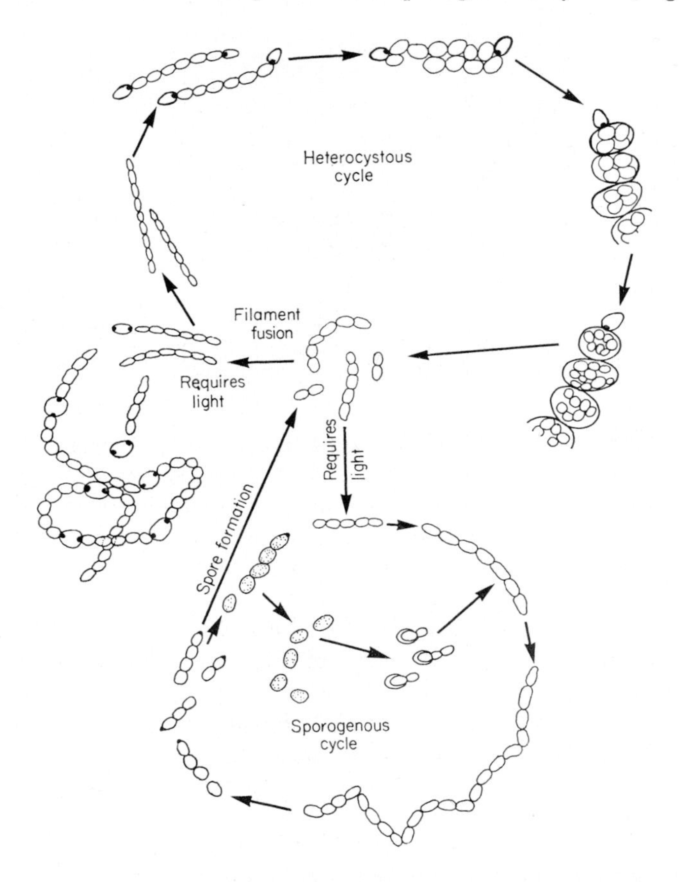

FIG. 11.39. Postulated alternation of sporogenous and heterocystous generations by *Nostoc muscorum* A. After N. Lazaroff and W. Vishniac (1962) *J. gen. Microbiol.* **28**, 203, Fig. 4. For details see text.

In the heterocystous cycle motile hormogonia are produced and eventually come to rest, the terminal cells differentiating into heterocysts. The intercalary cells continue to grow and by successive divisions each produces a cluster of four cells within the parent envelope. This is termed the "aseriate" stage. Development also proceeds to this stage when the alga is grown heterotrophically in the dark but further development is then arrested unless the cultures are illuminated. However, development continues in the dark after a short exposure to light as well as in continuous light. After activation by light, the clusters of cells give rise to short motile filaments. According to Lazaroff and Vishniac (1962) such filaments may join end to end in a sexual process of filament anastomosis which occasionally results in a spore-like structure. This unique observation has not yet been confirmed. The spores on germination give rise to a new motile filament. As the filaments grow and mature, their sheaths dissolve and new hormogonia emerge. The heterocystous cycle may be repeated.

The short motile filaments which develop from the four-celled clusters in the aseriate stage may alternatively pass through the sporogenous cycle. This consists of filament growth with differentiation of heterocysts and akinetes. The germling emerging from the akinete coat can continue its development through a heterocystous cycle.

Development of the alga arrested in the dark at the aseriate stage can be induced by exposure to red light (650 nm). Photoinduction was found later (Lazaroff, 1966) not to be an absolute requirement for development. Differentiation can also occur after a prolonged (2 months) incubation in total darkness. The role of light may thus be in the acceleration of a process which occurs slowly in the dark.

Robinson and Miller (1970) have reported a similar photocontrolled developmental cycle in *Nostoc commune*. The development of motile trichomes was blocked, however, at the aseriate stage not only by exposure to green light but also by white light from fluorescent lamps. Apart from this difference, red light promoted the formation of motile trichomes while green light reversed the photoactivation as in *N. muscorum*.

Lazaroff (1966) demonstrated the presence of allophycocyanin in *N. muscorum* and showed that its absorption spectrum matches well the action spectrum for photoactivation. However, he could find no single pigment which could be regarded as receptor for the photoreversal, which has its minimum between 500 and 600 nm. The photoreversibility of the process suggests the presence of a phytochrome-type photomorphogenetic pigment with photoconvertible forms having absorption maxima in the red and green spectral region respectively (Lazaroff, 1966; Robinson and Miller, 1970). Phytochromes are chemically similar to phycobiliproteins and phytochrome-like morphogenetic pigments have been demonstrated in several algae including the red alga, *Porphyra tenera* (Dring, 1967), but their presence in blue-green algae does not seem to have been proved conclusively yet.

Forest (1968) considers that the Lazaroff type cycles may occur in other

Nostoc species "by allowing for variations, principally deletions, truncations and/or prolongations of stages" and considers that it may also apply to other Nostocaceae such as species of *Anabaena* and *Cylindrospermum* and to members of the Rivulariaceae. Such developmental cycles could explain perhaps why algae such as *Aphanizomenon* although forming distinct colonies in nature lose their colonial habit when grown in culture. The cycles could also explain in part why different morphological variations of the same alga show different metabolic characteristics.

Kantz and Bold (1969) studied the life cycle of 25 isolates of the genera *Nostoc* and *Anabaena*. They found that the life cycles of the *Nostoc* species do not show the complexity described for *N. muscorum* by Lazaroff and Vishniac (1962, 1964) and that they can be divided into two groups. In the so-called "*piscinale*" type, the mature mass of filaments lacks a common outer layer of "pellicle" and reproduction occurs by the formation of motile hormogonia and by the germination of akinetes in addition to filament fragmentation. In the "*commune*" type, the colony is contained by a hyaline sheath with a firm pellicle, and the trichomes are tightly coiled. New colonies are produced by protrusions ("budding") of the ends of trichomes through the pellicle and by hormogonium formation accompanied by the rupture of the pellicle. While in *Nostoc* species the motile (reproductive) and non-motile (vegetative) phases occur regularly, clearly defining the life cycle, no such regular alternation of phases was observed in *Anabaena* species. In these, the motile stage is identical with the vegetative phase, and the non-motile phase develops merely as a response to unfavourable conditions.

Chapter 12

Freshwater ecology

Blue-green algae are widespread and often abundant in freshwater. Although they give the impression of being favoured by warm and nutrient-rich conditions, there are many exceptions to this generalization. Similarly, although they are found characteristically in neutral, and particularly in alkaline waters, certain species sometimes grow well in acid waters; *Chroococcus turgidus*, for example, is common in *Sphagnum* bogs of pH about 4. Blue-green algae may be most conspicuous as dense plankton blooms (Fig. 12.1—facing p. 256) but epilithic and epiphytic forms may be equally abundant in certain situations. They occur both in static and in running waters.

Planktonic freshwater forms

1. OCCURRENCE

These belong mostly to the Chroococcales and Nostocales. In the former, the genera *Coelosphaerium*, *Gomphosphaeria*, *Merismopedia* and *Microcystis* are the most commonly represented. Certain planktonic organisms which have been described as bacteria properly belong here; for example, *Pelogloea bacillifera*, a sometimes abundant plankton form which once was considered to be related to the bacterium *Chlorobium* (van Niel and Stanier, 1959), and *Synechococcus* sp. which are small unicells without gas vacuoles (Bailey-Watts *et al.*, 1968). The other planktonic Chroococcales have comparatively large spherical or irregular colonies of numerous cells embedded in mucilage. The common filamentous planktonic forms belong to the genera *Arthrospira*, *Oscillatoria*, *Lyngbya*, *Spirulina*, *Anabaena*, *Anabaenopsis*, *Aphanizomenon* and *Gloeotrichia*. The filaments may be free-floating, as in *Oscillatoria*, in bundles, as in *Aphanizomenon*, or in clumps surrounded by mucilage, as in *Anabaena*. Apart from *Synechococcus*, planktonic blue-green algae usually have gas vacuoles. *Anabaena*, *Anabaenopsis*, *Aphanizomenon* and *Gloeotrichia* spp. commonly produce akinetes.

Having gas vacuoles, these blue-green algae may collect at the water surface in calm weather and then give the impression of being the dominant and most productive plankton group. However, this is not necessarily so, for quantitative investigation shows that their total biomass is not exceptionally high and does not often exceed that of other types of planktonic algae present.

Pearsall (1932), who carried out a study of the composition of the phytoplankton of the English Lakes in relation to dissolved substances, was among the first to make generalizations about the factors favouring the growth of blue-green algae. He showed that the maximum proportion of blue-green algae in the plankton of a particular lake was related to the concentration of "albuminoid ammonia" (a crude measure of dissolved organic matter) which was found a month previously (Fig. 12.2) and also that blue-green algae tended to grow most rapidly when concentrations of inorganic nutrients such as nitrate and phosphate were lowest.

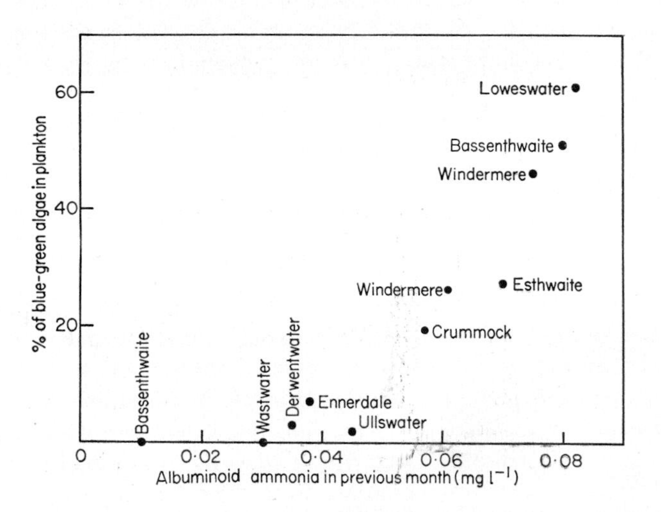

FIG. 12.2. Abundance of blue-green algae in the plankton of lakes in the English Lake District in relation to concentration of "albuminoid ammonia" (an approximate measure of dissolved organic matter) in the previous month. Data of W. H. Pearsall (1932) *J. Ecol.* **20**, 241–262.

The idea that growth of planktonic blue-green algae is favoured by dissolved organic matter has been supported by observations in different parts of the world (Singh, 1955; Brook, 1959; Vance, 1965; Horne and Fogg, 1970) although there are a few instances of blue-green algae being

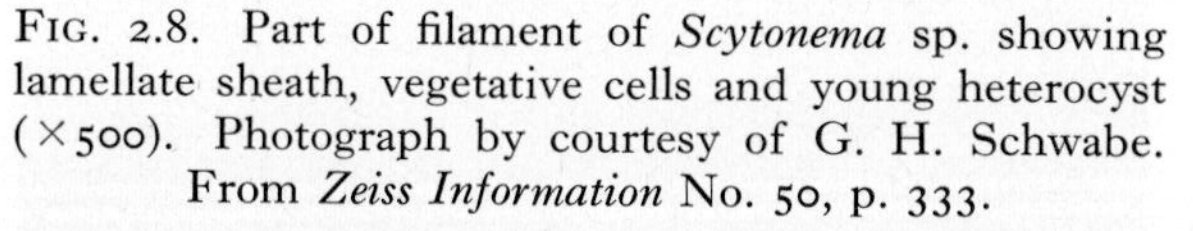

FIG. 2.8. Part of filament of *Scytonema* sp. showing lamellate sheath, vegetative cells and young heterocyst (×500). Photograph by courtesy of G. H. Schwabe. From *Zeiss Information* No. 50, p. 333.

FIG. 12.1. Surface bloom of *Microcystis aeruginosa* in a Scottish loch (photograph M. J. Daft and W. D. P. Stewart).

FIG. 12.8. A thermal stream in Yellowstone National Park showing blue-green algae in mid stream (mainly *Synechococcus*) and along the edge of the stream where they appear as a black leathery growth. From W. D. P. Stewart (1970) *Phycologia* **9,** 261–268.

FIG. 14.7. Bundles of *Trichodesmium erythraeum* from the Indian Ocean (×40) (photograph by courtesy of P. M. David).

dominant in waters with apparently low concentrations of organic substances. A species of *Anabaena*, for example is the main phytoplankton organism in Crater Lake, Oregon, which has no obvious source of organic contamination (Utterback *et al.*, 1942). From work described in other chapters it is apparent that many blue-green algae are obligate phototrophs and that no freshwater species require exogenous organic growth factors. In experiments in which columns of lake water, isolated *in situ* by plastic cylinders, were enriched with thiamine (vitamin B_1) it was found that whereas growth of plankton algae belonging to most groups was stimulated, that of blue-green algae appeared to be reduced (Hagedorn, 1971).

In establishing cultures in inorganic media, the presence of organic chelating agents is helpful (Allen, 1952) but blue-green algae are not exceptional in this respect and in any case produce extracellular products with complex-forming properties during the course of their growth (Fogg and Westlake, 1955). Studies in laboratory culture of *Anabaena flos-aquae* (Tischer, 1965) and *Oscillatoria rubescens* (Staub, 1961) have given no indication that these plankton species have an absolute requirement for any organic substance. Lange (1970) reported that various species of planktonic blue-green algae grew more rapidly and reached higher population densities in bacterized culture under carbon dioxide deficient conditions when an organic substrate such as sucrose was supplied. This stimulation is evidently due to bacterial production of carbon dioxide since it does not occur when the cultures are aerated. Although this effect may be important in dense blooms under static conditions, it is unlikely to provide a general explanation of the correlation of blue-green algae with dissolved organic matter because carbon dioxide is rarely severely limiting under lake conditions (Fogg, 1965). Since blue-green algae liberate relatively large amounts of extracellular organic material (Fogg, 1952) the possibility must be considered that their abundance in the plankton is the cause rather than the effect of high concentrations of dissolved organic matter. However, since dissolved organic matter in lake water normally exceeds by 5–20-fold that in phytoplankton cells, even in unpolluted waters (Hutchinson, 1957), this explanation seems unlikely. The correlation of abundance of blue-green algae with dissolved organic matter thus does not seem to depend on any simple direct mechanism.

Pearsall's other generalization, that blue-green algae grow best when nutrient concentrations are lowest, is supported by many observations (Hutchinson, 1967). In the case of planktonic blue-green algae which fix nitrogen, good growth can, of course, occur in the absence of combined nitrogen but species which apparently do not fix nitrogen, such as *Microcystis aeruginosa*, may be as abundant as nitrogen-fixing forms at times of nitrate deficiency. The reason for this may be that the algae store previously available nitrogen in the form of cyanophycin granules, or phycocyanin which they use under nitrogen-limiting conditions (see Stewart, 1972). Similarly, growth in waters low in phosphorus may be due to the algae utilizing phosphorus which they had accumulated previously as polyphosphate (Stewart and Alexander, 1971) (see p. 65). In general, growth maxima of blue-green algae tend to occur some weeks after nutrient concentrations have decreased. It is paradoxical that although they tend to develop at times of nutrient deficiency planktonic blue-green algae are nevertheless characteristic of waters receiving high nutrient inputs. Thus Vollenweider (1968) from a survey of the available information concluded that massive development of blue-green algae is likely if nutrient concentrations exceed 10 mg P m^{-3} and 200–300 mg N m^{-3} in the spring and/or the specific supply loading per unit area of lake reaches 0·2–0·5 g P m^{-2} per year and 5–10 g N m^{-2} per year. He did qualify this, however, by saying that many other factors are concerned and that the situation is complex and not yet properly understood.

2. SUCCESSION AND PERIODICITY

Attempts to identify the factors which favour the growth of planktonic blue-green algae have usually depended on correlation of the growth with general levels of various physico-chemical factors in the water. However, algal populations often show highly localized concentrations and generally the water bodies in which they occur show marked physical and chemical stratification. The factors governing bloom development might be clearer if the levels of factors obtaining in the precise environment in which growth occurs were considered. As a necessary preliminary to this an account of the periodicity of planktonic blue-green algae will be given.

The observations of Nauwerck (1963) on Lake Erken, a eutrophic lake of approximately 23 km^2 area and 9 m mean depth near Uppsala in Sweden, give a picture of a common type of succession and periodicity. Although most abundant in summer and autumn, blue-green algae were scarcely ever absent from the plankton. In 1957 colonies of *Coelosphaerium naegelianum* were nearly always present, even in the winter, when they were found concentrated near the bottom and, to a lesser extent, under the ice. Following the break-up of the ice at the end of April, a small species of *Eucapsis* developed strongly, reaching a concentration of 10^5 colonies per litre. Diatoms, chrysomonads, green algae and dinoflagellates predominated in the early summer and the main maximum of blue-green algae, consisting of *Anabaena flos-aquae*, *Coelosphaerium naegelianum* and *Aphanizomenon flos-aquae*, occurred in July. This maximum developed quickly and was concentrated in the top 4 m of the water. Whereas

Anabaena and *Gloeotrichia echinulata* which was present in smaller amount, were largely confined to the surface, *Aphanizomenon* was more generally distributed in depth. At this time, when there were about 10^3 colonies of *Anabaena* and 8×10^3 bundles of *Aphanizomenon* per litre, blue-green algae comprised over 60% of the total volume of the phytoplankton. In August blue-green algae were replaced by a maximum population of the dinoflagellate *Ceratium hirundinella*. In a second maximum in late summer and autumn other blue-green algae, including *Microcystis* spp., *Gomphosphaeria lacustris*, *Aphanothece* sp., *Aphanocapsa* sp. and *Chroococcus limneticus*, made their appearance although *Aphanizomenon* and *Coelosphaerium* predominated. The filamentous forms, *Oscillatoria* and *Lyngbya*, were of minor importance in Lake Erken in 1957.

A similar succession has been described by Reynolds (1971) for the Shropshire and Cheshire meres. Again *Microcystis* tends to develop later in the year, after *Anabaena* spp. and *Aphanizomenon*.

Among the factors which determine the abundance and succession of blue-green algae in the plankton nitrogen fixation deserves some consideration. Blue-green algae appear to be the main agents of fixation in the freshwater plankton and so far appreciable rates have been found *in situ* only when heterocystous species are present (see pp. 190 and 329). In terms of the total nitrogen income of a lake the contribution of nitrogen fixation by plankton algae is generally small but in particular water layers it may, at certain times, be the major source of combined nitrogen. Hutchinson (1967) has reported an instance of nitrogen fixation being the

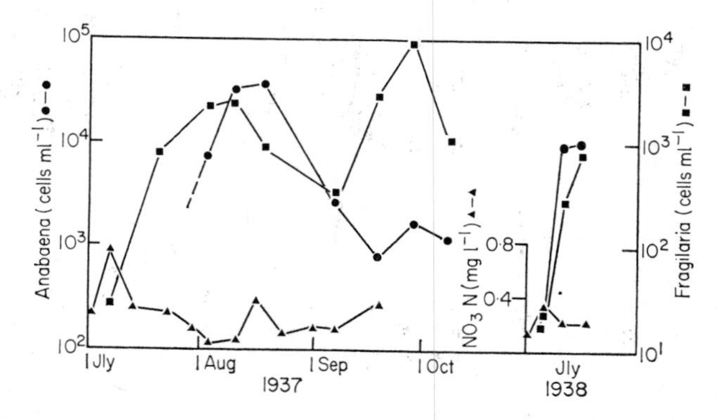

FIG. 12.3 Changes in populations of *Anabaena circinalis* (—●—) and of *Fragilaria crotonensis* (—■—) in Linsley Pond, Connecticut. In 1937 *Fragilaria* increased before *Anabaena* at a time when nitrate concentration (—▲—) was initially high. In 1938, with a lower nitrate concentration, *Anabaena* started to increase before *Fragilaria*. Redrawn after G. E. Hutchinson (1967), "A Treatise on Limnology", Vol. 2. (Wiley, New York).

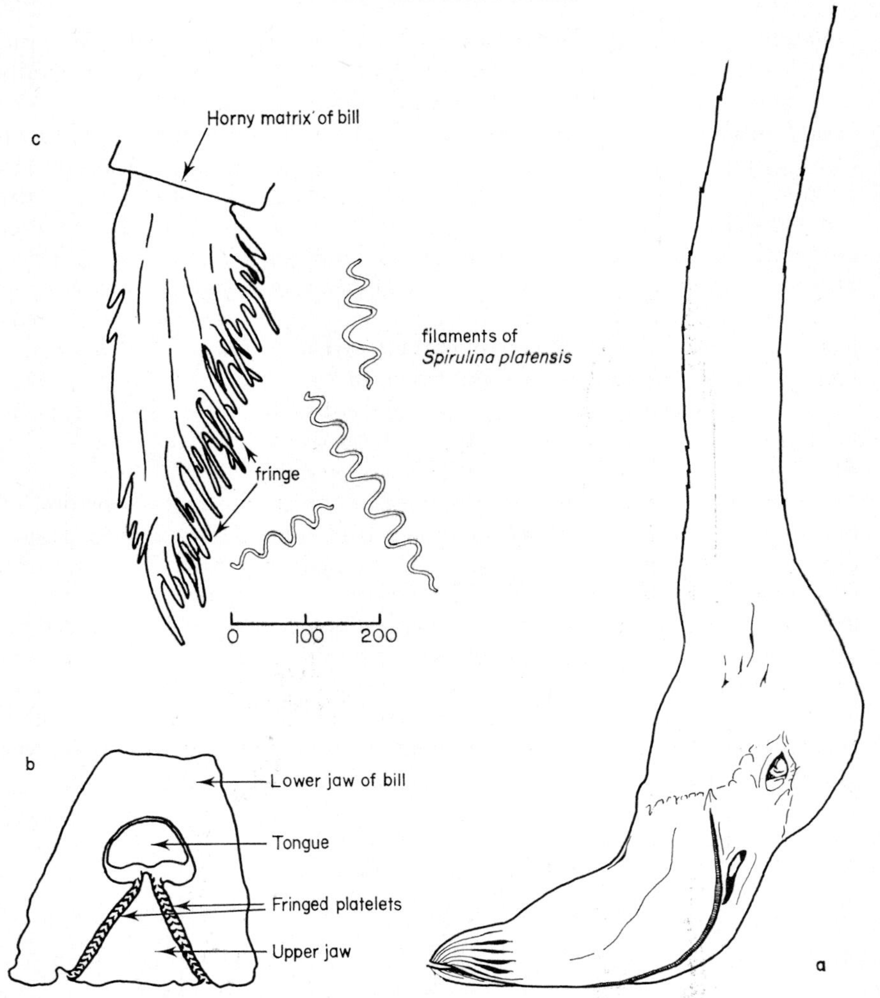

FIG. 12.4. The bill of the lesser flamingo *Phoeniconaias minor*—a mechanism for filter feeding on blue-green algae and diatoms. The bird feeds on planktonic algae by sweeping its head from side to side near the surface of the water as it walks forward. *Spirulina platensis*, a gas-vacuolate blue-green alga which forms surface waterblooms in the saline lakes of East Africa, is one of the principal food organisms. The carotenoids of the alga are deposited in the feathers of the bird and account for the pinkish-red coloration of its plumage. a. Head of the flamingo in the feeding position. The inverted bill is so angled that the upper jaw lies horizontally at the water surface. b. Cross section of the inverted bill. The piston-like tongue, which lies in a groove in the deep-keeled lower jaw, moves to and fro pumping water over the fringed platelets lining the gap between the two jaws. c. Fringed platelets of the jaw (which filter out the alga) and coiled filaments of the alga *Spirulina platensis* enlarged to the same scale. (Redrawn from the photographs and drawings of Jenkin, 1957.)

apparent determining factor in the succession of *Anabaena circinalis*, which almost certainly fixed nitrogen, and the diatom *Fragilaria crotonensis*. Both algae develop in Linsley Pond, Connecticut, at low nutrient concentrations and appear to have similar requirements. In the summer of 1937 *Fragilaria* developed first, at a time when the nitrate–nitrogen concentration was slightly over 0·7 mg l^{-1}, to be succeeded by *Anabaena* when this concentration had fallen to about 0·1 mg l^{-1}. In the following year when the nitrate-nitrogen concentration was lower, 0·3 mg l^{-1}, the *Anabaena* maximum preceded that of *Fragilaria* (Fig. 12.3).

Tolerance to high salt concentrations is a feature that enables blue-green algae to dominate the plankton of saline lakes whether these be of the chloride, sulphate or carbonate type. Lake Magadi in the Rift Valley, Kenya, for example, which is saturated with sodium carbonate (concentration up to 11·4%) and has a pH of around 10, has a dense growth of plankton all the year round consisting of *Oscillatoria*, *Merismopedia*, *Chroococcus*, *Synechocystis* and *Anabaena* species. *Arthrospira* is another genus characteristic of carbonate waters. Such blooms of blue-green algae provide the major source of food for the immense flocks of flamingoes which frequent the African lakes. In fact the bill of the flamingo is especially modified to permit feeding from material floating at the water surface (Fig. 12.4). The pink colour in the feathers of these birds is said to be derived from the carotenoids of the blue-green algae on which they feed. Growths of blue-green algae also occur in brine during salt manufacture. Singh (1961) mentions, for example, that *Anabaenopsis circularis* is abundant in brine containing 14–17% of salts from the Sambrha salt lake in Rajasthan.

3. STRATIFICATION AND BUOYANCY REGULATION

Species of *Oscillatoria* and *Lyngbya* illustrate particularly clearly the phenomenon of well defined population maxima at certain depths, a feature which is probably shown to some extent by most planktonic blue-green algae in thermally stratified waters. Of the many examples that have been reported that of *Oscillatoria rubescens* in Lake Lucerne, Switzerland, will be described here. This lake, which is 71 m deep, was once oligotrophic but following eutrophication with sewage effluents now regularly develops a bloom of this alga. Figure 12.5, drawn from the data of Zimmermann (1969), shows the seasonal cycle in 1966. In March when the lake was isothermal the population was uniformly distributed in the top 20 m. By May, when thermal stratification was beginning, a population maximum appeared at around 7·5 m. This maximum sank to 12·5 m in July and the *Oscillatoria* became rare in the upper 10 m. Thereafter the maximum rose to 5 m in September. Mixing of the water by a storm brought part of the population to the surface in October and by December the vertical distribution was once more uniform. During the period May to October the total population in the 20 m water column increased only slightly, from a mean of 0·448 to 1·105 mm^3 cell material l^{-1}. *Oscillatoria rubescens* is the

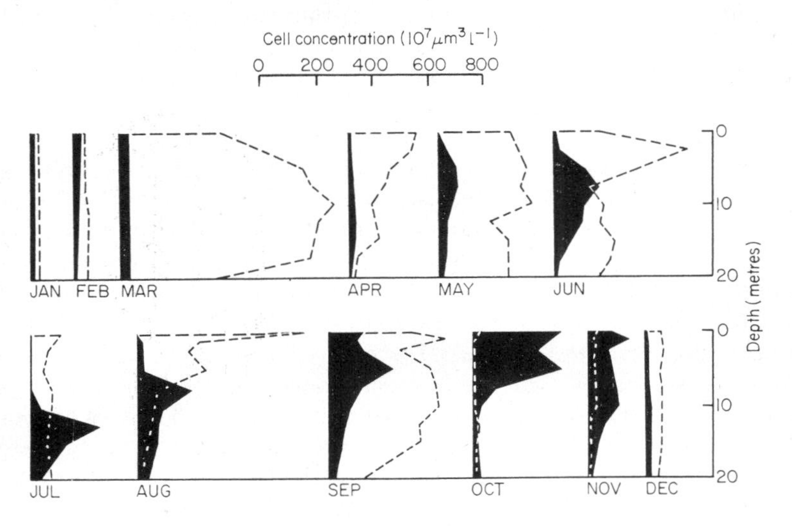

FIG. 12.5. Vertical distribution of *Oscillatoria rubescens* (black) and other phytoplankton (dotted lines) in the Vierwaldstättersee (Lake Lucerne) Switzerland, in 1966 (redrawn after U. Zimmermann (1969) *Schweiz. Z. Hydrol.* **31**, 1–58).

dominant plankton form in Lake Lucerne at all times of the year. Other blue-green algae occur to a much lesser extent, the diatom *Tabellaria fenestrata* and various flagellates being the other major forms. Although these other plankton algae often show a non-uniform vertical distribution they do not show consistent population maxima as the *Oscillatoria* does.

Brook *et al.* (1971), who have studied similar populations of blue-green algae in Minnesota lakes found that the maxima occur at constant depths at any one time regardless of the horizontal position, that is, they stretch as a level layer from shore to shore. After complete mixing of the water column enclosed in a plastic cylinder *in situ* the population maxima would re-establish after a few weeks at the same depth as before. The relative depths at which different species occur are remarkably constant from lake to lake.

In the tropics planktonic blue-green algae are usually abundant throughout the year. Lake George, Uganda, is an extensive eutrophic lake of 2 m average depth with no permanent stratification. Blue-green algae are the major components of the plankton and a dense bloom of *Microcystis* and *Anabaenopsis* spp. is present all the year. Chlorophyll *a* concentrations are about 250 mg m^{-3} and the euphotic zone is consequently limited to about 0·4 m. Although the water column is nearly isothermal, the population becomes concentrated at the surface in the morning and sinks to the bottom by evening in calm weather. Under windy conditions the phytoplankton becomes distributed evenly throughout the water body (Ganf, 1969).

It cannot be expected that any general hypothesis of the nature of the factors controlling the growth of planktonic blue-green algae will account for the variety of behaviour indicated in the previous paragraphs. Nevertheless it seems possible that the buoyancy-regulating mechanism provided by the gas vesicles may provide a key in many cases. As has been explained (p. 106) these structures develop most abundantly in the cells in dim light and collapse as a result of vigorous photosynthesis. We may therefore envisage that these opposing processes will result in the cells becoming poised at some intermediate depth in the water column as, in fact, observed for such species as *Oscillatoria rubescens*. Formation or collapse of gas vesicles is evidently sufficiently rapid to result in diurnal up-and-down movement, at least with species having large colonies with low form resistance to movement through water. Evidence that vertical movement of this sort does occur has been provided by Sirenko *et al.* (1968) for *Aphanizomenon flos-aquae* and by Ganf (1969) for *Microcystis aeruginosa*.

Buoyancy regulation might be of biological advantage in two ways. One might be that the alga becomes positioned at the point in some physico-chemical gradient most favourable for its growth. The other might be that the steepening of concentration gradients resulting from movement through the water produces "forced convection" (Munk and Riley, 1952) and so enhances uptake of limiting nutrients.

Various factors, such as light, temperature and concentrations of nitrate, phosphate and oxygen, important for the growth of blue-green algae show more or less marked gradients with depth. Eberley (1967) has pointed out that the effects of these factors are interrelated so that with limiting nutrient concentrations optimum growth conditions are attained at low light intensities and low temperatures. With higher concentrations of nutrients light and temperature optima might be expected to be increased so that the population would be better suited higher in the water column. Conversely, the population would find optimum conditions low in the water column if nutrient concentrations decreased. It is, in fact, generally observed that downward movement of population maxima occurs in the summer as light intensities increase and nutrient concentrations decrease. Thus Zimmermann (1969) found that as the *Oscillatoria rubescens* maximum in Lake Lucerne moved downwards from 7·5 m in June to 12·5 m in July (Fig. 12.5) the phosphate concentration fell from 1·9 to 0·6 μg P l^{-1} at these respective depths. Later the maximum rose to 5 m, where there was 1·8 μg P·l^{-1}, in September. During this period the light transmission of the water was nearly constant and the temperature in the epilimnion rose from 18 to 21°C.

Baker *et al.* (1969) found that *Oscillatoria agardhii* was able to photosynthesize more rapidly when exposed to much greater light intensities and temperatures than those obtaining *in situ* at the depth of the population maximum but it would be expected that in the absence of added nutrients these high rates could not be maintained. The sharp localization of these natural planktonic populations may perhaps be related to the limited metabolic flexibility of blue-green algae, resulting from lack of enzyme induction and repression (see p. 177). This lack of flexibility might necessitate closely circumscribed levels of environmental conditions for balanced metabolism and active growth. Another circumstance which may determine the optimum depth for growth of nitrogen-fixing species may be a differential effect of light intensity on carbon and nitrogen assimilation. Cobb and Myers (1964) found that photostimulation of nitrogen fixation by *Anabaena cylindrica* is saturated at lower intensities than is photosynthesis. It may be, therefore, that the correct C : N balance for growth cannot be achieved at high light intensities. In this respect it is perhaps of interest that the one planktonic alga which seems to survive best at the surface of lakes, *Microcystis aeruginosa*, appears not to fix nitrogen.

It is also noteworthy that population maxima tend to occur towards the bottom of the photic zone, where the light intensity is around 1 per cent of that at the surface. If the respiration rate is low such intensities may still suffice to support active growth by normal photosynthesis. In addition there is the possibility that growth is partially sustained by the photoassimilation of organic substances. As we have seen, many blue-green algae can grow in darkness on organic substrates and the ability to assimilate simple organic compounds in the light may be even more widespread (p. 155). The biological advantage of photoassimilation in dimly lit situations is that by starting with an already reduced source of carbon the limited assimilatory power from the photochemical reactions may be used to provide for a much greater amount of growth than would be possible with carbon dioxide as the carbon source. Since substrates such as glucose, acetate (Hobbie and Wright, 1965), and glycollate (Fogg *et al.*, 1969) are found in eutrophic waters in concentrations of the order of 0·05 mg l^{-1} it seems possible that photoassimilation could allow increased growth of plankton algae at the bottom of the photic zone and that this might partially explain the correlation between the abundance of planktonic blue-green algae and the concentration of dissolved organic matter. There is, however, no direct evidence, as yet, of this taking place under natural conditions.

Oxygen concentration, which often shows marked changes with depth, is probably another important interacting factor. As already mentioned (p. 167) metabolic functions such as nitrogen fixation and photosynthetic carbon fixation in *Anabaena flos-aquae* are inhibited by high levels of oxygen and it is likely that in the light blue-green algae in general grow more rapidly under microaerophilic than under fully aerobic conditions (Stewart and Pearson, 1970). Blooms of blue-green algae are certainly more prominent in lakes which tend to become depleted of oxygen in the hypolimnion in summer. Among the lakes studied by Zimmermann (1969), for example, the maximum population achieved by *Oscillatoria rubescens* is much greater (7·63 mm^3 cell material l^{-1}) in the Mauensee, which becomes anaerobic below 7 m in the summer, than in Lake Lucerne (1·10 mm^3 l^{-1}), the bottom waters of which never become depleted of oxygen. It is possible that herein is another possible explanation of the correlation between the abundance of planktonic blue-green algae and the concentration of dissolved organic matter, for oxygen depletion is likely to be greater when this is high. It may also be noted that the importance of associated bacteria in the development of blooms, which Lange (1970) has emphasized, depends on reduction of the oxygen concentration in the immediate vicinity of the alga by bacterial respiration as much as on increased carbon dioxide production. However, from the available data it is difficult to discern any clear relationship between the depth of a blue-green population maximum and the depth profile of oxygen concentration at any given moment. Sirenko *et al.*, (1968) reported a negative correlation between numbers of *Aphanizomenon* colonies in surface samples from a tank and the oxygen concentration in the water but the extensive data of Zimmermann (1969) on distribution of *Oscillatoria rubescens* and oxygen concentrations show no simple relationships. Any tendency for the population maximum to occur where low oxygen concentration favours growth appears to be obscured by oxygen production by photosynthesis of the algae themselves —indeed, population maxima of blue-green algae are sometimes associated with metalimnetic oxygen maxima (see, for example, Baker *et al.*, 1969).

In this discussion it has been assumed that buoyancy regulation is linked to growth in such a way as to maintain the alga in an optimum position with respect to gradients of different factors such as light intensity, temperature, nutrient concentrations and oxygen concentration. The integrative mechanism allowing such refined adjustment to a complex of environmental factors might conceivably prove to be a simple one. Thus, if nutrients are in short supply photosynthesis will result in accumulation of sugars, the rise in turgor will collapse gas vacuoles and the alga sink to lower light intensities. If, on the other hand, nutrients are in ample supply, the products of photosynthesis will largely be used for the manufacture of osmotically inactive cell material so that gas vacuoles will remain uncollapsed and the alga rise towards higher intensities (Dinsdale and Walsby, 1972). There is also some indication that the growth of planktonic blue-green algae does depend on the maintenance of a position in a stratified water column because the artificial circulation of a water body is sometimes particularly efficacious in suppressing their growth (Metropolitan Water Board, 43rd report; Wirth and Dunst, 1967). However, it must be added that artificial circulation also reduces the growth of other types of algae. If there is, in fact, some special effect it might be that the need of blue-green algae for closely circumscribed levels of certain environmental factors could account for their relative absence from the turbulent waters of the open oceans.

There can be little doubt on theoretical grounds (see Hutchinson, 1967) that movement relative to the water will promote the uptake of nutrients and that consequently those blue-green algae which show appreciable diurnal migration as a result of alternate formation and deflation of gas vacuoles should achieve greater growth than they would if they stayed suspended at a fixed depth. It may even be possible that such migration might carry colonies down into the nutrient-rich water below the photic zone in the evening to rise again, recharged, in the morning as gas vacuoles are reformed to carry out photosynthesis in the depleted water layers above. In some such way the paradox of planktonic blue-green algae apparently growing most vigorously when nutrients are in shortest supply may be resolved.

The discussion thus far has centred around forms such as species of *Oscillatoria* which normally remain concentrated at a depth of several metres. These occur as single filaments of small diameter having a high resistance to movement through the water and their consequent slow response to changes in buoyancy lends itself to precise poising at a particular depth. There remains the problem presented by the often striking surface scums of species of *Anabaena*, *Microcystis*, *Aphanizomenon* and *Gloeotrichia*. These species characteristically occur in large colonies which sink or rise through the water relatively rapidly (see p. 106). If a large colony forms gas vacuoles in dim light it may rise quickly in the water and, overshooting the optimum depth for photosynthesis, find itself in an inhibitory light intensity near the surface. Another possibility is that in periods of turbulence which do not permit stratification, cells may become over-vacuolated so that they rise rapidly to the surface in periods of calm following storm. At the surface sufficient photosynthesis to cause collapse of gas vesicles may not be possible and, indeed, Reynolds (1971) has found that gas vacuolation may actually increase in a surface scum so that damage to the cells through prolonged exposure to ultraviolet light or high oxygen concentrations may occur. It is a fact that surface scums in the Shropshire meres result from the concentration of colonies previously distributed in a water-layer several metres deep, rather than from growth at the surface (Fig. 12.6) and that cells taken from the surface are rarely viable (Reynolds, 1967, 1971). Fitzgerald (1967) observed that whereas colonies taken from several metres down grew readily in artificial culture, material taken at the surface rarely did.

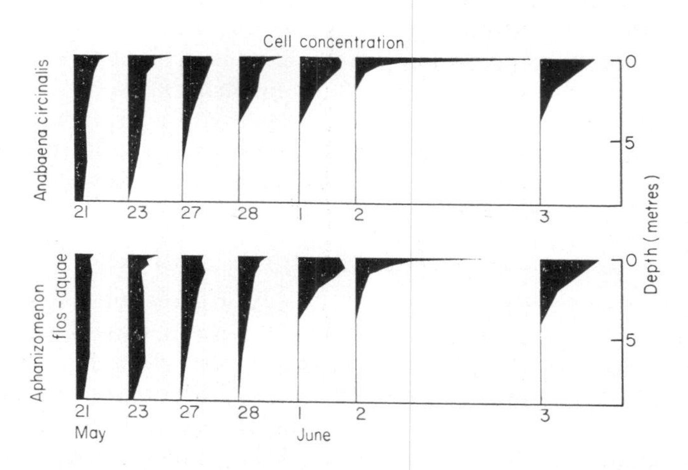

FIG. 12.6. Vertical distribution of *Anabaena circinalis* and *Aphanizomenon flos-aquae* in Crose Mere, Shropshire, England, in 1968 showing the progressive accumulation of the populations at the surface during a period of calm and their redistribution by wind. It will be noticed that *Anabaena* is consistently more buoyant than *Aphanizomenon* (data of C. S. Reynolds (1971) *Fld. Stud.* **3**, 409–432).

This view, that surface scums consist of moribund populations trapped in periods of calm, perhaps needs some qualification. The formation of large aggregates may have biological advantages, for example, immunity from grazing zooplankton, and these may offset the occasional mass-destruction. Furthermore, surface populations are not invariably moribund and as Walsby (1970) has pointed out, there are advantages in such scums. Access to atmospheric carbon dioxide is better and higher rates of photosynthesis may thus be possible in the eventuality of dissolved inorganic carbon becoming limiting. Since gas vacuoles are small and of low refractive index their light scattering powers are considerable, especially for shorter wavelengths (see p. 107), so that in a surface scum there must be a light shielding effect as postulated long ago by Lemmermann (1910) and, in particular, the penetration of ultraviolet rays will be slight. The formation of a quite thin film of colonies at the surface will greatly modify the light field below so that optimum conditions may occur nearer the surface.

Planktonic blue-green algae are also abundant in shallow unstratified lakes and small ponds as well as in stratified lakes. Such ponds and lakes are characterized by receiving high inputs of organic matter either through pollution or intense primary production. Village ponds and temple tanks in India are examples of the first sort whereas Lake George in Uganda exemplifies the latter. Although the water circulates freely, oxygen becomes depleted at the bottom during the night (Ganf, 1969). The vertical migration of the blue-green algae results in their exposure to anaerobic conditions for part of the 24 h cycle. Together with high light intensities, high temperature and high nutrient concentrations this appears to provide optimum conditions for growth of blue-green algae but the situation requires further investigation.

4. ROLE OF BIOLOGICALLY ACTIVE SUBSTANCES

It has been suggested that certain biologically active substances may be important in determining the succession of algae in natural waters. For example Lefèvre (1964) remarked that many green algae disappear when waterblooms of blue-green algae develop, and attributed their disappearance to toxins produced by the waterbloom. However, direct evidence for this is lacking, and the disappearance of suspended green algae could equally well be explained by the fact that they are shaded out by the dense layers of blue-green algae overlying them.

There is, nevertheless, good evidence that blue-green, among other planktonic algae, produce substances which may inhibit or stimulate their own growth or that of other algae. Harder (1917b) first showed that *Nostoc punctiforme*, an endophytic alga rather than a free-living, planktonic form, produces autotoxic substances. Jakob (1954) and Proctor (1957a, b) also found evidence of substances produced by blue-green algae which were toxic to other algae.

Other observations of the effects of filtrates from cultures on the growth of the same and different species suggest that production of autoinhibitors, antibiotics or growth-promoting substances by algae is a frequent phenomenon and the available evidence suggests that blue-green algae commonly produce inhibitors active against species belonging to a variety of different groups (see Lefèvre, 1964; Hutchinson, 1967; Fogg, 1971). Experiments in which different species were grown separated by a porous membrane in two limbs of a U-tube showed that *Microcystis* produced substances inhibitory to the growth of other algae such as *Chlamydomonas*, *Haematococcus*, *Navicula* and *Cryptomonas* (Vance, 1965). Blue-green algae are themselves sensitive to many antibiotics (Vance, 1966; Whitton, 1967b; see also p. 87) although resistant strains may develop rather quickly (Kumar, 1964). Direct evidence is so far lacking that any such substances are of importance in determining the position of blue-green algae in phytoplankton succession in the natural environment. Moreover, in no case has a substance been successfully isolated and characterized.

There is also a large number of reports concerning the production by blue-green algae of toxins which are active against organisms other than algae, including cattle, wildfowl, fish and man (see Gorham, 1964; Schwimmer and Schwimmer, 1964, 1968). However, despite the obvious importance of such observations from academic, medicinal and socio-economic view points, the substances involved have been little investigated and only in a small number of cases has the active principal been characterized.

The classical example is the "fast death factor" (FDF) produced by *Microcystis aeruginosa*. Many reports of algal toxicity have been associated with this alga. Hughes, Gorham and Zehnder (1958) isolated a strain, designated NRC-1, from which a toxin was extracted. It proved fatal to mice when injected intraperitoneally or administered orally and the symptoms produced are similar to those which had been described for animals which had ingested waterblooms containing *Microcystis*. Bishop *et al.* (1959) followed the activity through various fractionation procedures and found that it resided in one of five similar oligopeptides which could be separated by electrophoresis. The active peptide contained the amino acids aspartic acid, glutamic acid, D-serine, valine, ornithine, alanine and leucine in the approximate molar ratio of 1 : 2 : 1 : 1 : 1 : 2 : 2. Its high diffusibility through dialysis tubing indicated a low molecular weight and quite probably it is a decapeptide, although small multiples of ten residues cannot be ruled out. Dinitrophenylation revealed no N-terminal amino acid, suggesting a cyclic molecule. The toxicity of the molecule might reside in such a cyclic structure, or in the D-amino acid (serine), or in ornithine (a non-protein amino acid). The cultures from which this material was isolated contained several bacteria (and also, according to Stanier *et al.*, 1971, a small-celled *Synechococcus* having the microscopic appearance of a bacterium) but there was good evidence that the activity was produced by the *Microcystis* cells rather than by any of the contaminating organisms.

Less well documented cases have been reported of other bloom-forming blue-green algae, principally species of *Oscillatoria*, *Anabaena* and *Aphanizomenon*, being associated with toxin production. A "very fast-death factor", produced by cultures of *Anabaena flos-aquae*, capable of killing mice within 2 min, has also been investigated by Gorham and his collaborators at the National Research Council Laboratories in Canada (see Gorham *et al.*, 1964).

For some time it was wondered whether the toxicity associated with waterblooms in which *Aphanizomenon flos-aquae* was the preponderant alga was a product of this alga or some other (such as *Microcystis*) which also occurred in the samples tested. However, Sawyer *et al.* (1968) succeeded in isolating *A. flos-aquae* from such blooms and demonstrating that it was indeed the source of a potent toxin effective against crustaceans and fish. The toxin appears to be similar, if not identical, to Saxitoxin, a paralytic shellfish toxin produced by the marine dinoflagellate *Gonyaulax catenella* (see Jackim and Gentile, 1968).

Toxic substances are not limited in their occurrence to freshwater algae (see Habekost *et al.*, 1955). Recently Moikeha and Chu (1971) have extracted a toxin from material gathered from Hawaiian offshore blooms of *Lyngbya majuscula* which possesses antibacterial activity, and causes lysis of protozoa and death of rats when injected intravenously. The toxin, which has caused severe dermatitis in swimmers who have come into contact with the algal blooms, has not been characterized but appears to be a lipid substance judging by its solubility in organic solvents (Moikeha *et al.*, 1971).

The toxicity of waterblooms is very variable—few cases have been reported from Britain whereas they appear frequent in the United States, South Africa and New Zealand. The main explanation for this variability probably lies in the existence of toxic and non-toxic strains. Peary and Gorham (1966) have shown that light intensity and temperature also have marked effects on toxin production by *Anabaena flos-aquae*. In addition, the bloom must be concentrated if the animal is to absorb a lethal dose and, with *Microcystis*, the toxin must be released from the cells by decomposition if it is to be effective. Destruction or inactivation of the toxins may also affect their potency under natural conditions (Gorham, 1964). Where substances produced by blue-green algae are effective against other algae, which might compete for light or nutrients, or predatory organisms, their survival value is clear. Their toxicity to vertebrates and other organisms which present no obvious threat to algae, on the other hand, is perhaps fortuitous.

5. EUTROPHICATION AND CONTROL OF BLUE-GREEN ALGAE

With eutrophication of freshwaters taking place at an increasing rate in many parts of the world the problem of controlling the growth of planktonic blue-green algae is becoming acute. The ideal solution is to prevent the input of excessive amounts of nitrate, phosphate and organic matter into the waterbodies concerned, but this is rarely attainable since much of the input is from diffuse sources, such as leaching from agricultural land, rather than point sources which may be controllable. Treatment with copper sulphate to give a concentration in the water of about 0·2 mg l^{-1} has been used since the beginning of the century and is often effective in preventing the development of blooms without damaging higher plants or fish. This treatment sometimes fails, possibly because dissolved organic matter, including polypeptides liberated by the algae themselves, produce non-toxic complexes with the copper ion (Fogg and Westlake, 1955). In any case copper sulphate is only a temporary palliative and the wisdom of repeatedly adding a permanently poisonous substance to a lake or reservoir is questionable (Hasler, 1949). Organic substances, such as 2,3-dichloro-1, 4-napthoquinone, having selectively toxic action against blue-green algae have been found (Fitzgerald and Skoog, 1954), although there again, the extracellular products of the algae seem to reduce the toxicity of the poison (Whitton and MacArthur, 1967). Selective algicides of this sort do not appear to have been used to any great extent against cyanophycean water-blooms, possibly because of their high cost, but they are used to control the growths of algae in many swimming pools and in fish tanks. Artificial circulation has already been mentioned as a method which has proved effective in the reservoirs of the Metropolitan Water Board of London (p. 266). This technique has not yet been tested on a wide enough scale and it obviously would not be effective in the shallow unstratified type of water body which often develops the heaviest blooms. Since the biological success of planktonic blue-green algae probably depends largely on the possession of gas vacuoles and it is the buoyancy which they confer which causes the formation of the scums which are particularly a nuisance, inhibition or destruction of these structures might be an effective form of control. A specific antimetabolite inhibiting their formation might be found or it might prove practicable to use ultrasonic waves to destroy them since they are more sensitive to this than are many other biological structures. It has also been suggested that the pressures generated by detonating explosive devices might be used to collapse gas vacuoles in algal blooms and so to destroy their buoyancy (Walsby, 1968a). Experiments carried out on a small scale in the laboratory suggest that this should be feasible and pilot trials carried out by D. Menday and R. W. Edwards (unpublished) on a small reservoir in South Wales suggest that it could possibly provide an inexpensive means of treating blue-green algal blooms. Another possible technique is to add pathogens of blue-green algae to the waters (see Chapter 13).

The attached blue-green algae of the littoral zone of lakes

Attached blue-green algae may be as abundant in the littoral zone of a lake as planktonic forms are in its open waters. The occurrence of different species may be affected by a complex of factors including both the chemistry of the water and that of the substratum, the physical nature of the substratum, light penetration, degree of exposure to wave and ice action and variation in water level. Broadly speaking, attached blue-green algae are found in a wider variety of lake types than are the planktonic species. *Scytonema* species, for example, may be abundant in base-poor lakes and a benthic felt of *Phormidium* spp. predominates in Antarctic lakes; in both these types of water planktonic blue-green algae are absent.

The nature of the flora is largely determined by the general nature of the substratum and is classified into *epilithic*, growing on rock and stones; *epiphytic*, growing on larger plants; and *epipelic*, growing on, or in, bottom sediments.

Distinct zonation of epilithic algae has been described for several temperate lakes. In Lake Windermere in the English lake district, Godward (1937) found a spray zone dominated by blue-green algae including *Pleurocapsa fusca*, *Tolypothrix distorta*, var. *penicillata* and *Phormidium autumnale*. Other species present were *Homoeothrix fusca*, *Calothrix parietina*, *Schizothrix funalis*, *Pleurocapsa fluviatilis* and *Nostoc* sp. No doubt the predominance of the blue-green algae in this situation is related to their capacity to survive desiccation. The same species occurred below water level down to about 0·5 m (Fig. 12.7) but, here, together with green algae and diatoms. Between 2 and 3·5 m there was another distinct community including *Nostoc verrucosum*, *Dichothrix baueriana* var. *crassa* *Plectonema thomasinianum* var. *cincinnatum*. In the Traunsee, Austria, a more calcareous lake than Windermere, Kann (1959) recorded a similar zonation but a richer cyanophycean flora. She found exposed rocks were often covered up to 20 cm above mean water level with a black growth of *Gloeocapsa sanguinea* status *alpinus* and *Scytonema myochrous* status *crustaceus*. Towards the water line where they were more frequently wetted were *Nostoc sphaericum*, *Calothrix parietina*, *C. fusca*, *Scytonema myochrous* and *Chlorogloea microcystoides*. Below this was a zone dominated by *Tolypothrix distorta* var. *penicillata* and *S. myochrous* extending down to 10 cm below mean water level. Between 5 and 25 cm there was a *Rivularia haematites* community with intermixed *S. myochrous*, *T. distorta*, *Nostoc caeruleum*, *N. sphaericum*, the green alga *Mougeotia* and diatoms. This merged into the lowermost zone, which consisted of a *Schizothrix lacustris* community with *Tolypothrix tenuis*, *S. myochrous*, *Phormidium subfuscum* var. *joannianum*, *Chamaesiphon polymorphus*, *Tolypothrix distorta* and

Fig. 15.1. Growth of blue-green algae in a rice field in Burma (photograph, G. E. Fogg).

Fig. 16.5. Artificial stream carrying water from an alkaline hot spring in Yellowstone National Park and showing an extensive growth of blue-green algae, particularly *Mastigocladus* sp. (photograph W. D. P. Stewart).

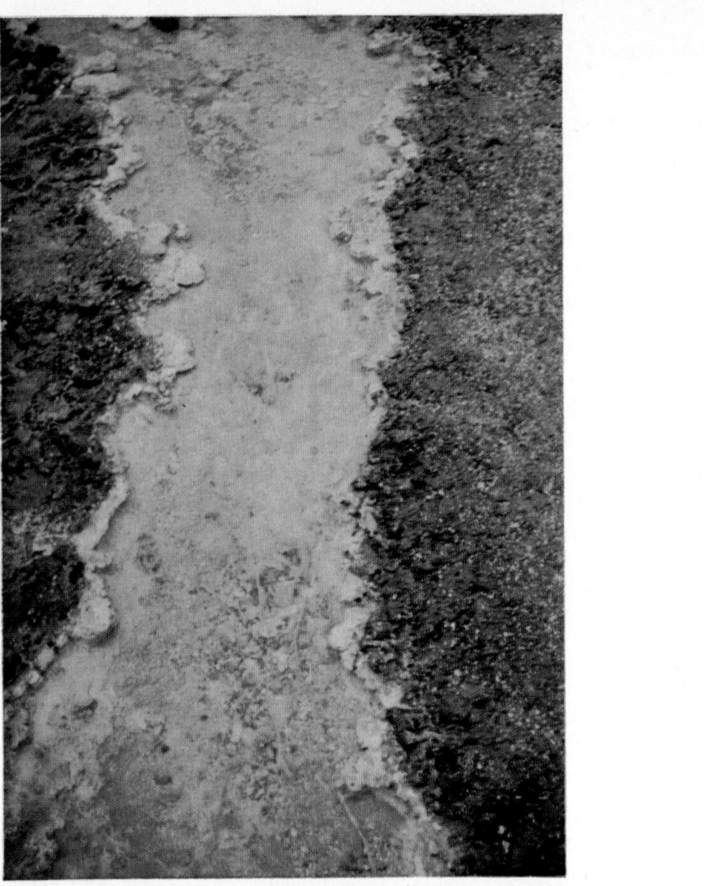

FIG. 16.6. Blackish crust of blue-green algae growing in a region bathed by stream on the side of an alkaline hot spring stream in Yellowstone National Park. The mid-stream temperature is within the range 60–70 °C. (Photograph W. D. P. Stewart).

FIG. 17.3. Colonies of *Geosiphon pyriforme* (×2.5). Photograph by courtesy of H. A. von Stosch.

Nostoc caeruleum, together with various green algae and diatoms. The relative importance of the different species showed seasonal variation.

No detailed study appears to have been made of the ecology of the epilithic blue-green algae in any tropical lake but it is probably fairly similar to that in temperate lakes.

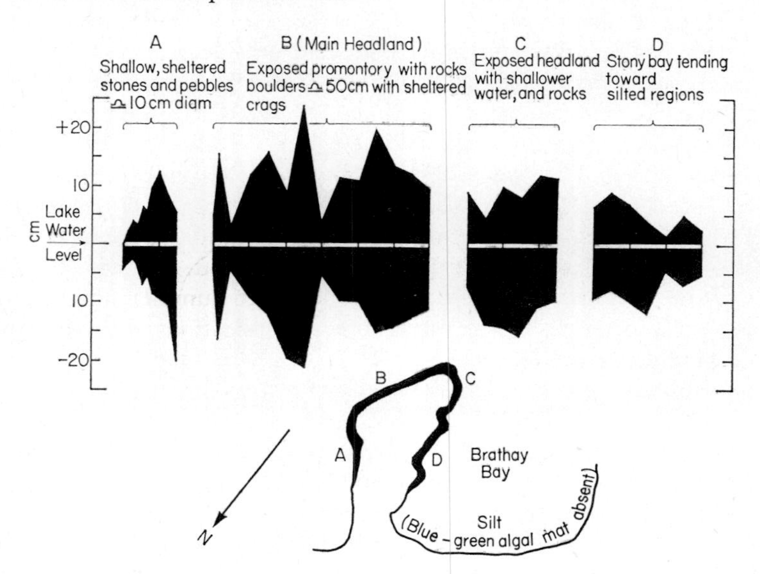

FIG. 12.7. Horizontal and vertical distribution of littoral blue-green algae (*Phormidium* spp.) on an exposed headland in the north basin of Windermere, English Lake District, in April 1970. The bars on the horizontal axis represent 10 m distances along the shore line shown in the sketch map below (unpublished data of R. Wood).

One of the most striking developments of epilithic blue-green algae is that of Antarctic lakes. Characteristically, a rusty red felt of *Phormidium* species covers the lake bottom from mean water level down to 1 or 2 m. *P. laminosum* and *P. autumnale* were identified by Fritsch (1917) as dominant species in samples from Cape Adare and Evans Cove but generally similar mats from Cape Evans and Cape Bird contained *Microcoleus lyngbyaceus* (*Lyngbya aestuarii*) or *Calothrix parietina* as dominants with *Schizothrix calcicola, Microcoleus vaginatus* and other species accompanying them (Zaneveld, 1969). The felt shows a distinct layering with a blue-green zone below the rusty-red surface. This may possibly be an adaptation which gives protection against ultraviolet rays and permits more efficient absorption of longer wavelengths. The temperature of the felt is commonly 1 or 2°C above that of the water immediately above it, a circumstance of obvious biological advantage (Goldman et al., 1963; Fogg and Horne, 1970). Pieces of detached felt melt their way through solid ice as they absorb radiation and, eventually reaching the surface, are dispersed by wind (Wilson, 1965).

Epiphytic algae are influenced by the specific characteristics of their host plant such as the length of life, the leaf arrangement, the surface and the exudates, as well as by water movement and other factors. Diatoms and green algae are usually the predominant forms in temperate lakes but certain blue-green algae are also characteristic. On submerged plants between 0 and 0·5 m in Lake Windermere, Godward (1937) found *Gloeotrichia* sp. and *Lyngbya perelegans* while on submerged plants between 3 and 6 m *Chamaesiphon cylindricus* var. *ampla*, *C. confervicola*, *Lyngbya purpurascens* and *L. purpurea* (?) were found. These last four forms are reddish, lilac or purplish in colour at these depths whereas when they grow in shallow water, as happens mainly in summer, they become pale blue-green or greyish. The physiological implications of such pigment changes were discussed on p. 150.

Epipelic blue-green algae include species forming mucilaginous colonies lying loosely on the surface, *Aphanothece stagnina* and *Nostoc pruniforme* for example, and motile forms, such as *Merismopedia* spp., *Oscillatoria* spp. and other filamentous species, which can actively maintain their position in a film on the sediment surface. In Windermere epipelic blue-green algae may go down to 16 m and show pigmentation similar to the deep water epiphytic species just mentioned. Round (1957) found that blue-green algae were more characteristic of the sediments of nutrient-rich than oligotrophic lakes in the English lake district. In a shallow tropical lake, Lake Chilwa in Malawi, on the verge of complete drying out Moss and Moss (1969) found *Oscillatoria* spp. abundant on all but dry mud surfaces. As in other situations these *Oscillatoria* mats tended to float to the water surface when buoyed up by bubbles of oxygen after periods of vigorous photosynthesis.

The quantitative study of benthic algae presents far greater difficulties than that of planktonic forms. Round (1961) found that in the English Lake District epipelic blue-green algae are present throughout the year and show maximum development in the spring. The time at which peak development occurs varies somewhat with depth, occurring later in the deeper situations as one might expect from consideration of light penetration. Growth of algae on the sediments was greatest at about 3 m, wave action removing the algae above this level and light intensity being limiting below. Moss (1968) has tabulated values for the chloropyll *a* contents of various benthic communities which give an idea of their relative biomass. Epipelic communities with filamentous blue-green algae dominant gave values between

15 and 51 mg chlorophyll *a* m^{-2} but if mats are formed the values are ten or more times higher. These values may be compared with 300 mg m^{-2} which has been estimated to be the maximum chlorophyll content of the photic zone when a dense bloom of phytoplankton is present (Talling, 1961). The contribution of benthic algae to total primary productivity is probably quite considerable even in a large lake. Measurements by the ^{14}C technique of photosynthesis by the benthic felt of Antarctic lakes by Goldman *et al.* (1963) and by Fogg and Horne (1970) showed high rates which suggest that it contributes far more than the phytoplankton.

Blue-green algae of rivers and streams

The complexity and variability of rivers and streams as habitats have discouraged systematic investigation and much less is known about the blue-green algae in them than about the blue-green algae in lakes.

Since plankton species normally have a generation time of 24 h or more they are liable to be washed out of a flowing system unless it is slow moving. Blue-green algae are not usually found in the plankton of small or medium sized rivers but may occur in larger ones such as the Nile. In the White Nile, near Khartoum, marked seasonal variations in phytoplankton have been recorded (Rzoska *et al.*, 1955). After high water and as current speeds diminish in September, diatoms and then blue-green algae appeared. The latter became dominant with *Anabaena flos-aquae* as the principal form although sometimes *Raphidiopsis curvata* or *Lyngbya limnetica* were abundant. In December current speeds increased again and these populations were washed out. It appears that the development of this plankton took place in the large lake-like storage basin of the Gebel Aulyia dam upstream and the annual cycle of plankton at Khartoum closely follows the cycle of water level in the basin.

Attached algae in rivers and streams may again be classified as epilithic, epiphytic or epipelic. Moving water promotes the uptake of nutrients and removal of waste products so that luxuriant growths are often achieved in rapids and certain species may have become adapted to such conditions. The currents limit their size, both in so far as they affect the stability of the substrate and as they disrupt the algal thallus. Picken (1936) found that in regions of relatively rapid flow the size of colonies of *Rivularia* were proportional to the size of the stones to which they were attached. In more slowly moving water thallus size was independent of stone size. The bulk of this algae increases more rapidly than does the area of its attachment so that the current limits the size of the thallus by tearing it away from the stone if it does not transport the stone to a slower part of the stream. The current obviously assists in the dissemination of attached forms. Blum (1963) has described how masses of *Oscillatoria* growing in quiet reaches are detached by being buoyed up by photosynthetically produced oxygen bubbles and are then carried downstream releasing individual filaments on the way.

A comparative study of the epilithic algae in various Austrian streams has been made by Kann (1966). Streams flowing over calcareous and non-calcareous rocks differ both quantitatively and qualitatively in this respect. The calcareous streams had about twice as many species as the other type and the growth, consisting of a mosaic of species, covered the rocks in a continuous visible sheet of about 1–2 mm thickness. On the non-calcareous rocks the growth was sparser, occurring in flecks and tufts and often only detectable by microscopic examination of scrapings. Species of *Chamaesiphon, Homoeothrix, Nostoc, Phormidium* and *Pleurocapsa* were present in both but certain species were characteristic of one or the other type. *Chamaesiphon polonicus, C. geitleri* and *C. pseudopolymorphus* were macroscopically evident in the calcareous streams whereas *C. fuscus* was the principal form in the non-calcareous type. There was considerable variation within each type. A stream of moderate alkalinity had a rich flora of blue-green algae but one of greater alkalinity in which tufa formation (biogenic deposition of calcium carbonate) was occurring had only three species of which one, *Phormidium umbilicatum*, was dominant. More species were found in rapidly flowing regions than in quieter parts of the stream. Light intensity also had an effect. Not only is the growth different in colour—grey-green in the shade, brown or olive-green in the sun—but certain species such as *Pleurocapsa minor, Chamaesiphon incrustans, C. polymorphus, Xenococcus kerneri, Oncobyrsa rivularis* and *Hyella fontana* var. *maxima*, are more prevalent in the shade, whereas others, *Homoeothrix* species for example, prefer well lighted regions. Some species growing on the underside of stones, such as *Schizothrix perforans* and *Aphanocapsa endolithica*, grow in minimal light intensity and show pigment adaptations similar to those discussed elsewhere (p. 274), being bright blue-green or violet in colour.

Epiphytic algae occur on angiosperms, bryophytes and on larger algae such as *Cladophora*. *Cladophora* species which are branched and have a rough non-mucilaginous surface, carry a great variety of algae among which are blue-green algae such as species of *Chamaesiphon, Oncobyrsa, Dermocarpa, Oscillatoria* and *Lyngbya*. Angiosperms and bryophytes have similar epiphytes and also other forms such as *Tolypothrix* and *Nostoc* (Round, 1965). Epipelic communities in rivers are particularly difficult to study and are liable to be complicated by epilithic and epiphytic algae which have become detached and which become intermixed with them.

Tufa formation is particularly associated with blue-green algae growing in calcareous streams (Fritsch, 1945; Gessner, 1959). Species of *Rivularia*,

Schizothrix, *Gloeocapsa*, *Petalonema* and *Phormidium* are chiefly concerned. Thermal blue-green algae are particularly active and the calcium carbonate deposit, called travertine, which they produce, usually brightly coloured by the algae within it, is a feature of hot-spring areas such as Yellowstone National Park. The calcium carbonate is deposited within the mucilage envelope, never within the cells. In *Rivularia* colonies, the calcite crystals may be deposited rhythmically in concentric zones. The process presumably depends basically on the precipitation of calcium carbonate as carbon dioxide is withdrawn by photosynthesis from water containing calcium bicarbonate, but since one species may cause it whilst under similar conditions another does not, there are evidently complications which are not yet understood. It is not essential for development as the species concerned can grow in media with very low concentrations of calcium.

Blue-green algae can be used as indicators of pollution in rivers. The classification of saprobic zones is complex and perhaps not altogether useful and it will suffice to note here that encrusting algal communities of *Chamaesiphon polonius* and *Calothrix* species are characteristic of unpolluted waters whereas *Oscillatoria chlorina* and *Spirulina jenneri* can survive in waters so heavily polluted with organic matter as to become deoxygenated. Various *Oscillatoria* and *Phormidium* spp. are characteristic of intermediate stages in organic pollution.

Blue-green algae of hot springs

The bright coloration of hot springs regions in all parts of the world, and particularly in the United States, Iceland, Italy and New Zealand, is due largely to the presence of blue-green algae and carotenoid-containing flexibacteria. Mention of thermal springs is made in the works of the ancient Greeks and Romans (see, for example, Frazer, 1914; Davis and De Wiest, 1966). Pliny the Elder wrote "Green plants grow in the hot springs of Padua" and Brock (1967b) reports that the hot springs are still there and are colonized by blue-green algae. In the environment of the hot spring where the water temperatures range from boiling point down to ambient temperature, which may be below freezing, relatively stable temperature gradients are established. Along these gradients different characteristic algal growths occur, the nature of which depends on the type of spring from which the streams originate rather than on the geographical location of the spring. In general three distinct types can be recognized: (1) alkaline springs, usually with a pH of around 9·0. These are the most usual and show the greatest algal biomass; (2) calcium carbonate springs depositing travertine which have a rather similar, although less well developed, algal flora and (3) acid springs in which the pH may be as low as

2·1 and in which blue-green algae are absent or rare but where the anomalous unicellular eukaryotic alga, *Cyanidium*, is usually predominant.

Alkaline streams have attracted most attention because it is in them that life occurs at higher temperatures than anywhere else on Earth. The uppermost temperature limit for life is often difficult to determine exactly because the gradients are steep and surges of water from springs may cause fluctuations of up to 10 °C. The generally accepted upper constant temperature limit for algal growth is 72–75 °C (Setchell, 1903; Nash, 1938; Peary and Castenholz, 1964; Brock, 1967b) and reports of growth at higher temperatures are probably erroneous (Castenholz, 1969). The algae which grow at these higher temperatures in alkaline waters are species of the unicellular cyanophycean genus *Synechococcus* (*S. elongatus* and *S. lividus*) and blue-green algae in those streams which have a relatively stable temperature gradient grow right up to the limit of their tolerance. Bacteria can, however, grow at higher temperatures. The maximum temperature for the growth of *Cyanidium caldarium* is 55–60°C (Castenholz, 1969; Doemel and Brock, 1970).

Total algal biomass in alkaline streams is usually greatest in the temperature range 51–56 °C in which *Mastigocladus laminosus*, *Phormidium laminosum* and *Synechococcus* spp. often predominate, intermingled with flexibacteria, bacterial rods and photosynthetic purple sulphur bacteria. The flexibacteria and simple filamentous algae such as *Phormidium* are difficult to distinguish in the field, and until recently, some flexibacteria were considered as species of *Phormidium*. Hot spring flexibacteria differ from the blue-green algae, however, in containing bacteriochlorophyll, not chlorophyll *a*, and in not evolving oxygen in photosynthesis (Pierson and Castenholz, 1971). Brock (1968) suggests that species referred to by Copeland (1936) as *Oscillatoria filiformis*, *Phormidium geysericole*, *P. subterraneum* and *P. bijahensis* are probably flexibacteria. At temperatures below 45° species of *Calothrix* are abundant and predominate together with thick leathery growths of colonial unicellular algae and diatoms. The floristic composition may vary, however, from spring to spring. For example, in Icelandic hot springs, *Synechococcus* species are absent although they occur in what seem to be comparable habitats in Yellowstone. It is found that chlorophyll (which reaches 700–800 mg m^{-2}), ribosenucleic acid and protein content are all greatest within the temperature range 51–56°. Figure 12.8 (facing p. 257) shows the upper reaches of one such Yellowstone stream.

Hot spring algae not only show a distinct zonation down the stream as temperature decreases but also a lateral zonation across the stream and a vertical zonation within the algal mat. The lateral zonation corresponds fairly closely to that found in the direction of flow except that the temperature gradient is compressed, and, as a result blackish growths of algae may be seen at the sides of the stream at temperatures of 45° only a few centimetres from mid-stream algae growing at temperatures about 30° higher.

Within the algal mat there is marked vertical zonation of micro-organisms, which is particularly well seen in thick mats. Usually there is a thin layer of blue-green algae at the surface, a wide band of flexibacteria beneath and, at the bottom, there are bacteria other than flexibacteria. These bacteria presumably depend for growth on the extracellular or lytic products released by the photosynthetic algae, assimilating them either heterotrophically or, as is probable for the flexibacteria, phototrophically (Pierson and Castenholz, 1971). Nitrogen-fixing species of blue-green algae probably provide combined nitrogen for the growth of other organisms (see p. 339).

Hot springs provide stable environments for ecosystems which may have remained unchanged, according to Brock (1967b) for thousands of years. The steady state biomass depends on the interactions of a variety of physical, chemical and biological factors. These may be classified into two main types: firstly, those which affect primary production rates and secondly those which superimpose themselves on the primary production rates. Light and temperature are examples of the first type, and flow rate and grazing are examples of the second type. Light is probably the most important of the physical factors which regulate production. For example if the light intensity is reduced markedly the algae fail to multiply at a sufficient rate to compensate for those washed away and the stream becomes bare. This happens, for example, in Iceland in streams during the months of almost complete darkness in the winter and can be induced artificially in the field simply by covering the test area. Brock and Brock (1968) have done this with natural populations of *Synechococcus*. Within one day of covering, the algal population had decreased. The decrease was exponential with time and within a week the growth had virtually disappeared. When light was re-admitted the algae re-appeared and grew exponentially. This technique offers a simple means of determining whether the algae present in a particular habitat are growing or not. If they are maintaining themselves by growth in a current they will get washed out when the area is darkened. The method can also be used to calculate the generation time of field populations of algae. In Brock's experiments the calculated generation time was about 40 h but Castenholz (1969) considered that with a 14 h-light day the generation time in the light should be much shorter, corresponding to 1·0–1·7 doublings per day. These values are rather similar to the 2 doublings per day which he obtained with *Synechococcus* at 65° in the laboratory. On a unit area basis the increase in organic matter ascribed to *Synechococcus* is 1–2 g m^{-2} day^{-1}.

Although temperature has a marked effect on species distribution it has only a small effect on primary production rates. This is because the algae adapt and metabolize most efficiently at the particular temperatures at which they occur. This has been shown in extensive studies on primary production using ^{14}C as tracer (see Brock, 1970a). Moreover, Castenholz (1969) has shown that filamentous blue-green algae may move by gliding to areas where temperature is optimum if the temperature where they are initially becomes sub-optimal (see p. 127). Some hot spring algae are truly thermophilic species, that is their optimum temperature range extends above 45° (Castenholz, 1969). Nevertheless, other hot spring algae can grow well at temperatures very different from those at which they were isolated. For example, Fogg (1951b) showed that a strain of *Mastigocladus laminosus* from New Zealand hot springs fixed nitrogen well at temperatures much lower than that at which it grew in the field. There is no exact information on how long a time these thermal algae take to adapt their photosynthetic rate to the new conditions.

The importance of herbivores in regulating the standing crop of algae has been appreciated by many workers. Schwabe (1966) was the first to suggest that the great accumulation of organic matter at the higher temperatures may be not so much the result of greater photosynthetic efficiency as of the absence of herbivores above about 45°. The most important herbivores in hot springs are probably ephydrid (Diptera) larvae which occur in large numbers below 45° and which, as ^{14}C studies show clearly (Brock *et al.*, 1969), browse on the algae present. Thus the algae which are most abundant at these lower temperatures may be those which are resistant to grazing. This may explain why total algal biomass is low at low temperatures; the decrease above the optimum is probably due to disorganization of the protoplasmic structure as discussed on p. 83.

Finally one may ask why blue-green algae should occur abundantly in thermal springs at all. The environment is extreme and lethal for most eukaryotic organisms. The total dissolved solids are often about 2000 mg l^{-1} compared with the values of less than 150 mg l^{-1} which are usual for lakes and streams and concentrations of sulphur and manganese may reach toxic levels. On the other hand, nitrate and nitrite are frequently undetectable and ammonium-nitrogen, although more prevalent, has been detected only in about half of over 860 hot springs tested (Castenholz, 1969). The occurrence of blue-green algae in hot springs evidently depends not only on their ability to adapt to high temperatures but also on their resistance to other unfavourable conditions and, at least below 55–60°C, on the capacity of some species to fix nitrogen.

Chapter 13

Pathogens of blue-green algae

Much attention has been paid to the factors regulating the growth of blue-green algae in various ecosystems, but these studies have been concerned mainly with the effects of chemical and physical conditions on growth and until the work of Canter and Lund on fungal parasites of algae (see Lund, 1957) the possible importance of biological interactions was scarcely considered. Now it is known that various micro-organisms including protozoa, fungi, bacteria and viruses all play roles in regulating algal growth in freshwater. Presumably, similar pathogens operate in other habitats but there is scarcely any information about this.

Fungal pathogens

All the fungi so far described on planktonic blue-green algae are chytrids, except one, which has been placed in the Blastocladiales (Canter, 1972). At the time this paper was sent for publication no bi-flagellate fungus was known to occur on these algae but more recently, in the autumn of 1970, a virulent bi-flagellate species was found on *Anabaena solitaria* in Windermere, the English Lake District (Canter, personal communication). Species of the planktonic blue-green genera *Anabaena*, *Aphanizomenon*, *Gomphosphaeria*, *Lyngbya*, *Microcystis* and *Oscillatoria* are all known to be attacked by chytridiaceous fungi. Examples infesting the genus *Anabaena* are *Rhizosiphon crassum* (Fig. 13.1a, b, c) which parasitizes *A. solitaria* and other *Anabaena* spp., and the closely related *R. anabaenae* (Canter 1951), originally described from Sweden as *Phlyctidium anabaenae* by Rodhe and Skuja (Skuja, 1948), which parasitizes *A. sphaerica*, *A. spiroides* and *A. macrospora* among others.

Many of the chytrids appear to have a host range limited within a single genus and some may even be confined to a specific structure such as a resting spore or heterocyst; *Rhizosiphon akinetum* apparently occurs only on akinetes and has been recorded on *Anabaena affinis* in the English Lakes and on *A. macrospora* in Czechoslavokia (Canter, 1954). Again, *Chytridium*

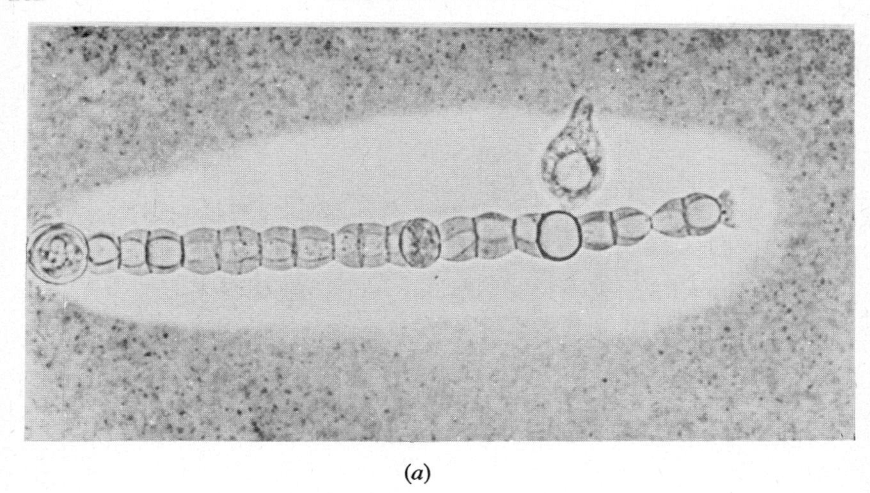

(*a*)

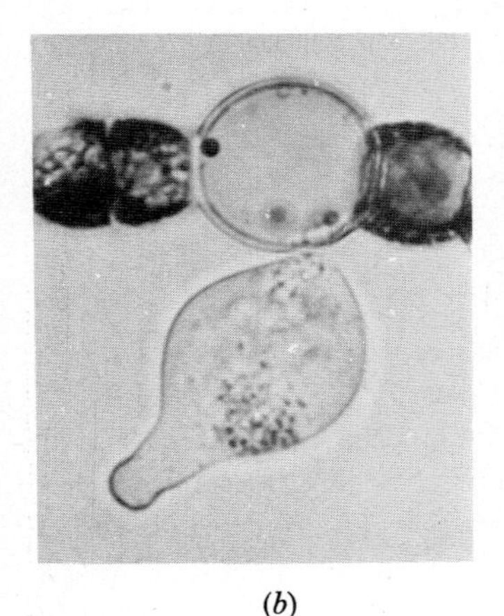

(*b*)

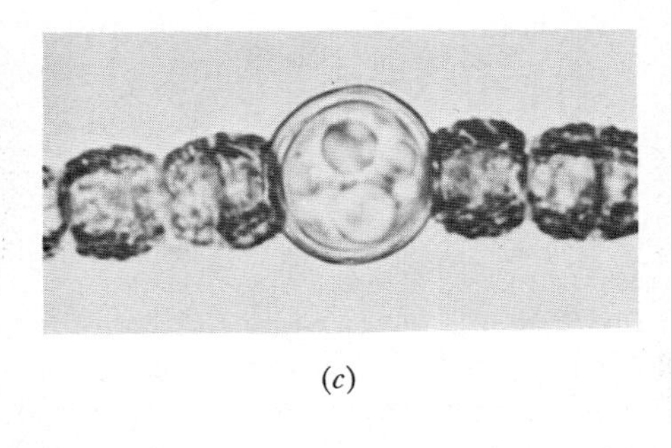

(*c*)

FIG. 13.1. *Rhizosiphon crassum:* (a) general view in negatively stained preparation showing sporangium, prosporangium, rhizoid passing along the algal filament and a resting spore (×660); (b) sporangium attached to swollen spore (×1300); (c) resting spore (×1300). By courtesy of H. M. Canter.

cornutum (Canter, 1963) is found only on the heterocysts of *Aphanizomenon*. *Blastocladiella anabaenae* (Fig. 13.2), which may belong to the Blastocladiales (Canter and Lund, 1968), occurs predominantly on *Anabaena flos-aquae* as well as on *Aphanizomenon flos-aquae*, *Anabaena circinalis* and *A. solitaria*.

It is not possible to deal here in detail with all the different life cycles of these parasites but one example, *Rhizosiphon crassum*, may be considered.

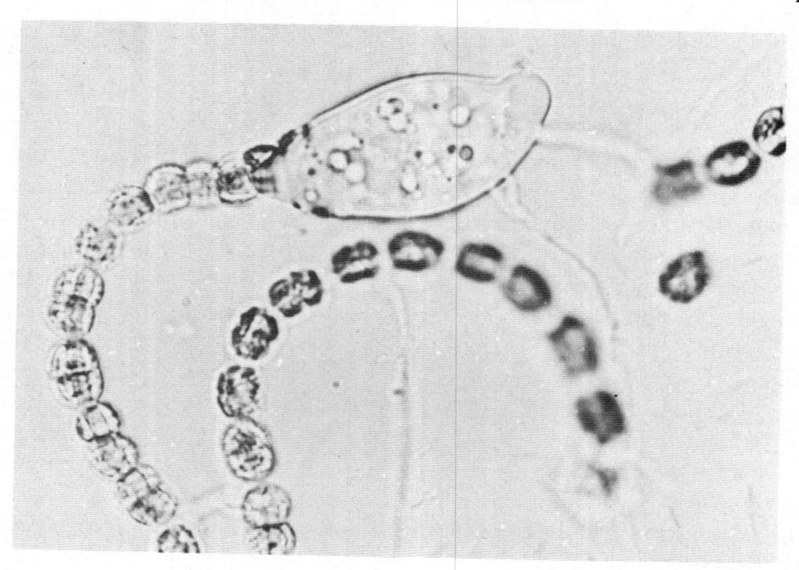

FIG. 13.2. *Blastocladiella anabaenae*: young sporangium with rhizoids passing through the cells of an *Anabaena* sp. (×1330). By courtesy of H. M. Canter and L. G. Willoughby (1964) *J. R. microsc. Soc.* **83**, 365, Plate 83, Fig. 15.

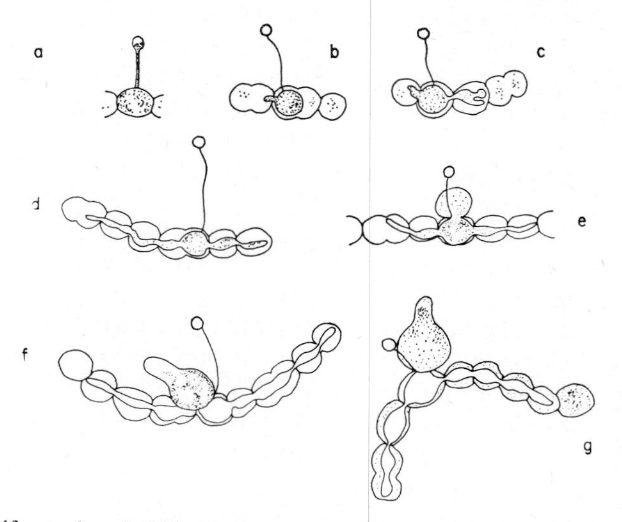

FIG. 13.3. Life cycle of *Rhizosiphon crassum*. (a) fungal zoospore attached to filament of *Anabaena* with contents passing into algal cell; (b)–(d), fungal rhizoidal system extending out from prosporangium; (e)–(g), production of external sporangium by fungus. Redrawn after H. M. Canter (1951) *Ann. Bot. Lond.* N.S. **15**, 129–156.

Its life cycle is shown diagrammatically in Fig. 13.3. According to Canter (1951) a fungal zoospore penetrates the algal mucilage sheath and encysts. From it, a fine thread grows toward and penetrates an *Anabaena* cell. The fungal contents then pass into the algal protoplast where they form a swelling, the prosporangium. This prosporangium enlarges and from it, usually on both sides, a tubular rhizoidal system develops which passes through the adjacent vegetative cells so that a variable number of algal cells, up to eleven, may be attacked by one fungal unit. Eventually the fungal protoplasm pushes out from the algal filament near the original point of infection and passes into a swelling which finally becomes a flask-shaped sporangium (Fig. 13.1b) from which many posteriorly uniflagellate zoospores are released apically. Thick walled spherical resting spores (Fig. 13.1c) may also develop from the same portion of the thallus as the prosporangium in the sporangial stage and it now seems almost certain that some sort of sexual process is involved in their formation (Canter, personal communication) although this is not shown in Fig. 13.3.

The complexity of biological interactions in freshwater ecosystems is such that these chytrids occurring on blue-green algae are themselves parasitisized by other chytrids. These hyperparasites are all very imperfectly known and consequently remain unnamed. They have been recorded on *Rhizophydium megarrhizum* and *R. subangulosum*, parasites of *Oscillatoria*; on *Rhizosiphon anabaenae* and on an unnamed parasite infecting *Microcystis* sp. Several algae are known to occur as epiphytes within the mucilage sheath of the blue-green alga *Gomphosphaeria naegeliana* (*Coelosphaerium naegelianum*) and they also may be parasitized by chytrids. *Rhizophydium hyalobryonis* occurs on *Hyalobryon mucicola* (Canter, 1951) while *Stylosphaeridium stipitatum* may be attacked by both *R. ephippium* (Canter, 1950) and *Entophlyctis molesta* (Canter, 1965).

The seasonal variation in fungal pathogens of blue-green algae is correlated directly with the abundance of the algae on which they occur and the fungi appear and disappear with the algae. The ultimate abundance of pathogens may determine to some extent the eventual decline of the host population. The fact that they are present only for short periods makes these fungi difficult to study in detail. A further difficulty is that, so far, it has not proved possible to isolate and grow them in laboratory culture.

Protozoa

The possible importance of protozoan grazing of phytoplankton has been recognized by Canter and Lund (1968). In the English lakes protozoa appear to prefer green algae and diatoms and seldom graze heavily on blue-green algae, but extensive grazing of the latter by protozoa occurs in some Scottish waters. This is an area where further studies on the interactions between primary productivity and secondary productivity are warranted.

Viruses

The suggestion that algal viruses may be the cause of otherwise inexplicable disappearances of algal blooms was made by Krauss (1961). In the same year Lewin (1960) described a phage attacking *Spirochaeta rosea*, which he regarded as being a non-photosynthetic blue-gree nalga. Two years later Safferman and Morris (1963) isolated for the first time a virus pathogenic to photosynthetic algae and since then viruses pathogenic to blue-green algae have been reported from Russia (Rubenchik *et al.*, 1966), India (Singh and Singh, 1967), Israel (Padan *et al.*, 1967), Sweden (Granhall and von Hofstein, 1969) and Scotland (Daft *et al.*, 1970).

Virus LPP-1, isolated originally by Safferman and Morris (1963) has been studied most extensively. It is effective against *Lyngbya*, *Phormidium* and *Plectonema* strains, hence the name LPP-1. Closely related viruses are G111 from Israel (Padan *et al.*, 1967), D-1 from Scotland (Daft *et al.*, 1970) and possibly LPP-4 and LPP-5 from India (Singh and Singh, 1967).

Morphologically LPP-1 is similar to a bacteriophage, with a distinct head and tail, and such viruses are termed *cyanophages*. The head is hexagonal, icosahedral and with an average diameter of 66 nm (Safferman and Morris, 1963). The tail is 20 nm long and not forked as was originally thought (Luftig and Haselkorn, 1967). Virus G111 has an icosahedral head, 56–60 nm diameter, and a tail length of 28–37 nm (Padan *et al.*, 1967) and D-1 has a head approximately 58·6 nm in diameter and a tail length of 20 nm (Fig. 13.4). Viruses LPP-1, G111 and D-1 are serologically related but they show no relationship to bacteriophages, T1, T2, T3 or T7.

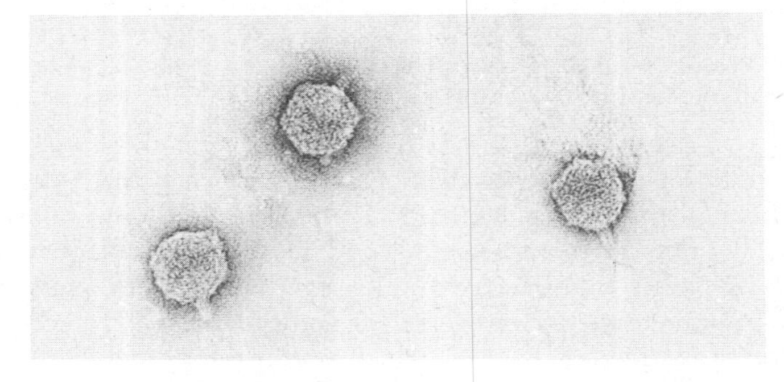

FIG. 13.4. Cyanophage D–1 showing icosahedral head and short tail (× 136,000) Electron micrograph from M. J. Daft, J. Begg and W. D. P. Stewart (1970), *New Phytol.* **69**, 1029, Plate 1.5.

Chemically, the LPP-1 virus is characterized by DNA with a guanine-cytosine ratio of 55–57% (Goldstein and Bendet, 1967) which comprises 40% of the virus, as calculated on the basis of phosphorus analyses (Goldstein *et al.*, 1967). The DNA is linear, 13·2 ± 0·5 nm in length, and the molecular weight is calculated to be 27×10^6 daltons (Luftig and Haselkorn, 1967). The molecular weight of the whole virus is $51 \pm 3 \times 10^6$ daltons, hydration values are 0·37 g water g^{-1} virus and all the essential amino acids are present (Goldstein *et al.*, 1967). The characteristics of a phage infecting the flexibacterium, *Saprospira grandis*, are similar to those of cyanophages and of phages attacking Eubacteriales (Lewin *et al.*, 1964).

LPP-1 and other viruses cause lyses in liquid culture and plaque formation on lawns of algae. The plaques (Fig. 13.5) which usually take

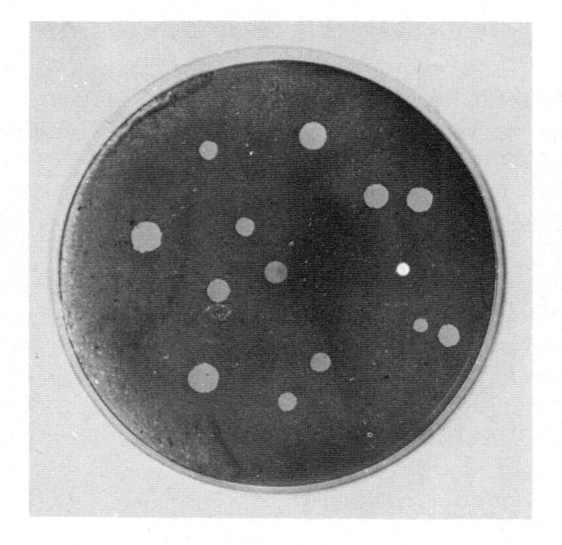

FIG. 13.5. Plaques formed by cyanophage D–1 on lawns of *Plectonema boryanum* strain 594. Photograph by M. J. Daft and W. D. P. Stewart.

3–7 days to develop, may vary widely in size. Safferman and Morris (1964) observed two distinct variants of LPP-1 which cause plaques ranging from less than 0·1 mm to more than 8·0 mm in diameter. Plaque numbers obtained by serial dilution techniques can be used to measure viral numbers in pure culture. However, Daft and Stewart (1971), who isolated from natural populations bacteria which produce morphologically similar plaques to those produced by viruses (compare Figs 13.5 and 13.8), suggest that the use of plaque numbers to obtain virus counts from mixed populations is liable to error.

The mode of action of algal viruses such as LPP-1 is that characteristic of phages in general. The virus first attaches itself to the wall of a vegetative cell and viral DNA is injected into the host via the tail, leaving behind "ghosts" which are empty virus heads still attached to the wall by their tail. In the host cell nucleoplasm the viral DNA multiplies and moves in between the photosynthetic lamellae where helices then form. Finally, these helices move into the virogenic stroma where they develop a protein coat and assume their typical shape. They are released eventually when the algal cells lyse (Smith *et al.*, 1967).

Viral attack has a marked effect on the physiology and metabolism of the susceptible algae. Respiratory and photosynthetic rates show little change during the first 24 h after infection with the virus. This may be equivalent to the latent period preceding increase in number of viral particles. Photosynthesis then decreases until at 120 h only 10% of the original activity remains. Respiration on the other hand increases to 300% of the uninfected control after 72 h and then falls to the control level at 120 h. This fall may be due to a decrease in the number of intact cells rather than to any specific decrease in activity per intact cell. Ginsburg *et al.* (1968) who studied the effect of viral attack on ^{14}C-labelled carbon dioxide uptake in the light found that on infection there was a rapid and complete cessation of activity. This was not due to cell lysis, nor to nitrogen starvation, and although ^{14}C carbon dioxide uptake was inhibited, cyanophage multiplication continued normally. These workers suggest that an inhibition of carbon dioxide uptake is an integral part of the infection process in *Plectonema*. Daft and Stewart (1971) in studies with D-1, found that

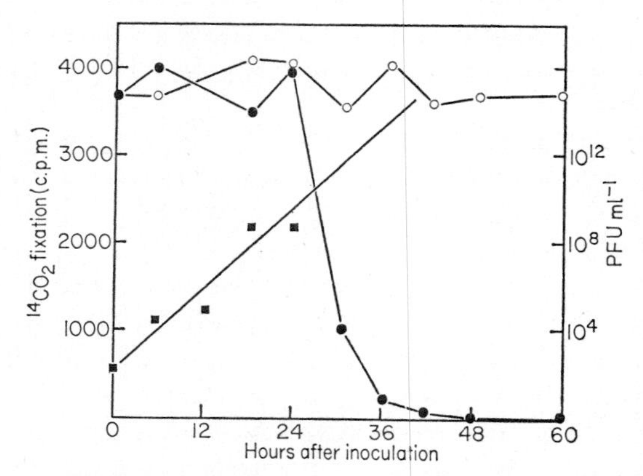

FIG. 13.6. Effect of cyanophage D–1 on uptake of ^{14}C from $NaH^{14}CO_3$ by *Anabaena flos-aquae* ■—■, virus titre; ●—●, ^{14}C uptake by infected algae, ○—○, ^{14}C uptake by alga without added virus. M. J. Daft and W. D. P. Stewart (unpublished).

although there was a rapid increase in viral titre with time, there was no inhibition of ^{14}C carbon dioxide uptake until after 24 h when, as the culture lysed, there was a sharp drop in the rate of uptake (Fig. 13.6).

Granhall and von Hoffstein (1969) demonstrated that acetylene reduction was inhibited rapidly when *Anabaena variabilis* was affected by a virus. This virus has not been isolated, but appears under the electron microscope to be a typical phage with a head 40–50 nm in diameter and a tail 10–20 nm long. Singh and Singh (1967) have also reported on phages effective against heterocystous algae. Phage C-1 is effective against some, but not all, strains of *Cylindrospermum* while phage AR-1 lyses *Anabaenopsis raciborskii* and *A. circularis*, both of which have heterocysts, and *Rhaphidiopsis indica* which does not. Neither phage attacks heterocysts, spores or gas vacuoles although vegetative cells are ruptured.

Lysis of the unicellular alga *Microcystis pulverea*, by what was probably a virus was reported by Rubenchik *et al.* (1966) and by Goryushin and Chaplinskaya (1966), but it was Safferman *et al.* (1969) who first isolated and characterized a virus active against unicellular blue-green algae. This virus SM-1, was isolated on *Synechococcus elongatus* and also causes lysis of a *Microcystis* species. Morphologically it is a polyhedron 88 nm in diameter without a tail and probably with double-stranded DNA. Safferman *et al.* (1972) reported on another virus, AS-1, which is also active against unicellular blue-green algae (*Synechococcus cedrorum* and *Synechococcus* sp. (*Anacystis nidulans*)) but not against filamentous blue-green algae. The virus-like SM-1 has been isolated from waste stabilization ponds although it appears to be rare there. It is the largest cyanophage yet isolated, with an icosohedral head 90 nm in diameter and with a contractile tail 243·5 nm in length. The DNA is double-stranded and has a guanine-cytosine ratio of 53–54%. This compares with a value of 55% for LPP–1 DNA (Goldstein and Bendet, 1967).

The distribution and seasonal variation of algal viruses has been studied in the United States (Safferman and Morris, 1964), in Israel (Padan and Shilo, 1968) and in Scotland (Daft *et al.*, 1970). This has depended in most studies on bringing the waters to be tested back to the laboratory and adding them to algal lawns, either directly (Safferman and Morris, 1964) or after concentration (Padan *et al.*, 1967). An alternative is to use a baiting method (Daft *et al.*, 1970). "Wefts" of test algae held between muslin are placed in the lake or sewage works, near an outflow if possible, and after 48 h are brought back to the laboratory where the algae are examined for pathogens. Both techniques have advantages and disadvantages. Using the former method it is possible to carry out quantitative studies on viral numbers, but there is a disadvantage in that the water mass which is sampled is usually very small. The latter method has the advantage in that the algae are exposed to large volumes of water but is qualitative only. Safferman and Morris (1967) found that LPP-1 was present in 11 out of the 12 waste stabilization ponds which they studied in the United States and recorded titres of up to 270 plaque-forming units per ml. They also studied 132 pond samples of which about 70% yielded viruses of the LPP type. In studies on seasonal variation over a 15 month period in five ponds they found that two of these always contained the virus. For some unknown reason, numbers decreased in the summer months when algal numbers increased. This suggests perhaps that at other times the algal viruses may have been responsible for the low algal numbers. In Israel, seasonal variation in LPP-type viruses has been studied by Padan and Shilo (1968) who found these viruses to be present in more than 90% of the waters taken from 40 fish ponds. Viral numbers were studied from October 1967 to July 1968 in five ponds using the plaque technique and they found, unlike Safferman and Morris (1967), that numbers increased dramatically in May when algal blooms also developed. The numbers present throughout the year ranged from 1 to 13,000 plaque-forming units per litre. The use of plaque counts as a measure of virus abundance in natural populations has some disadvantages because, as mentioned earlier, identical plaques may be produced by different agents. Nevertheless in serological studies in Scotland, Daft and Stewart (1971) detected a virus of the LPP type in each of over one hundred water samples tested.

Bacterial pathogens

The possible significance of bacterial pathogens in regulating the growth of blue-green algae in natural ecosystems has been appreciated only recently. Shilo (1966) was the first to point out that bacterial pathogens exist which cause rapid lysis of algae. Later Shilo reported (1970) on a number of gram-negative bacteria isolated from fish ponds in Israel which caused lysis of blue-green algae including *Synechococcus* sp. (*Anacystis nidulans*), *Coccochloris peniocystis*, *Synechococcus* sp., *Plectonema boryanum* and *Nostoc* species. Similarly Wu *et al.* (1968) reported briefly on a myxobacterium, probably a species of *Myxococcus*, which lysed some blue-green algae. Its activity was lost on subculturing.

More detailed studies on algal lysing bacteria have been reported by Stewart and Brown (1969), Shilo (1970) and Daft and Stewart (1971). The organisms are all members of the Myxobacterales, an order of aflagellate aerobic gram-negative bacteria which produce a silky growth and move by a slow gliding motion. The isolate studied by Stewart and Brown (1969) is 0·4–0·6 × 1·9–3·7 μm in size, has rounded ends, is aflagellate and without fruiting bodies. It digests cellulose but not chitin and is considered to be a species of *Cytophaga* (*Cytophaga* N5) although DNA base ratios which might confirm this identification were not determined. It digested all blue-green algae tested as well as species of green algae and various bacteria, apparently by producing extracellular lytic material. Isolate FP-1 studied by Shilo (1970) and isolates CP-1, CP-2, CP-3 and CP-4, isolated by Daft and Stewart (1971) are closely related myxobacteria. FP-1 is 3–9 × 0·6–1·0 μm in size but increases in length with age, while CP-1 the isolate studied by Daft and Stewart (1971) is on average 2·4 × 10·6 μm in size (Fig. 13.7). These gram-negative isolates did not develop spores under the growth conditions used and both contain carotenoid-like pigments. The GC ratio for FP-1 is approximately 70%, while the isolates CP-1–CP-4 have GC ratios of 65–69% (Table 13.I). Thus although

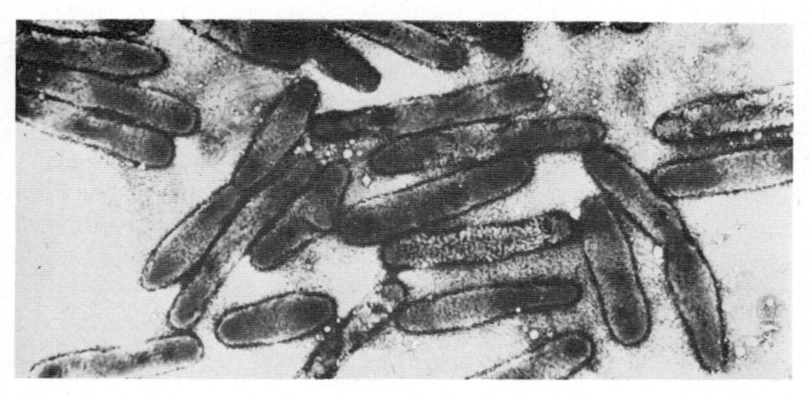

FIG. 13.7. Bacterium CP–1 (× 12,800). Note extracellular mucilaginous material and convoluted outer wall. Electron micrograph by M. J. Daft and W. D. P. Stewart (unpublished.)

TABLE 13.I. Guanine–cytosine ratios of various algal lysing bacteria

Organism	Mol % G + C	Reference
FP-1	70·0	Shilo, 1970
CP-1	68·9	Daft and Stewart, 1971
CP-2	68·4	Daft and Stewart, 1971
CP-3	65·3	Daft and Stewart, 1971
CP-4	66·8	Daft and Stewart, 1971
Myxobacter 44	69·4	Stewart and Brown, 1971
Myxobacter 45	71·3	Stewart and Brown, 1971
Myxobacter 46	69·5	Stewart and Brown, 1971
Myxobacter 3C	69·2	Stewart and Brown, 1971
Myxobacter AL-1	69·2	Stewart and Brown, 1971

fruiting bodies have not been found the GC ratios suggest that the isolates are more closely related to the fruiting myxobacteria than to *Cytophaga* species, which have GC ratios of 30–40%. A subsequent report on several myxobacteria, isolated from North America, which lyse blue-green algae, has come from Stewart and Brown (1971). In most respects they resemble the myxobacteria just described.

The plaques caused by myxobacteria resemble viral plaques in appearance (Fig. 13.8) although FP-1 plaques are slightly depressed in the agar.

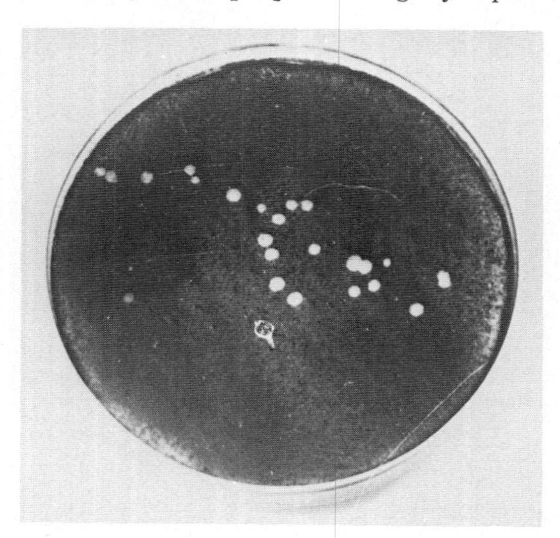

FIG. 13.8. Plaques formed by lytic bacterium CP–1 on lawns of *Nostoc ellipsosporum*. Photograph by M. J. Daft and W. D. P. Stewart (1971) *New Phytol.*, **70**, 819, Plate 1.7.

In general, one plaque is produced by one bacterium. Unlike viruses, these bacteria can be propagated in the absence of the host and, unlike the bacterium studied by Wu *et al.* (1968), their pathogenicity is not lost on subculturing. The algae lysed by FP-1 are *Nostoc* species 6305, *Plectonema boryanum*, *Oscillatoria prolifera*, *Spirulina platensis* and *S. tenuis* but *Anabaena cylindrica* and *Oscillatoria amphibia* as well as *Chlorella pyrenoidosa* and *Prymnesium parvum* are resistant. The CP isolates are effective against over 40 algal strains belonging to all orders of the Cyanophyceae although not all algal species or strains are necessarily susceptible. Lysis may be rapid and complete within 2–10 h but may take up to 3 days (Fig. 13.9). They also lyse various natural blooms taken from Scottish freshwaters (Fig. 13.10) as well as certain gram-positive and gram-negative bacteria. It is interesting to note that these bacteria are strictly aerobic

FIG. 13.9. Lysis of *Anabaena flos-aquae* by bacterium CP–1 in liquid culture. The flask on the left is a control without CP–1; the flask on the right was inoculated with CP–1 three days previously. Photograph by M. J. Daft and W. D. P. Stewart (unpublished).

FIG. 13.10. Lysis of a natural population of *Anabaena circinalis* by bacterium CP–1. Treatments from left to right are: control without added bacteria; bacteria added for 12 h; 24 h; and 48 h. The dark coloration of the liquid in the treated cultures is due to phycocyanin release. Note that CP–1 attack results in a loss in bouyancy of the algae. Photograph by M. J. Daft and W. D. P. Stewart (unpublished).

organisms while the blue-green algae evolve oxygen during photosynthesis. This may be of importance in natural ecosystems.

The mode of action of FP-1 and the CP isolates is rather similar. Shilo (1970) has shown by microscopical observation that direct contact between the ends of the bacteria and the susceptible algal cells is necessary for lysis to occur. She found that, within 20 min of contact, lysis of the infected cell resulted and that the bacterium then moved to another cell of the alga and the process was repeated. The occurrence of CP-1 attacking *Oscillatoria redekii* is shown in Fig. 13.11. As with FP-1, the bacteria attack end-on.

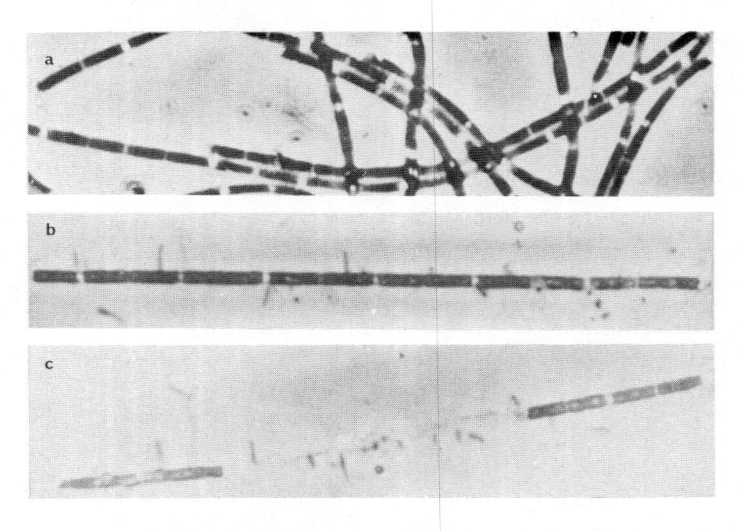

FIG. 13.11. Filaments of *Oscillatoria redekii* being attacked by bacterium CP–1 (a) control series without added bacterium; (b) bacteria attaching end-on to algal cells; (c) lysis of vegetative cells (× 1150). Photograph by M. J. Daft and W. D. P. Stewart (1973) *New Phytol.*, in press.

Lysis of the attacked vegetative cells then occurs, but heterocysts (Fig. 13.12) and spores are not lysed. After lysis all that remains of the algal cell is a complex of membranous material (Fig. 13.13). There is no evidence that the bacteria FP-1 and CP-1 produce extracellular lytic agents which accumulate in the medium. Daft and Stewart (1971), for example, showed that ^{14}C uptake from bicarbonate was inhibited within 2 h by the addition of intact cells to the culture, but not by filtrates from bacterial cultures alone. Effects on acetylene reduction were similar (Fig. 13.14). Stewart and Brown (1971) have found, however, that one of their strains does produce extracellular lytic material, so there is obviously some variation from strain to strain.

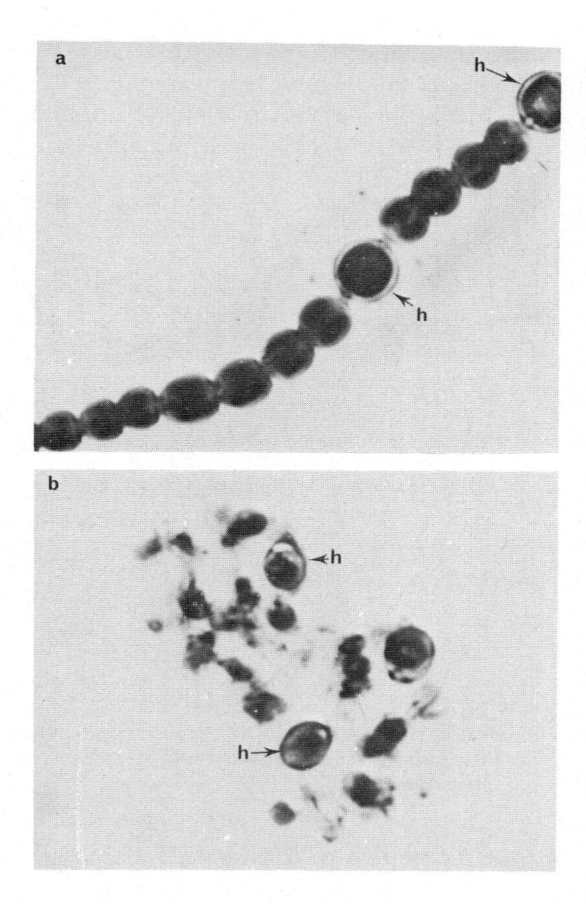

FIG. 13.12. Filaments of *Anabaena* sp. before (a) and after (b) lysis by bacterium CP–1. Note that only the heterocysts (h) remain clearly distinguishable after attack. From M. J. Daft and W. D. P. Stewart (1971), *New Phytol.* **70**, 819, Plate 2, 9–10.

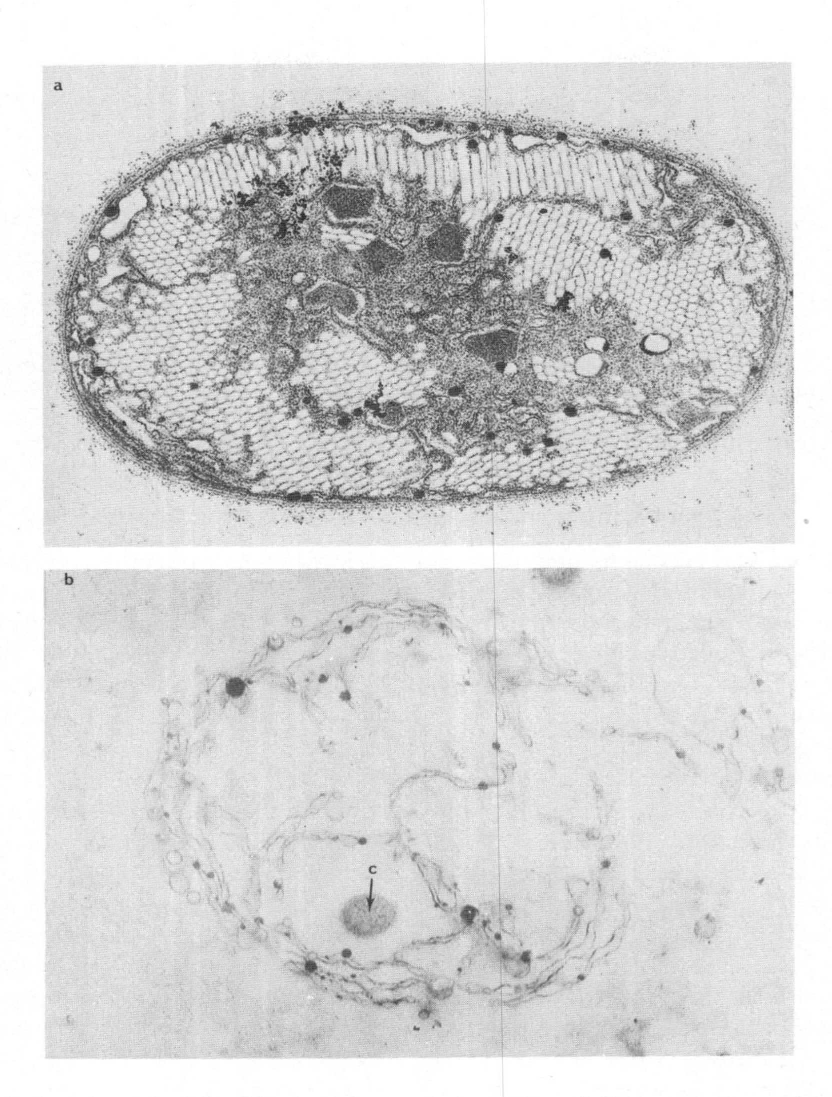

FIG. 13.13. A healthy cell of *Anabaena circinalis* (a) and the membranous remains (b) after lysis by bacterium CP–1 (c). (× 16,500), Electron micrographs M. J. Daft and W. D. P. Stewart (1973) *New Phytol.*, in press.

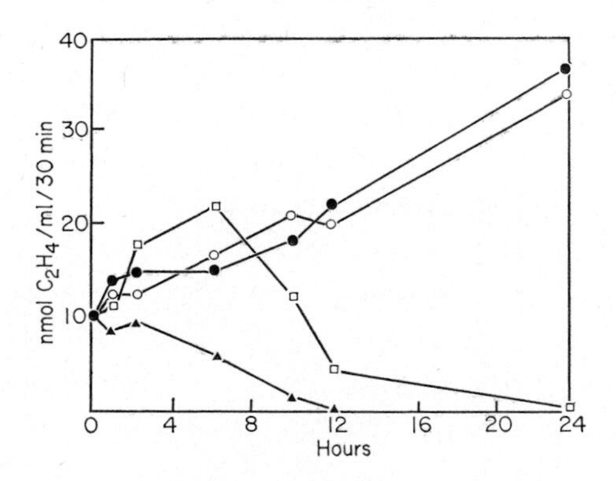

FIG. 13.14. The effect of bacterium CP–1 on acetylene reduction by *Nostoc ellipsosporum* ●, treated with sterile medium; ○, treated with CP-1 culture filtrate; □, treated with washed CP–1 cells; ▲, treated with CP–1 suspension. From M. J. Daft and W. D. P. Stewart (1971), *New Phytol.* **70**, 819, Fig. 5.

The practical use of pathogens in controlling blooms

The isolation and characterization of a variety of micro-organisms which lyse blue-green algae opens up interesting possibilities of the biological control of bloom formation in freshwaters. The myxomatosis virus was successful in controlling, at least for some time, rabbit populations in Australia and elsewhere, and it is possible that algal viruses could play a similar role in the control of water-blooms. The possible use of algal viruses is limited at the moment by the fact that few which are active against dominant bloom-forming species of algae have been isolated. The LPP group, which appear to be most widespread, affects only *Lyngbya*, *Phormidium* and *Plectonema* species and these seldom occur as water blooms. Of course this could be because they are so susceptible to the virus. The virus SM-1 which affects *Microcystis* may be more useful but it is difficult to isolate and is highly specific. Insufficient information has yet been accumulated on the viruses which affect heterocystous algae to decide what their potential as controlling agents may be. In addition to the disadvantage of algal viruses being specific to certain species, there is also the circumstance that resistant strains of algae arise readily. The use of bacteria may be a better proposition not only because these have a much wider host spectrum, but also because they cause algal lysis much more quickly and can be cultured in bulk, free from the host organism.

There have been few field trials on the effectiveness of these pathogens although the Russian workers, Goryushin and Chaplinskaya (1968) reported that they successfully cleared blooms of algae using viruses. Daft and Stewart (1971) have caused lysis of samples of natural blooms in the laboratory within 1–10 h using CP-1 and in field tests on small isolated sections of a reservoir noted the disappearance of an algal bloom within 48 h of the addition of CP-1 (Fig. 13.15). This result gives some promise that biological agents such as viruses and bacteria could serve in the future as bloom-control agents.

FIG. 13.15. Effect of bacterium CP–1 on artificial ponds isolated at the edge of a reservoir. Pond on the left has been treated with the bacterium for 48 h. Note disappearance of algae and cloudiness of water caused by phycocyanin release. Pond on right is untreated and *Microcystis aeruginosa* colonies occur as floating masses. From M. J. Daft and W. D. P. Stewart (unpublished).

Chapter 14

Marine blue-green algae

Blue-green algae are of widespread distribution in marine ecosystems although they are seldom dominant and often pass unnoticed to the casual observer. They are particularly abundant in estuarine and intertidal areas where they are subjected to extremes of environmental change, but on the other hand, the planktonic form *Trichodesmium* is particularly sensitive to fluctuations in the environment and is difficult to culture.

It is questionable whether most of the intertidal blue-green algae that have been described can be separated satisfactorily from freshwater forms. Some species certainly overlap. Fan (1956), for example, concluded from his taxonomic study of the genus *Calothrix* that the marine species *C. scopulorum* was a form of the freshwater alga *C. parietina* and Drouet (1968) in his monograph on the Oscillatoriaceae considers that, with a few exceptions such as *Oscillatoria erythraea* (*Trichodesmium* spp.), there is no true separation into freshwater and marine species.

Physiological studies also show that there may be considerable overlap. Fay and Fogg (1962) found, for example, that their freshwater isolate of *Chlorogloea fritschii* could, after a few days' adaptation to marine medium, grow as well in that as it did originally in freshwater. Also, Van Baalen (1962), who studied the sodium chloride requirement of marine isolates of *Coccochloris elabens*, *Agmenellum quadruplicatum* and *Microcoleus chthonoplastes*, found that although the optimum sodium chloride concentrations for the growth of *Coccochloris* and *Agmenellum* were 10‰ and 20‰ respectively, the two algae could still grow when only traces of sodium chloride were present. Stewart (1964a) reported similarly that his marine isolates of *Calothrix scopulorum* (Fig. 14.1) and *Nostoc entophytum* could grow well at salinities approaching those of freshwater. These laboratory investigations are supported by the field observations of Ercegovic (1930) who studied the algal flora present on rocks subjected to different salinities along the Dalmatian coast. He found that in areas subjected to open sea water the common species included *Calothrix scopulorum*, *Mastigocoleus testarum*, *Hyella* sp. and numerous Chroococcaceae, whereas on rocks at the mouth of a river where salinities were as low as 4·3‰ *Calothrix* persisted but *Mastigocoleus* and *Hyella* were absent. That is, *Calothrix* only appeared able to withstand low salinities. In upper rock pools, where salinity variations were wide (0–283‰) as a result of rain and evaporation, *Calothrix* disappeared.

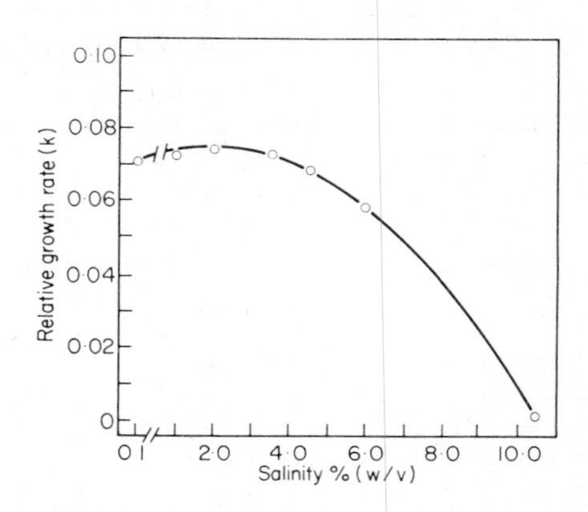

FIG. 14.1. Effect of salinity concentration on the growth of *Calothrix scopulorum*. After W. D. P. Stewart (1964) *J. gen. Microbiol.* **36**, 415.

Although some species grow at salinities approaching freshwater it is equally clear that for others a higher sodium chloride concentration is essential for growth. This was clear from the ecological observations of Ercegovic (see above) and has been confirmed by Van Baalen (1962). His strain of *Microcoleus chthonoplastes*, which in culture has a sodium chloride optimum of 20–25‰, does not grow when only traces of sodium chloride are present. Hof and Frémy (1933), who studied the flora of salt works divided the algae into two physiological groups which they termed *halotolerant* and *halophilic*. Halotolerant species are those which cannot grow at sodium chloride concentrations above 3 M (175·5‰) and *Calothrix scopulorum* is a good example of this type. In laboratory culture it can withstand, but cannot grow at, a salinity of 106‰ (Stewart, 1964a). Other examples are some salt works algae such as *Microcoleus chthonoplastes* and *Lyngbya aestuarii*. Halophilic species such as *Spirulina subsalsa*, which can grow at salt concentrations above 3 M, occur commonly in salt works and their ability to do so is probably helped by their prokaryotic cellular organization with its absence of large sap vacuoles. It is worth noting that certain eukaryotic algae such as *Dunaliella* sp. also occur in salt works and indeed the latter alga is more resistant to salt than are the blue-green algae. Hof and Frémy (1933) observed that although it was rare to obtain live blue-green algae from crude salt kept for more than three years, *Dunaliella* could be obtained from salt seven years old. Hof and Fremy summarize the literature on the algae of salt works.

Detailed laboratory studies on the general physiology of marine isolates were undertaken more recently by Van Baalen (1962). The most striking finding is that over 50% of the species examined have an absolute requirement for vitamin B_{12} (Table 14.I). No other vitamin requirement has been

TABLE 14.I Vitamin B_{12} requirement of various marine blue-green algae (after Van Baalen, 1962)

Species	No. of strains tested	Growth in absence of vitamin B_{12}
Agmenellum quadruplicatum (Menegh.) Bréb	1	+
Agmenellum quadruplicatum (Menegh.) Bréb	1	—
Anacystis marina (Hansg.) Dr. & Daily	1	+
Coccochloris elebans (Bréb) Dr. & Daily	1	—
Lyngbya aestuarii (Mert.) Lyngb.	1	+
Lyngbya lagerheimii (Möb.) Gom.	3	—
Microcoleus chthonoplastes (Mert.) Zanard.	2	+
Microcoleus tenerrimus Gom.	1	+
Oscillatoria amphibia Ag.	1	—
Oscillatoria subtilissima Kütz.	1	—

detected and in general the vitamin B_{12} analogue specificity is broad. This requirement for vitamin B_{12} is unique to marine blue-green algae and there is no corresponding vitamin B_{12} requirement by species from freshwater habitats. There is nothing peculiar about the nitrogen nutrition of these marine strains. Nitrate and ammonia are good nitrogen sources for growth, some fix nitrogen (see p. 340) and casamino acids support growth only for a short time, which suggests that only some of the amino acids serve as nitrogen sources. Urea is used but guanidine is not, and the ability of these algae to utilize organic nitrogen may account in part for their abundance in rocks and pools near bird colonies. This is the case for example on the Isle of May, off the Scottish east coast. One marine alga, a strain of *Lyngbya lagerheimii*, has been reported to grow, but only slowly, in the dark on glucose (Van Baalen, 1962). Cultures of marine blue-green algae, in addition to providing useful physiological data, could, if used by taxonomists, also help to avoid some of the confusion which different ecophenes of the same species cause in taxonomic studies of marine blue-green algae.

Intertidal and supralittoral algae

Blue-green algae are often abundant in intertidal and supralittoral zones of the seashore, where they form distinct narrow bands on vertical faces and broad bands on flat sheltered areas. Their abundance in these habitats is helped by their ability, compared with most other plants, to withstand the adverse conditions which prevail there so that they are at a competitive advantage. In salt marshes and marine muds where they are also abundant, they may benefit from the reducing conditions which sometimes occur, and from the availability of vitamin B_{12} produced by bacteria in these habitats.

1. ROCKY SHORES

Distinct crustose growths of endolithic and epilithic blue-green algae occur on porous rocks particularly in temperate zones. They are, for example, abundant in the spray zone of the chalk cliffs of southern England (Anand, 1937), the sandstone rocks of some Scottish coasts (Stewart, 1967a) and along the Dalmatian coast (Ercegovic, 1932). Free-living forms are rare on hard rocks such as slate, but lichens such as *Lichina confinis* and *L. pygmaea*, which contain blue-green algae as symbionts often occur on these rocks. Indeed, a supralittoral fringe containing blue-green algae or lichens is stated by Stephenson and Stephenson (1949) to be a universal occurrence.

The dominant epilithic algae on temperate rocky shores are often species of *Calothrix*, *Entophysalis*, *Gloeocapsa*, *Phormidium*, *Plectonema* and *Oscillatoria* (Anand, 1937; de Halperin, 1967, 1970; Den Hartog, 1959; Frémy, 1934; Lindstedt, 1943). This flora frequently extends throughout the littoral zone although it is most obvious in the supralittoral and upper littoral areas where it is not masked by a growth of seaweeds. The flora of rocky shores varies with the season. The *Calothrix* dominated flora occurs on many shores. In Scotland this is most profuse in autumn, winter and spring, but peels off from the supralittoral zone in summer as a result of desiccation (Fig. 14.2). On the other hand on parts of the south coast of England, such as at Wembury Point in Devon, *Calothrix* species are rare but the summer annual *Rivularia bullata* occurs as an extensive growth on barnacles at and below mid-tide level. Here distinct colonies first appear in early June and increase in size and number until late August, after which they decrease, and by mid-October have disappeared.

Common endolithic species in temperate areas are *Hyella caespitosa*, *Mastigocoleus testarum*, *Gomontia polyrhiza*, *Phormidium molle* and *Gloeocapsa* sp. which occur extensively in supralittoral calcareous rocks (Anand, 1937; Frémy, 1934) and others, such as *Mastigocoleus testarum* and *Hyella caespitosa*, often occur in the shells in intertidal marine molluscs (Nadson, 1900, 1932; Ercegovic, 1932; de Halperin, 1970). According to Le Campion-Alsumard (1970) some endolithic forms resembling *Gloeocapsa*, *Scopulonema* and *Hyella* species may be morphological variants of a single

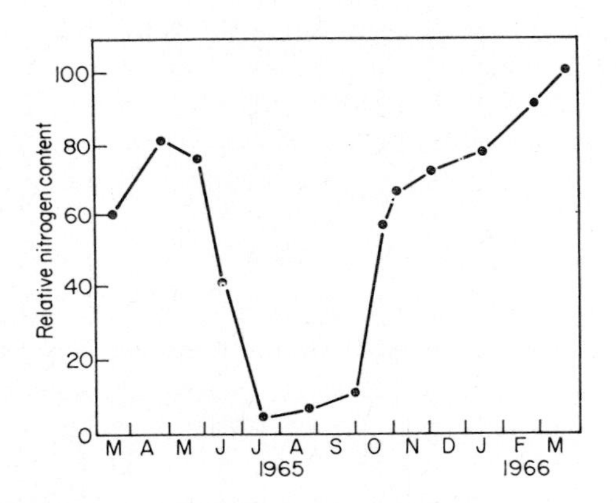

FIG. 14.2. Relative changes with season in the total nitrogen content of a *Calothrix* dominated supra-littoral fringe of a Scottish rocky shore. After W. D. P. Stewart (1967a) *Ann. Bot.* N.S. **31**, 385–406.

species *Entophysalis deusta*, but another widely distributed form, *Hormatonema*, is quite distinct. The mechanisms by which these algae penetrate shells and rocks is uncertain. Golubic (1969) considers that boring is a dissolution process carried out by terminal cells of the algae. The material secreted may be acid as has been suggested by early workers (*see* Fritsch, 1945) but this seems unlikely because in culture the production of extracellular acids is insufficient to lower the pH of the medium markedly. Chodat (1898), on the other hand, noted an alkaline reaction in the neighbourhood of algal filaments. This could be due to the pH of the immediate environment around the alga rising as carbon dioxide is removed from the water during photosynthesis. Another possibility is that the peptides and polypeptides which are generally secreted by blue-green algae (see p. 208) may have a solvent action by virtue of their chelating properties. However, if a simple solution mechanism operates it is surprising, perhaps, that all epilithic algae are not endolithic.

Blue-green algae are rare in the supralittoral and littoral regions of tropical shores because of the severe desiccation which results during low tide. However, *Nostoc* colonies have been found on coral reefs in the Pacific (Wood, 1965).

The few physiological studies carried out on rocky shore blue-green algae in pure culture have been concerned primarily with nitrogen fixation (Stewart, 1965, 1967a, and unpublished). *Calothrix scopulorum* grows well at salinities of 1–60‰ (Fig. 14.1) and has a pH optimum of 7–8 with activity decreasing rapidly below pH 6 and above 9 (Fig. 14.3). At pH 5

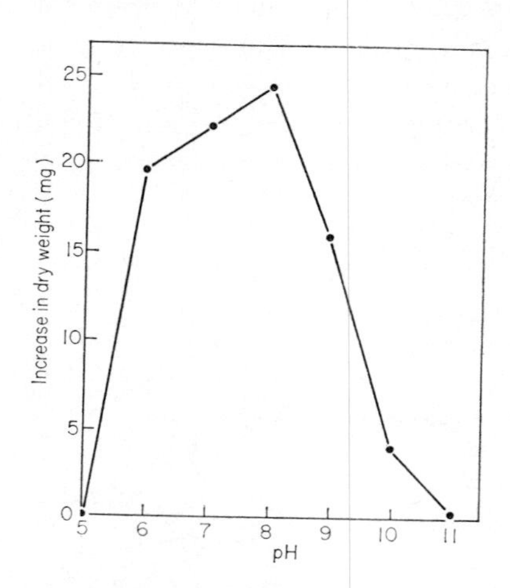

FIG. 14.3. Effect of pH on the growth of *Calothrix scopulorum* after 12 days growth in medium free of combined nitrogen (W. D. P. Stewart, unpublished).

the algae die. The temperature optimum is 30°C in the laboratory although fixation continues in the field at low rates at temperatures approaching 0°C. In fact, except on warm sunny days at low tide *Calothrix* is seldom likely to be exposed *in situ* to its optimum temperature, and when it is, it is probably affected adversely by the concomitant desiccating effects of the sun's rays. The optimum light intensity is approximately 8000 lux, a light intensity which is much lower than that often found in the field. The pigments which occur in the sheaths of *Calothrix* (see p. 17) may help to cut down the solar radiation reaching the algal protoplasm and it may be that in natural populations the most active algae will be found within the algal mat rather than at its surface.

2. SALT MARSHES

In salt marsh regions, where blue-green algae are often ubiquitous, there is a steady gradation from almost typically freshwater habitats to truly marine habitats. The algae, which play a role in the accretion of marshes by stabilizing the sand and mud and by building up humus, often penetrate the silt to depths of 5 mm and still retain their pigmentation. Those at the surface are sometimes brownish or reddish in colour. This is reminiscent of the excessive brownish pigmentation found in some salt marsh diatoms. It is probably due to an excessive development of phycoerytherin and perhaps carotenoids as well but the physiological reason for this is uncertain. Salt marsh algae have to withstand extremes of desiccation, salinity and redox values which may range from approximately −170 to +600 mV. This redox range is rather similar to that over which the purple sulphur bacteria (Thiorhodaceae) grow (Fig. 14.4) and this can be seen readily if

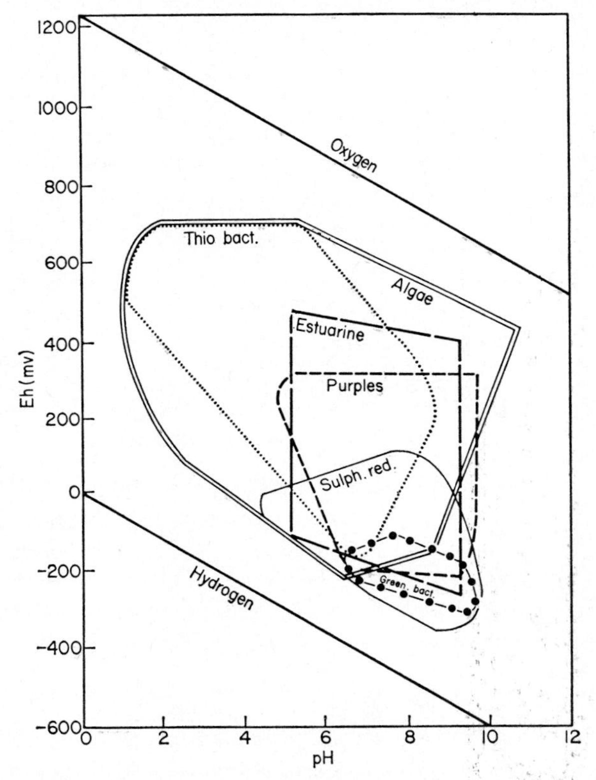

FIG. 14.4. Redox and pH ranges for the growth of various estuarine micro-organisms, including blue-green algae. After L. G. M. Baas-Becking and E. J. Ferguson Wood (1965) *Proc. Acad. Sci. Amst.* B, **58**, 173–181.

salt marsh mud is set up in a Winogradsky column. There the development of purple sulphur bacteria is followed, as conditions become less reducing, by a profuse growth of Cyanophyceae such as species of *Lyngbya*, *Microcoleus*, *Phormidium*, *Rivularia* and unicellular forms.

A remarkable feature of salt marsh algae, as pointed out by Chapman (1962), is that their zonation is rather similar in various temperate regions of the world. Most species usually occur, in small numbers at least, throughout the marsh (Carter, 1932, 1933; Stewart and Pugh, 1963) but nevertheless they do vary in the abundance at different levels and distinct patterns are observed.

At the top of the marsh and often on the vertical faces of escarpments *Rivularia biassolettiana*, *R. atra* and occasionally *Nostoc* predominate and may form colonies of up to an inch in diameter, just as they frequently do on soil at the top of the supralittoral fringe of rocky shores. These forms, with their thick gelatinous sheaths and small surface volume ratios, are particularly well suited to withstanding the desiccation to which they are subjected. Even so, on dry summer days the algae may become so dry, brittle and covered with salt that they are almost unrecognizable. Under such conditions the algae have a very low metabolic rate but on re-wetting the thalli metabolic activity may increase markedly within a few hours. This *Rivularia*-dominated zone is sparse in the salt marsh at Canvey, on the English east coast, where Carter (1932, 1933) carried out extensive studies (see Fig. 14.5), but nevertheless it is common in many marshes.

Below the *Rivularia* zone and in the more open parts of British salt marshes there is often an abundant development of filamentous forms such as the tightly interwoven filaments of *Phormidium*, the rope-like bundles of

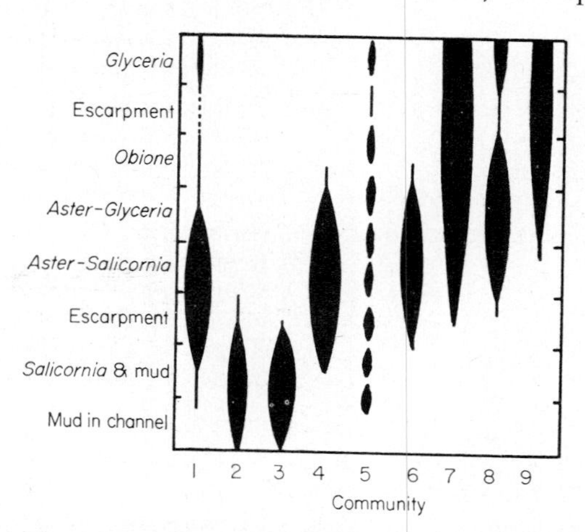

FIG. 14.5. The distribution of algal communities in relation to habitat in Canvey salt marshes (after Carter, 1933). 1, general Chlorophyceae community; 2, marginal diatom community; 3, marginal Cyanophyceae community; 4, *Ulothrix flacca* community; 5, *Enteromorpha minima–Rhizoclonium* community; 6, *Anabaena torulosa* community; 7, filamentous diatom community; 8, autumn Cyanophyceae community; 9, *Phormidium autumnale* community.

Microcoleus and thick sheathed algae such as *Gloeothece palea*. These algae are often most abundant near higher plants which provide protection from desiccation and excessive light. Often, particular algae are associated with specific higher plants, for example, the alga *Phormidium autumnale* with the angiosperm *Halimione portulacoides*. This is probably a result of both plants preferring similar environmental conditions rather than to any direct inter-dependency. The algae which occur in salt pans are quite distinct from those on the general salt marsh surface. In pans near the top Chlorophyceae predominate but in pans near the bottom these are replaced by blue-green algae such as *Anabaena torulosa* and *Oscillatoria* sp. which often form as a thick felt on the bottom.

TABLE 14.II. Algal communities and species of blue-green algae found commonly on Canvey salt marshes (Carter, 1933)

Community No.	Community description	Blue-green algae
1	General Chlorophyceae	*Microcoleus chthonoplastes, Lyngbya aestuarii, Rivularia* spp., *Gloeothece palea, Microcoleus tenerrima, Hydrocoleus lyngbyaceum, Lyngbya lutea, Nostoc commune, Spirulina subsalsa, Oscillatoria corallinae, Phormidium angustissimum, P. foveolarum, Plectonema phormioides*
2	Marginal diatoms	*Microcoleus acutirostris, M. chthonoplastes, Pseudoanabaena brevis, Holopedia sabulicola, Merismopedia revoluta, M. convoluta*
3	Marginal Cyanophyceae	*Oscillatoria sancta, O. corallinae, Phormidium angustissimum, P. tenue, Spirulina major, S. subsalsa, S. subtilissima, Microcoleus chthonoplastes, Oscillatoria formosa, O. laetevirens, Phormidium foveolarum*
4	*Ulothrix flacca*	—
5	*Enteromorpha minima–Rhizoclonium*	—
6	*Anabaena torulosa*	*Anabaena torulosa, Nodularia harveyana*
7	Filamentous diatoms	—
8	Autumn Cyanophyceae	*Oscillatoria bonnemaisonii, O. nigro-viridis, O. brevis, Microcoleus chthonoplastes, Lyngbya confervoides*
9	*Phormidium autumnale*	*Phormidium autumnale*
10	*Rivularia–Phaeococcus*	*Rivularia atra, Phormidium molle, Phormidium autumnale*

The general pattern of algal development in relation to habitat found on the Canvey marshes by Carter (1933) is shown in Fig. 14.5 and the species which may occur are listed in Table 14.II. Community 1 is composed mainly of green algae, and blue-green algae, although common, are a minor component. Community 2 and community 3 species occur particularly along the margins of the channels, but at different times. Community 6, of *Anabaena torulosa* and *Nodularia harveyana* often forms almost pure patches, several centimetres in diameter, on the marsh surface. In community 8 *Oscillatoria* species predominate and in community 9 *Phormidium autumnale* forms an almost pure sheet, except for a few diatoms. Blue-green algae are usually absent from communities 4, 5 and 7 but these have been included for completeness.

The seasonal development of blue-green algae in salt marshes follows a fairly consistent pattern (Fig. 14.6) with development reaching a maximum

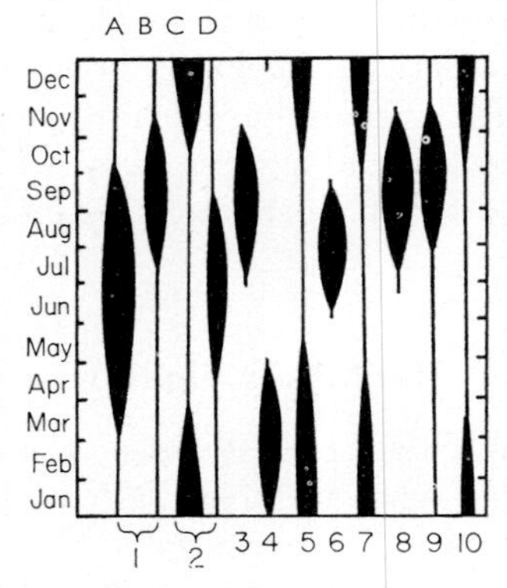

FIG. 14.6. Seasonal variation in algal communities in Canvey salt marshes (after Carter 1933). 1, general Chlorophycean community (A, Chlorophyceae component; B, Cyanophyceae component); 2, marginal diatom community (C, components with a winter maximum; D, components with a summer maximum,); 3, marginal Cyanophyceae community; 4, *Ulothrix flacca* community; 5, *Enteromorpha minima–Rhizoclonium* community; 6, *Anabaena torulosa* community; 7, filamentous diatom community; 8, autumn Cyanophyceae community; 9, *Phormidium autumnale* community; 10, *Rivularia–Phaeococcus* community.

in late summer and autumn. In spring and early summer the blue-green algae, although present, often become overgrown by species of *Vaucheria* and *Enteromorpha*.

Benthic algae

While littoral and supralittoral blue-green algae occur most extensively in temperate waters, benthic algae are more common in clear warmer seas.

There are exceptions of course. Pintner and Provasoli (1958) studied *Phormidium persicinum*, a red pigmented blue-green alga which was isolated from a depth of 20 m off the coast of Norway. *Calothrix confervicola* and *Symploca hydnoides* have been found down to 5 m below low water mark in the Rio de Arosa, Spain, and *Lyngbya majuscula* down to 30 m in the same locality (Donze, 1968). Bunt *et al.* (1970) observed rich growths of several blue-green algae, mainly Oscillatoriaceae, at depths of 20–25 m off the Bahamas and South Eastern Florida. The commonest species were *Schizothrix calcicola, S. mexicana, S. tenerrima, Porphyrosiphon notarisii, P. kurzii, Microcoleus lyngbyaceus* and *Spirulina subsalsa*. Price (1971) noted that in lagoon areas on Aldabra in the Indian Ocean growths of *Oscillatoria submembranacea* occur in the sublittoral beneath overhanging coral which produce a shaded environment. Deep water benthic forms show chromatic adaptation (see p. 148). For example, *Lyngbya sordida*, which is blue-green or brownish when growing in surface illumination is pink when it occurs at a depth of 20–30 m. This is because of a preponderance of *c*-phycoerythrin over *c*-phycocyanin (Feldmann, 1959). More information on benthic blue-green algae is required and perhaps with the interest in SCUBA diving which exists now this information may soon be forthcoming.

Planktonic algae

In contrast to their abundance in freshwater habitats, there are few truly planktonic blue-green algae in the sea. The commonest are species of *Trichodesmium*. This genus is characterized by having the reddish pigmented filaments aggregated into rafts or bundles of 6–25 filaments, which float towards the surface of the water and which are easily visible to the naked eye (Fig. 14.7, facing p. 257). Sailors give these the name of "sea sawdust". Desikachary (1959) describes three species from the Indian Ocean: *Trichodesmium erythraeum, T. thiebautii* and *T. hildenbrandii* and Wood (1965) records a species *T. rubescens*. However Drouet (1968) in his taxonomic revision of the Oscillatoriaceae classifies all *Trichodesmium* species as *Oscillatoria erythraea*.

Trichodesmium is a genus of tropical waters and grows where the surface water temperature is above 25°C and the salinity near 35‰. It is abundant in the Indian Ocean, the tropical Atlantic Ocean, the Red Sea (to which it is believed to have given its name) and in the Pacific Ocean (see Bowman and Lancaster, 1965; Wood, 1965; Ramamurthy, 1970). It occurs usually in long windrows which vary from a few feet to several miles in length and Wood (1965) records *Trichodesmium* blooms occupying up to 52,000 km^2 off the western coast of Australia. Blooms of this size however are rare. In summer the alga extends to sub-tropical and even temperate regions. It is sometimes found, for example, extending northward from the Sargasso Sea to the continental shelf. Farran (1932) and Stewart and Boalch (unpublished) found it off the south coast of Ireland, and Wood (1965) records a species, *Trichodesmium rubescens*, off the south of New Zealand and even in Antarctic waters. It is not known whether the alga is meta bolically active in these colder waters and one of the reasons for our lack of knowledge is that it is extremely difficult to keep the alga in culture and thus experiment with it.

The environmental factors which initiate bloom formation are uncertain but intriguing. Wood (1965) considers that upwelling and rainfall may be important in providing nutrients and that this may help the bloom to develop. The abundance of *Trichodesmium* in coastal waters of high salinity and in oceanic waters into which large rivers such as the Amazon flow suggests that the nutrients it requires are derived from run-off from the land. Ramamurthy (1970), who reported that gibberellic acid stimulates the growth of *Trichodesmium*, suspects that the availability of growth substances may be important but this requires rechecking as the concentrations he used were much higher than those likely to be encountered in natural seawater. Vitamin B_{12}, which is required by the red pigmented *Phormidium persicinum* (Pinter and Provasoli, 1958), by many other marine Oscillatoriaceae (see p. 300), and which is produced abundantly by estuarine bacteria (see Starr, 1956) may be required for growth of *Trichodesmium*.

Despite the fact that *Trichodesmium* occurs in the plankton, Feldmann (1932) considers it to be a bottom living form which occasionally floats to the water surface. Thus the red pigmentation could be explained in terms of chromatic adaptation (see p. 148) but some of the waters where *Trichodesmium* occurs are so deep as to make the suggestion that it is really a benthic alga unlikely. *Trichodesmium*, like freshwater planktonic blue-green algae, possesses gas vacuoles (Stevens and Van Baalen, 1970) which presumably play a part in regulating buoyancy. Its biological success may therefore be related to an ability to locate itself at optimum depth in the water column (see p. 263). The difficulty encountered in culturing *Trichodesmium* suggests that surface blooms are largely moribund.

Trichodesmium blooms in general are non-toxic, unlike the red tides of the dinoflagellate *Gymnodinium breve*. Ramamurthy (1970) reported no ill effects after an "accidental drenching" in a bloom of the alga. Indeed, the same author suggests (but does not provide convincing data in support of this) that *Trichodesmium* produces anti-bacterial substances. The alga has been reported to cause fish mortality by the clogging of gills (Desikachary, 1959) and certain fish such as the adult *Chanos chanos* avoid it, although strangely its larvae are said to feed on it (Wood, 1965). Perhaps it is the offensive odour of some *Trichodesmium* blooms which the adult fish dislike.

In the nutrient depleted surface waters of areas such as the Sargasso Sea, *Trichodesmium* occurs even when the phosphate-phosphorus and nitrate plus nitrite levels are reduced to 1·5–3·1 μg P l^{-1} and 1·4–7·0 μg N l^{-1} respectively. It is metabolically active there although the agreement between replicates obtained both in nitrogen fixation (Dugdale *et al.*, 1964) and photosynthesis (McLeod *et al.*, 1962) studies is not good. The occurrence of different coloured colonies, which may range from grey, through yellow, green, purple and red, suggests corresponding variation in metabolic activity. The capacity to fix nitrogen may enable *Trichodesmium* to thrive in waters low in combined nitrogen (see p. 193) but nitrate, ammonium nitrogen, glycine and urea are all assimilated by the alga (Dugdale and Goering, 1967).

Although most attention has been paid to *Trichodesmium*, Bernard and Leçal (1960) report that species of *Nostoc* are important components of the marine plankton of tropical and sub-tropical waters, being found abundantly at depths of 600–1000 m in the Arabian Sea, the Mediterranean, the Indian Ocean, the Atlantic Ocean and even in sub-antarctic waters. Three forms have been reported. The commonest, which is 3–5 μm in diameter, has few heterocysts and is of variable morphology, but resembles the freshwater species *Nostoc planktonicum*. A second, smaller form occurs off the Algerian coast and in the Atlantic, while the third species has cells 7–10 μm in diameter. These forms have not yet been reported or investigated by other workers, and until this is done one must only remark that this is surprising when so many regular samples are being taken by a variety of workers in these waters. A third genus which has been recorded frequently in the plankton of the Indian Ocean and other warmer seas is *Katagnymene*. Drouet (1968) classifies this genus together with *Trichodesmium* species as *Oscillatoria erythraea*, and he is probably correct in doing so. The alga is characterized by having isolated trichomes, which are not very different in size from *Trichodesmium* and are surrounded by a soft diffluent sheath. In the North Atlantic a planktonic species of *Anabaena* has been reported (R. Johnston, pers. comm.) but little is known of its distribution or physiology. Another rare genus in the plankton is *Calothrix* which Allen (1963) reports as occurring sometimes in the plankton up to 20 miles off the coasts of California and Mexico. It is likely that these are filaments of an epilithic *Calothrix* which are washed out to sea rather than filaments of a truly planktonic species.

Chapter 15

Terrestrial ecology

Blue-green algae have a world-wide distribution in terrestrial habitats, occurring predominantly on the surface of soils, stones, rocks and trees, particularly in moist, neutral or alkaline areas. Despite this, and the attention given by early workers (e.g. Bristol-Roach, 1927; Petersen, 1935; Fritsch, 1936) their ecology has been studied less than that of heterotrophic micro-organisms such as soil fungi and bacteria, and this has given rise sometimes to a fallacious impression that they are unimportant soil micro-organisms (see, for example, Parkinson *et al.*, 1972).

Terrestrial algae, in fact, play a major role as primary colonizers which help in the establishment of other members of the soil flora and in the accumulation of humus. They do this in at least four main ways. Firstly, they bind sand and soil particles and prevent erosion. This is helped by their gelatinous sheaths and by the fact that many, such as species of *Porphyrosiphon*, which covers large eroded areas in Brazil (Drouet, 1937), *Microcoleus*, *Plectonema*, *Schizothrix* and *Scytonema*, all form closely intertwined rope-like bundles in and among the soil particles. Secondly, they help to maintain moisture in the soil. Booth (1941) found in studies in Oklahoma that soil with an algal covering had a moisture content of 8·9% compared with 1·3% in the absence of algae. Thirdly, blue-green algae are important as contributors of combined nitrogen (see Chapter 16). Fourthly, it has been suggested by several Indian and Russian workers that blue-green algae help higher plant growth by supplying growth substances (Venkataraman and Neelakantan, 1967; Mishustin and Shil'nikova, 1971).

Methods of estimating algal abundance

Although algal growths are readily distinguishable with the naked eye in many habitats there is considerable difficulty in measuring their abundance quantitatively. Three main methods are used but none is completely satisfactory. Plating and dilution techniques are sometimes used. These have an advantage in that one can enumerate the different species present but the kinds of algae isolated do depend to some extent on the culture conditions provided. Furthermore, all the algae which develop are counted equally and therefore distinctions cannot be made between those that were metabolically active in the soil and those that may have been present as inactive forms such as spores. There is also the practical difficulty that algae with mucilage sheaths are often aggregated together and do not form a homogeneous suspension easily. Also, the technique is of limited use with filamentous algae where the number of colonies which develop varies according to the degree to which the filaments fragment during the dilution procedure.

A second method uses pigment analysis as a measure of algal biomass. Here the pigments are extracted in acetone and absorption by the extracted pigment is measured at 490 nm (the absorption peak of myxoxanthin). Singh (1961) has used this technique to measure the abundance of *Cylindrospermum licheniforme* in rice fields, but usually it is not particularly suitable in most instances with field material because, in the extraction procedure, coloured materials such as humic acids and chlorophyll degradation products may be extracted as well. This method gives no indication of the algal species present; it simply is a measure of total algal biomass.

A third method is to make a direct microscopic examination of the soil flora. This can be done using either ordinary light microscopy or fluorescence microscopy. According to Tchan (1953) the algal cells appear reddish using the latter technique due to the fluorescence of chlorophyll. We have not found this modification particularly satisfactory and it does not readily allow algae of different species or classes to be distinguished. An alternative technique is to place coverslips on top of the soil for a period, then remove these and examine the algae which have grown on the surface of the glass (Lund, 1947). Although microscopic examination is excellent for distinguishing the different forms, its accurate quantitative application is difficult.

Numbers of terrestrial algae

The earliest detailed studies on soil algae were carried out by Bristol-Roach (1927) who, like some subsequent workers, did not always distinguish between blue-green algae and other algal groups. In general the numbers of blue-green algae recorded are extremely varied from habitat to habitat, and even within different areas of the same habitat. Furthermore, there are marked seasonal fluctuations in the flora. An average value perhaps is about 20,000–50,000 blue-green algae per gram of soil. Tchan and

Beadle (1955) found low numbers (an average of 1000 and a maximum of 3000 blue-green algae per gram of soil) in arid Australian soils. Jurgensen and Davey (1968) recorded up to 97,000 nitrogen-fixing algae per gram in the top 5 mm of a tree nursery soil in North Carolina (see Table 16.II) and up to 6000 per gram of soil in Alaskan tundra soils. Singh (1961) calculated 18,000,000 filaments of *Cylindrospermum licheniforme* per gram of soil in maize fields in India. Other workers (e.g. Duvigneaud and Symoens, 1948; Behre, 1953; Muzafarov, 1953) have listed the numbers of species which they found in various soils. A comparison of these data is of limited quantitative significance here because the numbers recorded vary greatly depending on the taxonomic treatments used.

Geographical distribution

Blue-green algae are of world-wide distribution and, in general, the more inhospitable the habitat, the more likely it is that blue-green algae will be important components.

In polar regions species of *Nostoc* are among the important terrestrial algae and a variety of species belonging mainly to the genera *Schizothrix*, *Oscillatoria*, *Lyngbya*, *Dichothrix*, *Phormidium* and *Stigonema*, have been recorded. On top of alkaline marble areas on Ross Island, South Victoria Land, free-living *Nostoc* forms an "algal peat", 10–15 cm deep (Holm-Hansen, 1963b) and in the South Orkney Islands, *Nostoc* is abundant on amphibolite areas and on silt between the stone stripes produces by frost-heaving (Fogg and Stewart 1968; Horne, 1972). *Nostoc* also occurs in symbiotic state in Antarctic lichens either as primary phycobiont, as in *Collema*, or as secondary phycobiont, as in *Stereocaulon*.

In temperate areas blue-green algae are seen less readily with the naked eye but nevertheless they can be isolated from almost all neutral or alkaline soils (see Lund, 1967). For example, in a survey of Scottish soils carried out by Stewart and Harbott (unpublished) over 80% of the soils collected developed a copious algal growth when kept in the laboratory for 4–8 weeks, although only in about 10% of these could the algae be observed with the naked eye upon collection. The algae appear to be more abundant in cultivated soils than in uncultivated ones (Tiffany, 1951). Shtina (1961, 1963, 1964b, 1965a, b) has carried out extensive studies on the abundance of blue-green algae in Russian soils. She found, for example, that *Nostoc* and *Scytonema* were particularly important components of the fertile steppe land, and she also observed (Shtina and Roizin, 1966) that the algal flora could be used as an indicator of the soil type and condition in various habitats. In cultivated soils species of *Cylindrospermum* and *Anabaena cylindrica* were usually most abundant, in damp virgin land *Fischerella muscicola*, *Nostoc commune*, *N. sphaericum*, *Tolypothrix tenuis* and *Scytonema hofmanii* were common, and in very wet virgin lands *Anabaena variabilis*, *Calothrix* species and *Nostoc muscorum* predominated.

In sub-tropical and tropical regions the blue-green algae in arid soils and in rice paddy soils have received most attention. Arid soils are probably the most inhospitable of all habitats where blue-green algae grow because the temperatures are high and water is severely limiting. However, as we have already seen (p. 85), these algae are especially resistant to such adverse conditions and it is therefore not surprising to find that they are in many cases the dominant components of the microflora. Shields and Durrell (1964), who summarized much of the relevant data, noted how coverings of Cyanophyceae form continuous layers over hundreds of acres of badly eroded land in the south eastern United States, while Shields and Drouet (1962) also record that algal crusts form the dominant cover in parts of the deserts in Nevada. In such areas the mean annual soil temperatures are high, for example, 20–22°C in Arizona (Cameron, 1962), 24°C in the Takyr soils of the Soviet Union (Sdobnikova, 1958). The maximum temperatures at the surface are much higher. The algae of arid regions occur particularly in five types of habitat: (1) the few permanent bodies of water (and there must be a few indeed when the mean annual rainfall may only be 3 in. per annum); (2) in intermittent streams and transitory pools; (3) in cracks and depressions in the soil surface; (4) under pebbles, rocks and stones, and (5) in lichen symbiosis.

The dominant algae present in American arid soils are species of *Schizothrix* with species of *Phormidium*, *Scytonema* and *Protosiphon* also being abundant. *Microcoleus* and *Nostoc* are found frequently in soils from India (Mitra, 1951), Algeria (Behre, 1953) and Australia (Moewus, 1953). Cameron (1960) records *Nostoc* as present in about 50% of the soil samples from arid areas in the United States while species of Chroococcaceae have been recorded beneath translucent pebbles and stones in various parts of the world (see Lund, 1967).

Despite the attention paid to other tropical environments, interest has focused predominantly on rice paddy soils. In all there are about one hundred million square kilometres of paddy fields, some of which are in southern Europe and the United States, but about 95% are in India and the Far East. Usually rice is grown on land which is submerged under 10 cm or so of water for 60–90 days during the growing season and then allowed to dry to facilitate harvest. The warm conditions demanded by rice, the availability of nutrients, the reducing conditions in the soil and algal ability to withstand desiccation all favour growth of algae. Pandey (1965), for example, reports that over 70% of the algal species present in Indian paddy soils are blue-green algae; they are also abundant in Japan

(Watanabe, 1966), Italy (Materassi and Balloni, 1965) and in other paddy soils (Fig. 15.1, facing p. 272).

Singh (1961) described the periodicity of blue-green algae in paddy fields in the vicinity of Banaras, India. Early in the rainy season, which lasts from the end of June to the middle of October, the soil surface becomes covered with a thick patchy growth of algae. Numerous different species are found, among them *Aphanothece pallida*, *Microcoleus chthonoplastes*, *Aulosira fertilissima*, and species of *Scytonema*, *Nostoc* and *Cylindrospermum*. The algae continue to grow until the end of July and meanwhile the paddy fields are puddled, mixing in the growth with the top 20 cm or so of soil. After about a fortnight when the mud has settled and the rice seedlings have been transplanted, *Microcoleus*, a good soil binder, appears together with various species of *Anabaena* spp., *Cylindrospermum*, *Tolypothrix* and *Fischerella*. *Aulosira fertilissima*, which is perhaps the most important nitrogen-fixing species, develops more slowly but by the middle of September forms an extensive brownish-yellow gelatinous growth. This is dominant for about three months and forms a papery layer on the soil surface when the fields dry. The ecology of the paddy field environment has been relatively little studied but the effects of cultivation on the physico-chemical conditions, the effect of these on the growth of different species and the interrelations between the various micro-organisms and the rice plants are all undoubtedly complex. Knowledge of these would help in making more practical use of the blue-green algae. It should be mentioned that they are also of importance in paddy fields as soil conditioners, in suppressing growth of mosquitoes (Griffin and Rees, 1956) and, perhaps, in supplying growth substances since their beneficial effect on rice plants appears to be greater than could be attributed solely to nitrogen fixation (Venkataraman and Neelakantan, 1967; Mishustin and Shil'nikova, 1971).

Factors affecting algal distribution

The cosmopolitan distribution of blue-green algae is evidence of their ability to withstand and grow under a wide variety of environmental conditions. In the various habitats there are several factors, which, to different degrees depending on the particular environment, affect the growth and development of terrestrial algae. Some, such as nutrients and temperature, have been considered elsewhere. Others are:

1. SUBSTRATE

Blue-green algae often develop abundantly on damp, porous and particularly on calcareous rocks. The early reports of lithophytic blue-green algae were summarized by Fritsch (1936). He drew attention to their abundance in damp mountain regions in temperate countries such as Scotland. There, gelatinous colonies of *Gloeocapsa*, *Gloeothece*, *Aphanocapsa* and *Nostoc* species are common on rocks near mountain streams. Lime encrusted colonies of *Rivularia* often occur on rock surfaces subjected to sea spray and in the tropics extensive growths of *Scytonema* cause blackening of the rocks. This happens for example in Angola where the Pedras Negras owe their name to the phenomenon. Similarly blue-green algae are prominent on the inselbergs in Nigeria and on sandstone cliffs in Natal and Rio de Janiero. Blue-green algae colonize bare lava readily; Behre and Schwabe (1970) record their presence on the island of Surtsey off Iceland, five years after its formation, and Treub (1888) noted that on the volcanic island of Krakatoa blue-green algae were early colonizers. On other tropical islands such as the coral atoll of Aldabra in the Indian Ocean Whitton (1971) records that blue-green algae, and in particular heterocystous species, are abundant on the otherwise bare rocks. Singh (1961) discusses the occurrence of blue-green algae on rocks in India where *Scytonema crassum*, *Schizothrix lenormandiana* and *Symploca dubia* are prominent on calcareous rocks and *Stigonema minutum* is common on siliceous rocks. Blue-green algae in general are found rarely on hard rocks such as slate. As well as natural rock surfaces, buildings in the tropics usually have a blackish covering of blue-green algae. This is not only unsightly (Fig. 4.3) but has been considered as the cause of the rotting of stone and deterioration of antiquities. Metal surfaces, if non-toxic, are also colonized; the gilded surface of the Shwedagon pagoda in Rangoon requires frequent cleaning from what is evidently a growth of blue-green algae.

2. LIGHT

Blue-green algae as a group are primarily photoautotrophic microorganisms (see p. 157) and cores through the soil show that they are usually located in the top 0·5 mm layer of the soil. For example, in paddy soils the algae occur particularly as a surface scum (Singh, 1961) and Stewart and Harbott (unpublished) observed that in Scottish grasslands over 90% of the observed algae were present in the top 1·0 cm of soil (see also Chapter 16). Jurgensen and Davey (1968) obtained essentially similar results in nursery soils in North Carolina. Viable cells of blue-green algae may occur, however, at depth in the soil. Fehér (1948) recorded that many algae occurred at depths of 15–20 cm below the soil surface and, in fact, considered that this was a particularly active zone of algal activity. Other workers have not obtained similar findings and it is more likely that the large algal numbers which are sometimes observed at depths are transported vertically there by rain or perhaps by earthworms (Petersen, 1935) and are probably relatively inactive. Vaidya (1964), who investigated soils from the vicinity of New Delhi, found blue-green algae at depths of 5·5 m and suggested that they were carried there by ants. Certain algae do grow abundantly just below the soil surface, however. Fogg (unpublished) found that in the white quartzite sands on the coasts of the Falkland Islands a distinct zone of blue-green algae was frequently present about 2 mm below the surface, and Stewart (unpublished) noted an abundant and healthy growth of blue-green algae, particularly *Nodularia* species just below the sand surface on areas of the Scottish west coast. These sub-surface growths are healthy in appearance with an abundance of phycocyanin. In this situation they are protected from the adverse effects of excessive light intensity and desiccation (see below) and sufficient light probably penetrates between the soil or sand particles to allow them to grow photosynthetically. Certain soil algae such as *Chlorogloea fritschii* and *Cylindrospermum gorakhporense* can grow heterotrophically although their growth rate under such conditions is slow (see p. 162). As Lund (1967) has said, "facultative heterotrophy seems to be of biological value in maintaining populations buried in the soil but is of little importance in production as a whole. In the surface layers where vertical movement of the algae may occur depending on the environmental conditions at the surface, it is autotrophy rather than heterotrophy which generates the necessary energy for these movements."

Slight lateral movement in soils may be brought about by gliding motility but wind is probably the main agent of algal dispersion over wider areas. Lateral movement in ungrazed prairie soil in Oklahoma was studied by Forest *et al.* (1959). They removed soil cores, replaced them after sterilization, then after an interval compared the algal composition of the sterilized areas with the unsterilized surrounding soil. They found that there were no special pioneer forms and that it was the dominant species from the surrounding soil which became most abundant on previously sterilized soil.

3. HYDROGEN-ION CONCENTRATION

Hydrogen-ion concentration is an important factor which directly or indirectly affects the distribution and abundance of terrestrial blue-green algae. Brannon (1945) concluded from studies in Florida that under conditions when light and moisture were not limiting, pH was the major physico-chemical factor affecting algal growth. It is doubtful whether a sweeping statement of this kind is entirely justified, but nevertheless it is probably a useful generalization. Cyanophyceae prefer neutral or alkaline conditions (see p. 138), and this accounts to some extent for their abundance in alkaline "Usar" lands in India (Singh, 1950), in Russian "Takyr" soils (see, for example, Sdobnikova, 1958; Bolyshev, 1952) and in alkaline paddy soils. Indeed one of the reasons for the abundance of blue-green algae in arid soils may be their universal alkalinity. This preference for neutral and alkaline areas is evident from the fact that in the field, as in the laboratory, there is little growth of blue-green algae below pH 6. Stewart and Harbott found in Scotland (Fig. 15.2) that the maximum

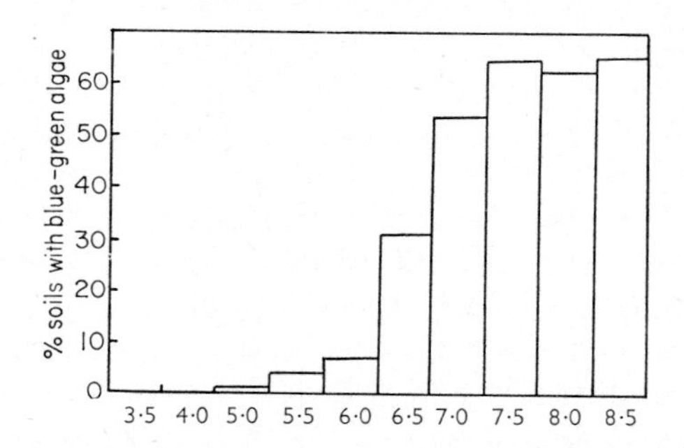

FIG. 15.2. The percentage number of Scottish soils of different pH showing the presence of blue-green algae after incubation for 30 days in the laboratory (W. D. P. Stewart and B. J. Harbott, unpublished).

growths of blue-green algae occurred in soils of pH 7·0–8·5 and that the algae were rare at pH 5 or below. Granhall (1970) noted similar distributions in Swedish soils. A few species however do occur in acid soils. For example species of *Hapalosiphon* and *Chroococcus* occur abundantly in localized areas such as in certain acid bog-lands; these algae presumably have an efficient internal pH-regulating mechanism which maintains the pH in the region of 7–8. Jurgensen and Davey (1968) found no growth of *Nostoc* below pH 5·0 in North America.

A good example of the importance of pH within a particular environment comes from the work of Fogg and Stewart (1968) and Horne (1972) on Signy Island, in Antarctica. They found that *Nostoc*, either free-living or in symbiotic association in the lichens *Collema* and *Stereocaulon*, occurred only on neutral or alkaline substrata and was absent from acid areas. Furthermore, in *Stereocaulon*, *Nostoc*-containing cephalodia developed only in neutral or alkaline areas and were absent from *Stereocaulon* in acid habitats.

It is difficult to distinguish whether the reported effects of pH are due directly or solely to changes in hydrogen-ion concentration or whether other physico-chemical factors are also involved. For example at high pH levels the poor growth of blue-green algae may be due to soluble iron being precipitated out of solution as ferric hydroxide which thus becomes largely unavailable. Also, low pH levels could lead to an unavailability of molybdenum which is essential for nitrogenase and nitrate reductase. It is also uncertain whether the beneficial effect noted on liming paddy soils is due solely to the increased availability of calcium or whether much of it is due to the increase in pH of the soil which also occurs. The finding that many blue-green algae appear to be calcicoles may simply be related to their preference for neutral or alkaline conditions.

4. DESICCATION

Soil algae have a tremendous capacity to withstand desiccation. In tropical and arid desert regions such as Arizona, the algae dry up and become inactive during the dry season, but then during the monsoon period in Indian paddy soils, for example, or after short periods of rain in American arid deserts, profuse growths may develop from these previously desiccated algae. In areas such as parts of the Antarctic, or the arctic tundra, or the Russian takyrs, the soils freeze over completely in winter, then become flooded as the snow and ice melts, and this is followed during summer by a period of intense desiccation due to strong sunlight and high winds. There is little doubt that in such soils the availability of moisture is the overriding factor affecting the metabolism of algae occurring there.

The importance of moisture is seen most markedly in the experiments of Singh (1950). In the alkaline infertile "Usar" lands of northern India, he introduced a simple technique of taking plots of less than an acre in area and surrounding these with an earth embankment. These retained the water, pools were formed, and a rich natural population of blue-green algae developed. *Microcoleus* species were the first colonizers and were followed by species of *Scytonema*, *Porphyrosiphon*, *Campylonema* and *Cylindrospermum*, and finally by *Nostoc commune*, a species which "extends for many miles in extensive areas of 'Usar' soils" (Singh, 1961). This growth of algae resulted in soil which had a higher nitrogen content and better water-retaining properties than before and which supported good growth of rice and sugar cane on these areas which otherwise would have been infertile.

Similarly, in Italian paddy soils, Materassi and Balloni (1965) found that blue-green algae comprised about 70% of the flora under waterlogged conditions and about 30% under dry conditions.

Whitton (1971) made interesting observations on the effect of moisture on algae on the island of Aldabra in the Indian Ocean. He found that there, also, blue-green algae developed profusely in the pools and small bodies of water which occurred during the wet season. He found colonies of *Wollea* particularly common together with various other colonial Nostocaceae. Species of *Spirulina*, *Lyngbya* and *Oscillatoria* were abundant. Whitton also noted an interesting distribution of algae on certain of the rocks, which appeared to be governed in part by moisture. On convex rock surfaces there was little algal cover whereas in small concavities the algae, which were mainly members of the Scytonemataceae and Chroococcales, developed as a dark blue-green turf. Some small depressions, in addition, were often filled with colonies of *Nostoc commune* which imbibe water rapidly.

A circumstance which may be determined in part by desiccation, is that in many soils, both in the tropics and in temperate regions, blue-green algae are often more abundant in cultivated soils than in uncultivated ones (Tiffany, 1951). Shtina (1959) also found that in Russia, blue-green algae developed best under perennial crops. This could be due not only to the increased availability of nutrients in the former but probably also to the importance of higher plant cover in affording protection from desiccation. Rich growths of blue-green algae also occur in non-arable soils which remain moist. For example, Stewart (1965) found profuse growths of *Nostoc* (Fig. 15.3) among long grass, but not in the more open regions, of a sand dune slack and Whitton (1971) found that *Nostoc commune* var. *flagelliforme* was often particularly common among grass in coconut groves on Aldabra.

Other examples of the importance of moisture are the frequent occurrence of *Rivularia* colonies on rock faces above the supralittoral zone of the seashore, which have fresh water passing over them, and the abundance in many damp mountain regions in Scotland of gelatinous colonies of *Gloeocapsa*, *Gloeothece*, *Aphanocapsa* and of *Nostoc*, particularly near mountain streams (see Fritsch, 1936).

The habit of the surface flora helps to protect the algae from extremes of desiccation. These forms are usually filamentous with intertwined trichomes and coalescent sheaths forming close-knit bundles. This is found in the strands of *Microcoleus*, the sheets of *Phormidium* and the bundles of *Oscillatoria*. In others, such as *Nostoc*, *Rivularia* and the colonial unicellular forms such as *Gloeocapsa*, the algae become aggregated into gelatinous globose colonies. These mucilaginous colonies lose water slowly and have a remarkable capacity to take up water quickly. Durrell and Shields (1961), for example, showed that algal crust organisms may absorb 12–13 times their sheath volume in six minutes. The rapidity of uptake is advantageous because, in arid areas, rain, which not only supplies water for metabolism *per se*, but also nutrients dissolved out of the soil, may fall only for a short time. The fact that uptake into the sheath is a passive process is important

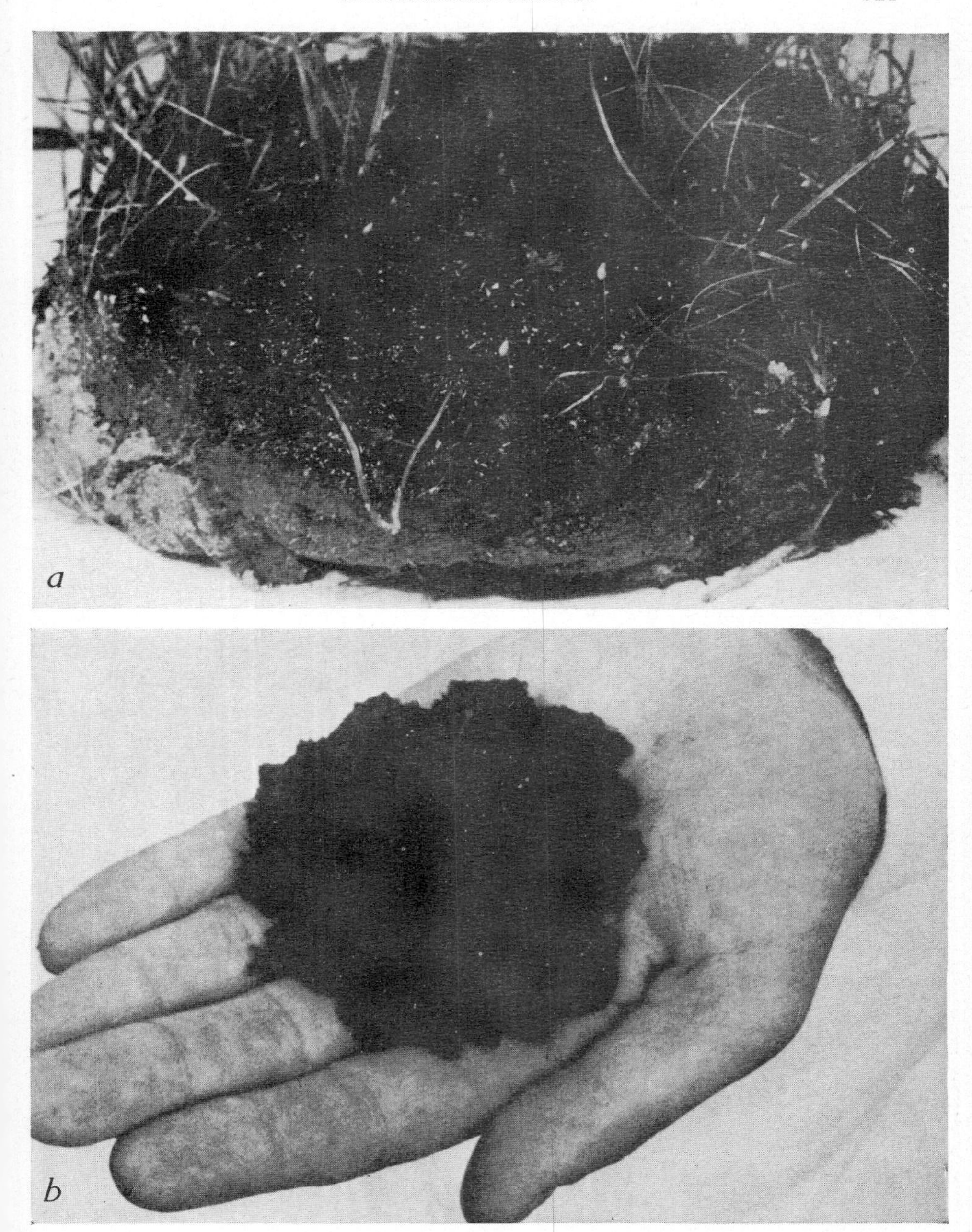

FIG. 15.3. (a) Sample of sand dune slack soil from Blakeney Point, Norfolk in August 1965 showing a thick gelatinous cover of blue-green algae (×0·75); (b) growth of *Nostoc* scraped off a 160 cm² area of sand-dune slack soil from among long grass in May 1966 (× 0·7). From W. D. P. Stewart (1967) *Ann. Bot.* N.S. **31**, 385, Plate 1.

when it is appreciated that when water becomes available the algae may not be in a vigorous metabolic state. Not all algal sheaths show a capacity to absorb water rapidly. Old filaments of *Calothrix* and certain Scytonemataceae actually require soaking for several hours before rapid uptake occurs. This is why taxonomists often soak their herbarium species in dilute detergent before examining them. Non-wettable algae depend on atmospheric moisture for water and occur frequently below translucent pebbles where the relative humidity is high.

5. THE OXIDATION–REDUCTION LEVEL OF THE SOIL

Soil microbiologists have in the past attributed the stimulatory effect of waterlogging of soils on algal growth mainly to an increased availability of water. However, various laboratory workers over the years have appreciated the beneficial effect of low redox levels for the growth of blue-green algae (see Stewart and Pearson, 1970) and indeed small amounts of sodium sulphide have often been added to the culture medium to enhance the growth of blue-green algae (Allen, 1952). Thus in the natural environment part of the beneficial effect of waterlogging may be through creating lower redox levels. Singh (1961), in fact, reported on some such correlation although he appeared not to recognize its possible significance. For example he reported that at the start of the rainy season the redox potential (Eh_7) was high (510–480 mV) at a depth of 2 cm but that this decreased with waterlogging and that "in August during the time of puddling, the potential stabilised on the reducing side. At this stage there was a high crop of algae which again resulted in a slight increase in potential ($Eh_7 = 300$ mV)." Similarly Stewart (unpublished) observed reducing conditions in sand-dune slacks when abundant growths of *Nostoc* occurred and Baas-Becking and Wood (1955) have studied the importance of pH and redox levels on the growth of blue-green algae and other micro-organisms in estuarine muds. Although the latter studies were not strictly on terrestrial habitats, they emphasize the importance of redox potential because as Fig. 15.3 shows, particularly good growth of blue-green algae occurs under reducing conditions.

Chapter 16

The ecology of nitrogen fixation by algae

In discussing the ecology of blue-green algae in the previous chapters mention of nitrogen fixation has been inevitable. However, there remain various aspects of this ecologically important property which are best dealt with in one place.

Blue-green algae are of world-wide distribution and are often abundant. There is a general correlation between rates of nitrogen fixation under aerobic conditions and the abundance of heterocystous algae in lakes, soils and most maritime habitats, but under microaerophilic or reducing conditions the position is less clear as some non-heterocystous algae also fix nitrogen (see p. 193). The extent to which they do this in the field has not yet been established. Several reviews on ecological aspects of nitrogen fixation by algae are available (Singh, 1961; Fogg and Stewart, 1965; Stewart, 1966, 1969, 1970a, 1973d).

Tropical and rice paddy soils

Particular attention has been focused on the role of nitrogen-fixing species in rice paddy fields (see Chapter 15) for in this situation their ecological significance may be immense. Many of the world's paddy soils receive no artificial fertilizer, and when one appreciates that over half the world's population lives on rice as a staple diet, the importance of nitrogen-fixing algae in sustaining rice crops can readily be appreciated.

Extensive studies on the distribution of nitrogen-fixing blue-green algae in paddy fields and other tropical soils have been carried out by Watanabe over a 30-year period (see Watanabe and Yamamoto, 1971). He concludes from analyses of soil samples from Japan, South and East Asia, Africa, and India that nitrogen-fixing species are widely distributed but are more abundant in soils south of 30° latitude than north of it (11·8 and 1·7% of the total flora respectively). That is, they are more common in tropical and sub-tropical soils than in temperate areas. Apart from Watanabe's studies, information on nitrogen-fixing algae in the tropics and sub-tropics is fragmentary, simply because of the large land area to be covered. Much work has been done on Indian soils (see Singh, 1961) and there is information for Japan (see Kobayashi *et al.*, 1967; Watanabe, 1966), Australia (Bunt, 1961a), Italy (Materassi and Balloni, 1965,) America (see Shields and Durrell, 1964), Russia (see Shtina, 1965a) and the Philippines (MacRae and Castro, 1967; Yoshida *et al.*, 1973). The general consensus of opinion is that nitrogen-fixing algae are widely distributed in paddy soils except in Australia, where Bunt (1961a) found that in North Queensland nitrogen-fixing algae made up only a small proportion of the flora. Possibly this was because heavy dressings of ammonium sulphate were applied to the soils, or because high levels of copper sulphate (up to 5 ppm) were added to the irrigation water. This emphasizes one aspect of nitrogen fixation by paddy soil blue-green algae. They are probably least important as nitrogen sources in countries with well developed agricultural practices.

The dominant nitrogen-fixing algae are usually species of *Nostoc*, *Anabaena*, *Tolypothrix*, *Aulosira* and *Scytonema* which develop most profusely, together with photosynthetic bacteria, during the wet season (Materassi and Balloni, 1965; Kobayashi *et al.*, 1967). Yoshida *et al.* (1973), for example, isolated 308 strains of blue-green algae from soils in the Philippines and of these 267 were identified as *Nostoc*, 38 as *Anabaena* and 1 each as *Stigonema*, *Tolypothrix* and *Scytonema*.

The ability of the paddy soil microflora to fix nitrogen has been demonstrated convincingly by MacRae and Castro (1967) at the International Rice Research Centre at Laguna in the Philippines. They exposed paddy soils in the greenhouse to an atmosphere enriched with $^{15}N_2$, observed uptake of the isotope and calculated nitrogen fixation rates equivalent to 10–55 kg nitrogen ha^{-1} ann^{-1}. Yoshida *et al.* (1973) using a similar technique, also at Laguna, obtained nitrogen fixation rates equivalent to 40–80 kg nitrogen ha^{-1} ann^{-1}. Nitrogen fixation and algal abundance were greatest at the soil surface (0–2 cm) and there was negligible nitrogenase activity in the dark. This suggests that algae and photosynthetic bacteria were responsible for most, if not all, of the observed nitrogen fixation. These workers also conclude that the nitrogen fixed probably becomes available for rice growth when the algae lyse.

Data obtained with $^{15}N_2$ show, therefore, the nitrogen-fixing potential of paddy soils. The next problem is how to obtain reliable data on nitrogen fixation in the field. To date these are not available, mainly because until recently there were no good, simple, techniques which could be used in *in situ* studies. Values obtained using diverse techniques range from 1 to 50 kgNha^{-1} ann^{-1} (see Singh, 1961). Furthermore it is not always possible to compare values obtained because the results have often been expressed differently by different workers. Unfortunately this unhappy state of affairs still exists as data presented on subsequent pages will show. The introduction of the acetylene reduction techniques for measuring nitrogenase *in situ* in paddy soils should provide more reliable data.

A direct, and in some ways more useful, method of demonstrating the importance of paddy soil algae is to compare the growth of the rice plant in the presence and absence of the algae. Beneficial effects have been shown to occur in pot culture (see, for example, De and Sulaiman, 1950; Watanabe *et al.*, 1951; Ley, 1959; Singh, 1961; Iha *et al.*, 1965) and rice yields have been increased up to 600% on the addition of blue-green algae in the laboratory (Allen, 1956). The latter findings cannot be related readily to the less favourable conditions which occur in the field but nevertheless they do emphasize the possible benefit to the rice crop. This has been confirmed in actual field trials by workers such as Singh (1961) who found that the addition of *Aulosira fertilissima* gave a 114% increase in yield in the field. Increases, usually between 5 and 30%, have also been noted in China (Ley, 1959), Egypt (Nawawy *et al.*, 1958) and Russia (Perminova, 1964; Shtina, 1965a). Subrahmanyan *et al.* (1965) using an inoculum of four species increased the yield by 30% in India and some of their data are shown in Table 16.I.

TABLE 16.I. Yield of paddy grain and straw (kg ha^{-1}) in fields with and without blue-green algae in 1963 (after Subrahmanyan *et al.*, 1965)

	Grain Yield		Straw Yield	
Treatment	− Algae	+ Algae	− Algae	+ Algae
Untreated	1797	2195	1994	2275
$(NH_4)_2SO_4$	2463	—	2630	—
Lime, phosphate, Mo	3001	3979	3145	4327

$(NH_4)_2SO_4$ supplied as 20 kg N ha^{-1}; lime as 1000 kg ha^{-1}; phosphate as 20 kg P_2O_5 ha^{-1} and molybdenum as 0·28 kg ha^{-1} of sodium molybdate; blue-green algae as a mixture of *Nostoc*, *Anabaena*, *Scytonema* and *Tolypothrix* supplied as 200 g dry weight ha^{-1}.

On the basis of such field trials and promising laboratory results using *Tolypothrix tenuis*, Watanabe (1961) investigated the possibility of making large quantities of blue-green algae available for use as fertilizer. He mass-cultured the alga on moist, sterile, porous gravel and found that it could be stored in this way in air-tight bags for up to three years without loss of activity. When this material was broadcast on to the soil at a level of 1–5 kg dry weight of alga per hectare when the rice plants were bedded out, he obtained beneficial results, although the response was slow at first. Inoculated soils showed gains in plant yield over untreated controls of 2% in the first year, 8% in the second year, 15% in the third year, 20% in the fourth year and 11% in the fifth year. Watanabe considers that fertilization in this way is equivalent to an ammonium sulphate application of 29 kg nitrogen ha^{-1}. There are problems however with this technique: it is expensive to produce the algae, they sometimes fail to grow on the soils to which they are added and sometimes the whole algal inoculum is consumed rapidly by soil invertebrates. For these reasons and in particular because it cannot compete commercially with inorganic nitrogen fertilizer the technique has not found application.

Mishustin and Shil'nikova (1971) relate how the water fern *Azolla* which contains an *Anabaena* as symbiont (see p. 349) has been used successfully to increase rice yield in Vietnam. This technique was also reported on by Bortels (1940) and according to Sung-Hong Hien (1957) was introduced into Vietnam by a peasant called Ba Heng who recommended adding 15–20 kg of *Azolla* per hectare. This method was apparently so successful that when Bà Heng died a pagoda was built to the "Goddess of *Azolla*".

In the arid desert soils of the United States there are many known, or possible, nitrogen-fixing algae such as *Scytonema hofmanni*, *S. arcangelii*, *Anabaena levanderi*, *Nostoc* sp. and species of *Dichothrix*, *Tolypothrix*, *Rhabdoderma*, *Microchaete* and coccoid algae (Cameron and Fuller, 1960). Mayland *et al.* (1966) demonstrated, under simulated field conditions and using $^{15}N_2$ as tracer, that nitrogen fixation by desert crust algae occurred at a constant rate for 520 days and that the nitrogen fixed amounted to 10·9 kg nitrogen (hectare of crust)$^{-1}$ ann^{-1}. They showed further (Mayland and McIntosh, 1966) that the nitrogen which is fixed becomes available at a steady rate to shoots of *Artemisia* plants.

These crusts, of course, do not extend over the entire soil surface in nature but nevertheless such nitrogen fixation contributes appreciably to the nitrogen economy of these soils which, in Arizona, cover 20 million acres and are important rearing grounds for the beef cattle of the state.

In the south central arid zone of Australia nitrogen fixation by lichens is important. Rogers *et al.* (1966) tested twelve of the most common species and found that three of these incorporated $^{15}N_2$ after a six-day exposure period at 21°C. Uptake was highest in *Collema coccophorus* which has *Nostoc* as phycobiont. It was low but significant in *Lecidia crystalifera* and was marginal (0·0073 atom % excess) in *Parmelia conspersa*. The latter two lichens contain a green algae (*Protococcus* sp.) as phycobiont so that the low levels of fixation observed may have been due to associated bacteria.

Temperate soils

Although many studies have been carried out on the distribution of blue-green algae in temperate soils (see Chapter 15) few have been concerned specifically with nitrogen-fixing algae. One reason for this has been the difficulty in enumerating potential nitrogen-fixing species. It is a time-consuming process to obtain pure cultures and without these, tests for nitrogenase activity, whatever the technique used, are equivocal (see p. 311). The best method is to consider all heterocystous species as being capable of fixing nitrogen. As stressed earlier this underestimates nitrogen-fixing forms as some non-heterocystous species fix nitrogen. The data of Jurgensen and Davey (1968) may be typical for aerobic nitrogen-fixing algae. They found that in Canadian soils there were up to approximately 6000 cells of nitrogen-fixing algae (*Nostoc* sp.) per gram of soil, although up to 100,000 cells per gram were found elsewhere. These values are as high as in some tropical soils (Kobayashi *et al.*, 1967). Shtina (1963, 1964) and Gollerbakh and Shtina (1969) have shown that in Russian sod-podzolic soils potential nitrogen-fixing algae, particularly species of *Nostoc*, *Anabaena*, and *Cylindrospermum* are often abundant, but in arable land they are less important. Kuchkarova (1962) reports that in certain parts of Russia blooms of blue-green algae are collected from waters and ditches and added as fertilizer to crops such as cotton.

The various factors which affect blue-green algae in general (see p. 134) also affect nitrogen fixation. Desiccation is particularly important and Beck (1968) considered that shortage of water was the most important factor limiting nitrogen fixation in the field. Henriksson (1971) concluded, however, from extensive studies in Sweden that the high rates of nitrogen fixation which occur when the soil is damp more than compensate for the lack of fixation during dry periods. The pattern of pH dependency follows that in blue-green algae in general (see p. 138) and nitrogen fixation is largely light dependent, occurring primarily in the surface layers (Stewart, 1965; Granhall and Henriksson, 1969; Jurgensen and Davey, 1968; Table 16.II). Some species fix nitrogen in the dark although the rates are much lower than in the light (see p. 162). The effects of temperature on nitrogen fixation by field algae are controversial. Henriksson (1971) found that temperature was unimportant in Sweden, but Scottish soils which show no nitrogen-fixing activity in the field in winter develop nitrogenase activity when incubated in the laboratory at 15°C but not at 4°C.

The available information on the quantities of nitrogen fixed by algae in temperate soils suggests that the levels may be higher than hitherto supposed. Granhall and Henriksson (1969) studying Swedish soils obtained nitrogen fixation rates of 0·1–1·0 mg nitrogen (g soil)$^{-1}$ day^{-1} during the

TABLE 16.II. Abundance of nitrogen-fixing blue-green algae at various depths in a forest nursery soil (after Jurgensen and Davey, 1968)

Depth (cm)	Soil moisture (%)	N_2-fixing algae (g^{-1} soil)
0·0–0·5	13·2	96,500
0·5–2·0	8·2	20,500
5·0–6·0	7·6	2100
9·0–10·0	8·2	100

Sampling date 9.2.65.

summer months in areas where heterocystous algae were abundant. Henriksson (1971), who studied over 1000 soil samples from many localities in Sweden, estimated annual fixation rates, in an agricultural field where *Nostoc* was abundant, of 15–51 kg ha^{-1} ann^{-1} and 4–44 kg ha^{-1} ann^{-1} in a lakeside meadow containing *Nostoc*, *Anabaena*, *Cylindrospermum* and *Calothrix* species. Highest rates were found in damp areas, as Paul *et al.* (1971) also found in observations on *Nostoc* in Canadian grasslands. Henriksson concludes that the high rates which she found result, to some extent, from the fact that in the summer in Sweden there is little darkness and long periods of twilight so that light is seldom a limiting factor. Stewart and Harbott (unpublished) surveyed a variety of soils in Scotland for nitrogenase activity by algae using the acetylene reduction technique and observed reduction rates which were equivalent to up to 40 μg N m^{-2} h^{-1} in arable land and up to 140 μg N m^{-2} h^{-1} in pasture during the summer of 1970.

Arctic and Antarctic regions

Until recently most attention has been focused on the tropical and sub-tropical deserts of the world and there were few data on nitrogen fixation by blue-green algae in frigid desert regions. However the position has changed and increasing interest is being paid to the sources of nitrogen in these ecosystems. In the Arctic, potential nitrogen-fixing species of *Nostoc* have been recorded from tundra in Alaska (V. A. Alexander, pers. comm.) and Canada (Jurgensen and Davey, 1968) and *Scytonema hofmanni* is present in volcanic soils in Alaska (Cameron *et al.*, 1965). Alexander and Schell (unpublished) have carried out extensive studies on *in situ* nitrogen fixation in tundra ecosystems and their findings have emphasized the importance of nitrogen fixation by algae there. They consider that nitrogen fixation is a major source of nitrogen input into the tundra and may occur at a rate of about 28 μg N m^{-2} h^{-1}. Blue-green algae, particularly in lichen association, are probably the most important sources of this biologically fixed nitrogen.

The importance of nitrogen-fixing algae in the Antarctic had been realized a few years earlier. Holm-Hansen (1963, 1964) showed that unialgal cultures of various *Nostoc* species from South Victoria Land fixed nitrogen, while Fogg and Stewart (1968) in *in situ* tests on Signy Island, South Orkney Islands, detected nitrogen fixation in 14 of the 42 sites sampled. On all but one of the sites giving positive results *Nostoc commune*, or lichens containing *Nostoc*, were visible to the naked eye. Nitrogen fixation continued throughout the Antarctic summer and topography and relief of the habitats where the algae occurred were important factors since they determined the microclimate of the algae. Although nitrogen fixation occurred at temperatures which did not exceed 4°C, activity was highest in moist and sheltered areas such as among plant tufts where around mid-day, in bright sunshine, temperatures increased to 10°C or higher. From the data obtained it was found that the Q_{10} for nitrogen fixation was approximately 6, a value similar to that observed in laboratory studies with *Anabaena cylindrica* by Fogg and Than-Tun (1960). In the Antarctic, just as in other arid areas, desiccation is an important factor governing nitrogen fixation rates. In the studies of Fogg and Stewart (1968) the only *Nostoc*-containing sample tested which did not fix nitrogen was a desiccated sample of *Collema*. The mean rates of nitrogen fixation found were approximately 0·48 μg nitrogen fixed (mg nitrogen)$^{-1}$ day^{-1} for *Nostoc* and 0·39 μg nitrogen fixed (mg nitrogen)$^{-1}$ day^{-1} for *Collema*. Horne (1972) has estimated that *Nostoc* and *Collema* may, together, fix up to 2·4 mg nitrogen m^{-2} ann^{-1} on the more favourable sites on Signy Island.

Aquatic habitats

Probably the most extensive *in situ* studies on nitrogen fixation have been in aquatic ecosystems. This may be due to some extent to the fact that blooms of nitrogen-fixing algae are often a nuisance in freshwaters, particularly if they are becoming polluted, but also it may be because techniques for measuring nitrogen fixation can be applied most easily to water samples.

Blooms of nitrogen-fixing species have been reported from waters in many countries and states including: Alaska (Billaud, 1967), Pennsylvania (Dugdale and Dugdale, 1962), Wisconsin (Goering and Neess, 1964; Stewart *et al.*, 1967; Stewart *et al.*, 1971; Rusness and Burris, 1970), the English Lake District (Horne and Fogg, 1970), Holland (Horne, 1969) and

TABLE 16.III. Maximum recorded rates of nitrogen fixation for surface waters of various lakes *in situ* in the light (after Fogg, 1971)

Lake and date	Lake type[a]	Principal spp. of blue-green algae present	Total organic N ($mg\ l^{-1}$)	Nitrogen fixed ($\mu g\ l^{-1}\ day^{-1}$)	Nitrogen fixed (% of total N day^{-1})	Reference
Sanctuary Pennsylvania, Aug. '59	Temperate, 3rd class, polluted	*Anabaena flos-aquae*, *A. circinalis*, *A. spiroides*,	3·6	125	3·5	Dugdale and Dugdale, 1962
Mendota, Wisconsin, Aug. '61	Temperate, dimictic 2nd class, polluted	*Gloeotrichia echinulata*	0·72	8·5	1·18	Goering and Neess, 1964
Wingra, Wisconsin, July '61	Temperate, 3rd class, polluted	*Microcystis aeruginosa*, *Anabaena* sp.	2·1	12	0·55	Goering and Neess, 1964
Smith, Alaska, June '64	Sub-arctic, dimictic 2nd class	*Anabaena flos-aquae*	(0·41)[b]	2·88	(0·7)	Billaud, 1967
Windermere, N. basin, England, June, '66	Temperate, dimictic 2nd class	*Anabaena flos-aquae*, *Oscillatoria* spp.	0·164	0·098	0·060	Horne and Fogg, 1970
Windermere, S. basin, England, Oct. '65	Temperate, dimictic 2nd class	*Anabaena solitaria*, *A. flos-aquae*, *Oscillatoria* spp.	0·47	2·82	0·060	Horne and Fogg, 1970

Esthwaite Water, England, Aug. '65	Temperate, dimictic 2nd class	*Aphanizomenon flos-aquae*, *Anabaena flos-aquae*, *A. circinalis*	0·255	0·244	0·096	Horne and Fogg, 1970
Loch Leven, Scotland, May, July, Sept. '68	Temperate, dimictic 2nd class	*Synechococcus* sp., *Oscillatoria* sp., *Anabaena* sp.	0·186	0	0	Horne, unpublished
Tjeukemeer, Holland, Sept. '66	Temperate, 3rd class	*Aphanizomenon* sp., *Oscillatoria* sp.	2·1	14·9	0·66	Horne, unpublished
George, Uganda, March '68	Tropical 3rd class	*Microcystis* sp., *Anabaena* sp.	2·2	4·1	0·19[c]	Horne, unpublished
McIlwaine, Rhodesia, March '68	Warm monomictic 2nd class, polluted	*Microcystis* sp.	1·275	0	0	Horne, unpublished
Kariba, Rhodesia, March '68	Warm monomictic 2nd class	*Oscillatoria* sp.	0·30	0	0	Horne, unpublished

[a] See Hutchinson 1957.
[b] Particulate N only.
[c] Extrapolated from hourly rate on basis of diurnal variation in photosynthesis.

Scotland (Stewart, 1972a, see also Table 16.III). In the tropics and subtropics, nitrogen fixation may be associated with *Anabaena flos-aquae* in the White Nile (Prowse and Talling, 1958) and Blue Nile (Telling and Rzoska, 1967) and with *Anabaenopsis raciborskii* in fish-ponds in Malaya (G. A. Prowse, pers. comm.).

Dugdale *et al.* (1959), using $^{15}N_2$ as tracer, were the first to obtain conclusive data which showed nitrogen fixation in lake water. In a subsequent study in Pennsylvania, Dugdale and Dugdale (1962) studied the variation in nitrogen fixation in the surface waters and found that significant light-dependent nitrogen fixation occurred from July until September with maximum fixation rates of about 130 μg N l^{-1} day^{-1} occurring in August. Moderate rates of fixation were subsequently recorded for Lake Mendota by Goering and Neess (1964) who, like Dugdale and Dugdale, observed low rates of dark nitrogen fixation on the addition of carbohydrates to the water. The rates of nitrogen fixation which Billaud (1967) observed in Alaska during a bloom of *Anabaena* which occurred for a 2-week period during the height of the summer season were 1–3 mg nitrogen m^{-2} h^{-1}. These are as high as any found in temperate regions (Fogg, 1971). In Lake George, Uganda, Horne and Viner (1971) recorded rates of 0·44–1·66 μg nitrogen l^{-1} h^{-1} which were apparently maintained throughout the year and this contributed approximately 44 kg nitrogen ha^{-1} ann^{-1} to the lake. This was probably the major source of nitrogen input into the ecosystem (Table 16.IV). Thus data on *in situ* nitrogen fixation rates from Alaska to equatorial Africa all testify to the abundance and importance of nitrogen fixation by algae in aquatic ecosystems.

TABLE 16.IV. Nitrogen balance sheet for a tropical lake (Lake George, Uganda) (after Horne and Viner, 1971)

N gain ann^{-1}		N loss ann^{-1}	
Source	mg N	Source	mg N
N_2 fixation	1280	Denitrification	?
Stream inflow[a]	322	Stream plankton outflow	3180
Rainfall	277	Sedimentation	655
Hippopotamus excreta	76–99	Commercial fish export	50–75

[a] Does not include surface run-off.

The biggest source of error in lake studies, as in others, is in sampling since the distribution of the algae is usually far from uniform and there are great fluctuations in algal periodicity and metabolic rate. The extent of these fluctuations can be seen in diurnal studies on Green Bay, Wisconsin. Figure 16.1 shows that, in the surface waters, nitrogenase activity fluctuated throughout the day at 0 m and 2 m, where, in addition, the abundance of algae (estimated here as total nitrogen) also varied. It should be

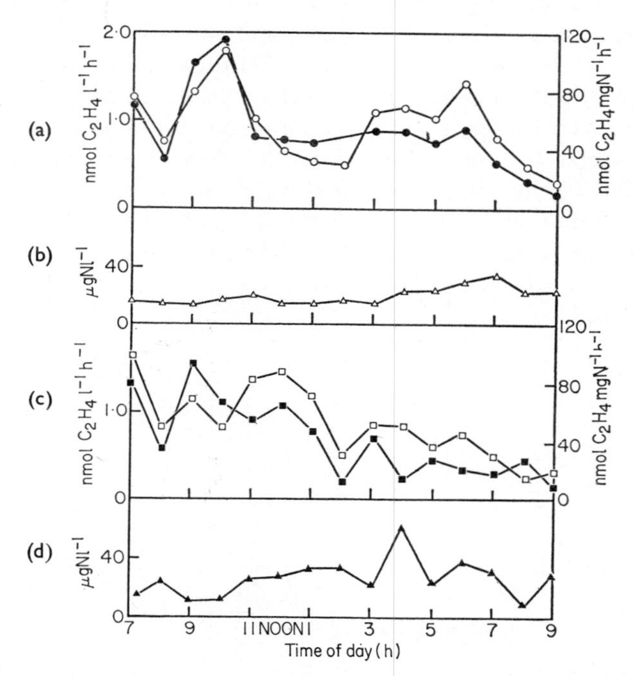

FIG. 16.1. Daily variation in acetylene reduction rates in Green Bay, Wisconsin on October 6th, 1969 (a) acetylene reduction l^{-1} h^{-1} (○—○) and acetylene reduction (mg nitrogen)$^{-1}$ h^{-1} (●—●) in the surface waters; (b) total nitrogen in the surface waters (△—△); (c) acetylene reduction l^{-1} h^{-1} (□—□) and acetylene reduction (mg nitrogen)$^{-1}$ h^{-1} (■—■) at 2m; (d) total nitrogen at 2m (▲—▲) (after Stewart *et al.*, 1971).

stressed that these results serve only to emphasize the fluctuations which may occur and that the pattern of fluctuation varies markedly from lake to lake and from day to day. There is also variation in nitrogenase activity with depth and usually nitrogenase activity is highest at, or just below, the surface, as Table 16.V shows. However in some lakes, particularly those where algae are scarce, highest levels of nitrogenase activity may occur in the sediments (Brezonik and Harper, 1969).

Particular attention has been paid to the importance of nitrogen fixation by algae in waters of differing degrees of eutrophication. In their initial studies in the eutrophic Lake Mendota using the acetylene reduction technique, Stewart *et al.* (1967) noted that fixation was restricted largely to the surface waters, was light dependent and was correlated directly with the abundance of *Gloeotrichia* in the samples. Howard *et al.* (1970) subsequently demonstrated algal nitrogenase activity in the surface waters and bacterial nitrogenase activity in the sediments of Lake Erie. Rusness and Burris (1970), who surveyed various Wisconsin lakes in 1968, found negligible acetylene reduction in the surface waters of oligotrophic lakes, although values as high as 2·2 μg nitrogen l^{-1} h^{-1} were detected in eutrophic waters. In studies on Wisconsin lakes of differing degrees of eutrophication in the summer of 1969 much lower levels of nitrogenase activity were found, with a maximum of 0·2 μg nitrogen l^{-1} h^{-1} occurring in the eutrophic southern Green Bay (Stewart *et al.*, 1971). Horne and

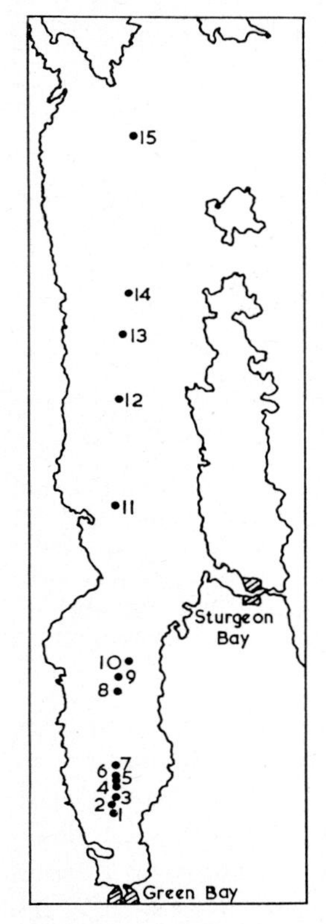

FIG. 16.2. Map of Green Bay, Northern Wisconsin showing the position of the eutrophic southern waters (sampling sites 1–10) and the oligotrophic northern basin (sampling sites 11–15). See Table 16.V. (Redrawn from Stewart *et al.*, 1971, *New Phytol.* **70**, 497–509).

Fogg (1970) consider that in the English lakes, as a result of the opposing effects of organic nitrogen and nitrate concentrations, nitrogen fixation attains a maximum in the intermediate stages of eutrophication.

The effects of eutrophication may even be seen within a particular body of water. Green Bay in Northern Wisconsin (Fig. 16.2) is oligotrophic in its northern basin and eutrophic in its southern basin and, as Table 16.V shows, nitrogenase activity per unit volume of lake water is much higher in the southern waters than in the northern waters. Even in the southern waters the highest rates of acetylene reduction occur near urban areas, such as the city of Green Bay, where eutrophication is maximal.

TABLE 16.V. Acetylenere duction per unit volume in surface waters the of eutrophic southern Green Bay and oligotrophic northern Green Bay in July–August, 1969 (after Stewart *et al.*, 1971)

Sampling area	Station	nmoles C_2H_4 l^{-1} h^{-1}
Southern Green Bay	1	8·6
	2	16·6
	3	6·9
	4	19·7
	5	7·6
	6	2·8
	7	1·6
	8	0·5
	9	0·7
	10	0·6
Northern Green Bay	11	0·6
	12	0·3
	13	0·2
	14	0·3
	15	0·1

See Fig. 16.2 for positions of sampling sites.

Two studies provide estimates of the total nitrogen input into mesotrophic and eutrophic lakes. In studies on Esthwaite Water, a mesotrophic lake in the English lake district, Horne and Fogg (1970) measured nitrogen fixation over a 17 month period using ^{15}N as a tracer. They found that light-dependent nitrogen fixation occurred during the period March-October in 1965 and in 1966 and this was correlated with the abundance of

TABLE 16.VI. Acetylene reduction with depth in various areas of Green Bay

Depth (m)	Sampling station	nmoles C_2H_4 $l^{-1}h^{-1}$	Sampling station	nmoles C_2H_4 $l^{-1}h^{-1}$	Sampling station	nmoles C_2H_4 $l^{-1}h^{-1}$
0	18	8·6	43	0·63	28	0·05
2		3·2		0·16		0·02
5		1·6		0·11		0·00
0	19	16·6	44	1·40	29	0·06
2		11·4		0·56		0·02
5		1·9		0·10		0·01
0	20	6·9	49	0·48	30	0·07
2		11·4		0·48		0·13
5		5·4		0·05		0·07

The dates of sampling were as follows: 10 July 1969 (station 43); 11 July 1969 (stations 44 and 49); 17 July 1969 (stations 18, 19 and 20) and 1 August, 1969 (stations 28, 29 and 30). Each value is the mean of triplicates. There was no appreciable activity at 15 m at any of the sampling stations. These sampling stations are not shown in Fig. 16.2. After Stewart *et al.*, 1971.

heterocystous algae (Fig. 16.3). Fixation rates were highest when the levels of organic nitrogen in the water were high and when the levels of nitrate nitrogen were below 0·3 mg l^{-1}, but there was no significant negative correlation of the rate of fixation with nitrate concentration. The average

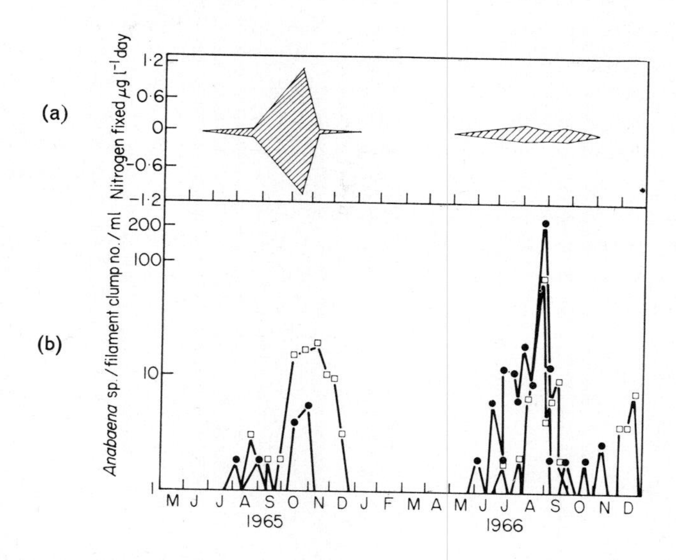

FIG. 16.3. (a) Nitrogen fixation and (b) variation in numbers of *Anabaena flos-aquae* (●—●) and *Anabaena solitaria* (□—□) in Windermere south basin in 1965–6 (after Horne and Fogg, 1970).

rate of fixation in the top 6 m was calculated as 0·006–0·13 g nitrogen m^{-2} ann^{-1}. This compares with 0·11–0·29 g nitrogen m^{-2} ann^{-1} for the less eutrophic waters of Windermere south basin and 0·04 for the oligotrophic Windermere north basin. In Lake Mendota the average rate of nitrogen fixation calculated by Stewart *et al.* (1971) for the top 5 m on the basis of measurements taken at weekly intervals at 0, 2 and 5 m during most of the ice-free season in 1969 was 2·6 kg ha^{-1} ann^{-1} (0·26 g m^{-2} ann^{-1}).

The factors which cause increased growth of nitrogen-fixing algae in eutrophic waters are complex and interacting. It is often suggested that inorganic nitrogen and phosphorus are the two most important nutrients causing eutrophication. However, as emphasized by Stewart *et al.* (1970) and Stewart and Alexander (1971), it is questionable whether the input of inorganic nitrogen into fresh waters is important if phosphorus and other nutrients are available in quantity. Under these conditions blooms of nitrogen-fixing algae may still occur, even if there is no combined nitrogen in the waters, and the major difference between nitrogen-depleted and nitrogen-rich waters is probably in floristic composition rather than in differences in the total biomass of algae.

Stewart *et al.* (1970) have shown that nitrogenase activity increases markedly when phosphorus-starved *Anabaena flos-aquae* is supplied with available phosphorus either as inorganic phosphate or as phosphorus-containing detergent. There is a marked stimulation of activity within 15 min by as little as 3–10 $\mu g\ l^{-1}$ of phosphorus. Natural populations do not always respond in this way and this may be because, when they are growing in phosphorus rich waters, blue-green algae accumulate and store phosphorus in the form of polyphosphate (Jensen, 1968, 1969) and they use this for subsequent growth in phosphorus-depleted waters. The rapid response, which occurs in the light and in the dark, may be due to ATP limiting nitrogenase activity in phosphorus-depleted cells. There is a net synthesis of ATP on adding phosphorus to phosphorus-depleted cells (Fig. 16.4).

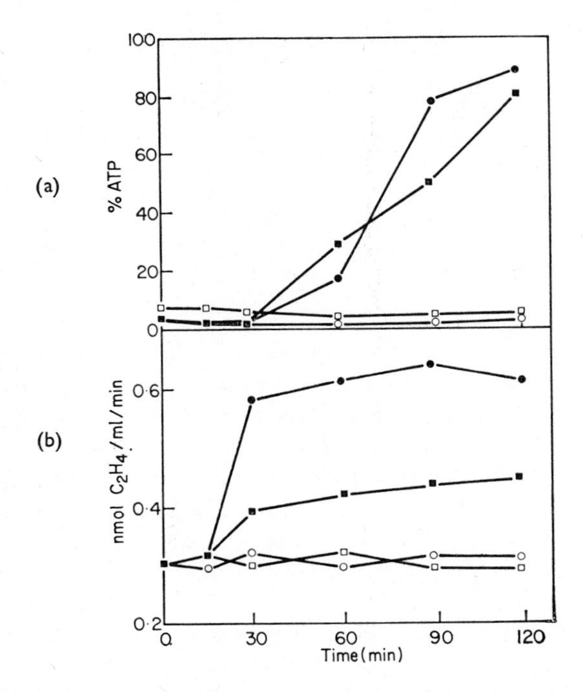

Fig. 16.4. The effect of added phosphorus on (a) the endogenous ATP and (b) on acetylene reduction by *Anabaena flos-aquae*, (●—●), light + phosphorus; (■—■), dark + phosphorus; (○—○), light − phosphorus; (□—□), dark − phosphorus (after Stewart and Alexander, 1971).

Nitrogen fixation in streams and flowing water has received little attention. Potential nitrogen-fixing algae, including species of *Calothrix*, *Rivularia* and *Tolypothrix*, are nevertheless common lithophytes on pebbles and rocks both in temperate (West and Fritsch, 1927) and tropical (Singh, 1961) streams. The most extensive studies so far have been on streams originating from hot springs. The combined nitrogen levels in the calcareous waters in which hot spring blue-green algae predominate are low (seldom more than 1·0 ppm of ammonium nitrogen and negligible levels of nitrate nitrogen, see Castenholz, 1969; Stewart, 1970b). Thus in these waters nitrogen-fixing species appear to be at a competitive advantage and indeed heterocystous algae are particularly abundant. For example, *Mastigocladus*, a nitrogen-fixing genus (Fogg, 1951b) is, according to Copeland (1936), second only to *Phormidium* in its abundance in Yellowstone National Park and when one considers that certain organisms hitherto referred to the genus *Phormidium* are probably flexibacteria, it may be that *Mastigocladus* is an even more important component of the flora. Other common heterocystous forms are species of *Calothrix*, *Rivularia*, and, possibly *Dichothrix*, all of which are usually restricted to waters above pH 6.

The *in vivo* thermal stability of algal nitrogenase is remarkable since in *in vitro* experiments it is inactivated on heating at 60°C for 10 min (Robinson and Stewart, unpublished). The effect of temperature on nitrogen fixation by algae in hot springs has been studied by Billaud (1967), Stewart, (1968, 1970b) and Stewart *et al.* (in press). Billaud (1967) found high rates of $^{15}N_2$ incorporation by a *Mastigocladus*-dominated flora at 45°C in thermal regions in Alaska. Stewart (1968, 1970b) also concluded from studies in the Lower Geyser basin area of Yellowstone National Park that nitrogen fixation, probably by *Mastigocladus*, was particularly important and observed fixation at temperatures as high as 54°C. Table 16.VII shows the relationship between temperature, nitrogen fixation and algae present in these alkaline streams. There was no evidence of nitrogen fixation at temperatures above 70°C where pink, filamentous bacteria predominated or in the temperature range 60–70°C where *Synechococcus* was abundant. Nitrogen fixation was appreciable at 25–54°C. At the latter temperature *Mastigocladus* was the most abundant heterocystous form whereas in the temperature range of 25–40°C species of *Calothrix* and *Dichothrix* occur as blackish crusts on the sinter together with diatoms and non-filamentous blue-green algae.

A more critical assessment of the highest temperatures at which nitrogen fixation occurs in alkaline hot springs has been carried out by Stewart *et al.* (unpublished) using an artificial stream built in Yellowstone National Park by Dr. R. Wiegert of the University of Athens, Georgia (Fig. 16.5, facing p. 272). The flora which developed on the wooden boards was dominated by

TABLE 16.VII. Nitrogen fixation and prominent micro-organisms in an alkaline hot spring stream in Yellowstone National Park (from Stewart, 1970b, and unpublished)

Temperature	Prominent micro-organisms	μg N fixed (mg total N)$^{-1}$ day^{-1}
80	Pink filamentous bacteria	0·0
70	*Synechococcus*	0·0
54	*Phormidium*, non-heterocystous filamentous blue-green algae, flexibacteria and *Mastigocladus*	0·5
46	*Phormidium*, non-heterocystous filamentous blue-green algae, flexibacteria and *Mastigocladus*	3·2
38	Rivulariaceae and non-filamentous blue-green algae	5·5
30	Rivulariaceae	4·9

Mastigocladus and there nitrogenase activity was detected at a temperature of 60°C. This is the highest temperature at which nitrogen fixation has been recorded on the surface of the Earth. Furthermore, as emphasized by Brock (1970b) in relation to studies on other metabolic processes, the algae are particularly well adapted to the temperatures at which they occur. For example in transfer studies carried out in the artificial stream it was shown that when an alga which had been grown at a particular temperature was transferred to a temperature only a few degrees higher there was an inhibition of nitrogenase activity. For example material grown at 48°C stopped reducing acetylene when transferred to 55°C although samples taken at 55°C reduced acetylene at 58°C.

Although the distribution of algae which are submerged continuously in hot spring streams is determined mainly by temperature, moisture is an important factor governing the metabolism of those algae which live on the sides of the streams exposed to steam arising from the hot water. There the algae, which form a blackish crust, are more prominent on the sides on to which the steam is chiefly blown (see Fig. 16.6, facing p. 273). These algae are most active in the morning when they are moistened by the large quantities of steam which accumulate there, but in the afternoon as evaporation increases, the algae dry out and stop reducing acetylene.

Marine environments

Nitrogen-fixing blue-green algae are less common in marine environments than in freshwaters. The only planktonic marine alga which has been studied extensively in connexion with nitrogen fixation is the red pigmented alga *Trichodesmium*, a member of the Oscillatoriaceae which occurs in thermally stratified waters in the tropics. (see p. 308).

Trichodesmium was first reported to fix nitrogen by Dugdale *et al.* (1961) who studied natural populations in the tropical Atlantic. Since then they have extended their work to the Arabian Sea, the Pacific Ocean and the Southern Atlantic Ocean and have accumulated a large amount of data which shows that high rates of nitrogen fixation are associated with *Trichodesmium* blooms, and that the alga rather than associated bacteria probably is responsible for the observed fixation (Dugdale *et al.*, 1964; Goering *et al.*, 1966; Dugdale and Goering, 1967). The rates vary markedly from station to station and this probably reflects the different physiological states of the algae used in the tests. Dugdale *et al.* (1964) reported that there were two main colony types, both of which fixed nitrogen. One comprised green bundles and the other brownish tufts and both occur at depths down to 40 m. The rates of fixation observed may be as high as 2 μg nitrogen fixed l^{-1} h^{-1} with average values in spring and summer of 0·05 and 0·15 μg nitrogen l^{-1} h^{-1} respectively. These nitrogen fixation rates may be up to 10 times higher than the rate of removal of ammonia from the sea by algae and fixation, which is light dependent, must represent a considerable input of new nitrogen into the ecosystem.

To date there has been no good evidence of nitrogen fixation by pure cultures of *Trichodesmium* but this has probably been due largely to the difficulty in culturing the alga. Bunt *et al.* (1970) have applied the acetylene reduction technique to almost pure cultures and obtained good evidence of an active nitr enase. They also tested benthic marine algae occurring at depths of 20–2 m off the Bahamas and south eastern Florida. The organisms there were mainly members of the family Oscillatoriaceae and although these did not reduce acetylene when incubated with or without added oxygen, an associated *Anabaena* species from the same habitat did, providing the gas phase contained no added oxygen.

In marine intertidal habitats, sand-dune systems and salt marshes nitrogen-fixing blue-green algae often comprise an important part of the flora. Stewart (1971a) lists those species from marine environments for which there is evidence of nitrogen fixation and these include, in addition to those mentioned earlier, the following intertidal or spray-zone forms: *Calothrix scopulorum*, *C. aeruginea*, *C. confervoides*, *Anabaena torulosa*, *A. variabilis*, *Microchaete* sp., *Nodularia harveyana*, *N. spumigena*, *Nostoc entophytum*, *N. linckia*, *Nostoc* sp., *Rivularia atra*, *R. biassolettiana* and *R. bullata*.

The most detailed studies have been carried out on *C. scopulorum*, a species which is common in the supralittoral fringe of many shores both in north and south temperate regions. Nitrogen fixation *in situ* shows a marked seasonal variation on Scottish rocky shores with the rates being highest in April and in late August and lowest in winter, when temperatures are low, and in high summer, when as a result of desiccation the algae peel off the rock surface. The annual fixation rate corresponds to 2·5 g nitrogen fixed m^{-2} ann^{-1} which is about one tenth of that fixed on good agricultural land by a legume crop (Stewart, 1967a). In sand-dune ecosystems where *Nostoc* is often abundant high rates of nitrogen fixation also occur in summer when the temperatures are high and where excessive desiccation does not occur (Stewart, 1967b). Webber (1967) found that about half of the prominent forms in a Massachusetts salt marsh are heterocystous algae and are thus potentially capable of fixing nitrogen.

Chapter 17

Symbiosis

Blue-green algae occur in symbiotic association more frequently than any other algal group with the possible exception of the Chlorophyceae. These associations, listed in Table 17.I, appear, at least superficially,

TABLE 17.I. Associations of blue-green algae with other plant and animal groups

Group	Genera	Algal genera
Fungi	Ascomycetes in lichens	*Calothrix*, *Chroococcus*, *Dichothrix*, *Gloeocapsa*, *Hyella*, *Nostoc*, *Scytonema*, *Stigonema*
	Phycomycete in *Geosiphon*	*Nostoc*
Filamentous algae	*Enteromorpha*	*Calothrix*
Diatoms	*Rhizosolenia*	*Microchaete*
Bryophytes	*Anthoceros*, *Blasia*, *Cavicularia*,	*Nostoc* or/and *Anabaena*
Ferns	*Azolla*	*Anabaena azollae*
Gymnosperms	*Bowenia*, *Ceratozamia*, *Cycas*, *Dioon*, *Encephalartos*, *Macrozamia*, *Stangeria*, *Zamia*	*Nostoc* or/and *Anabaena*
Angiosperms	*Gunnera*	*Nostoc* (*punctiforme?*)
Protozoa	*Cyanophora paradoxa*, *Cryptella cyanophora*, *Glaucocystis nostochinearum*, *Paulinella chromatophora*, *Peliana cyanea*	various "cyanelles"

to be of two main types. There are those such as lichens and the *Azolla–Anabaena* association where the algae are clearly extracellular with respect to the other partner, and there are others such as the cyanelles, which appear to be endosymbiotic blue-green algae. Usually the associations are not obligate and the algae can be isolated in culture. This poses the question of what biological advantage it is for these algae, which can grow perfectly well in a free-living state, to engage in a symbiotic association in which they may become modified to such an extent, as in some lichens, as to be virtually unrecognizable under the light microscope. This question, together with a recent revival of the theory of Mereschkowsky (1905) that the chloroplasts of higher plants may have arisen from endophytic blue-green algae (see Chapter 18) have aroused particular interest in associations of blue-green algae with other plant groups.

There are three main approaches to the study of blue-green algae in symbiosis. Firstly, the algae can be isolated in pure culture and their physiology investigated. It was once difficult to obtain pure cultures of symbiotic algae (*phycobionts*), but Bowyer and Skerman (1968) have developed a simple method of subculture for isolating and purifying those of *Gunnera* and *Cycas* which may find application with other systems as well. Studies with isolated phycobionts which, contrary to some earlier reports, grow rapidly in pure culture, are essential for investigations such as those to show whether or not the algae fix nitrogen or to determine their pigment composition. A disadvantage of this type of approach, however, is that the physiology of the isolated phycobiont may be different from that in the symbiotic state. As there is often great morphological variation between the free-living and symbiotic phycobionts, for example that of *Peltigera canina* has few heterocysts when growing symbiotically whereas heterocysts are common in the isolated phycobiont (Griffiths *et al.*, 1972), it would be surprising if there was not an attendant physiological variation as well. Thus a second line of approach, which is most useful for physiological studies of the symbiosis, is to use the intact plants, or portions such as discs cut out of lichens. A third method which has yielded interesting and often confirmatory information is the characterization of the endophytes *in situ* by electron microscopy. This has proved particularly useful in studies on cyanelles (see p. 351).

Associations with fungi

The association of algae with fungi to form lichens is probably the commonest of all symbiotic associations. In most lichens there is a single algal component which is a green alga, a blue-green alga, or, as in species of *Verrucaria*, a member of the Xanthophyceae. In a few cases such as *Lobaria pulmonaria* and *L. retigera* which contain, as phycobionts, a *Coccomyxa* and a *Nostoc* species respectively, the thalli are indistinguishable in external morphology in spite of the presence of different phycobionts (Asahina, 1937). Almost all the fungal partners (*mycobionts*) are ascomycetes but certain *Fungi imperfecti* and basidiomycetes also occur in lichens (see Ahmadjian, 1967). There are approximately 17,000 species of lichens which have been described and blue-green algae occur in about 8% of these (Fogg, 1956; Bond, 1959; Ahmadjian, 1967). The commonest blue-green algae are species of *Nostoc* but *Stigonema, Scytonema, Calothrix, Dichothrix, Gloeocapsa, Chroococcus* and *Hyella* all occur as lichen phycobionts (Ahmadjian, 1967). It is of interest that species belonging to the first five of these genera fix nitrogen, as do certain strains of *Gloeocapsa* (see p. 194).

Most lichens are composed of one kind of alga and one kind of fungus only but certain species contain two algal species, and when this happens one is always a blue-green alga. Exceptionally, as in *Solorina*, the blue-green alga and green alga form well-defined layers in the thallus. More usually the secondary phycobiont (generally a species of *Nostoc* or *Scytonema*) occurs in special structures called *cephalodia*. In most lichens these are protuberances which arise on the thallus surface, but occasionally, as in *Peltigera apthosa*, the cephalodia are internal.

Some earlier workers such as Cengia-Sambo (1926) reported on the presence of *Azotobacter* in lichen species and considered that a three-partnered symbiosis of alga, fungus and *Azotobacter* existed. Panosyan and Nikogosyan (1966) also found *Azotobacter* associated with lichens in Armenia but they were not found by Krasilnikov (1949) in the 40 species of Russian lichens which he examined. The role of *Azotobacter* in lichens thus seems dubious. It is probable that *Azotobacter* occurs as a contaminant on lichen surfaces, just as do yeasts, fungi and other microbes. The possible importance of *Azotobacter* in lichen associations has been discussed by Scott (1956) and Ahmadjian (1967).

Despite the widespread abundance and undoubted success of the lichen association, no one has yet satisfactorily achieved the synthesis of a typical lichen thallus from the isolated partners. Ahmadjian (1962, 1963) was partially successful with *Acrospora fuscata*, and later with *Cladonia cristella* was able to re-establish an abnormally pigmented thallus with ascogial filaments and ascogenous hyphae but no asci (Ahmadjian, 1966). Such results give good reason to believe that the reproducible resynthesis of typical lichen thalli will soon be achieved. In nature regeneration may result more from the dispersion of pieces of the symbiotic system rather than from synthesis from the isolated partners.

Although the isolated partners can grow rapidly in culture the growth rates of the composite organisms are remarkably slow. *Peltigera* species, which are generally reputed to be the fastest growing lichens, increase in size by only 2–3 cm per year. Nevertheless the phycobionts are often particularly active in the symbiotic state. Millbank (1972) has shown, for example, that the nitrogenase activity of the *Peltigera* phycobiont, which comprises only 2·7% of the lichen dry weight, may be several times higher than that of the isolated phycobiont (see also p. 210). Photosynthesis by the phycobiont, as measured by ^{14}C incorporation from carbon dioxide, is also rapid (see Smith *et al.*, 1970). The phycobiont is thus capable of playing an important role by providing for the mycobiont fixed carbon and nitrogenous products also, if it fixes nitrogen.

Carbohydrate transfer from the phycobiont to the mycobiont has been studied extensively by Smith and his collaborators (see Smith *et al.*, 1970). In species such as *Lobaria scrobiculata*, *Peltigera polydactyla*, *P. praetextata* and *P. horizontalis*, which contain *Nostoc* as phycobiont, there is a rapid transfer (within 60 s) of fixed carbon from the alga to the fungus and over a 4 h experimental period in the light approximately 40% of all the carbon fixed passed into the fungus. Typical results over a 2 h period are shown in Fig. 17.1. Within 2 min of being fixed the *Nostoc* carbon was liberated as

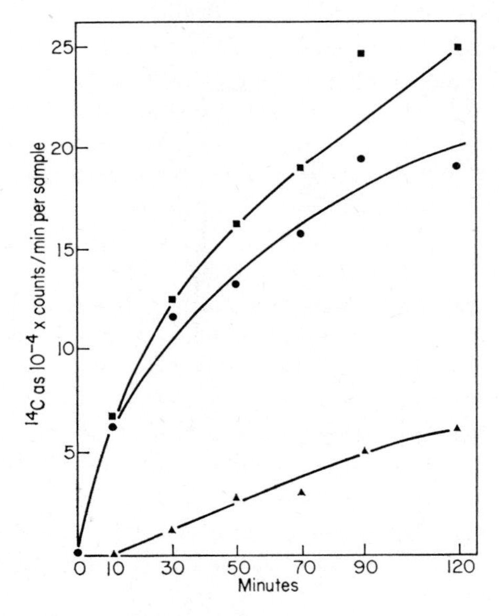

FIG. 17.1. Transfer of ^{14}C labelled photosynthate from the algal layer (●—●) to the medulla (▲—▲) of *Peltigera polydactyla*. (■—■) represents the total ^{14}C label. After Smith and Drew (1965). *New Phytol* **64**, 195, Fig. 1.

glucose and converted by the fungus into mannitol and assimilated. The phycobiont does not assimilate mannitol. Not all the extracellular carbohydrates liberated by the phycobiont are available to the mycobiont. For example, Henriksson (1964) found that the polysaccharides produced in old cultures of *Collema* are largely unavailable to the fungus; these may resemble the carbohydrates released on autolysis by other algal groups (Guillard and Wangersky, 1958; Marker, 1965). In addition to glucose, other extracellular products such as polysaccharides, nitrogenous products, thiamine, riboflavin, nicotinic acid, pantothenic acid and biotin are all produced by the *Nostoc* of *Collema tenax* (Henriksson, 1964).

How the fungus modifies the algae so that such large amounts of carbohydrate and combined nitrogen are liberated is not known, but electron microscope observations show certain morphological adaptations of the mycobiont which may facilitate transfer (Tschermak, 1941; Plessl, 1963). Many mycobionts, for example, produce distinct thin walled haustoria which invaginate the algal cell walls and penetrate the protoplasm (Moore and McAlear, 1960; Fig. 17.2). Whether this is a true penetration or whether it is simply an invagination of the host plasmalemma is uncertain, but the host cells often die. This is not surprising when as many as five haustoria have been found within one algal cell (Tschermak, 1941; Plessl, 1963). In other lichens such as *Peltigera aphthosa* haustoria are not formed, perhaps because the walls of both partners are often very thin. In *Collema*, where the association is loose and the alga retains its typical filamentous form, haustoria are generally not produced and the hyphae simply grow between the algal filaments and into the algal sheaths. However, on rare occasions and for some unknown reason, the hyphae become parasitic, penetrate the *Nostoc* cells and kill them.

With carbohydrates, nitrogenous products and vitamins all being available, the benefits of the symbiosis to the fungus are evident, but the advantages to the alga are less obvious. In the symbiotic state the alga may be protected from environmental extremes such as desiccation, but free-living blue-green algae are particularly resistant to such extremes and it is unlikely that an increased ability to withstand desiccation is the only benefit. Furthermore the mycobiont may produce substances which are inhibitory to the phycobiont (Henriksson, 1964). It seems that in addition to protecting the alga from environmental extremes the symbiotic state provides a particularly favourable environment for cellular metabolism of the algae. Stewart and Pearson (1970) emphasized how nitrogen-fixing algae prefer conditions of low oxygen tension for nitrogen fixation and how high pO_2 levels inhibit activity. They consider that in lichen thalli respiration by the mycobiont could prevent the accumulation of high oxygen levels. This could partially explain, perhaps, the high rates of nitrogen fixation and possibly of other metabolic processes shown by the phycobiont in the symbiotic state.

Although one tends to think only of lichens when considering symbioses of blue-green algae and fungi, there is another interesting, but largely neglected, type of symbiosis between them which is exemplified by *Geosiphon pyriforme* (Fig. 17.3, facing p. 273). Here a blue-green alga

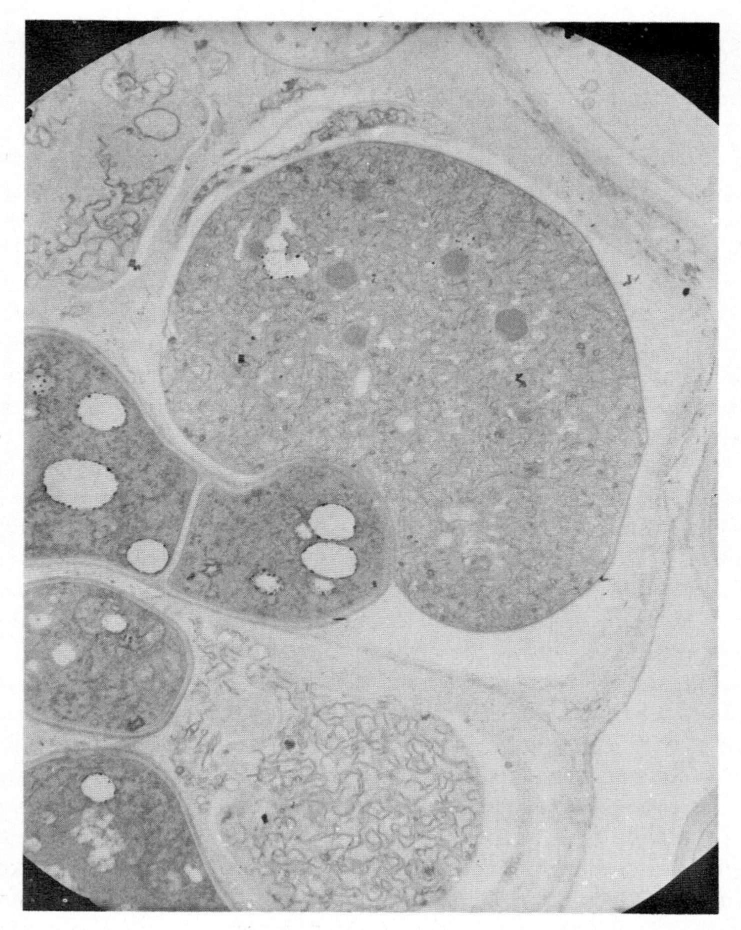

FIG. 17.2. Electron micrograph of a section of the lichen *Heppia lutosa*, in which the phycobiont is *Scytonema* sp., showing a thin-walled haustorium of the mycobiont penetrating an algal cell; ×10,000. Photograph by courtesy of V. Ahmadjian and M. Fuller.

(probably *Nostoc sphaericum*) occurs intracellularly within a phycomycetous fungus. The symbiosis, which has been studied using electron microscopy by Schnepf (1964) is not obligatory. Indeed both organisms propagate themselves independently and the association becomes re-established when the fungal wall dissolves away on contact with the free-living *Nostoc*. The fungal plasmalemma then invaginates so that an intracellular vesicle containing the alga develops within the fungal hyphae. The system has not been investigated physiologically.

Associations with bryophytes and pteridophytes

Nostoc species occur in symbiotic association with several bryophytes such as species of *Anthoceros*, *Blasia* and *Cavicularia*. In *Anthoceros* there are small mucilage-filled cavities on the undersurface which open through narrow pores to the outside. These become filled with *Nostoc* cells and the surrounding *Anthoceros* cells then increase in size to form slight protuberances. These are induced only by *Nostoc* filaments and do not occur when *Oscillatoria* filaments, or diatoms, enter thec avities (Myer, 1966). Some species of *Anthoceros*, such as *A. laevis*, may not produce cavities and there *Nostoc* occurs on the thallus surface. Rather similar cavities are produced in *Cavicularia*, and in *Blasia* where the algae comprise about 1% of the total volume of the thallus (Bond and Scott, 1955).

The algae which occur in association with bryophytes are not specific. Takesige (1937) found that *Blasia* plants could be infected by *Nostoc* isolated either from *Cavicularia* or from cycad root nodules (see below). Pankow and Maertens (1964) showed, in fact, that the *Blasia pusilla* endophyte is the common free-living nitrogen-fixing species *Nostoc sphaericum*. Bond and Scott (1955) concluded from studies on the intact system using ^{15}N as tracer that the nitrogen fixed by the blue-green algae probably becomes available to the liverwort. Such a transfer could benefit the hepatics considerably in the nitrogen-poor habitats where they frequently occur.

There is only one well-known symbiosis between a fern and a blue-green alga, namely the water fern *Azolla–Anabaena* association. Here the alga occurs in cavities on the dorsal lobe of the leaf. Again the association is not obligatory but it is the normal state of affairs, perhaps because akinetes of *Anabaena* are generally found in the megasori of the host. In the presence of combined nitrogen the *Azolla* grows well in the absence of *Anabaena* (Huneke, 1933; Wildemann, 1934). However the *Anabaena* fixes nitrogen (Bortels, 1940) and in medium free from combined nitrogen this can provide all the nitrogen required for the healthy growth of both the alga and fern. In pure culture the endophyte liberates large quantities of extracellular nitrogenous compounds and this is probably the means whereby nitrogen is transferred to the fern. It would be interesting to know whether this symbiotic association occurred during the Jurassic and Cretaceous periods when, according to Bond (1967), *Azolla* was distributed widely.

Association with gymnosperms and angiosperms

Among the gymnosperms nitrogen-fixing root nodules seem only to occur in the cycads. There are about 90 species of cycads, distributed in about eight genera, and root nodules have been recorded on about a third of these. At the present time cycads are restricted to the warmer parts of the southern hemisphere but there is good evidence from fossils that in the Jurassic and Cretaceous periods they were of widespread and dominant distribution both in the northern and southern hemispheres (see Walton, 1953).

The root nodules occur as coralloid masses superficially resembling those of nodulated non-leguminous angiosperms such as *Alnus*, being dichotomously branched and ageotropic. Nodules containing algae occur only at, or near, the soil surface. Reinke (1879) was the first to observe that the alga occurs as a distinct green zone in the nodule cortex, as can be seen readily with the naked eye when the nodule lobes are sliced. According to Watanabe (1924) the nodules are free from algae when first formed and those which become infected persist whereas those which do not die. There is some uncertainty as to the exact identities of the algal symbionts but they are referred to generally as species of *Nostoc* or *Anabaena* (Reinke, 1879; Winter, 1935). In fact it has been reported (Takesige, 1937) that three endophytic algae, two *Nostoc* and one *Anabaena* species may occur within the same nodule. It is more likely that these are ecophenes of the same species (see, for example, Forest, 1968). The report that a species of the green alga *Chlorococcum* occurs as a symbiont (Fernandez and Bhat, 1945) has not been substantiated. All the blue-green algal endophytes so far isolated fix nitrogen and recent studies (Bowyer and Skerman, 1968) do not confirm the earlier view of Winter (1935) that the endophytes have largely lost their ability to grow photosynthetically. Fixation rates by the intact nodules are comparable to those of non-leguminous angiosperms such as *Alnus* (Bergersen *et al.*, 1965). Some aspects of nitrogen fixation by cycad nodules have been mentioned earlier (p. 210).

Gunnera, a member of the Haloragidaceae, is perhaps the only angiosperm genus which normally has blue-green algae as symbionts. There are about 30 *Gunnera* species and these are widely distributed in the southern hemisphere, particularly in aquatic and marshy habitats. The algal partner, usually regarded as *Nostoc punctiforme* (Harder, 1917b; Winter, 1935; Schaede, 1951), occurs in wart-like swellings on the leaf bases in *G. macrophylla*, in which the algal mat is large enough to be dissected out easily. In *G. manicata* these swellings usually occur in the stem. The mechanism of penetration has been considered in detail by Schaede (1951). Unlike most other *Nostoc* associations the *Gunnera* endophyte appears to be intracellular and is surrounded by a thin cytoplasmic membrane of the host. It probably grows mixotrophically, that is, photosynthetic carbon dioxide fixation, photo-assimilation of organic substrates and oxidative assimilation as well, are all possible, the importance of each at a particular time depending on the environmental conditions (Harder, 1917b; Winter, 1935). It has been shown, using ^{15}N as a tracer, that nitrogen fixed in the swellings containing the alga is transferred to other parts of the plant (Table 17.II). *Nostoc punctiforme* has been reported as occurring in the root

TABLE 17.II. ^{15}N enrichment of various parts of *Gunnera arenaria* plants after exposure to $^{15}N_2$ (after Silvester and Smith, 1969)

Exposure time (h)	Atom % ^{15}N excess Plant part			
	Node cluster	Leaf	Internode	Root
1·5	0·400	0·010	0·006	0·004
3	0·613	0·022	0·018	0·016
5	1·184	0·080	0·144	0·039
9	2·231	0·463	0·182	0·184

nodules of *Trifolium alexandrinum* (Bhaskaran and Venkataraman, 1958) but this association appears to be rare.

Associations with other algae and protozoa

Blue-green algae, or structures resembling them, occur in association with a small number of protozoan and algal species but whether the associations are in all cases "symbiotic", that is, mutually beneficial, is doubtful. Pascher (1914b) introduced the term *Syncyanosen* to cover the general phenomenon of blue-green algae living in association with another member of the Protista and later (Pascher, 1929a) distinguished between *Ectocyanosen*, in which the blue-green alga occurs extracellularly and *Endocyanosen* in which they occur intracellularly. It may be questioned whether the Ectocyanosen are anything more than casual associations and we will consider further only the Endocyanosen. Pascher (1914b, 1929b) described the former fully. Korschikoff (1924) was the first to describe an association between a flagellate and an intracellular blue-green alga-like structure and Pascher (1929a) termed these intracellular structures *Cyanellen*. In general the term Endocyanosen refers to any blue-green alga-like inclusion, whereas the term cyanelle is usually restricted to inclusions which resemble unicellular blue-green algae. There is considerable doubt as to whether the cyanelles are blue-green algae rather than chloroplast-like organelles within the protozoan. They occur in certain freshwater Protozoa only and many of the known associations are listed in Table 17.I.

Initial studies on organisms containing cyanelles (see Geitler, 1959) were concerned with general morphology, and forms which were studied in some detail include *Paulinella chromatophora*, a rhizopod with two intracellular *Synechococcus*-like inclusions (Lauterborn, 1895; Geitler, 1927; Pascher, 1929b), *Peliana cyanea*, a colourless flagellate with 1–6 *Synechococcus*-like inclusions arranged parietally (Pascher, 1929b) and *Cryptella cyanophora* which again contains blue-green algae-like unicells

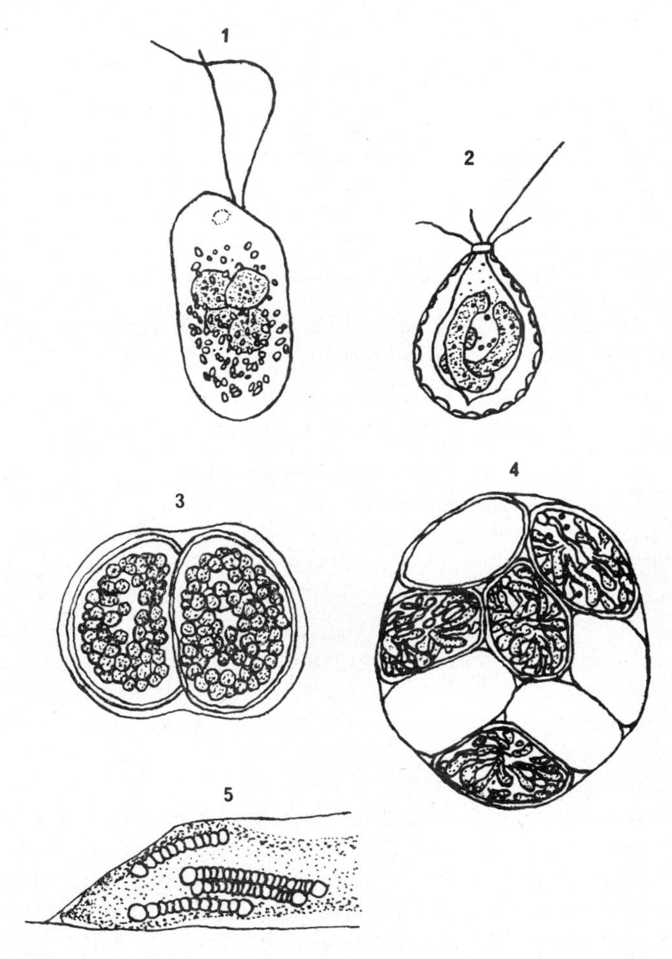

FIG. 17.4. Protista with blue-green alga-like symbionts: (1) *Peliana cyanea* Pascher; (2) *Paulinella chromatophora* Lauterb.; (3) *Cyanoptysche gloeocystis* Pascher (all after Pascher); (4) *Glaucocystis nostochinearum* (after Geitler); (5) *Rhizosolenia* sp. containing *Richelia intracellularis* (after Schmidt).

(Pascher, 1931). More recently electron microscope observations have been carried out (see Hall and Claus, 1963, 1967; Schnepf, 1965; Schnepf and Koch, 1966; Schnepf *et al.*, 1966). Certain of the cyanelles have been isolated and cultured and some of their physiological characteristics investigated (Provasoli and Pintner, 1953). Forms on which detailed information is now available include *Cyanophora paradoxa* and *Glaucocystis nostochinearum* (see Figs 17.4–17.6).

Cyanophora is the generic name given to a cryptomonad which contains what appears to be an intracellular blue-green alga belonging to the Chroococcales. Several species have been described; Hall and Claus (1963) studied one which they called *Cyanophora paradoxa*, but which differs in certain points from the original description of this species. In this symbiosis (see Fig. 17.5) the unicellular alga is surrounded intracellularly by a thin limiting membrane only and the four-layered wall which is characteristic of free-living blue-green algae is absent. The chromatoplasm of the endosymbiont shows the typical lamellar arrangement found in blue-green algae and in the chloroplasts of Rhodophyceae but the centroplasm is atypical in being surrounded by a halo-like region. Hall and Claus (1963) suggest that this endophyte might be a forerunner of a eukaryote-like cell. Because it differs from free-living blue-green algae in various respects such as in the absence of a typical cell wall, and in having a modified centroplasm, Hall and Claus (1963) classify these cyanelles as a new species of blue-green algae, *Cyanocyta korschikoffiana* and place it in a new family Cyanocytaceae, in the Chroococcales. It has been isolated in culture by Provasoli and Pintner (1953) and shown to have a vitamin B_{12} requirement. The isolated form contains typical blue-green algal pigments and develops cyanophycin granules which are absent from the cyanelle in the symbiotic state.

Glaucocystis nostochinearum is another instance of a distinct association between a blue-green algal-like inclusion and an apochlorotic alga which superficially resembles *Oocystis* (Chlorophyceae). The discovery by Schnepf *et al.* (1966) that the host has rudimentary flagella raises the question, however, of whether it is in fact an *Oocystis*. Robinson and Preston (1971b) in an investigation of the fine structure of *Glaucocystis* found an unusual type of plasmalemma reminiscent of that of the dinoflagellates but no features characteristic of the Chlorophyceae. Its taxonomic position is thus very uncertain. In *Glaucocystis* the cyanelles are rod-shaped and resemble free-living *Aphanothece*, *Synechococcus* or *Rhabdoderma* cells (see Figs 17.4, 17.6). They are 0·7–10 μm in length × 1·0–1·6 μm in width and divide by binary fission. They may be distributed at random within the host or they may be arranged radially. As with the cyanelles of *Cyanophora* the typical cell wall is missing. The cyanelle thus appears to have become modified from a typical free-living alga in such a way as to maximize the interchange of substances between the endosymbiont and its host, suggesting perhaps that this is a long-established composite association rather than two distinct organisms which have recently come

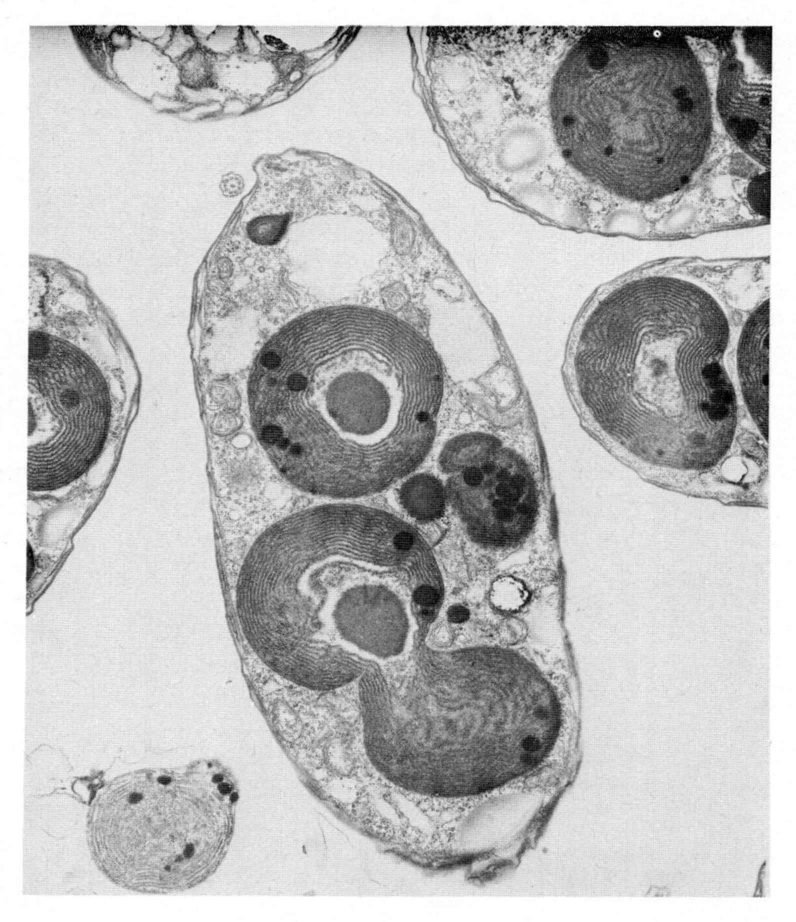

FIG. 17.5. The cyanelle *Cyanocyta korschikoffiana* in the cryptomonad *Cyanophora paradoxa* (× 12,000). By courtesy of W. T. Hall and G. Claus.

in contact by chance. Indeed Robinson and Preston (1971a) consider cyanelles as permanent organelles of the host cell which function as the chloroplasts do in other algae and plants and (1971b) are inclined to doubt whether they are cyanophycean in nature. The absence of a wall makes the endosymbiont particularly susceptible to osmotic shock, however, and this is one of the reasons why cyanelles are sometimes difficult to culture. Hall and Claus (1967) classify the cyanelle as a new species *Skujapelta nuda* in the new family Skujapeltaceae in the order Chroococcales (Fig. 17.6).

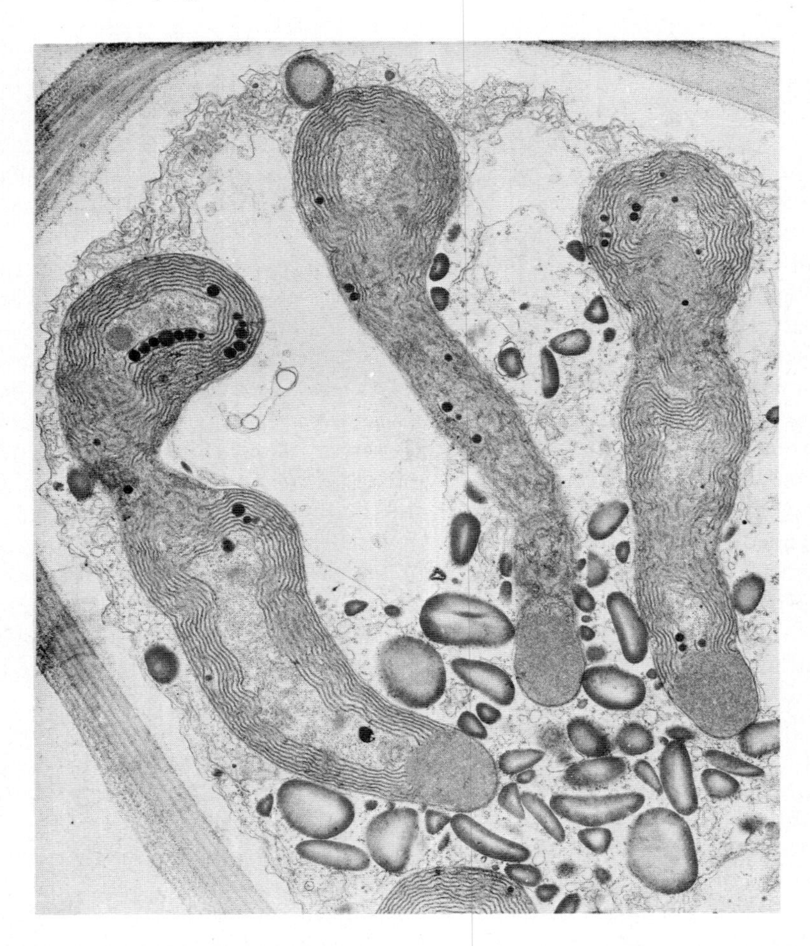

FIG. 17.6. The cyanelle *Skujapelta nuda* in *Glaucocystis nostochinearum* (× 12,000). By courtesy of W. T. Hall and G. Claus (1967), *J. Phycol.* **3**, 37, Fig. 3.

The benefit of the association to the host protistan is obvious in *Paulinella* (Fig. 17.4). This rhizopod does not feed holozoically and shows a positive phototactic response when cyanelles are present, although not otherwise (Pascher, 1929a). The cyanelle apparently provides the organic material for the composite cell. If it did not the rhizopod would die of starvation (or from rupture as a result of increased growth of the cyanelles!). Occasionally lysis of the cyanelles is observed and this is perhaps a method whereby cyanelle numbers can be controlled. Lysis of all cells seldom occurs except occasionally just prior to reproduction of the rhizopod. As Droop (1963) remarked, "One does not, in general, kill the goose that lays the golden egg".

Continuity of the symbiosis is usually achieved by synchronization of reproduction of the cyanelle and the host, although not always in the same way. In *Peliana* (Fig. 17.4), for example, the alga and flagellate usually divide simultaneously with both daughter flagellates receiving the same number of cyanelles. In *Paulinella* cell division of the protozoan and the *Synechococcus* is linked, but out of phase. When the rhizopod divides each cell receives one algal cell and then this divides so that each adult protozoa contains two cyanelles.

In certain marine diatoms of the genus *Rhizosolenia*, blue-green alga-like filaments which resemble a species of *Microchaete* and which have been classified as a blue-green alga, *Richelia intracellularis*, sometimes occurs intracellularly (Fig. 17.4). Here the diatom, usually but not always, has poorly developed chromatophores and the *Richelia* assumes the role of the photosynthetic organelles. Drum and Pankratz (1963) observed cellular inclusions in the freshwater diatoms *Rhopalodia gibba* and *R. gibberula* under the electron microscope. Although these resemble unicellular blue-green algae in some respects it is not certain that this is in fact their true nature. Lami (1958) has also reported on the occurrence of *Calothrix* within filaments of *Enteromorpha*. This association, which is uncommon, requires further study.

The mode of penetration of endosymbionts into the host algae is not well understood and in the case of some cyanelles it is questionable whether they ever occur in a free-living state. Blue-green algae do not possess cellulases which would be necessary to lyse the wall of a green alga, so that it is more likely that they get in either by ingestion or else by entering at some stage in the life of the host when a rigid cell wall is lacking. The latter is the manner in which *Richelia* becomes associated with *Rhizosolenia*. The possible importance of cyanelles and endophytic blue-green algae as indicators of the possible path of evolution of higher plant chloroplasts is considered in the next chapter.

Associations with metazoa

There are several reports of associations between what appear to be strains of the unicellular blue-green alga *Aphanocapsa* and various sponges.

The association was reported on first by Feldmann (1933) in the sponges *Petrosia ficiformis* and *Ircinia variabilis* and is now known in others as well, including *Chondrilla nucula*, *Pellina semitubulosa* and *Verongia aerophoba* (see Sarà, 1971). The alga forms a dense layer in the cortex of the sponge, where it grows extracellularly and intracellularly and reproduces. It provides mutual shading for both organisms and as a result the sponge is able to grow at higher light intensities when the alga is present than when it is absent. Sarà (1971) has carried out an electron microscope study of the *Aphanocapsa–Ircinia* association and considers that two species of *Aphanocapsa* are present. The algae, unlike cyanelles, have typical cell walls and Sarà concludes that carbon compounds for the sponge are provided both as a result of the alga secreting material and also by disintegration of the alga.

Chapter 18

Evolution and phylogeny

The position of blue-green algae in the hierarchy of living organisms was discussed at the beginning of this book and various points bearing on their taxonomic relationships have been considered in subsequent chapters. It remains to summarize their possible inter-relations with other plant groups and views on their possible origin and evolution.

Relationships to other plant groups

It is clear that they resemble the bacteria more than any other plant group. The chief points of similarity between the two groups are as follows:

1. Cellular organization: together with the bacteria, the blue-green algae stand in sharp contrast to the rest of living organisms in having a prokaryotic cellular organization (p. 35). As well as in the absence of membrane-bound organelles and other general prokaryotic features there is resemblance in the details of the structure of gas vesicles, which are only found in certain bacteria and blue-green algae and nowhere else (p. 97).

2. Cell wall structure and chemistry: the fine structure as seen in electron micrographs of sections of the walls of blue-green algae bears a general resemblance to that of bacteria, particularly that of the gram-negative kinds (p. 71) in that both have a mucopeptide as a major cell wall component. This contains α-ϵ-diaminopimelic acid and muramic acid (p. 74) and is susceptible to lysozyme treatment (p. 74).

3. Cell division: this process, which is accomplished by diaphragm-like ingrowths from the periphery towards the centre, is superficially at least, similar in the two classes and distinct from other modes of division found in the higher plants and animals (p. 214).

4. Ribosomes: there is good evidence that there are two forms of these bodies, characterized by different sedimentation properties in the ultracentrifuge and by the RNA species which they contain, and that one type

is characteristic of bacteria, blue-green algae, higher plant chloroplasts and mitochondria, whereas the other is found only in the cytoplasm of higher plants and in animals (p. 59).

5. Poly-β-hydroxybutyric acid: this reserve product occurs in bacteria and in blue-green algae but is absent from eukaryotic organisms (p. 69).

6. Nitrogen fixation: the capacity to assimilate elemental nitrogen is found only in certain free-living bacteria and blue-green algae, and in symbiotic systems containing bacteria or blue-green algae, and has not been demonstrated convincingly in any eukaryotic organism without symbiotic micro-organisms (p. 5).

7. Survival at high temperatures: in general bacteria and blue-green algae are capable of survival and growth at temperatures about 20°C higher than the highest tolerated by eukaryotes (p. 84).

8. Susceptibility to antibiotics: many antibiotics, such as penicillin and streptomycin, inhibit the growth of some bacteria and blue-green algae more severely than that of other micro-organisms (p. 88).

9. Genetic recombination: in both bacteria and blue-green algae this is accomplished by parasexual processes and not by sexual reproduction involving syngamy and meiosis as in eukaryotes (p. 90).

The Prokaryota therefore are a well-defined group and there is some justification for considering the blue-green algae as a specialized group of bacteria. On the other hand it is equally clear that there are important differences between the two groups, including the following:

1. The blue-green algae show much greater structural diversity than the bacteria. Multicellular organisms with protoplasmic connexions between the cells are found among blue-green algae but not amongst the bacteria with the possible exception of the flexibacteria, which are of uncertain taxonomic position (see p. 174). The branched multiseriate filamentous thallus of algae such as species of *Stigonema* (p. 31) is much more elaborate than anything found in the bacteria. The bacteria do not as a rule form distinct colonies under natural conditions whereas most blue-green algae are aggregated into colonies of more or less characteristic form and structure. Only in a few instances, for example with *Merismopedia* in the blue-green algae and *Thiopedia* amongst the bacteria, is there any parallelism.

2. Differentiation of the cells is dissimilar: blue-green algae do not produce bacterial-type endospores and there are no structures resembling the heterocysts or akinetes of the blue-green algae to be found in bacteria.

3. Blue-green algae never produce flagella of any sort whereas the bacteria, although they do not produce the "9 + 2" stranded type, characteristic of the Eukaryota, include many forms with single-stranded flagella. It is, however, possible that bacterial flagella may be homologous with the contractile fibrils which are postulated as occurring at the surface of some gliding blue-green algae (p. 117, but see Bisset, 1973).

4. Although there is probably a close similarity between the basic photochemical mechanism of the photosynthetic bacteria and of blue-green algae, the pigments which mediate them are chemically distinct and, whereas blue-green algae are capable of utilizing water as the ultimate hydrogen donor with consequent evolution of oxygen, no bacteria are capable of doing this (p. 152).

5. The fatty acid compositions of blue-green algae are different from those of bacteria. Polyunsaturated fatty acids are absent from the photosynthetic bacteria and the mono-unsaturated acid present is mostly the Δ-11 C_{18} vaccenic acid, which is also found in non-photosynthetic bacteria. Polyunsaturated fatty acids are present in the morphologically more complex forms and the monoenoic acid found is the Δ-9 compound oleic acid (Holton *et al.*, 1968).

There are, furthermore, some resemblances which seem to be of significance between the Cyanophyceae and a eukaryote group, the Rhodophyceae (Allsopp, 1969; Klein, 1970). The most obvious are in the photosynthetic apparatus. The principal pigment, chlorophyll *a*, is the same in both groups and β-carotene and phycobiliproteins are common accessory pigments. As indicated above (p. 54) the phycobiliproteins, while not absolutely identical in these two algal groups, are closely related, according to immunological studies (Berns, 1967) and are arranged in similar structures, the phycobilisomes. The phycobiliproteins of the Cryptophyceae, the only other group where they occur, are immunologically distinct and do not form phycobilisome aggregates. There are, of course, similar proteinaceous pigments, phytochromes, which occur widely in higher plants and in some algae, having a similar, though distinct, bilin prosthetic group. The thylakoids also show a resemblance in occurring singly rather than being closely stacked, as in higher plants and other algal groups. Furthermore, it is striking that of all the eukaryote groups the Rhodophyceae is the only one having no flagellated forms and showing no indication that such organelles were ever possessed in the ancestry of the group. There are further resemblances in chemical features which may or may not be significant. The major reserve polysaccharide of blue-green algae has a highly branched molecule consisting mainly of α-1,4-linked glucose units and resembling the amylopectin of higher plants (p. 62). Rhodophycean starch is very similar (Percival and McDowell, 1967). The trisaccharide, trehalose, which is otherwise uncommon in the plant kingdom, occurs in blue-green algae and in certain red algae (Meeuse, 1962).

Although such similarities exist it is important to stress that these should not be interpreted too strongly in terms of phylogeny. Indeed, it would be naïve to suppose that the blue-green algae evolved directly from organisms resembling a present-day bacterium or that present-day red algae are derived directly from an ancestral blue-green alga. Klein (1970), for example, who pointed out that the unicellular eukaryotic hot spring alga, *Cyanidium caldarium*, has many characteristics intermediate between blue-green and red algae, was properly cautious in not suggesting a direct phylogenetic link.

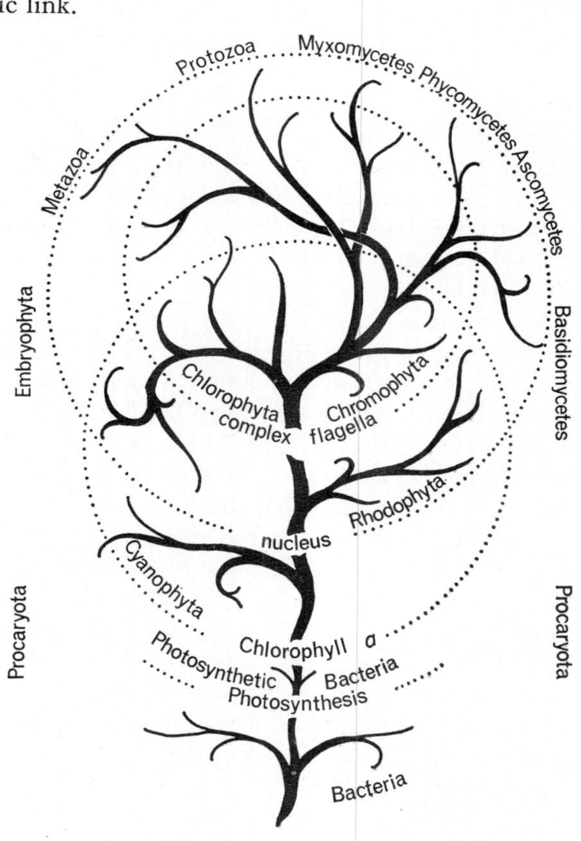

FIG. 18.1. Diagram showing the relationships of the major groups of organisms according to the distribution of chlorophyll *a* and certain cytological characters (redrawn after T. Christensen (1964) *In* "Algae and Man", ed. D. F. Jackson. Plenum Press, New York, pp. 59–64).

It is possible, nevertheless, that the three groups could represent successive stages of evolution, with the appearance of chlorophyll *a*, marking an important advance between the bacteria and blue-green algae and the appearance of the membrane-enclosed nucleus being the important step between the latter group and the Rhodophyceae. Beyond this the acquisition of complex flagella marks off the cell type from which higher plants and animals were probably evolved. It is likely that such complex features as a photosynthetic mechanism based on chlorophyll *a*, a nucleus and an 11-stranded flagellum have each arisen independently and successfully only once in evolution and the idea that such features might have been combined by symbiotic fusion of primitive organisms is touched on at the end of this chapter. A diagram summarizing the classification of lower organisms in these terms has been given by Christensen (1964; see Fig. 18.1).

Phylogeny and the fossil record

A great deal of attention has been paid since the early 1950s to the study of fossil blue-green algae and over three-quarters of the known records have been obtained since then. These fossils have been found predominantly in Precambrian rocks, which have been studied extensively by workers interested in the origin of life on Earth because it is there that the earliest micro-organism-like structures have been found. The degree of preservation in some of the recently discovered fossils is so good that characteristics, not only of gross morphology, but also of cell type, sheath etc. can often be discerned. A well preserved fossil blue-green alga is the heterocystous alga *Kidstoniella fritschii* which was present in Middle Devonian material from the Rhynie Chert in Aberdeenshire, Scotland (Fig. 18.2). This material is about 380 million years old and is thus of relatively recent origin, but well-preserved fossil blue-green algae extend well back into the Precambrian period. The Precambrian period, in fact, because of the frequent occurrence of blue-green algae then has been termed by Schopf (1973) "the age of blue-green algae" just as other eras have been termed "the age of reptiles" and "the age of mammals".

The oldest reliable microfossils are those from the Fig Tree series of South Africa ($3{\cdot}1 \times 10^9$ years old). Some of these, which may or may not have been blue-green algae, have been referred to as "algae-like" (Pflug, 1967; Schopf and Barghoorn, 1967; Pflug *et al.*, 1969). They are never associated with traces of higher plants or animals, and do not appear to be contaminants or artefacts. Oehler and Schopf (1971) have shown that in their degree of preservation and mineralogic setting they are comparable to blue-green algae artificially fossilized in crystalline silica at moderately elevated temperatures and pressure.

Although one cannot be certain that such "alga-like" unicellular structures were indeed blue-green algae it is known that in the Bulawayan series of the Early Precambrian which is about 3·0–2·8 × 10^9 years old, stromatolite-like structures have been recognized. Stromatolites are

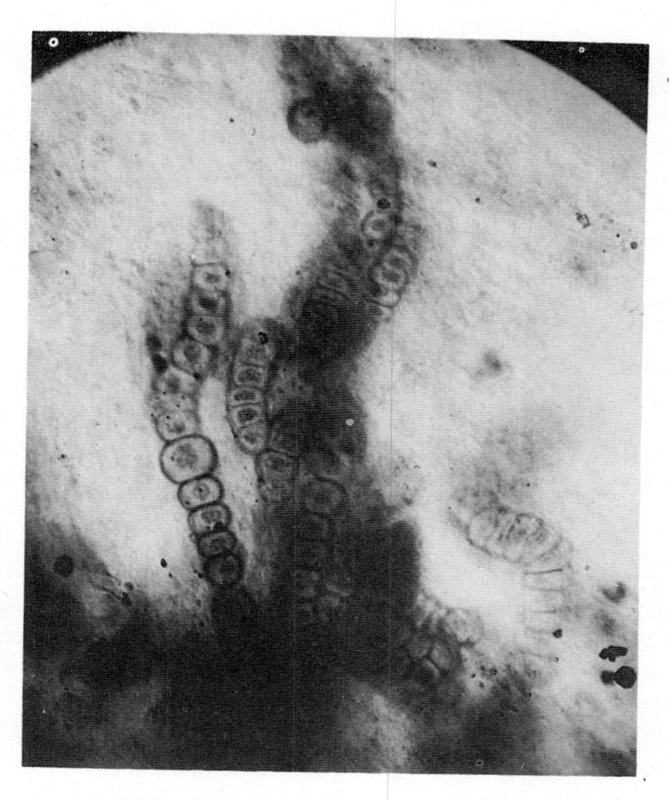

FIG. 18.2. *Kidstoniella fritschii*, a fossil blue-green alga with heterocysts from the Rhynie Chert of Aberdeenshire (Middle Devonian, about 380 million years old). By courtesy of W. N. Croft and E. A. George, 1959, *Bull. Br. Mus.* (*Nat. Hist.*), *Geology* **3**, Fig. 27.

macroscopic calcareous structures which closely resemble deposits produced by present-day blue-green algae in supra-tidal and intertidal environments in tropical seas. Although the Bulawayan stromatolites cannot be said with certainty to contain blue-green algae, comparisons with present-day structures which contain filamentous blue-green algae suggests strongly that filamentous blue-green algae had evolved by that time. If, indeed, the "alga-like" micro-fossils are blue-green algae it implies that the most primitive blue-green algae were probably unicellular forms. There is controversy as to whether the capacity to photosynthesize had evolved then. Oro and Nooner (1967) on the basis of carbon dating studies consider that it had not. On the other hand Schopf (1973) points out that the arrangement of present-day stromatolites with successive stacking, on top of each other, is due to the algae growing towards the light, and as this arrangement is also found in early Precambrian stromatolites it does suggest that a phototrophic mode of life may have existed then. He also points out that these micro-organisms could, of course, possibly be non-oxygen evolving flexibacteria which contain bacteriochlorophyll (Pierson and Castenholz, 1971).

If there is doubt about the presence of unicellular blue-green algae in the Early Precambrian, there can be little about their presence in the Middle Precambrian ($2{\cdot}5$–$1{\cdot}7 \times 10^9$ years ago). By then chroococcalean-like blue-green algae were abundant. For example, in the Gunflint formation in Canada, forms which appear to be unicellular sheathless, unicellular sheathed, and colonial types were present (Barghoorn and Tyler, 1965; Cloud, 1965). By the Late Precambrian ($1{\cdot}7$–$0{\cdot}6 \times 10^9$ years ago) fossil colonial blue-green algae including forms resembling *Eucapsis* were present (Fig. 18.3; Schopf, 1973) and by 1×10^9 years ago forms resembling living Entophysalidaceae and Pleurocapsaceae had appeared (Vologdin and Kordé, 1965; Schopf, 1973).

Filamentous blue-green algae, although absent from the Fig Tree series apparently had evolved soon thereafter, because as we have seen above stromatolites which probably contained filamentous algae were present $3{\cdot}0$–$2{\cdot}8 \times 10^9$ years ago. In the Middle Precambrian well-developed Nostocaceae-like and *Oscillatoria*-like forms can be recognized easily. It seems, according to Schopf (1973), that Nostocacean-like blue-green algae may have developed before *Oscillatoria*-like types but that by the Late Precambrian, the latter were the more prominent. Excellently preserved Late Precambrian fossil blue-green algae have been obtained from the Bitter Springs formation in North Central Australia (Schopf and Barghoorn, 1969). There, *Oscillatoria*-like types were abundant and well-developed and *Anabaena*-like heterocystous algae such as *Archaeonema longicellularis* were present. Less is known from the fossil record of the origins of complex filamentous types resembling living Stigonematales and Scytonemataceae as they have not been recorded from the Precambrian. It would be unwise to speculate on their origin until more information becomes available. In summation, therefore, there is fossil evidence which indicates that unicellular blue-green algae were the most primitive, that filamentous forms developed early in the evolution of blue-green algae, that Nostoceae-like filamentous types may have been more

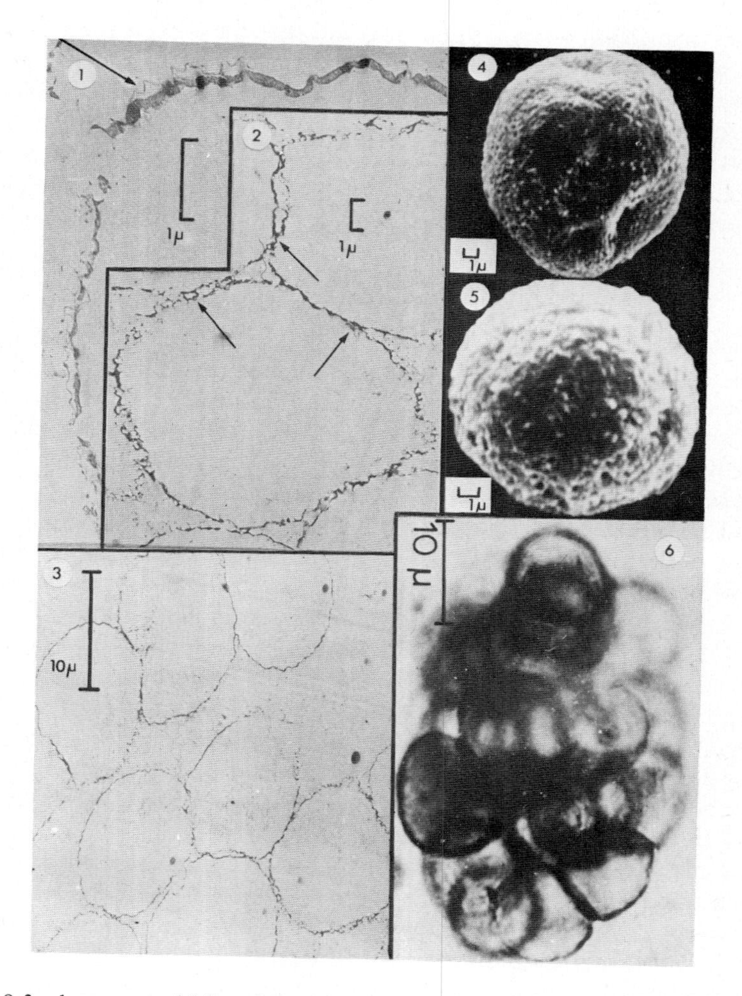

FIG. 18.3. A presumed blue-green alga, *Myxococcoides minor* Schopf, from black chert of the Late Precambrian Bitter Springs Formation of central Australia (*ca.* 825 million years old). Figs 1–3, transmission electron micrographs of ultra-thin sections of epoxy-embedded algal colonies, freed from the chert matrix by treatment with hydrofluoric acid. The arrow in Fig. 1 points to an outermost membranous cell wall layer adjacent to a relatively thick, granular, more electron-dense layer. The apparent fusion of the walls of adjacent cells indicated by the arrows in Figs 2 and 3 is probably a preservational artefact. Figs 4–5, scanning electron micrographs of isolated cells from acid-resistant residues. Fig. 6, optical photomicrograph of an algal colony in a petrographic thin section, photographed by transmitted light. By courtesy of J. W. Schopf (1970). *J. Palaeontol.* **44**, 1.

primitive than *Oscillatoria*-like filaments and that there is no good evidence yet for complex Stigonematales or Scytonemataceae until the Devonian. However, as has been emphasized by Schopf (1973), who should be consulted for further details and references, any postulated phylogeny based on fossil material is biased by many factors including the quantity, quality and geological distribution of the fossil evidence available. As more evidence becomes available the pattern suggested above may have to be modified.

Geochemical and biochemical evolution

Associated with morphological fossils just described are chemical fossils such as the hydrocarbons prytane and pristane and L-amino acids. These are generally believed to be of biological origin rather than the result of secondary contamination of the rocks and the range found in Precambrian sediments (Gelpi *et al.*, 1970) is similar to that found in contemporary blue-green algae. Analysis by mass spectrometry of the ratio of the isotopes ^{12}C and ^{13}C in such organic material shows a slightly greater amount of ^{12}C than in the parent rock. This is an indication of the occurrence of some such biological process as photosynthesis, which discriminates in favour of the lighter isotope, about $2 \cdot 5 \times 10^9$ years ago (Echlin, 1966).

Thus there is geochemical evidence which supports the view that photosynthetic micro-organisms had certainly evolved by the Middle Precambrian.

It is generally supposed, from geochemical evidence, that the atmosphere of the primitive Earth was reducing, with substances such as water, hydrogen, methane, ammonia and some hydrogen sulphide present, but with no carbon dioxide or oxygen. Oparin (1957) suggested that simple organic compounds such as hydrocarbons first arose on the Earth by reactions between water and carbides in the hot rocks, and that these hydrocarbons then combined with ammonia and water to give further organic compounds. Miller (1953) demonstrated experimentally that after passing an electrical discharge through a mixture of methane, ammonia, hydrogen and water, organic acids, traces of hydrogen cyanide and the amino acids glycine and aspartic acid could be recovered. Oparin further drew attention to the fact that when certain substances of high molecular weight, such as proteins, are mixed in solutions they sometimes form structures called *coacervates*, i.e. microscopic droplets which develop a skin composed of some of the larger molecules. These droplets absorb water and divide and structures such as these may have played an important part in the development of life. It is supposed that by a process of chemical survival of the fittest, such substances or structures gave rise to more complicated self-replicating units which formed the basis of the first living systems. It is implicit in this theory that these organisms were not photosynthetic but heterotrophic, deriving their energy by anaerobic conversions of pre-existing organic substances (see Oparin, 1957).

The formation of organic substances would, however, decrease with time as hydrogen continued to escape from the earth's gravitational field and as the heterotrophic micro-organisms used up the supply of simple organic materials. At the point when the atmosphere lost its excess of free hydrogen, oxygen would begin to accumulate through the decomposition of water by ultraviolet light. Ozone, derived from oxygen, absorbs ultraviolet rays strongly so that the process would be self-regulating around an oxygen concentration of well below 1%, but the amount of ultraviolet radiation reaching the surface of the Earth and available for producing organic compounds would be decreased.

Pigments capable of initiating chemical reactions after excitation with visible radiation may then have become important. Among these, porphyrins, which are easily formed from the simple organic substances likely to have been present and which are excellent photosensitizers in laboratory experiments, probably played a key role. Most of the photochemical reactions which occurred would have taken place with an overall loss in potential chemical energy and the first crucial step in the evolution of photosynthesis was probably the appearance of a magnesium–porphyrin derivative of the chlorophyll type, which, in a suitable environment, could promote a photochemical reaction which was not followed by retrograde back-reactions and which accumulated potential chemical energy in stable compounds.

The complicated photosynthetic system found in modern plants most probably evolved in stages and it is likely that the first photosynthetic organisms possessed only a single photosystem, capable of producing high energy phosphate by means of radiant energy, that is, by a process akin to cyclic photophosphorylation. This could enable the photoassimilation of preformed organic substances present in the environment to take place, in much the same way as in present day photosynthetic bacteria and algae. There would be essentially no carbon dioxide fixation and, of course, oxygen would not be evolved. It is unlikely that even this simplest type of photosynthetic mechanism would have evolved more than once and the great chemical similarity between chlorophyll *a*, and bacteriochlorophyll, as well as other similarities in their photosynthetic mechanisms, suggest that the blue-green algae and photosynthetic bacteria evolved from a common ancestral type. The preference of living Cyanophyceae for minimal oxygen concentrations and their great resistance to ultraviolet radiation is an indication, perhaps, that the group made its first appearance at this stage of biochemical evolution.

The paths along which further evolution of photosynthesis took place have been suggested by Olson (1970). Slower production and greater utilization of abiotically formed organic substances probably led to the success of a mutant which could use as carbon sources the more oxidized compounds, including finally carbon dioxide itself, which would be evolved in respiration. These substances could be reduced via a non-cyclic electron transport system. The necessary electron donors were probably substances such as hydrogen and hydrogen sulphide but as these were used up there may have been selection for systems using weaker reducing agents, such as ammonia and nitrite. Photosynthetic bacteria, perhaps because of shading by larger algae, adapted to utilize the longer wavelengths not absorbed by chlorophyll *a*, but thereby sacrificed about 0·4 eV of free energy stored by the primary photochemical reaction and thus prevented themselves from using water itself as an electron donor. In the blue-green algae, however, the second photosystem was finally modified to extract electrons from water and to dispose of the strongly oxidizing residue by evolution of molecular oxygen. Some representatives, however, have retained to this day the ability to use hydrogen and hydrogen sulphide as electron donors in photosynthesis.

Geochemical evidence suggest that the Earth's atmosphere became oxidizing between 1200 and 2700 million years ago (Sagan, 1967), and the theory that photosynthesis was the main source of the oxygen seems well-founded. If, as seems likely, the blue-green algae were the photosynthetic organisms responsible then they may have brought about the most profound change that the face of the Earth has ever seen. Anaerobic mechanisms are inefficient in providing energy for life processes and appear incapable of sustaining elaborate forms of life, for present day obligate anaerobes are all unicellular. Oxygen-based respiratory mechanisms are much more efficient and made possible the evolution of organisms of greater activity, size and complexity. More elaborate organisms, such as protozoa, corals and sponges, appeared in abundance just before the Cambrian era (which began about 600 million years ago) which suggests that dissolved oxygen was available then. There were perhaps two different oxygen regimes. The first was when the atmosphere contained less than 1% of oxygen and shading by ozone was still insufficient to reduce lethal ultraviolet radiation to a level at which life was possible, except under water, which absorbs these wavelengths. The second was when the concentration reached about 10% oxygen. Then, sufficient ozone would be present to reduce the ultraviolet rays to a level at which it was possible for life to emerge from water and colonize the land.

Ample supplies of ammonia were present on the primitive Earth but with the onset of oxidizing conditions this was probably converted to nitrate and subject to further conversion to elemental nitrogen by the process of denitrification. The high proportion of nitrogen-fixing species in the Cyanophyceae suggests that they may have undergone their great evolutionary development at a time when supplies of combined nitrogen were depleted (Hutchinson, quoted in Wilson, 1951). There is geological evidence (Rutten, 1962) that free oxygen was present in the middle Pre-cambrian; the presence of heterocystous algae (Schopf, 1973) is a further indication that oxygen was present then.

Phylogeny and living Cyanophyceae

Phylogeny within the Cyanophyceae is problematical. Despite the different views of systematists as to what constitutes genera, families and species, there is general agreement that the coccoid algae are the most primitive and that the filamentous forms are derived (Geitler, 1925; Tilden, 1935; Elenkin, 1936; Fritsch, 1942). Fritsch (1942) considered that among the filamentous types, heterocystous forms have been derived from non-heterocystous forms and that the Stigonematales with their heterotrichous habit are the most advanced. Tilden (1933) also thought that the heterocystous algae are the most advanced filamentous forms, and that the Rivulariaceae, the Nostocaceae and the Stigonemataceae, all of which are heterocystous families, have been derived independently from non-filamentous forms. On the other hand, Geitler (1925) and Elenkin (1936) considered that non-heterocystous algae have been derived from heterocystous forms. Desikachary (1959) has discussed in more detail the phylogenetic views of many of the early workers.

In any attempt to delineate evolutionary trends in the blue-green algae there are at least three possible approaches. Firstly, one can consider the fossil evidence, as we have already done. Secondly, one can look for morphological intermediates, that is living forms which have features intermediate between different living species. This is the approach of the classical taxonomists such as Geitler (1925, 1932, 1942), Fritsch (1942) and Elenkin (1936). Thirdly, biochemical differences and similarities between living forms can be compared.

Examples of possible links between the unicellular and colonial types and simple filamentous forms are *Chlorogloea fritschii*, possibly *Nostoc* spp., and *Johannesbaptista pellucida* (Fig. 2.2j). The latter species consists of discoid cells which originate from a single cell and which become aggregated into a single series within a narrow colourless mucilage sheath. From a type such as this, which is classified in the Pleurocapsales by Fritsch (1945), it would be easy to derive simple non-heterocystous filamentous types resembling species of *Phormidium* or *Lyngbya*. *Chlorogloea fritschii*, an alga first considered to be a member of the Chroococcales by Mitra (1950), is now known to be a *Nostoc* species because it has a filamentous stage with heterocysts as well as unicells and a pseudoparenchymatous stage (see p. 19). It thus has affinities with the Chroococcales, Pleurocapsales and Nostocales. Similarly *Nostoc muscorum* has unicellular, pseudoparenchymatous, and non-heterocystous filamentous stages as well as typical heterocystous filaments (Lazaroff and Vishniac, 1961, 1962). By considering these forms and the filamentous mutants of unicellular types considered above (p. 33) it is easy to see how particular morphological types may have been derived from one another or from a common ancestor.

Possible interrelations between filamentous types can be envisaged when one considers *Plectonema* species, which resemble the Oscillatoriaceae, in that they have no heterocysts, and heterocystous Scytonemataceae, in that they show a similar type of branching. There is also a similarity between forms such as *Nodularia* which has simple unbranched filaments with heterocysts and *Lyngbya* which has no heterocysts.

Information about biochemical differences and similarities between various blue-green algae is scanty. Edelman *et al.* (1967) determined the DNA composition of a large number (29 strains) of various blue-green algae. They established distinct ranges of mean percentage guanine + cytosine (GC) content for Oscillatoriaceae, Nostocaceae and Scytonemataceae, and discussed their data with regard to taxonomy and phylogenetic significance. Stanier *et al.* (1971) investigated in more detail the DNA base composition of unicellular algae, finding that this group is particularly heterogeneous in this respect and that DNA composition is poorly related to the existing classification of the group. The most striking biochemical difference between different morphological groups is the presence or absence and abundance of polyunsaturated fatty acids. These acids are present in most filamentous blue-green algae including *Calothrix*, *Anabaena*, *Chlorogloea* and *Oscillatoria* types but are absent or present in low concentrations only in unicellular algae such as *Synechococcus* sp., *Anacystis nidulans*, *Anacystis marina* and *Synechococcus cedrorum*. They are absent or low in what Kenyon and Stanier (1970) call *Gloeocapsa* type 1 and are absent in the two strains of the filamentous genus *Spirulina* which they examined. The presence of the polyunsaturated γ-linolenic acid has been reported, however, in *Spirulina platensis* by Nichols (1970) who draws attention to the fact that this acid is a precursor of arachdonic acid, which occurs in some higher algae. He suggests that this may be of evolutionary significance with one group of blue-green algae evolving from types such as *Anabaena* which have α-linolenic acid and others from *Spirulina* type which have γ-linolenic acid. The content of polyunsaturated acids in different types of *Gloeocapsa* is correlated with the DNA base ratios (the group which has high amounts of polyunsaturated fatty acids having values of 48 mol % guanine plus cytosine and those with low amounts having values of 35) and with features of fine structure (Stanier *et al.*, 1971). This seems to suggest that there is more than one distinct evolutionary line among the unicellular blue-green algae.

From the above summary it is clear that there are few firm pointers as to the pathways of evolution within the blue-green algae. Collectively these suggest that the unicellular forms evolved first and that evolution of filamentous forms was almost simultaneous or occurred shortly thereafter. The possible origin of filamentous forms from unicellular forms may perhaps be investigated most readily by comparing the naturally-occurring variants of extant unicellular and filamentous algae, the filamentous mutants of unicellular types noted by Ingram and Van Baalen (1970), and biochemical differences.

Symbiotic origin of chloroplasts

Limited by their prokaryotic cell structure, the evolution in morphological complexity of the blue-green algae seems to have been slow and they were soon outstripped by the eukaryotic type whose existence they had made possible. The fundamental biochemical similarities between prokaryotes and eukaryotes leave little doubt that the latter evolved from the former but the way in which this happened is debatable and may perhaps always remain so because there are no certain intermediate types between the Prokaryota and Eukaryota.

It may be that a eukaryotic type of cell arose from a prokaryote such as a blue-green alga by partitioning off of specialized regions of the protoplasm. Thus, enclosure of the other photosynthetic lamellae by the outermost one or by some other membraneous cell component might have given rise endogenously to a chloroplast. Alternatively, there is the possibility, first suggested by Mereschkowsky (1905), that the chloroplasts of plant cells were derived exogenously by blue-green algae becoming established endosymbiotically within cells of another sort. This theory has recently attracted considerable attention as new evidence in its support has been put forward. Sagan (1967) has further developed the hypothesis to explain the origin of eukaryotic cells by supposing that their three organelles—mitochondria, chloroplasts and the basal bodies of flagella—were each separately derived from once free-living Prokaryota.

As we saw in the previous chapter, blue-green algae show a remarkable propensity for symbiosis and there are undoubted cases of them being incorporated into the cells of other organisms as endosymbionts. Beyond this, there are structures such as the cyanelles of *Glaucocystis*, which can be cultured separately after extraction from the cell containing them and have some resemblance to blue-green algae, but which have become so modified, by reduction of the cell wall and in other ways, as to make identification uncertain. It is tempting to think that the end-product of such a sequence might be a photosynthetic organelle incapable of independent growth, i.e. a chloroplast.

There are, in fact, many resemblances between blue-green algae and chloroplasts. These have been considered in detail by Kenyon and Stanier (1970), Echlin (1970) and Carr and Craig (1970). Both are bounded by semi-permeable membranes about 7 nm in thickness and contain thylakoids with membranes of 7–8 nm thickness. Both contain DNA which is not organized into chromosomes but present as 2·5 nm diameter fibrils not associated with histones. The total amount of DNA present in *Anacystis nidulans* and *Anabaena variabilis* cells is of the same order as that in chloroplasts from *Euglena* and from two higher plant species, although much higher than in *Acetabularia* chloroplasts. On the detailed biochemical level, apart from the similarity in the photosynthetic pigments and products, there are also other striking correspondences. The major ribonuclear component in plant cells is of 80 S size, as in other Eukaryota, but chloroplasts contain only 70 S ribosomes—which are the size of those found in Prokaryota (see p. 61). Furthermore, the 80 S ribosomes of Eukaryota contain one molecule of 25–28 S RNA hydrogen bonded to one molecule each of 5·8 S, 18 S and 5 S RNA, whereas the 70 S ribosomes of bacteria, blue-green algae and chloroplasts contain only 3 RNA molecules; one each of 23 S, 16 S and 5 S RNA (Payne and Dyer, 1972). Chloramphenicol, which inhibits protein synthesis by prokaryotic cells, has been found to bind to ribosomes from chloroplasts and to those from *Oscillatoria* but not to eukaryotic ribosomes. Moreover, the specificity of chloramphenicol isomers is the same for prokaryotic and chloroplast ribosomes (Carr and Craig, 1970).

There are, however, some facts which are not easy to reconcile with the endosymbiotic view of the origin of the chloroplast. Although chloroplasts have all the genetic apparatus for protein synthesis it appears that they are not entirely autonomous in this respect, as perhaps might be expected if they were derived from once free-living cells, but that they depend to some extent on messenger RNA of nuclear origin. Using the inhibitor lincomycin, which is highly specific for protein synthesis by chloroplast and prokaryotic ribosomes, Ellis and Hartley (1971) have shown that, of all the chloroplast enzymes they investigated, only ribulosediphosphate carboxylase synthesis was carried out by chloroplast ribosomes. Others, even though located exclusively in chloroplasts, depend on cytoplasmic ribosomes for synthesis. However, in the course of evolution of symbiotic cells into chloroplasts, it may well be that their autonomy would be restricted and part of the genetic control of their development taken over by the nucleus of the host cell. A more weighty objection to the theory is that the DNA base ratios of chloroplasts show a wide variation from 25 to 60 mol % of guanine plus cytosine, which seems to imply the unlikely possibility that chloroplasts of different species have originated from different prokaryotes (Carr and Craig, 1970).

On balance, it seems probable, but certainly not proven, that the chloroplasts of higher plants originated endosymbiotically. It may be possible to test this hypothesis experimentally but in the words of Woodcock and Bogorad (1971), "it seems as difficult to reach a final positive decision as to do the experiment to show that blue-green algae and bacteria arose as escapees from an intracellular community". Nevertheless, the idea that the chloroplasts of higher plants are derived from free-living organisms similar to the present day blue-green algae is one which appears to illuminate hitherto dark areas of the evolutionary scene. It seems possible that the blue-green algae are something more than a lowly branch of the evolutionary tree and represent a pervading and vital component which has made possible the highest forms of plant life.

References

AHLBORN, F. (1895) Über die Wasserblüte *Byssus flos-aquae* und ihr Verhalten gegen Druck. *Verh. naturwiss. Ver. Hamb.* **3**, 25–36.

AHMAD, M. R. and WINTER, A. (1968) Studies on the hormonal relationships of algae in pure culture. 1. The effect of indole-3-acetic acid on the growth of blue-green and green algae. *Planta* **78**, 277–286.

AHMADJIAN, V. (1962) Investigations on lichen synthesis. *Am. J. Bot.* **49**, 277–283.

AHMADJIAN, V. (1963) The fungi of lichens. *Scient. Am.* **208**, 122–132.

AHMADJIAN, V. (1966) Lichens. *In* "Symbiosis" (S. M. Henry, ed.) Vol. 1, pp. 35–97. Academic Press, New York.

AHMADJIAN, V. (1967) A guide to the algae occurring as lichen symbionts: isolation, culture, cultural physiology and identification. *Phycologia* **6**, 127–160.

ALLEN, C. F., FRANKE, H. and HIRAYAMA, O. (1967) Identification of a plastoquinone and two naphthoquinones in *Anacystis nidulans* by NMR and mass spectroscopy. *Biochem. biophys. Res. Commun.* **5**, 562–568.

ALLEN, M. B. (1952) The cultivation of Myxophyceae. *Arch. Mikrobiol.* **17**, 34–53.

ALLEN, M. B. (1956) Photosynthetic nitrogen fixation by blue-green algae. *Sci. Monthly* **83**, 100–106.

ALLEN, M. B. (1963) Nitrogen fixing organisms in the sea. *In* "Symposium on Marine Microbiology" (C. H. Oppenheimer, ed.) pp. 85–92. C. C. Thomas, Springfield, Illinois.

ALLEN, M. B. and ARNON, D. I. (1955a) Studies on nitrogen-fixing blue-green algae. I. Growth and nitrogen fixation by *Anabaena cylindrica* Lemm. *Pl. Physiol., Lancaster* **30**, 366–372.

ALLEN, M. B. and ARNON, D. I. (1955b) Studies on nitrogen-fixing blue-green algae. II. The sodium requirement of *Anabaena cylindrica*. *Physiologia Pl.* **8**, 653–660.

ALLEN, M. M. (1968a) Simple conditions for growth of unicellular blue-green algae on plates. *J. Phycol.* **4**, 1–4.

ALLEN, M. M. (1968b) Photosynthetic membrane system in *Anacystis nidulans*. *J. Bact.* **96**, 836–841.

ALLEN, M. M. (1968c) Ultrastructure of the cell wall and cell division of unicellular blue-green algae. *J. Bact.* **96**, 842–852.

ALLEN, M. M. (1972) Mesosomes in blue-green algae. *Arch. Mikrobiol.* **84**, 199–206.

ALLEN, M. M. and SMITH, A. J. (1969) Nitrogen chlorosis in blue-green algae. *Arch. Mikrobiol.* **69**, 114–120.

ALLEN, M. M. and STANIER, R. Y. (1968a) Growth and division of some unicellular blue-green algae. *J. gen. Microbiol.* **51**, 199–202.

ALLEN, M. M. and STANIER, R. Y. (1968b) Selective isolation of blue-green algae from water and soil. *J. gen. Microbiol.* **51**, 203–209.

ALLISON, F. E. and MORRIS, H. J. (1930) Nitrogen fixation by blue-green algae. *Science, N.Y.* **71**, 221–223.

ALLISON, F. E., HOOVER, S. R. and MORRIS, H. J. (1937) Physiological studies with the nitrogen fixing alga *Nostoc muscorum*. *Bot. Gaz.* **98**, 433–463.

ALLISON, R. K., SKIPPER, H. E. REID, M. R., SHORT, W. A. and HOGAN, G. L. (1953) Studies on the photosynthetic reaction. I. The assimilation of acetate by *Nostoc muscorum*. *J. biol. Chem.* **204**, 197–205.

ALLISON, R. K., SKIPPER, H. E., REID, M. R., SHORT, W. A. and HOGAN, G. L. (1954) Studies on the photosynthetic reaction. II. Sodium formate and urea feeding experiments with *Nostoc muscorum*. *Pl. Physiol., Lancaster* **29**, 164–168.

ALLSOPP, A. (1968) Germination of hormocysts of *Scytonema javanicum* and the function of blue-green algal heterocysts. *Nature, Lond.* **220**, 810.

ALLSOPP, A. (1969) Phylogenetic relationships of the procaryota and the origin of the eucaryotic cell. *New Phytol.* **68**, 591–612.

ALMODOVAR, L. R. (1963) The fresh-water and terrestrial Cyanophyta of Puerto Rico. *Nova Hedwigia* **5**, 429–435.

ANAND, P. L. (1937) An ecological study of the algae of the British chalk-cliffs. I. *J. Ecol.* **25**, 153–188.

ANKEL, H. and TISCHER, R. G. (1969) UDP–D–glucoronate 4'-epimerase in blue-green algae. *Biochim. biophys. Acta* **178**, 415–419.

ARNOLD, W. and OPPENHEIMER, J. R. (1950) Internal conversion in the photosynthetic mechanism of blue-green algae. *J. gen. Physiol.* **33**, 423–435.

ARNON, D. I. (1958) The role of micronutrients in plant nutrition with special reference to photosynthesis and nitrogen assimilation. *In* "Trace Elements" (C. A. Lamb, O. G. Bentley and J. M. Beattie, eds.) pp. 1–32. Academic Press, New York.

ARNON, D. I. (1965) Ferredoxin and photosynthesis. *Science, N.Y.* **149**, 1460–1470.

ASAHINA, Y. (1937) Über den taxonomischen Wert der Flechtenstoffe. *Bot. Mag., Tokyo* **51**, 759–764.

BAAS-BECKING, L. G. M. and WOOD, E. J. F. (1965) Biological processes in the estuarine environment. II. Ecology of the sulphur cycle. *Proc. Acad. Sci. Amst. B* **58**, 173–181.

BAHAL, M. (1969) Effect of 2,4–dinitrophenol on heterocyst development in *Anabaena ambigua*. *Phykos* **8**, 11–17.

BAHAL, M. and TALPASAYI, E. R. S. (1970a) Heterocyst differentiation in *Anabaena ambigua* Rao. III. Effect of thiol inhibitors. *Indian Biologist* **2**, 56–65.

BAHAL, M. and TALPASAYI, E. R. S. (1970b) Effect of azide and glycine on heterocyst development in *Anabaena ambigua* Rao. *Indian J. exp. Biol.* **8**, 237–238.

BAILEY-WATTS, A. E., BINDLOSS, M. E. and BELCHER, J. H. (1968) Freshwater primary production by a blue-green alga of bacterial size. *Nature, Lond.* **220**, 1344–1345.

BAKER, A. L., BROOK, A. J. and KLEMER, A. R. (1969) Some photosynthetic characteristics of a naturally occurring population of *Oscillatoria agardhii* Gomont. *Limnol. Oceanogr.* **14**, 327–333.

BARGHOORN, E. S. and TYLER, S. A. (1965) Microorganisms from the Gunflint chert. *Science, N.Y.* **147**, 563–577.

BATTERTON, J. C. and VAN BAALEN, C. (1968) Phosphorus deficiency and phosphate uptake in the blue-green alga *Anacystis nidulans. Can. J. Microbiol.* **14**, 341–348.

BATTERTON, J. C. and VAN BAALEN, C. (1971) Growth responses of blue-green algae to sodium chloride concentration. *Arch. Mikrobiol.* **76**, 151–165.

BAZIN, M. J. (1968) Sexuality in a blue-green alga: genetic recombination in *Anacystis nidulans. Nature, Lond.* **218**, 282–283.

BECK, S. (1963) Licht und elektronenmikroskopische Untersuchungen an einer sporenbildender Cyanophycee aus dem Formenkreis von *Pleurocapsa fuliginosa* Hauck. *Flora, Jena* **153**, 194–216.

BECK, T. (1968) "Mikrobiologie des Bodens". BLV, München, Basel and Wien.

BEHRE, K. (1953) Cyanophyceen über rieselten Felsen, von Herm Vaillant vornehmlich, in Algerian gesammelt. *Bull. Soc. Hist. Afr. Nord* **44**, 209–227.

BEHRE, K. and SCHWABE, G. H. (1970) Auf Surtsey Island in Sommer 1968 nachgewiesene nicht marine Algen. *Schr. Naturwiss. Ver. Schlesw.-Holst. Sonderband*, 31–100.

BEIJERINCK, M. W. (1901) Über oligonitrophile Mikroben. *Zentbl. Bakt. ParasitKde* (*Abt II*) **7**, 561–582.

BERGERON, J. A. (1963) Studies on the localization, physicochemical properties and function of phycocyanin in *Anacystis nidulans. In* "Photosynthetic Mechanisms of Green Plants" (B. Kok and A. T. Jagendorf, eds.) pp. 527–536. Natn. Acad. Sci. U.S.A.—Natn. Res. Council Publ.

BERGERSEN, F. J. (1970) The quantitative relationship between nitrogen fixation and the acetylene reduction assay. *Aust. J. biol. Sci.* **23**, 1015–1025.

BERGERSEN, F. J., KENNEDY, G. S. and WITTMANN, W. (1965) Nitrogen fixation in the coralloid roots of *Macrozamia communis. Aust. J. biol. Sci.* **18**, 1135–1142.

BERNARD, F. and LEÇAL, J. (1960) Plancton unicellulaire recolté dans l'ocean Indien par le Charcot (1950) et le Norsel (1955–1956). *Bull. Inst. océanogr. Monaco.* **1166**, 1–59.

BERNS, D. S. (1967) Immunochemistry of biliproteins. *Pl. Physiol., Lancaster* **42**, 1569–1586.

BERNS, D. S. and EDWARDS, M. R. (1965) Electron micrographic investigations of C-phycocyanin. *Archs Biochem. Biophys.* **110**, 511–516.

BHASKARAN, S. and VENKATARAMAN, G. S. (1958) Occurrence of blue-green algae in the nodules of *Trifolium alexandrinum*. *Nature, Lond.* **181**, 277–278.

BIENZENO, C. B. and KOCH, J. J. (1938) The buckling of a cylindrical tank of variable thickness under external pressure. *In* Proc. 5th Internat. Congr. Appl. Mechanics.

BIGGINS, J. (1967a) Preparation of metabolically active protoplasts from the blue-green alga, *Phormidium luridum*. *Pl. Physiol., Lancaster* **42**, 1442–1446.

BIGGINS, J. (1967b) Photosynthetic reactions by lysed protoplasts and particle preparations from the blue-green alga, *Phormidium luridum*. *Pl. Physiol., Lancaster* **42**, 1447–1456.

BIGGINS, J. (1969) Respiration in blue-green algae. *J. Bact.* **99**, 570–575.

BILLAUD, V. A. (1967) Aspects of the nitrogen nutrition of some naturally occurring populations of blue-green algae. *In* "Environmental Requirements of Blue-Green Algae" (A. F. Bartsch, ed.) pp. 35–53. U.S. Dept. Interior, Federal Water Control Admin., Corvallis, Oregon.

BIRDSEY, E. C. and LYNCH, V. H. (1962) Utilization of nitrogen compounds by unicellular algae. *Science, N.Y.* **137**, 763–764.

BISALPUTRA, T., BROWN, D. L. and WEIER, T. E. (1969) Possible respiratory sites in a blue-green alga *Nostoc sphaericum* as demonstrated by potassium tellurite and tetranitro-blue tetrazolium reduction. *J. ultrastruct. Res.* **27**, 182–197.

BISHOP, C. T., ADAMS, G. A. and HUGHES, E. O. (1954) A polysaccharide from the blue-green alga, *Anabaena cylindrica*. *Can. J. Chem.* **32**, 999–1004.

BISHOP, C. T., ANET, E. F. L. J. and GORHAM, P. R. (1959) Isolation and identification of the fast-death factor in *Microcystis aeruginosa* NRC–1. *Can. J. Biochem. Physiol.* **37**, 453–471.

BISHOP, N. I. (1966) Partial reactions of photosynthesis and photoreduction. *A. Rev. Pl. Physiol.* **17**, 185–208.

BISSET, K. (1973) This "prokaryotic-eukaryotic" business. *New Scientist* **57**, 296–298.

BLUM, J. L. (1963) The influence of water currents on the life functions of algae. *Ann. N.Y. Acad. Sci.* **108**, 353–358.

BOARDMAN, N. K. (1970) Physical separation of the photosynthetic photochemical systems. *A. Rev. Pl. Physiol.* **21**, 115–140.

BOLD, H. C. (1942) The cultivation of algae. *Bot. Rev.* **8**, 70–138.

BOLYSHEV, N. N. (1952) The origin and evolution of the soils of takyrs. *Pochovovedenie* **21**, 403–417.

BOND, G. (1959) Nitrogen metabolism in plants. The incidence and importance of biological fixation of nitrogen. *Advmt Sci., Lond.* **15**, 382–386.

BOND, G. (1967) Fixation of nitrogen by higher plants other than legumes. *A. Rev. Pl. Physiol.* **18**, 107–126.

BOND, G. and SCOTT, B. D. (1955) An examination of some symbiotic sytems for fixation of nitrogen. *Ann. Bot.* **19**, 67–77.

BONE, D. H. (1971a) Relationship between phosphates and alkaline phosphatase of *Anabaena flos-aquae* in continuous culture. *Arch. Mikrobiol.* **80**, 147–153.

BONE, D. H. (1971b) Kinetics of synthesis of nitrogenase in batch and continuous culture of *Anabaena flos-aquae*. *Arch. Mikrobiol.* **80**, 242–251.

BOOTH, W. E. (1941) Algae as pioneers in plant succession and their importance in erosion control. *Ecology* **22**, 38–46.

BORESCH, K. (1913) Die Färbung von Cyanophyceen und Chlorophyceen in ihrer Abhängigkeit vom Stickstoffgehalt des Substrates. *Jb. wiss. Bot.* **52**, 145–185.

BORESCH, K. (1919) Über die Einwirkung färbigen Lichtes auf die Färbung von Cyanophyceen. *Ber. dt. bot. Ges.* **37**, 25–39.

BORNET, E. and FLAHAULT, C. (1886–1888) "Revision des Nostocacees Heterocystees". Reprint 1959. J. Cramer, Weinheim.

BORNET, E. and THURET, G. (1880) "Notes Algologiques". II. G. Masson, Paris.

BORTELS, H. (1940) Über die Bedeutung des Molybdäns für stickstoffbindende Nostocaceen. *Arch. Mikrobiol.* **11**, 155–186.

BORZI, A. (1878) Nachträge zur Morphologie und Biologie der Nostochaceen. *Flora, Jena* **61**, 465–471.

BORZI, A. (1916) Studi sulle Mixoficee. II. Stigonemaceae. *Nuovo G. bot. ital.* **23**, 559–588.

BOTHE, H. (1970) Photosynthetische Stickstoffixierung mit einem zellfreien Extrakt aus der Blaualge *Anabaena cylindrica*. *Ber. dt. bot. Ges.* **83**, 421–432.

BOUILHAC, R. (1897) Sur le culture du *Nostoc punctiforme* en présence de glucose. *C. r. hebd. Séanc. Acad. Sci., Paris* **125**, 880–882.

BOUILHAC, R. (1901) Sur la végétation du *Nostoc punctiforme* en présence de differénts hydrates de carbone. *C. r. hebd. Séanc. Acad. Sci., Paris* **133**, 55–57.

BOWEN, C. C. and JENSEN, T. E. (1965) Fine structure of gas vacuoles in blue-green algae. *Am. J. Bot.* **52**, 641.

BOWMAN, T. E. and LANCASTER, L. J. (1965) A bloom of the planktonic blue-green alga *Trichodesmium erythraeum* in the Tonga Islands. *Limnol. Oceanogr.* **10**, 291–293.

BOWYER, J. W. and SKERMAN, V. B. D. (1968) Production of axenic cultures of soil-borne and endophytic blue-green algae. *J. gen. Microbiol.* **54**, 299–306.

BRADLEY, S. and CARR, N. G. (1971) The absence of a functional photosystem II in heterocysts of *Anabaena cylindrica*. *J. gen. Microbiol.* **68**, xiii–xiv.

BRAND, F. (1903). Morphologisch-physiologische Betrachtungen über Cyanophyceen. *Beih. bot. Zbl.* **15**, 31–64.

BRANNON, M. S. (1945) Factors affecting growth and distribution of Myxophyceae in Florida. *Proc. Florida Acad. Sci.* **8**, 296–303.

BRANTON, D. (1966) Fracture faces of frozen membranes. *Proc. natn. Acad. Sci. U.S.A.* **55**, 1048–1056.

BREMNER, J. M., CHENG, H. H. and EDWARDS, A. P. (1965) Assumptions and errors in nitrogen–15 tracer research. *Rep. FAO/IAE Technical Meeting, Brunswick-Volkenrode*, pp. 429–432. Pergamon Press, Oxford.

BREZONIK, P. L. and HARPER, C. L. (1969) Nitrogen fixation in some anoxic lacustrine environments. *Science, N.Y.* **164**, 1277–1279.

BRISTOL-ROACH, B. M. (1927) On the algae of some normal English soils. *J. agric. Sci., Camb.* **17**, 563–588.

Brock, M. L., Wiegert, R. G. and Brock, T. D. (1969) Feeding by *Paracoenia* and *Ephydra* (Diptera: Ephydridae) on the microorganisms of hot springs. *Ecology* **50**, 192–200.

Brock, T. D. (1967a) Relationship between standing crop and primary productivity along a hot spring thermal gradient. *Ecology* **48**, 566–571.

Brock, T. D. (1967b) Life at high temperatures. *Science, N.Y.* **158**, 1012–1019.

Brock, T. D. (1968) Taxonomic confusion concerning certain filamentous blue-green algae. *J. Phycol.* **4**, 178–179.

Brock, T. D. (1970a) High temperature systems. *A. Rev. Ecol. System.* **1**, 191–220.

Brock, T. D. (1970b) Photosynthesis by algal epiphytes of *Utricularia* in Everglades National Park. *Bull. mar. Sci.* **20**, 952–956.

Brock, T. D. and Brock, M. L. (1968) Measurement of steady-rate growth of a thermophilic alga directly in Nature. *J. Bact.* **95**, 811–815.

Brody, M. and Vatter, A. E. (1959) Observations on cellular structures of *Porphyridium cruentum*. *J. biophys. biochem. Cytol.* **5**, 289–294.

Brook, A. J. (1959) The waterbloom problem. *Proc. Soc. Water Treat. Exam.* **8**, 133–137.

Brook, A. J., Baker, A. L. and Klemer, A. R. (1971) The use of turbidimetry in studies of the population dynamics of phytoplankton populations with special reference to *Oscillatoria agardhii* var. *isothrix*. *Verh. int. Ver. theor. angew. Limnol.* **19**, 244–252.

Brown, A. H. (1953) The effect of light on respiration using isotopically enriched oxygen. *Am. J. Bot.* **40**, 719–729.

Brown, A. H. and Webster, G. C. (1953) The influence of light on the rate of respiration of the blue-green alga *Anabaena*. *Am. J. Bot.* **40**, 753–758.

Brown, D. L. and Bisalputra, T. (1969) Fine structure of the blue-green alga *Nostoc sphaericum:* the structured granule. *Phycologia* **8**, 119–126.

Brown, F., Cuthbertson, W. F. J. and Fogg, G. E. (1956) Vitamin B_{12} activity of *Chlorella vulgaris* Beij. and *Anabaena cylindrica* Lemm. *Nature, Lond.* **177**, 188.

Brownell, P. F. and Nicholas, D. J. D. (1967) Some effects of sodium no nitrate assimilation and N_2 fixation in *Anabaena cylindrica*. *Pl. Physiol., Lancaster* **42**, 915–921.

Brunthaler, J. (1909) Der Einfluss äusserer Faktoren auf *Gloeothece rupestris* (Lyngb.) Born. *Sber. Akad. Wiss. Wien, Mat.-Nat. Kl. I.* **118**, 501–573.

Bryce, T. A., Welti, D., Walsby, A. E. and Nichols, B. W. (1972) Monohexoside derivatives of long-chain polyhydroxyl alcohols; a novel class of glycolipid specific to heterocystous algae. *Phytochemistry* **11**, 295–302.

Buchanan, B. B., Bachofen, R. and Arnon, D. I. (1964) Role of ferredoxin in the reductive assimilation of CO_2 and acetate by extracts of the photosynthetic bacterium *Chromatium*. *Proc. natn. Acad. Sci. U.S.A.* **52**, 839–847.

Büchel, K. H., Röchling, H., Baedelt, H., Gerhardt, B. and Trebst, A. (1967) Hemmung der Photosynthese in *Anacystis* durch Alkylbenzimidazole. *Z. Naturf.* **22b**, 535–537.

BUCKLAND, B. and WALSBY, A. E. (1971) A study of the strength and stability of gas vesicles isolated from a blue-green alga. *Arch. Mikrobiol.* **79**, 327–337.

BULEN, W. A. and LECOMTE, J. R. (1966) The nitrogenase system from *Azotobacter:* two enzyme requirements for N_2 reduction, ATP dependent H_2 evolution and ATP hydrolysis. *Proc. natn. Acad. Sci. U.S.A.* **56**, 979–986.

BÜNNING, E. and HERDTLE, H. (1946) Physiologische Untersuchungen an thermophilen Blaualgen. *Z. Naturf.* **1**, 93–99.

BUNT, J. S. (1961a) Nitrogen-fixing blue-green algae in Australian rice soils. *Nature, Lond.* **192**, 479–480.

BUNT, J. S. (1961b) Blue-green algae—Growth. *Nature, Lond.* **192**, 1274–1275.

BUNT, J. S., COOKSEY, K. E., HEEB, M. A., LEE, C. C. and TAYLOR, B. F. (1970) Assay of algal nitrogen fixation in the marine subtropics by acetylene reduction. *Nature, Lond.* **227**, 1163–1164.

BURKHOLDER, P. R. (1933) Movement in the Cyanophyceae: Effect of pH upon the movement of *Oscillatoria. J. gen. Physiol.* **16**, 875–881.

BURKHOLDER, P. R. (1934) Movement in the Cyanophyceae. *Q. Rev. Biol.* **9**, 438–459.

BURKHOLDER, P. R. (1963) Some nutritional relationships among microbes of sea sediments and waters. *In* "Symposium on Marine Microbiology" (C. H. Oppenheimer, ed.) pp. 133–150. C. C. Thomas, Springfield, Illinois.

BURNS, R. C., HOLSTEN, R. D. and HARDY, R. W. F. (1970) Isolation by crystallization of the Mo–Fe protein of *Azotobacter* nitrogenase. *Biochem. biophys. Res. Commun.* **39**, 90–99.

BURRIS, R. H. (1969) Progress in the biochemistry of nitrogen fixation. *Proc. R. Soc.* B. **172**, 317–437.

BURRIS, R. H., EPPLING, F. J., WAHLIN, H. B. and WILSON, P. W. (1943) Detection of nitrogen fixation with isotopic nitrogen. *J. biol. Chem.* **148**, 349–357.

BURRIS, R. H. and WILSON, P. W. (1957) Methods for measurement of nitrogen fixation. *In* "Methods in Enzymology" (S. P. Colowick and N. O. Kaplan, eds.) Vol. 4 pp. 355–366. Academic Press, New York.

CALVIN, M. and LYNCH, V. (1952) Grana-like structure of *Synechococcus cedrorum. Nature, Lond.* **169**, 455–456.

CAMERON, R. E. (1960) Communities of soil algae occurring in the Sonoran Desert in Arizona. *J. Arizona Acad. Sci.* **3**, 85–88.

CAMERON, R. E. (1962) Species of *Nostoc* Vaucher occurring in the Sonoran Desert in Arizona. *Trans. Am. microsc. Soc.* **81**, 379–384.

CAMERON, R. E. and BLANK, G. B. (1966) Desert algae: soil crusts and diaphanous substrata as algal habitats. *Jet Propulsion Laboratory, Techn. Rep. no. 32–971*, pp. 1–41, Pasadena, California.

CAMERON, R. E. and FULLER, W. H. (1960) Nitrogen fixation by some soil algae in Arizona soils. *Soil Sci. Soc. Am. Proc.* **24**, 353–356.

CAMERON, R. E., MORELLI, F. A. and BLANK, G. B. (1965) Soil studies—desert microflora. VI. Abundance of microflora in an area of soil at White Mountain Range, California. *Jet Propulsion Laboratory, Space Programs Summary No. 37–32*, Vol. 4.

CANABAEUS, L. (1929) Über die Heterocysten und Gasvakuolen der Blaualgen und ihre Beziehung zueinander. *In* "Pflanzenforschung" (R. Kolkowitz, ed.) Vol. **13** pp. 1–48. Fischer, Jena.

CANTER, H. M. (1950) Fungal parasites of the phytoplankton. I. Studies on British Chytrids, X. *Ann. Bot.* **14**, 263–289.

CANTER, H. M. (1951) Fungal parasites of the phytoplankton. II. Studies on British Chytrids, XII. *Ann. Bot.* **15**, 129–156.

CANTER, H. M. (1954) Fungal parasites of the phytoplankton. III. Studies on British Chytrids, XIII. *Trans. Br. mycol. Soc.* **37**, 111–133.

CANTER, H. M. (1963) Concerning *Chytridium cornutum* Braun. *Trans. Br. mycol. Soc.* **46**, 208–212.

CANTER, H. M. (1965) Studies on British Chytrids, XXIV. *Entophlyctis molesta* sp. nov., parasitic on *Stylosphaeridium stipitatum* (Bachm.) Geitler et Gimesi from the plankton. *Jl R. microsc. Soc.* **84**, 549–557.

CANTER, H. M. (1972) A guide to the fungi occurring on planktonic blue-green algae. *In* "Taxonomy and Biology of Blue-green Algae" (T. V. Desikachary, ed.) pp. 145–158. University of Madras.

CANTER, H. M. and LUND, J. W. G. (1968) The importance of Protozoa in controlling the abundance of planktonic algae in lakes. *Proc. Linn. Soc. Lond.* **179**, 203–219.

CANTER, H. M. and WILLOUGHBY, L. G. (1964) A parasitic *Blastocladiella* from Windermere plankton. *Jl R. microsc. Soc.* **83**, 365–372.

CARR, N. G. (1966) The occurrence of poly–β–hydroxybutyrate in the blue-green alga, *Chlorogloea fritschii. Biochim. biophys. Acta* **120**, 308–310.

CARR, N. G. (1967) Aspects of enzymic control in blue-green algae. *Abstr. 4th Meet. Fedn Europ. Biochem. Soc.* p. 89. Universitetsforlaget, Oslo.

CARR, N. G. (1969) Growth of phototrophic bacteria and blue-green algae. *In* "Methods in Microbiology" (J. R. Norris and D. W. Ribbons, eds.) Vol. 3B, pp. 53–77. Academic Press, London and New York.

CARR, N. G. and CRAIG, I. W. (1970) The relationship between bacteria, blue-green algae and chloroplasts. *In* "Phytochemical Phylogeny" (J. B. Harbourne, ed.) pp. 119–143. Academic Press, London and New York.

CARR, N. G., EXELL, G., FLYNN, V., HALLAWAY, M. and TALUKDAR, S. (1967) Minor quinones of some Myxophyceae. *Archs Biochem. Biophys.* **120**, 503–507.

CARR, N. G. and HALLAWAY, M. (1965) Reduction of phenolindo–2,6–dichlorophenol in dark and light by the blue-green alga, *Anabaena variabilis. J. gen. Microbiol.* **39**, 335–344.

CARR, N. G. and HALLAWAY, M. (1966) Quinones of some blue-green algae. *In* "Biochemistry of Chloroplasts" (T. W. Goodwin, ed.) Vol. 1, pp. 159–163. Academic Press, New York.

CARR, N. G., HOOD, W. and PEARCE, J. (1969) Control and intermediary metabolism of blue-green algae. *In* "Progress in Photosynthesis Research" (H. Metzner, ed.) Vol. 3, pp. 1565–1569. Tübingen.

CARR, N. G. and PEARCE, J. (1966) Photoheterotrophism in blue-green algae. *Biochem. J.* **99**, 28–29P.

CARTER, H. J. (1856) Notes on the freshwater Infusoria of the Island of Bombay. I. Organisation. *Ann. Mag. nat. Hist.* (Ser. II) **18**, 115–132.

CARTER, N. (1932) A comparative study of the alga flora of two salt marshes. Part I. *J. Ecol.* **20**, 341–370.

CARTER, N. (1933) A comparative study of the alga flora of two salt marshes Part II. *J. Ecol.* **21**, 128–208.

CARTER, P. W., HEILBRON, I. M. and LYTHGOE, B. (1939) The lipoproteins and sterols of the algal classes. *Proc. R. Soc.* B **128**, 82–109.

CASTENHOLZ, R. W. (1967a) Environmental requirements of thermophilic blue-green algae. *In* "Environmental Requirements of Blue-green Algae" (A. F. Bartsch, ed.) pp. 55–79. U.S. Dept. Interior, Federal Water Poll. Control Admin., Corvallis, Oregon.

CASTENHOLZ, R. W. (1967b) Aggregation in a thermophilic *Oscillatoria. Nature, Lond.* **215**, 1285–1286.

CASTENHOLZ, R. W. (1968) The behavior of *Oscillatoria terebriformis* in hot springs. *J. Phycol.* **4**, 132–139.

CASTENHOLZ, R. W. (1969) Thermophilic blue-green algae and the thermal environment. *Bact. Rev.* **33**, 476–504.

CENGIA-SAMBO, M. (1926) Ancora della pdisimbiosi nei licheni ad algha cianoticee. I. Betteri simbionti. *Atti Soc. Ital. Nat. Museo Storia Nat. Milano* **64**, 191–195.

CHAO, L. and BOWEN, C. C. (1971) Purification and properties of glycogen isolated from a blue-green alga, *Nostoc muscorum. J. Bact.* **105**, 331–338.

CHAPMAN, D. J., COLE, W. J. and SIEGELMAN, H. W. (1967) The structure of phycoerythrobilin. *J. Am. Chem. Soc.* **89**, 5976–5977.

CHAPMAN, J. A. and SALTON, M. R. J. (1962) A study of several blue-green algae in the electron microscope. *Arch. Mikrobiol.* **44**, 311–322.

CHAPMAN, V. J. (1962) "The Algae". Macmillan, London.

CHATTON, E. (1937) "Titre et travaux scientifiques". Séte, Sottano.

CHENIAE, G. M. and MARTIN, I. E. (1969) Photoreactivation of manganese catalyst in photosynthetic oxygen evolution. *Pl. Physiol., Lancaster* **44**, 351–360.

CHEUNG, W. Y. and GIBBS, M. (1966) Dark and photometabolism of sugars by a blue-green alga, *Tolypothrix tenuis. Pl. Physiol., Lancaster* **41**, 731–737.

CHODAT, R. (1898) Etudes de la biologie lacustre. *Bull. Herb. Boissier* **6**, 49–77.

CHRISTENSEN, T. (1964) The gross classification of the algae. *In* "Algae and Man" (D. F. Jackson, ed.) pp. 59–64. Plenum Press, New York.

CLARK, R. L. and JENSEN, T. E. (1969) Ultrastructure of akinete development in a blue-green alga, *Cylindrospermum* sp. *Cytologia* **34**, 439–448.

CLOUD, P. E. (1965) Significance of the Gunflint (Precambrian) microflora. *Science, N.Y.* **148**, 27–35.

COBB, H. D. and MYERS, J. (1962) Experimentally produced variation in the C/N ratio of nitrogen-fixing blue-green algae. *Pl. Physiol., Lancaster*, Suppl. **37**, VII.

COBB, H. D. and MYERS, J. (1964) Comparative studies of nitrogen fixation and photosynthesis in *Anabaena cylindrica*. *Am. J. Bot.* **51**, 753–762.

COHEN-BAZIRE, G., KUNISAWA, R. and PFENNIG, N. (1969) Comparative study of the structure of gas vacuoles. *J. Bact.* **100**, 1049–1061.

COHEN-BAZIRE, G. and LEFORT-TRAN, M. (1970) Fixation of phycobiliproteins to photosynthetic membranes by glutaraldehyde. *Arch. Mikrobiol.* **71**, 245–257.

COHN, F. (1853) Untersuchungen über die Entwicklungsgeschichte mikroskopischer Algen und Pilze. *Nova Acta Acad. Caesar. Leop. Carol.* **24**, 103–256.

COHN, F. (1871–72) Grundzüge einer neuen natürlichen Anordnung der Kryptogamischen Pflanzen. *Jb. schles. Ges. vaterl. Kult.* **49**, 83–89.

COLLYER, D. M. and FOGG, G. E. (1955) Studies on fat accumulation by algae. *J. exp. Bot.* **6**, 256–275.

COPELAND, J. J. (1936) Yellowstone thermal Myxophyceae. *Ann. N.Y. Acad. Sci.* **36**, 1–229.

CORRENS, C. (1897) Über die Membran und die Bewegung der Oscillarien. *Ber. dt. bot. Ges.* **15**, 139–148.

COSTERTON, J. W. F. (1960) Cytological studies on the Schizophyceae and related organisms. Ph.D. Thesis, University of Western Ontario, London, Canada.

COSTERTON, J. W. F., MURRAY, R. G. E. and ROBINOW, C. F. (1961) Observations on the motility and the structure of *Vitreoscilla*. *Can. J. Microbiol.* **7**, 329–339.

COUPIN, H. (1923) Quelques rémarques sur la locomotion des Oscillaries. *C. r. hebd. Séanc. Acad. Sci., Paris* **176**, 1491–1493.

COX, R. M. (1966) Physiological studies on nitrogen fixation in the blue-green alga *Anabaena cylindrica*. *Arch. Mikrobiol.* **53**, 263–276.

COX, R. M. and FAY, P. (1967) Nitrogen fixation and pyruvate metabolism in cell-free preparations of *Anabaena cylindrica*. *Arch. Mikrobiol.* **58**, 357–365.

COX, R. M. and FAY, P. (1969) Special aspects of nitrogen fixation by blue-green algae. *Proc. R. Soc.* B **172**, 357–366.

COX, R. M., FAY, P. and FOGG, G. E. (1964) Nitrogen fixation and photosynthesis in a subcellular fraction of the blue-green alga *Anabaena cylindrica*. *Biochim. biophys. Acta* **88**, 208–210.

CRAIG, I. W. and CARR, N. G. (1967) Ribosomes from the blue-green alga *Anabaena variabilis*. *Biochem. J.* **103**, 64P.

CRAIG, I. W. and CARR, N. G. (1968) Ribosomes from the blue-green alga *Anabaena variabilis*. *Arch. Mikrobiol.* **62**, 167–177.

CRAIG, I. W., LEACH, C. K. and CARR, N. G. (1969) Studies with deoxyribonucleic acid from blue-green algae. *Arch. Mikrobiol.* **65**, 218–227.

CRESPI, H. L., MANDEVILLE, S. E. and KATZ, J. J. (1962) The action of lysozyme on several blue-green algae. *Biochem. biophys. Res. Commun.* **9**, 569–573.

CROFT, W. N. and GEORGE, E. A. (1959) Blue-green algae from the Middie Devonian of Rhynie, Aberdeenshire. *Bull. Br. Mus. nat. Hist., Geology* **3**, 339–353.

CROSSETT, R. N., DREW, E. A. and LARKUM, A. W. D. (1965) Chromatic adaptation in benthic marine algae. *Nature, Lond.* **207**, 547–548.

DAFT, M. J., BEGG, J. and STEWART, W. D. P. (1970) A virus of blue-green algae from freshwater habitats in Scotland. *New Phytol.* **69**, 1029–1038.

DAFT, M. J. and STEWART, W. D. P. (1971) Bacterial pathogens of freshwater blue-green algae. *New Phytol.* **70**, 819–829.

D'ARCY THOMPSON, W. (1917) "On Growth and Form". University Press, Cambridge.

DA SILVA, E. J. and JENSEN, A. (1971) Content of α–tocopherol in some blue-green algae. *Biochim. biophys. Acta* **239**, 345–347.

DAVIS, E. B., TISCHER, R. G. and BROWN, L. R. (1966) Nitrogen fixation by the blue-green alga, *Anabaena flos-aquae* A–37. *Physiologia Pl.* **19**, 823–826.

DAVIS, L. C., SHAH, V. K., BRILL, W. J. and ORME-JOHNSON, W. H. (1972) Nitrogenase II. Changes in the EPR signal of component I (iron-molybdenum protein) of *Azotobacter vinelandii* nitrogenase during repression and de-repression. *Biochim. biophys. Acta* **256**, 512–523.

DAVIS, S. N. and DE WIEST, R. J. M. (1966) "Hydrogeology". Wiley, New York.

DAVSON, H. and DANEILLI, J. F. (1943) "The permeability of natural membranes". University Press, Cambridge.

DE, P. K. (1939) The role of blue-green algae in nitrogen fixation in rice fields. *Proc. R. Soc.* B **127**, 121–139.

DE, P. K. and MANDEL, L. N. (1956) Fixation of nitrogen by algae in rice soils. *Soil Sci.* **81**, 453–458.

DE, P. K. and SULAIMAN, M. (1950) Fixation of nitrogen in rice soils by algae as influenced by crop, CO_2 and inorganic substances. *Soil Sci.* **70**, 137–151.

DEMETER, O. (1956) Über Modifikationen bei Cyanophyceen. *Arch. Mikrobiol.* **24**, 105–133.

DEN HARTOG, C. (1959) "The epilithic algal communities occurring along the coast of the Netherlands". North Holland Publ. Co., Amsterdam.

DE PUYMALY, A. (1957) Les hétérocystes des algues bleues: leur nature et leur rôle. *Botaniste* **41**, 209–270.

DESIKACHARY, T. V. (1959) "Cyanophyta". Indian Council of Agricultural Research, New Delhi.

DE SOUZA, N. J. and NES, W. R. (1968) Sterols: isolation from a blue-green alga. *Science, N.Y.* **162**, 363.

DHARMAWARDENE, M. W. N., STEWART, W. D. P. and STANLEY, S. O. (1972) Nitrogenase activity, amino acid pool patterns and amination in blue-green algae. *Planta* **108**, 133–145.

DHARMAWARDENE, M. W. N., HAYSTEAD, A. and STEWART, W. D. P. (1973) Glutamine synthetase of the nitrogen-fixing alga *Anabaena cylindrica*. *Arch. Mikrobiol.* **90**, 281–295.

DIETRICH, W. and BIGGINS, J. (1968) Respiratory system of *Leucothrix mucor*. *Pl. Physiol., Lancaster* **43**, S–30.

DIETRICH, W. E. and BIGGINS, J. (1971) Respiratory mechanisms in the Flexibacteriaceae: terminal oxidase systems of *Saprospira grandis* and *Vitreoscilla* species. *J. Bact.* **105**, 1083–1089.

DILWORTH, M. J. (1966) Acetylene reduction by nitrogen-fixing preparations from *Clostridium pasteurianum*. *Biochim. biophys. Acta* **127**, 285–294.

DINSDALE, M. T. and WALSBY, A. E. (1972) The interrelations of cell turgor pressure, gas-vacuolation and buoyancy in a blue-green alga. *J. exp. Bot.* **23**, 561–570.

DODD, J. D. (1960) Filament movement in *Oscillatoria sancta* (Kuetz) Gomont. *Trans. Am. microsc. Soc.* **79**, 480–485.

DOEMEL, W. N. and BROCK, T. (1970) The upper temperature limit of *Cyanidium caldarium*. *Arch. Mikrobiol.* **72**, 326–332.

DOETSCH, R. N. and HAGEAGE, G. J. (1968) Motility in prokaryotic organisms: problems, points of view, and perspectives. *Biol. Rev.* **43**, 317–362.

DÖHLER, G. and BRAUN, F. (1971) Untersuchung der Beziehung zwischen extracellularer Glykolsaüre-Ausscheidung und der photosynthetischen CO_2–Aufname bei der Blaualge *Anacystis nidulans*. *Planta* **98**, 357–361.

DONZE, M. (1968) The algal vegetation of the Ria de Arosa (N.W. Spain). *Blumea* **16**, 159–192.

DONZE, M., HAVEMAN, J. and SCHIERECK, P. (1972) Absence of Photosystem 2 in heterocysts of the blue-green alga *Anabaena*. *Biochim. biophys. Acta* **256**, 157–161.

DRAWERT, H. (1949) Zellmorphologische und zellphysiologische Studien an Cyanophyceen. I. Literaturübersicht und Versuche mit *Oscillatoria borneti* Zukal. *Planta* **37**, 161–209.

DRAWERT, H. and METZNER, I. (1956) Fluoreszenz– und elektronenmikroskopische Beobachtungen an *Cylindrospermum* und einigen anderen Cyanophyceen. III. Zellmorphologische und zellphysiologische Studien an Cyanophyceen. *Ber. dt. bot. Ges.* **69**, 291–301.

DRAWERT, H. and METZNER, I. (1958) Fluoreszenz– und elektronenmikroskopische Untersuchungen an *Oscillatoria borneti* Zukal. V. Zellmorphologische und zellphysiologische Studien an Cyanophyceen. *Z. Bot.* **46**, 16–25.

DRAWERT, H. and TISCHER, I. (1956) Über Redox-Vorgänge bei Cyanophyceen unter besonderer Berücksichtigung der Heterocysten. *Naturwissenschaften* **43**, 132.

DREWES, K. (1928) Über die Assimilation des Luftstickstoffs durch Blaualgen. *Zentbl. Bakt. ParasitKde* (Abt II) **76**, 88–101.

DREWS, G. (1955) Zur Frage der TTC–Reduktion durch Cyanophyceen. *Naturwissenschaften* **42**, 646.

DREWS, G. (1959) Beiträge zur Kenntnis des phototaktischen Reaktionen der Cyanophyceen. *Arch. Protistenk.* **104**, 389–430.

DREWS, G. and GIESBRECHT, P. (1965) Die Thylakoidstrukturen von *Rhodopseudomonas* spec. *Arch. Mikrobiol.* **52**, 242–250.

DREWS, G. and MEYER, H. (1964) Untersuchungen zum chemischen Aufbau der Zellwände von *Anacystis nidulans* und *Chlorogloea fritschii*. *Arch. Mikrobiol.* **48**, 259–267.

DREWS, G. and NIKLOWITZ, W. (1956) Beiträge zur Cytologie der Blaualgen. II. Zentroplasma und granuläre Einschlüsse von *Phormidium uncinatum*. *Arch. Mikrobiol.* **24**, 147–162.

DREWS, G. and NIKLOWITZ, W. (1957) Beiträge zur Cytologie der Blaualgen. III. Untersuchungen über die granuläre Einschlüsse der Hormogonales. *Arch. Mikrobiol.* **25**, 333–351.

DRING, M. J. (1967) Effect of daylength on growth and respiration of the Conchocelis-phase of *Porphyra tenera*. *J. mar. biol. Ass. U.K.* **47**, 501–510.

DROOP, M. R. (1963) Algae and invertebrates in symbiosis. *In* "Symbiotic Associations" (P. S. Nutman and B. Mosse, eds.) pp. 171–199. *13th Symp. Soc. gen. Microbiol.* University Press, Cambridge.

DROOP, M. R. (1967) A procedure for routine purification of algal cultures with antibiotics. *Br. phycol. Bull.* **3**, 295–297.

DROOP, M. R. (1969) Algae. *In* "Methods in Microbiology" (J. R. Norris and D. W. Ribbons, eds.) Vol. 3B, pp. 269–313. Academic Press, London and New York.

DROUET, F. (1937) The Brazilian Myxophyceae I. *Am. J. Bot.* **24**, 598–608.

DROUET, F. (1951) Cyanophyta. *In* "Manual of Phycology" (G. M. Smith, ed.) pp. 159–166. Chronica Botanica Co., Waltham, Mass.

DROUET, F. (1962) Gomont's ecophenes of the blue-green alga, *Microcoleus vaginatus*. *Proc. Acad. nat. Sci. Philad.* **114**, 191–205.

DROUET, F. (1964) Ecophenes of *Microcoleus chthonoplastes*. *Revue algol.* **7**, 315–324.

DROUET, F. (1968) "Revision of the Classification of the Oscillatoriaceae". *Monogr. Acad. nat. Sci. Philad.* **15**.

DROUET, F. and DAILY, W. (1956) "Revision of the Coccoid Myxophyceae". *Butler Univ. bot. Studies* **12**.

DRUM, R. W. and HOPKINS, J. T. (1966) Diatom locomotion: an explanation. *Protoplasma*, **62**, 1–33.

DRUM, R. W. and PANKRATZ, S. (1963) Fine structure of an unusual cytoplasmic inclusion in the diatom genus, *Rhopalodia*. *Protoplasma* **60**, 141–149.

DUA, R. D. and BURRIS, R. H. (1963) Stability of nitrogen-fixing enzymes and the reactivation of a cold-labile enzyme. *Proc. natn. Acad. Sci. U.S.A.* **50**, 169–175.

DUGDALE, R. C., DUGDALE, V., NEES, J. and GOERING, J. (1959) Nitrogen fixation in lakes. *Science, N.Y.* **130**, 859–860.

DUGDALE, R. C. and GOERING, J. J. (1967) Uptake of new and regenerated forms of nitrogen in primary productivity. *Limnol. Oceanogr.* **12**, 196–206.

DUGDALE, R. C., GOERING, J. J. and RYTHER, J. H. (1964) High nitrogen fixation rates in the Sargasso Sea and the Arabian Sea. *Limnol. Oceanogr.* **9**, 507–510.

DUGDALE, R. C., MENZEL, D. W. and RYTHER, J. H. (1961) Nitrogen fixation in the Sargasso Sea. *Deep Sea Res.* **7**, 298–300.

DUGDALE, V. A. and DUGDALE, R. C. (1962) Nitrogen metabolism in lakes: role of nitrogen fixation in Sanctuary Lake, Pennsylvania. *Limnol. Oceanogr.* **7**, 170–177.

DUGDALE, V. A. and DUGDALE, R. C. (1965) Nitrogen metabolism in lakes. III. Tracer studies of the assimilation of inorganic nitrogen sources. *Limnol. Oceanogr.* **10**, 53–57.

Dunn, J. H. and Wolk, C. P. (1970) Composition of the cellular envelopes of *Anabaena cylindrica*. *J. Bact.* **103**, 153–158.
Durrell, L. W. and Shields, L. M. (1961) Characteristics of soil algae relating to crust formation. *Trans. Am. microsc. Soc.* **80**, 73–79.
Duvigneaud, P. and Symoens, J. J. (1948) Exploration du Parc National Albert. Mission J. Lebrun (1937–1938). Fascicule 10. Cyanophycées. pp. 1–35. Institut des Parcs Nationaux du Congo Belge, Brussels.
Duysens, L. N. M. (1952) "Transfer of Excitation Energy in Photosynthesis". Ph.D. Thesis, University of Utrecht.
Duysens, L. N. M. (1956) The flattening of the absorption spectrum of suspensions as compared to that of solutions. *Biochim. biophys. Acta* **19**, 1–12.
Eberly, W. R. (1967) Problems in the laboratory culture of planktonic blue-green algae. *In* "Environmental Requirements of Blue-green Algae" (A. F. Bartsch, ed.) pp. 7–34. U.S. Dept. Interior, Federal Water Poll. Control Admin., Corvallis, Oregon.
Echlin, P. (1964a) The fine structure of the blue-green alga *Anacystis montana* f. *minor* grown in continuous illumination. *Protoplasma* **58**, 439–457.
Echlin, P. (1964b) Intra-cytoplasmic membranous inclusions in the blue-green alga, *Anacystis nidulans*. *Arch. Mikrobiol.* **49**, 267–274.
Echlin, P. (1966) Origins of photosynthesis. *Sci. J.* (April) 1–7.
Echlin, P. (1970) The photosynthetic apparatus in prokaryotes and eukaryotes. *In* "Organization and Control in Prokaryotic and Eukaryotic Cells" (H. P. Charles and B. C. J. G. Knight, eds.) pp. 221–248. *20th Symp. Soc. gen. Microbiol.* University Press, Cambridge.
Echlin, P. and Morris, I. (1965) The relationship between blue-green algae and bacteria. *Biol. Rev.* **40**, 143–187.
Edelman, M., Swinton, D., Schiff, J. A., Epstein, H. T. and Zeldin, B. (1967) Deoxyribonucleic acid of the blue-green algae (Cyanophyta). *Bact. Rev.* **31**, 315–331.
Edwards, M. R., Berns, D. S., Ghiorse, W. C. and Holt, S. C. (1968) Ultrastructure of the thermophilic blue-green alga, *Synechococcus lividus* Copeland. *J. Phycol.* **4**, 283–298.
Edwards, M. R. and Gantt, E. (1971) Phycobilisomes of the thermophilic blue-green alga *Synechoccocus lividus*. *J. Cell Biol.* **50**, 896–900.
Elenkin, A. A. (1936) "Monographia Algarum Cyanophycearum Aquidulcium et Terrestrium Infinibus URSS Inventarum". Vol. 1. Acad. Nauk URSS, Moscow and Leningrad.
Eley, J. H. (1971) Effect of carbon dioxide concentration on pigmentation in the blue-green alga *Anacystis nidulans*. *Pl. Cell Physiol., Tokyo* **12**, 311–316.
Ellis, R. J. and Hartley, M. R. (1971) Sites of synthesis of chloroplast proteins. *Nature, Lond.* **233**, 193–196.
Elo, J. E. (1937) Vergleichende Permeabilitätsstudien, besonders an niederen Pflanzen. *Ann. Bot. Soc. Zool. Bot. Fennicae Vanamo* **8**, 1–108.
Emerson, R. (1927) "Über die Wirkung von Blausaure, Schwefelwasserstoff und Kohlenoxyd auf die Atmung Verschideener Algen". Dissertation, Friedrich Wilhelms Universität, Berlin.

EMERSON, R. (1958) The quantum yield of photosynthesis. *A. Rev. Pl. Physiol.* **9**, 1–24.

EMERSON, R. and LEWIS, C. M. (1942) The photosynthetic efficiency of phycocyanin in *Chroococcus* and the problem of carotenoid participation in photosynthesis. *J. gen. Physiol.* **25**, 579–595.

ENGELMANN, T. W. (1883) Farbe und Assimilation. *Bot. Ztg.* **11**, 1–13.

ERCEGOVIC, A. (1930) Sur la tolerance des Cyanophycées vis-a-vis des variations brusques de la salinité de l'eau de mer. *Acta bot. Inst. bot. Univ. Zagreb* **5**, 48–56.

ERCEGOVIĆ, A. (1932) Etudes écologiques et sociologiques des Cyanophycées lithophytes de la côte Yougoslave de l'Adriatique. *Bull. int. Acad. Yougosl. Sci. Zagreb* **26**, 33–56.

ERWIN, J. and BLOCH, K. (1963) Polyunsaturated fatty acids in some photosynthetic organisms. *Biochem. Z.* **338**, 496–511.

EVANS, M. C. W., HALL, D. O., BOTHE, H. and WHATLEY, F. R. (1968) The stoichiometry of electron transfer by bacterial and plant ferredoxins. *Biochem. J.* **110**, 485–489.

EVERTON, M. and LORDS, J. L. (1967) Respiration and photosynthesis in the thermophilic blue-green alga *Schizothrix calcicola*. *Proc. Utah Acad. Sci. Arts Lett.* **44**, 416.

EYSTER, C. (1952) Necessity of boron for *Nostoc muscorum*. *Nature, Lond.* **170**, 755.

FAN, K. C. (1956) Revision of the genus *Calothrix Ag.* *Revue algol.* **2**, 154–178.

FARRAN, G. P. (1932) The occurrence of *Trichodesmium thiebautii* off the south coast of Ireland. *Rapp. Proc. Verb. Cons. int. Explor. Mer.* **77**, 60–64.

FAY, P. (1962) "Studies on growth and nitrogen fixation in *Chlorogloea fritschii* Mitra". Ph.D. Thesis, University of London.

FAY, P. (1965) Heterotrophy and nitrogen fixation in *Chlorogloea fritschii*. *J. gen. Microbiol.* **39**, 11–20.

FAY, P. (1969a) Metabolic activities of isolated spores of *Anabaena cylindrica*. *J. exp. Bot.* **20**, 100–109.

FAY, P. (1969b) Cell differentiation and pigment composition in *Anabaena cylindrica*. *Arch. Mikrobiol.* **67**, 62–70.

FAY, P. (1970) Photostimulation of nitrogen fixation in *Anabaena cylindrica*. *Biochim. biophys. Acta* **216**, 353–356.

FAY, P. (1973) Some aspects of heterotrophy by blue-green algae. *Br. phycol. J.* **8**, 209.

FAY, P. and COX, R. M. (1966) Decarboxylation performed by particulate fractions of two nitrogen-fixing blue-green algae. *Biochim. biophys. Acta* **126**, 402–404.

FAY, P. and COX, R. M. (1967) Oxygen inhibition of nitrogen fixation in cell-free preparations of blue-green algae. *Biochim. biophys. Acta* **143**, 562–569.

FAY, P. and FOGG, G. E. (1962) Studies on nitrogen fixation by blue-green algae III. Growth and nitrogen fixation in *Chlorogloea fritschii* Mitra. *Arch. Mikrobiol.* **42**, 310–321.

Fay, P. and Kulasooriya, S. A. (1972) Tetrazolium reduction and nitrogenase acitivity in heterocystous blue-green algae. *Arch. Mikrobiol.* **87**, 341–352.

Fay, P. and Kulasooriya, S. A. (1973) A simple apparatus for the continuous culture of photosynthetic micro-organisms. *Br. phycol.* J. **8**, 51–57.

Fay, P., Kumar, H. D. and Fogg, G. E. (1964) Cellular factors affecting nitrogen fixation in the blue-green alga *Chlorogloea fritschii. J. gen. Microbiol.* **35**, 351–360.

Fay, P. and Lang, N. J. (1971) The heterocysts of blue-green algae I. Ultrastructural integrity after isolation. *Proc. R. Soc.* B **178**, 185–192.

Fay, P., Stewart, W. D. P., Walsby, A. E. and Fogg, G. E. (1968) Is the heterocyst the site of nitrogen fixation in blue-green algae? *Nature, Lond.* **220**, 810–812.

Fay, P. and Walsby, A. E. (1966) Metabolic activities of isolated heterocysts of the blue-green alga *Anabaena cylindrica. Nature, Lond.* **209**, 94–95.

Fechner, R. (1915) Die Chemotaxis der Oscillarien und ihre Bewegungserscheinungen überhaupt. *Z. Bot.* **7**, 289–364.

Fehér, D. (1948) Researches on the geographical distribution of soil microflora. II. The geographical distribution of soil algae. *Erdész. Kisérl.* **48**, 57–93.

Feldmann, J. (1932) Sur la biologie des *Trichodesmium* Ehrenburg. *Revue algol.* **6**, 357–358.

Feldmann, J. (1933) Sur quelques Cyanophycées vivant dans les tissues des éponges de Banyuls. *Archs. Zool. exp. gén.* **75**, 381–404.

Feldmann, J. (1959) Sul l'ecologie des Rhodophycées et des Cyanophycées marines dans ses rapports avec les problemes de leur photosynthèse. *Bull. Soc. fr. Physiol. veg.* **5**, 161–167.

Fernandez, F. and Bhat, J. V. (1945) A note on the association of *Chlorococcum humicolum* in the roots of *Cycas revolta. Curr. Sci.* **14**, 235.

Fewson, C. A., Al-Hafidh, M. and Gibbs, M. (1962) Role of aldolase in photosynthesis. I. Enzyme studies with photosynthetic organisms with special reference to blue-green algae. *Pl. Physiol., Lancaster* **37**, 402–406.

Findley, D., Holton, R. W. and Herndon, W. R. (1968) Effect of light and temperature on the developmental cycle of *Chlorogloea fritschii. J. Phycol.* **4**, Suppl. 10.

Findley, D. L., Walne, P. L. and Holton, R. W. (1970) The effects of light intensity on the ultrastructure of *Chlorogloea fritschii* Mitra grown at high temperature. *J. Phycol.* **6**, 182–188.

Fitzgerald, G. P. (1967) Discussion. *In* "Environmental Requirements of Blue-green Algae" (A. F. Bartsch, ed.) pp. 97–102. U.S. Dept. Interior, Federal Water Poll. Control Admin., Corvallis, Oregon.

Fitzgerald, G. P. and Skoog, F. (1954) Control of blue-green algae blooms with 2,3–dichloronaphthoquinone. *Sewage ind. Wastes* **26**, 1136–1140.

Fitz-James, P. C. (1960) Participation of the cytoplasmic membrane in the growth and spore formation of bacilli. *J. biophys. biochem. Cytol.* **8**, 507–528.

Fogg, G. E. (1941) The gas vacuoles of the Myxophyceae (Cyanophyceae). *Biol. Rev.* **16**, 205–217.

FOGG, G. E. (1942) Studies on nitrogen fixation by blue-green algae. I. Nitrogen fixation by *Anabaena cylindrica* Lemm. *J. exp. Biol.* **19**, 78–87.

FOGG, G. E. (1944) Growth and heterocyst production in *Anabaena cylindrica* Lemm. *New Phytol.* **43**, 164–175.

FOGG, G. E. (1949) Growth and heterocyst production in *Anabaena cylindrica* Lemm. II. In relation to carbon and nitrogen metabolism. *Ann. Bot.* **13**, 241–259.

FOGG, G. E. (1951a) Growth and heterocyst production in *Anabaena cylindrica* Lemm. III. The cytology of heterocysts. *Ann. Bot.* **15**, 23–35.

FOGG, G. E. (1951b) Studies on nitrogen fixation in blue-green algae. II. Nitrogen fixation by *Mastigocladus laminosus* Cohn. *J. exp. Bot.* **2**, 117–120.

FOGG, G. E. (1952) The production of extracellular nitrogenous substances by a blue-green alga. *Proc. R. Soc.* B **139**, 372–397.

FOGG, G. E. (1956a) Nitrogen fixation by photosynthetic organisms. *A. Rev. Pl. Physiol.* **7**, 51–70.

FOGG, G. E. (1956b) The comparative physiology and biochemistry of the blue-green algae. *Bact. Rev.* **20**, 148–165.

FOGG, G. E. (1965) "Algal Cultures and Phytoplankton Ecology". University of Wisconsin Press, Madison and Milwaukee.

FOGG, G. E. (1966) The extracellular products of algae. *Oceanogr. Mar. Biol. Rev.* **4**, 195–212.

FOGG, G. E. (1968) "Photosynthesis". English Universities Press, London.

FOGG, G. E. (1969a) Survival of algae under adverse conditions. *In* "Dormancy and Survival" (H. W. Woolhouse, ed.) pp. 123–142. *23rd Symp. Soc. exp. Biol.* University Press, Cambridge.

FOGG, G. E. (1969b) The physiology of an algal nuisance. *Proc. R. Soc.* B **173**, 175–189.

FOGG, G. E. (1971) Extracellular products of algae in freshwater. *Arch. Hydrobiol. Beih. Ergegn. Limnol.* **5**, 1–25.

FOGG, G. E. (1971) Nitrogen fixation in lakes. *In* "Biological Nitrogen Fixation in Natural and Agricultural Habitats". (T. A. Lie and E. G. Mulder, eds.) pp. 393–401. *Pl. Soil*, Special volume.

FOGG, G. E. (1972) Organic substances: plants. *In* "Marine Ecology" (O. Kinne, ed.) Vol. 1(3) pp. 1551–1563. Wiley-Interscience, London and New York.

FOGG, G. E., EAGLE, D. J. and KINSON, M. E. (1969) The occurrence of glycollic acid in natural waters. *Verh. int. Ver. theor. angew. Limnol.* **17**, 480–484.

FOGG, G. E. and HORNE, A. J. (1970) The physiology of antarctic freshwater algae. *In* "Antarctic Ecology" (M. W. Holdgate, ed.) Vol. 2, pp. 632–638. Academic Press, London and New York.

FOGG, G. E. and STEWART, W. D. P. (1965) Nitrogen fixation in blue-green algae. *Sci. Progress* **53**, 191–201.

FOGG, G. E. and STEWART, W. D. P. (1968) *In situ* determinations of biological nitrogen fixation in Antarctica. *Br. Antarct. Surv. Bull.* **15**, 39–46.

Fogg, G. E. and Than-Tun (1958) Photochemical reduction of elementary nitrogen in the blue-green alga *Anabaena cylindrica*. *Biochim. biophys. Acta* **30**, 209.

Fogg, G. E. and Than-Tun (1960) Interrelations of photosynthesis and assimilation of elementary nitrogen in a blue-green alga. *Proc. R. Soc.* B **153**, 111–127.

Fogg, G. E. and Westlake, D. F. (1955) The importance of extracellular products of algae in freshwater. *Verh. int. Ver. theor. angew. Limnol.* **12**, 219–232.

Fogg, G. E. and Wolfe, M. (1954) The nitrogen metabolism of the blue-green algae (Myxophyceae). *In* "Autotrophic Micro-organisms" (B. A. Fry and J. L. Peel, eds.) pp. 99–125. *4th Symp. Soc. gen. Microbiol.* University Press, Cambridge.

Forest, H. (1968) The approach of a modern algal taxonomist. *In* "Algae, Man and the Environment" (D. F. Jackson, ed.) pp. 185–199. University Press, Syracuse, N.Y.

Forest, H. S., Willson, P. L. and England, R. B. (1959) Algal establishment on sterilized soil replaced in an Oklahama prairie. *Ecology* **40**, 475–477.

Forrest, H. S., Van Baalen, C. and Myers, J. (1957) Occurrence of pteridines in a blue-green alga. *Science, N.Y.* **125**, 699–700.

Forrest, H. S., Van Baalen, C. and Myers, J. (1958) Isolation and identification of a new pteridine from a blue-green alga. *Archs Biochem. Biophys.* **78**, 95–99.

Fowden, L. (1962) Amino acids and proteins. *In* "Physiology and Biochemistry of Algae" (R. A. Lewin, ed.) pp. 189–209. Academic Press, New York and London.

Frank, B. (1889) Über den experimentellen Nachweis der Assimilation freien Stickstoffs durch erdbodenbewohnende Algen. *Ber. dt. bot. Ges.* **7**, 34–42.

Frank, H., Lefort, M. and Martin, H. H. (1962) Elektronenoptische und chemische Untersuchungen an Zellwänden der Blaualge *Phormidium uncinatum*. *Z. Naturf.* **17**, 262–268.

Frazer, J. G. (1914) "Adonis Attis Osiris". MacMillan, London.

Fredrick, J. F. (1951) Preliminary studies on the synthesis of polysaccharides in the algae. *Physiologia Pl.* **4**, 621–626.

Fredrick, J. F. (1971) Storage polyglucan-synthesizing isozyme patterns in the Cyanophyceae. *Phytochemistry* **10**, 395–398.

Frémy, P. (1929) Les Myxophycées de l'Afrique équatoriale francaise. *Archs Bot., Caen 3,* Memoire no. 2.

Frémy, P. (1934) Les Cyanophycées des côtes d'Europe. *Mém. Soc. natn. Sci. nat. math. Cherbourg* **41**, 1–236.

Frenkel, A., Gaffron, H. and Battley, E. H. (1950) Photosynthesis and photoreduction by the blue-green alga, *Synechococcus elongatus* Näg. *Biol. Bull.* **99**, 157–162.

Frey-Wyssling, A. and Stecher, H. (1954) Über den Feinbau des *Nostoc*-Schleimes. *Z. Zellforsch. mikrosk. Anat.* **39**, 515–519.

Fritsch, F. E. (1904) Studies on Cyanophyceae. III. Some points in the reproduction of *Anabaena*. *New Phytol.* **3**, 216–228.

Fritsch, F. E. (1917) Freshwater algae. *In* "British Antarctic (Terra Nova) Expedition, 1910. Natural History Report. Botany." Part 1, pp. 1–16. British Museum (Natural History).

Fritsch, F. E. (1936) The role of the terrestrial alga in nature. *In* "Essays in Geobotany in Honor of William Albert Setchell", pp. 195–217. University of California Press.

Fritsch, F. E. (1942) The interrelations and classification of the Myxophyceae (Cyanophyceae). *New Phytol.* **41**, 134–148.

Fritsch, F. E. (1943) Studies in the comparative morphology of the algae. III. Evolutionary tendencies and affinities among Phaeophyceae. *Ann. Bot.* **7**, 63–87.

Fritsch, F. E. (1944) Present-day classification of algae. *Bot. Rev.* **10**, 233–277.

Fritsch, F. E. (1945) "The Structure and Reproduction of the Algae". Vol. 2. University Press, Cambridge.

Fritsch, F. E. (1951) The heterocyst: a botanical enigma. *Proc. Linn. Soc. Lond.* **162**, 194–211.

Fuhs, G. W. (1958a) Untersuchungen an Ultradünnschnitten von *Oscillatoria amoena* (Kütz) Gomont. *Protoplasma* **49**, 523–540.

Fuhs, G. W. (1958b) Über die Natur der Granula in Cytoplasma von *Oscillatoria amoena* (Kütz) Gomont. *Öst. bot. Z.* **104**, 531–551.

Fuhs, G. W. (1958c) Enzymatischer Abbau der Membranen von *Oscillatoria amoena* (Kütz.) Gomont mit lysozym. *Arch. Mikrobiol.* **29**, 51–52.

Fuhs, G. W. (1958d) Bau, Verhalten und Bedeutung der kernäquivalenten Strukturen bei *Oscillatoria amoena* (Kütz.) Gomont. *Arch. Mikrobiol.* **28**, 270–302.

Fuhs, G. W. (1963) Cytochemisch-elektronenmikroskopische Lokalisierung der Ribonukleinsäure und des Assimilats in Cyanophyceen. *Protoplasma* **56**, 178–187.

Fuhs, G. W. (1964) Die Wirkung der Uranylsalzen auf die Struktur des Bakteriennucleoids. *Arch. Mikrobiol.* **49**, 383–404.

Fuhs, G. W. (1968) Cytology of blue-green algae: light microscopic aspects. *In* "Algae, Man and the Environment" (D. F. Jackson, ed.) pp. 213–233. University Press, Syracuse, N.Y.

Fujita, Y. and Myers, J. (1965) Hydrogenase and NADP-reduction reactions by a cell-free preparation of *Anabaena cylindrica*. *Archs Biochem. Biophys.* **111**, 619–625.

Fulco, L. Karfunkel, P. and Aaronson, S. (1967) Effect of lysozyme (muramidase) on marine and freshwater blue-green algae. *J. Phycol.* **3**, 51–52.

Gaffron, H. and Fager, E. W. (1951) The kinetics and chemistry of photosynthesis. *A. Rev. Pl. Physiol.* **2**, 87–114.

Gaidukov, N. (1902) Über den Einfluss farbigen Lichtes auf die Färbung lebender Oscillarien. *Abh. dt. Akad. Wiss. Berl.* **5**, 1–36.

Gaidukov, N. (1923) Zur Frage nach der komplementären chromatischen Adaptation. *Ber. dt. bot. Ges.* **41**, 356–361.

Gallon, J. R., LaRue, T. A. and Kurz, W. G. W. (1972) Characteristics of nitrogenase activity in broken cell preparations of the blue-green alga *Gloeocapsa* sp. LB 795. *Can. J. Microbiol.* **18**, 327–332.

Ganf, G. G. (1969) "Physiological and ecological aspects of the phytoplankton of Lake George, Uganda." Ph.D. Thesis, University of Lancaster, England.

Gantt, E. and Conti, S. F. (1966) Granules associated with the chloroplast lamellae of *Porphyridium cruentum*. *J. Cell. Biol.* **29**, 423–434.

Gantt, E. and Conti, S. F. (1969) Ultrastructure of blue-green algae. *J. Bact.* **97**, 1486–1493.

Gantt, E., Edwards, M. R. and Provasoli, L. (1971) Chloroplast structure of the Cryptophyceae. Evidence for phycobiliproteins within intrathylakoidal spaces. *J. Cell. Biol.* **48**, 280–290.

Geitler, L. (1921) Versuch einer Lösung des Heterocysten-Problems. *Sber. Akad. Wiss. Wien, mat.-nat.* Kl. I. **130**, 223–245.

Geitler, L. (1925) Synoptische Darstellung der Cyanophyceen in morphologischer und systematischer Hinsicht. *Beih. bot. Zbl.* II. **41**, 163–294.

Geitler, L. (1927) Bemerkungen über *Paulinella chromatophora*. *Zool. Anz.* **73**, 333–334.

Geitler, L. (1932) Cyanophyceae. *In* "Kryptogamenflora von Deutschland, Österreich und der Schweiz" (L. Rabenhorst, ed.) Vol. 14. Akademische Verlagsgesellschaft, Leipzig.

Geitler, L. (1936) Schizophyceen. *In* "Handbuch der Pflanzenanatomie" (W. Zimmermann and O. Ozenda, eds.) Vol. 6, part 1. Bornträger, Berlin.

Geitler, L. (1942) Schizophyta: Klasse Schizophyceae. *In* "Die Natürlichen Pflanzenfamilien" (A. Engler and K. Prantl, eds.) 2nd ed. Vol. 1b, pp. 1–232.

Geitler, L. (1951) Über rechtwinkelige Scheidung von Scheidewänden und dreidimensionale Zellverbände. *Öst. bot. Z* **98**, 171–186.

Geitler, L. (1959) Syncyanosen. *In* "Handbuch der Pflanzenphysiologie" (W. Ruhland, ed.) Vol. 11, pp. 530–545. Springer Verlag, Berlin-Göttingen-Heidelburg.

Gelpi, E. Schneider, H., Mann, J. and Oro, J. (1970) Hydrocarbons of geochemical significance in microscopic algae. *Phytochemistry* **9**, 603–612.

Gentile, J. H. and Maloney, T. E. (1969) Toxicity and environmental requirements of a strain of *Aphanizomenon flos-aquae* (L.) Ralfs. *Can. J. Microbiol.* **15**, 165–173.

Gerhardt, B. and Santo, R. (1966) Photophosphorylierung in einem zellfreien System aus *Anacystis*. *Z. Naturf.* **21b**, 673–678.

Gerhardt, B. and Wiessner, W. (1967) On the light-dependent reactivation of photosynthetic activity by manganese. *Biochem. biophys. Res. Commun.* **28**, 958–964.

Gerloff, G. C., Fitzgerald, G. P. and Skoog, F. (1950a) The isolation, purification and culture of blue-green algae. *Am. J. Bot.* **37**, 216–218.

Gerloff, G. C., Fitzgerald, G. P. and Skoog, F. (1950b) The mineral nutrition of *Coccochloris peniocystis*. *Am. J. Bot.* **37**, 835–840.

Gerloff, G. C., Fitzgerald, G. P. and Skoog, F. (1952) The mineral nutrition of *Microcystis aeruginosa*. *Am. J. Bot.* **39**, 26–32.

Gessner, F. (1959) "Hydrobotanik. II. Stoffhaushalt." VEB Deutscher Verlag der Wissenschaften, Berlin.

Gibbs, M. (1962) Respiration. *In* "Physiology and Biochemistry of Algae" (R. A. Lewin, ed.) pp. 61–90. Academic Press, New York and London.

Gibbs, M., Ellyard, P. W. and Latzko, E. (1968) Warburg effect: control of photosynthesis by oxygen. *In* "Comparative Biochemistry and Biophysics of Photosynthesis" (K. Shibata, A. Takamiya, A. T. Jagendorf and R. C. Fuller, eds.) pp. 387–399. University of Tokyo Press, Tokyo and University Park Press, State College, Pennsylvania.

Gibbs, M., Latzko, E., Harvey, M. J., Plaut, Z. and Shain, Y. (1970) Photosynthesis in the algae. *In* "Phylogenesis and Morphogenesis in the Algae" (J. F. Fredrick and R. M. Klein, eds.) *Ann. N.Y. Acad. Sci.* **175**, 541–545.

Giesy, R. M. (1964) A light and electron microscope study of interlamellar polyglucoside bodies in *Oscillatoria chalybia*. *Am. J. Bot.* **51**, 388–396.

Ginsburg, D., Padan, E. and Shilo, M. (1968) Effect of cyanophage infection on CO_2 photoassimilation in *Plectonema boryanum*. *J. Virol.* **2**, 695–701.

Glade, R. (1914) Zur Kenntnis der Gattung *Cylindrospermum*. *Beitr. Biol. Pfl.* **12**, 295–344.

Glauert, A. M. and Glauert, R. H. (1958) Araldite as an embedding medium for electron microscopy. *J. biophys. biochem. Cytol.* **4**, 191–194.

Glazer, A. N. and Cohen-Bazire, G. (1971) Subunit structure of the phycobiliproteins of blue-green algae. *Proc. natn. Acad. Sci. U.S.A.* **68**, 1398–1401.

Glazer, A. N., Cohen-Bazire, G. and Stanier, R. Y. (1971) Comparative immunology of algal biliproteins. *Proc. natn. Acad. Sci. U.S.A.* **68**, 3005–3008.

Godward, M. B. E. (1937) An ecological and taxonomic investigation of the littoral algal flora of Lake Windermere. *J. Ecol.* **25**, 496–568.

Godward, M. B. E. (1962) Invisible radiations. *In* "Physiology and Biochemistry of Algae" (R. A. Lewin, ed.) pp. 551–566. Academic Press, New York and London.

Goedheer, J. C. (1968) On the low-temperature fluorescence spectra of blue-green algae and red algae. *Biochim. biophys. Acta* **153**, 903–906.

Goedheer, J. C. (1969) Energy transfer from carotenoids to chlorophyll in blue-green, red and green algae and greening bean leaves. *Biochim. biophys. Acta* **172**, 252–265.

Goering, J. J., Dugdale, R. C. and Menzel, D. W. (1966) Estimates of *in situ* rates of nitrogen uptake by *Trichodesmium* sp. in the tropical Atlantic Ocean. *Limnol. Oceanogr.* **11**, 614–620.

Goering, J. J. and Neess, J. C. (1964) Nitrogen fixation in two Wisconsin lakes. *Limnol. Oceanogr.* **9**, 535–539.

Goldman, C. R., Mason, D. T. and Wood, B. J. B. (1963) Light injury and inhibition in Antarctic freshwater phytoplankton. *Limnol. Oceanogr.* **8**, 313–322.

Goldstein, D. A. and Bendet, I. J. (1967) Physical properties of the DNA from the blue-green algal virus LPP-1. *Virology* **32**, 614–618.

Goldstein, D. A., Bendet, I. J., Lauffer, M. A. and Smith, K. M. (1967) Some biological and physicochemical properties of blue-green algal virus LPP-1. *Virology* **32**, 601–613.

Gollerbakh, M. M. and Shtina, E. A. (1969) "Pochvennye Vodorosli" (Soil Algae) Idz. Nauka, Leningrad.
Golubić, S. (1969) Distribution, taxonomy and boring patterns of marine endolithic algae. *Am. Zool.* **9**, 747–751.
Gomont, M. (1888) Recherches sur le enveloppes des Nostocacées filamenteuses. *Bull. Soc. bot. Fr.* **35**, 204–236.
Gomont, M. (1892) Monographie des Oscillariées. *Annls Sci. nat.* (*Bot.*) **15**, 263–368 and **16**, 91–264.
Goodwin, T. W. (1965) Distribution of carotenoids. *In* "Chemistry and Biochemistry of Plant Pigments" (T. W. Goodwin, ed.) pp. 143–174. Academic Press, London and New York.
Goodwin, T. W. and Taha, M. M. (1951) A study of the carotenoids echinenone and myxoxanthin with special reference to their probable identity. *Biochem. J.* **47**, 244–249.
Gorham, P. R. (1964) Toxic algae. *In* "Algae and Man" (D. F. Jackson, ed.) pp. 307–336. Plenum Press, New York.
Gorham, P. R., McLachlan, J. S., Hammer, U. T. and Kim, W. K. (1964) Isolation and culture of toxic strains of *Anabaena flos-aquae* (Lyngb.) de Bréb. *Verh. int. Ver. theor. angw. Limnol.* **15**, 796–804.
Goryunova, S. V. and Rzhanova, G. N. (1964) Vital excretions of nitrogen-containing substances in *Lyngbya aestuarii* and their physiological role. *In* "Biology of the Cyanophyta" (V. D. Fedorov and M. M. Telitchenko, eds.) pp. 118–138. University Press, Moscow.
Goryushin, V. A. and Chaplinskaya, S. M. (1966) Existence of viruses of blue-green algae. (Ukranian) *Mikrobiol. Zh. Akad. Nauk. Ukr. RSR* **28**, 94–97.
Goryushin, V. A. and Chaplinksaya, S. M. (1968) The discovery of viruses lysing blue-green algae in the reservoirs of the River Dneper. *In* "Tsvetenie Vody" (A. V. Topachevsky, ed.) pp. 171–174. (Russian) Naukova Dumka, Kiev, USSR.
Granhall, U. (1970) Acetylene reduction by blue-green algae isolated from Swedish soils. *Oikos* **21**, 330–332.
Granhall, U. and Henriksson, E. (1969) Nitrogen-fixing blue-green algae in Swedish soils. *Oikos* **20**, 175–178.
Granhall, U. and Hofsten, A. von (1969) The ultrastructure of a cyanophage attack on *Anabaena variabilis*. *Physiologia Pl.* **22**, 713–722.
Grant, B. R. (1968) The effect of carbon dioxide concentration and buffer system on nitrate and nitrite assimilation by *Dunaliella tertiolecta*. *J. gen. Microbiol.* **54**, 327–336.
Gregory, P. H., Hamilton, E. D. and Sreeramulu, T. (1955) Occurrence of the alga *Gloeocapsa* in the air. *Nature, Lond.* **176**, 1270.
Griffin, G. D. and Rees, D. M. (1956) *Anabaena unispora* and other blue-green algae as possible mosquito control factors in Salt Lake County, Utah. *Proc. Utah Acad. Sci. Arts Lett.* **33**, 101–103.
Griffiths, H. B., Greenwood, A. D. and Millbank, J. W. (1972) The frequency of heterocysts in the *Nostoc* phycobiont of the lichen *Peltigera canina* Willd. *New Phytol.* **71**, 11–13.

GRILLI, M. (1964) Infrastutture di *Anabaena azollae* vivente nelle foglioline di *Azolla caroliniana*. *Annali Microbiol. Enzimol.* **14**, 69–90.

GUILLARD, R. R. L. (1962) Salt and osmotic balance. *In* "Physiology and Biochemistry of Algae" (R. A. Lewin, ed.) pp. 529–540. Academic Press, New York and London.

GUILLARD, R. R. L. and WANGERSKY, P. J. (1958) The production of extracellular carbohydrates by some marine flagellates. *Limnol. Oceanogr.* **3**, 449–454.

HABEKOST, R. D., FRAZER, I. M. and HALSTEAD, R. W. (1955) Observations on toxic marine algae. *J. Wash. Acad. Sci.* **45**, 101–103.

HAGEDORN, H. (1960) Elektronenmikroskopische Untersuchungen an Blaualgen. *Naturwissenschaften* **47**, 430.

HAGEDORN, H. (1971) Experimentelle Untersuchungen über den Einfluss des Thiamins auf die natürliche Algenpopulation des Pelagials. *Arch. Hydrobiol.* **68**, 382–399.

HALFEN, L. N. and CASTENHOLZ, R. W. (1970) Gliding in a blue-green alga: a possible mechanism. *Nature, Lond.* **225**, 1163–1165.

HALFEN, L. N. and CASTENHOLZ, R. W. (1971a) Gliding motility in the blue-green alga *Oscillatoria princeps*. *J. Phycol.* **7**, 133–145.

HALFEN, L. N. and CASTENHOLZ, R. W. (1971b) Energy expenditure for gliding motility in a blue-green alga. *J. Phycol.* **7**, 258–260.

HALFEN, L. N. and FRANCIS, G. W. (1972) The influence of culture temperature on the carotenoid composition of the blue-green alga *Anacystis nidulans*. *Arch. Mikrobiol.* **81**, 25–35.

HALL, W. T. and CLAUS, G. (1962) Electron microscope studies on ultrathin sections of *Oscillatoria chalybea* Mertens. *Protoplasma* **54**, 355–368.

HALL, W. T. and CLAUS, G. (1963) Ultrastructural studies on the blue-green algal symbiont in *Cyanophora paradoxa* Korschikoff. *J. Cell Biol.* **19**, 551–563.

HALL, W. T. and CLAUS, G. (1965) The fine structure of the coccoid blue-green alga, nom. prov. *Synechococcus oceanica*. *Revta Biol., Lisboa* **5**, 63–74.

HALL, W. T. and CLAUS, G. (1967) Ultrastructural studies on the cyanelles of *Glaucocystis nostochinearum* Itzigsohn. *J. Phycol.* **3**, 37–51.

HALLDAL, P. (1958) Pigment formation and growth in blue-green algae in crossed gradients of light intensity and temperature. *Physiologia Pl.* **11**, 401–420.

HALLIER, V. W. and PARK, R. B. (1969) Photosynthetic light reactions in chemically fixed *Anacystis nidulans*, *Chlorella pyrenoidosa* and *Porphyridium cruentum*. *Pl. Physiol., Lancaster* **44**, 535–539.

HALPERIN, DE D. R. (1967) Cianoficeas marinas de Puerto Deseado II (Provincia de Santa Cruz, Argentina). *Darwinia* **14**, 273–354.

HALPERIN, DE D. R. (1970) Cianoficeas marina del Chubut (Argentina) I. Golfo San Hose, Golfo Nuevo Y Alrededores de Rawson. *Physis* **30**, 33–96.

HANSGIRG, A. (1883) Bemerkungen über die Bewegungen der Oscillarien. *Bot. Ztg* **41**, 831–843.

HARDER, R. (1917a) Über die Beziehung der Keimung von Cyanophyceensporen zum Licht. *Ber. dt. bot. Ges.* **35**, 58–64.

HARDER, R. (1917b) Ernährungsphysiologische Untersuchungen an Cyanophyceen, hauptsächlich dem endophytischen *Nostoc punctiforme*. *Z. Bot.* **9**, 145–245.

HARDER, R. (1918) Über die Bewegung der Nostocaceen. *Z. Bot.* **10**, 177–243.

HARDER, R. (1920) Über die Reaktionen freibeweglicher pflanzlicher Organismen auf plötzliche Änderungen der Lichtintensität. *Z. Bot.* **12**, 353–462.

HARDY, R. W. F. and BURNS, R. C. (1968) Biological nitrogen fixation. *A. Rev. Biochem.* **37**, 331–358.

HARDY, R. W. F., HOLSTEN, R. D., JACKSON, E. K. and BURNS, R. C. (1968) The acetylene-ethylene assay for N_2 fixation: laboratory and field evaluation. *Pl. Physiol., Lancaster* **43**, 1185–1207.

HARVEY, W. H. (1846–1851) "Phycologia Britannica". London.

HASLER, A. D. (1949) Antibiotic aspects of copper treatment of lakes. *Trans. Wisc. Acad. Sci. Arts Lett.* **39**, 97–103.

HATFIELD, D. L., VAN BAALEN, C. and FORREST, H. S. (1961) Pteridines in blue-green algae. *Pl. Physiol., Lancaster* **36**, 240–243.

HATTORI, A. (1962a) Light induced reduction of nitrate, nitrite and hydroxylamine in a blue-green alga, *Anabaena cylindrica*. *Pl. Cell Physiol., Tokyo* **3**, 355–369.

HATTORI, A. (1962b) Adaptive formation of nitrate reducing system in *Anabaena cylindrcia*. *Pl. Cell Physiol., Tokyo* **3**, 371–377.

HATTORI, A. (1970) Solubilization of nitrate reductase from the blue-green alga *Anabaena cylindrica*. *Pl. Cell Physiol., Tokyo* **11**, 975–978.

HATTORI, A. and FUJITA, Y. (1959a) Formation of phycobilin pigments in a blue-green alga, *Tolypothrix tenuis*, as induced by illumination with coloured lights. *J. Biochem., Tokyo* **46**, 521–524.

HATTORI, A. and FUJITA, Y. (1959b) Effect of pre-illumination on the formation of phycobilin pigments in a blue-green alga, *Tolypothrix tenuis*. *J. Biochem., Tokyo* **46**, 1259–1261.

HATTORI, A. and MYERS, J. (1966) Reduction of nitrate and nitrite by subcellular preparations of *Anabaena cylindrica* I. Reduction of nitrite to ammonia. *Pl. Physiol., Lancaster* **41**, 1031–1036.

HATTORI, A. and MYERS, J. (1967) Reduction of nitrate and nitrite by subcellular preparations of *Anabaena cylindrica* II. Reduction of nitrate to nitrite. *Pl. Cell Physiol., Tokyo* **8**, 327–337.

HATTORI, A. and UESUGI, I. (1968a) Purification and properties of nitrite reductase from the blue-green alga *Anabaena cylindrica*. *Pl. Cell Physiol.* **9**, 689–699.

HATTORI, A. and UESUGI, I. (1968b) Ferredoxin-dependent photoreduction of nitrate and nitrite by subcellular preparations of *Anabaena cylindrica*. *In* "Comparative Biochemistry and Biophysics of Photosynthesis" (K. Shibata, A. Takamiya, A. T. Jagendorf and R. C. Fuller, eds.) pp. 201–205. University of Tokyo Press, Tokyo and University Park Press, State College, Pennsylvania.

HAUPT, W. (1965) Perception of environmental stimuli orientating growth and movement in lower plants' *A. Rev. Pl. Physiol.* **16**, 267–290.

HAXO, F. T. (1960) The wavelength dependence of photosynthesis and the role of accessory pigments. *In* "Comparative Biochemistry of Photoreactive Systems" (M. B. Allen, ed.) pp. 339–359. Academic Press, New York and London.

HAXO, F. T. and BLINKS, L. R. (1950) Photosynthetic action spectra of marine algae. *J. gen. Physiol.* **33**, 389–422.

HAYSTEAD, A., ROBINSON, R. and STEWART, W. D. P. (1970) Nitrogenase activity in extracts of heterocystous and non-heterocystous blue-green algae. *Arch. Mikrobiol.* **75**, 235–243.

HAYSTEAD, A. and STEWART, W. D. P. (1972) Characteristics of the nitrogenase system from the blue-green alga *Anabaena cylindrica. Arch. Mikrobiol.* **82**, 325–336.

HEALEY, F. P. (1968) The carotenoids of four blue-green algae. *J. Phycol.* **4**, 126–129.

HEGLER, R. (1901) Untersuchungen über die Organisation der Phycochromaceenzellen. *Jb. wiss. Bot.* **36**, 229–354.

HEILBRON, I. M. and LYTHGOE, B. (1936) The carotenoid pigments of *Oscillatoria rubescens. J. chem. Soc.* 1376–1380.

HENNINGER, M. D., BHAGAVAN, H. N. and CRANE, F. L. (1965) Comparative studies on plastoquinones. I. Evidence for three quinones in the blue-green alga *Anacystis nidulans. Archs Biochem. Biophys.* **110**, 69–74.

HENRIKSSON, E. (1964) "Studies in the Physiology if the Lichen *Collema*". *Acta Univ. upsal.* **38**, 1–13.

HENRIKSSON, E. (1971) Algal nitrogen fixation in temperate regions. *In* "Biological Nitrogen Fixation in Natural and Agricultural Habitats", (T. A. Lie and E. G. Mulder, eds.) pp. 415–419. *Plant and Soil*, Special volume.

HERDMAN, M., FAULKNER, B. M. and CARR, N. G. (1970) Synchronous growth and genome replication in the blue-green alga *Anacystis nidulans. Arch Mikrobiol.* **73**, 238–249.

HÉRISSET, A. (1952) Influence de la lumière sur la fixation biologique de l'azote par le *Nostoc commune. Bull. Soc. Chim,. biol.* **34**, 532–537.

HERTZBERG, S. and LIAAEN-JENSEN, S. (1966a) The carotenoids of blue-green algae. I. The carotenoids of *Oscillatoria rubescens* and an *Arthrospira* sp. *Phytochemistry* **5**, 557–563.

HERTZBERG, S. and LIAAEN-JENSEN, S. (1966b) The carotenoids of blue-green algae. II. The carotenoids of *Aphanizomenon flos-aquae. Phytochemistry* **5**, 565–570.

HERTZBERG, S. and LIAAEN-JENSEN, S. (1967) The carotenoids of blue-green algae. III. A comparative study of mutachrome and flavacin. *Phytochemistry* **5**, 1119–1126.

HERTZBERG, S. and LIAAEN-JENSEN, S. (1969a) The structure of myxoxanthophyll. *Phytochemistry* **8**, 1259–1280.

HERTZBERG, S. and LIAAEN-JENSEN, S. (1969b) The structure of oscillaxanthin. *Phytochemistry* **8**, 1281–1292.

HESS, U. (1962) Über die hydratabhängige Entwicklung und die Austrocknungresistenz von Cyanophyceen. *Arch. Mikrobiol.* **44**, 189–218.

Hitch, C. J. B. and Stewart, W. D. P. (1973) Nitrogen fixation by lichens in Scotland. *New Phytol.* (in press).

Hoare, D. S. and Hoare, S. L. (1966) Feedback regulation of arginine biosynthesis in blue-green algae and photosynthetic bacteria. *J. Bact.* **92**, 375–379.

Hoare, D. S., Hoare, S. L. and Moore, R. B. (1967) The photoassimilation of organic compounds by autotrophic blue-green algae. *J. gen. Microbiol.* **49**, 351–370.

Hoare, D. S., Hoare, S. L. and Smith, H. J. (1969) Assimilation of organic compounds by blue-green algae and photosynthetic bacteria. *In* "Progress in Photosynthesis Research" (H. Metzner, ed.) Vol. 3, pp. 1570–1573. Tübingen.

Hoare, D. S., Ingram, L. O., Thurston, E. L. and Walkup, R. (1971) Dark heterotrophic growth of an endophytic blue-green alga. *Arch. Mikrobiol.* **78**, 310–321.

Hoare, D. S. and Moore, R. B. (1965) Photoassimilation of organic compounds by autotrophic blue-green algae. *Biochim. biophys. Acta* **109**, 622–625.

Hobbie, J. E. and Wright, R. T. (1965) Competition between planktonic bacteria and algae for organic solutes. *Mem. Ist. ital. Idrobiol.* **18** Suppl., 175–185.

Hoch, G., Owens, O. H. and Kok, B. (1963) Photosynthesis and respiration. *Archs Biochem. Biophys.* **101**, 171–180.

Höcht, H., Martin, H. H. and Kandler, O. (1965) Zur Kenntnis der chemischen Zusammensetzung der Zellwand der Blaualgen. *Z. Pflanzenphysiol.* **53**, 39–57.

Hof, T. and Frémy, P. (1933) On Myxophyceae living in strong brines. *Recl Trav. bot. néerl.* **30**, 140–162.

Hofstein, A. V. and Pearson, L. C. (1965) Chromatin distribution in Cyanophyceae. *Hereditas* **53**, 212–220.

Holm-Hansen, O. (1963a) Effect of varying residual moisture content on the viability of lyophilized algae. *Nature, Lond.* **198**, 1014–1015.

Holm-Hansen, O. (1963b) Algae: Nitrogen fixation by Antarctic species. *Science, N.Y.* **139**, 1059–1060.

Holm-Hansen, O. (1964) Isolation and culture of terrestrial and freshwater algae of Antarctica. *Phycologia* **4**, 43–51.

Holm-Hansen, O. (1967) Factors affecting the viability of lyophilized algae. *Cryobiology* **4**, 17–23.

Holm-Hansen, O. (1968) Ecology, physiology and biochemistry of blue-green algae. *A. Rev. Microbiol.* **22**, 47–70.

Holm-Hansen, O. and Brown, G. W. (1963) Ornithine cycle enzymes in the blue-green alga *Nostoc muscorum*. *Pl. Cell Physiol., Tokyo* **4**, 299–306.

Holm-Hansen, O., Gerloff, G. C. and Skoog, F. (1954) Cobalt as an essential element for blue-green algae. *Physiologia Pl.* **7**, 665–675.

Holton, R. W., Blecker, H. H. and Onore, H. (1964) Effect of growth temperature on the fatty acid composition of blue-green algae. *Phytochemistry* **3**, 595–602.

Holton, R. W., Blecker, H. H. and Stevens, T. S. (1968) Fatty acids in blue-green algae: possible relation to phylogenetic position. *Science, N.Y.* **160**, 545–547.

HOLTON, R. W. and FREEMAN, A. W. (1965) Some theoretical considerations of the gliding movement of blue-green algae. *Am. J. Bot.* **52**, 640.

HOLTON, R. W. and MYERS, J. (1963) Cytochromes of a blue-green alga: extraction of a *c*-type with a strongly negative redox potential. *Science, N.Y.* **142**, 234–235.

HOLTON, R. W. and MYERS, J. (1967a) Water-soluble cytochromes from a blue-green alga. I. Extraction, purification and spectral properties of cytochromes *c* (549, 552 and 554, *Anacystis nidulans*). *Biochim. biophys. Acta* **131**, 362–374.

HOLTON, R. W. and MYERS, J. (1967b) Water-soluble cytochromes from a blue-green alga. II. Physicochemical properties and quantitative relationships of cytochromes *c* (549, 552 and 554, *Anacystis nidulans*). *Biochim. biophys. Acta.* **131**, 375–384.

HOOD, W. and CARR, N. G. (1967) A single glyceraldehyde-3-phosphate dehydrogenase active with NAD and NADP in *Anabaena variabilis. Biochim. biophys. Acta* **146**, 309–311.

HOOD, W. and CARR, N. G. (1968) Threonine deaminase and acetolactate synthetase in *Anabaena variabilis. Biochem. J.* **109**, 4P.

HOOD, W. and CARR, N. G. (1971) Apparent lack of control by repression of arginine metabolism in blue-green algae. *J. Bact.* **107**, 365–367.

HOOD, W., LEAVER, A. G. and CARR, N. G. (1969) Extracellular nitrogen and the control of arginine biosynthesis in *Anabaena variabilis. Biochem. J.* **114**, 12P–13P.

HOPWOOD, D. A. and GLAUERT, A. M. (1960) The fine structure of the nuclear material of a blue-green alga, *Anabaena cylindrica* Lemm. *J. biophys. biochem. Cytol.* **8**, 813–823.

HORNE, A. J. (1969) "Nitrogen and carbon fixation in aquatic ecosystems." Ph.D. Thesis, University of Dundee.

HORNE, A. J. (1972) The ecology of nitrogen fixation in Signy Island, South Orkney Islands. *Brit. Antarct. Survey Bull.* **27,** 1—18.

HORNE, A. J. and FOGG, G. E. (1970) Nitrogen fixation in some English lakes. *Proc. R. Soc.* B. **175**, 351–366.

HORNE, A. J. and VINER, A. B. (1971) Nitrogen fixation and its significance in tropical Lake George, Uganda. *Nature, Lond.* **232**, 417–418.

HORTON, A. A. (1968) NADH oxidase in blue-green algae. *Biochem. biophys. Res. Commun.* **32**, 839–845.

HOUGH, L., JONES, J. K. N. and WADMAN, W. H. (1952) An investigation of the polysaccharide components of certain fresh-water algae. *J. chem. Soc.* 3393–3399.

HOWARD, D. L., FREA, J. I., PFISTER, R. M. and DUGAN, P. R. (1970) Biological nitrogen fixation in Lake Erie. *Science, N.Y.* **169**, 61–62.

HOWLAND, G. P. and RAMUS, J. (1971) Analysis of blue-green and red algal ribosomal RNA-s by gel electrophoresis. *Arch. Mikrobiol.* **76**, 292–298.

HUGHES, E. O., GORHAM, P. R. and ZEHNDER, A. (1958) Toxicity of a unialgal culture of *Microcystis aeruginosa. Can. J. Microbiol.* **4**, 225–236.

HUNEKE, A. (1933) Beitrage zur Kenntnis der Symbiose zwischen *Azolla* und *Anabaena. Beitr. Biol. Pfl.* **20**, 315–341.

HUTCHINSON, G. E. (1957) "A Treatise on Limnology" Vol. 1. Geography, physics and chemistry. John Wiley & Sons, New York and London.

HUTCHINSON, G. E. (1967) "A Treatise on Limnology" Vol. 2. Introduction to lake biology and the limnoplankton. John Wiley & Sons, New York and London.

HWANG, S. W. and HORNELAND, W. (1965) Survival of algal cultures after freezing by controlled and uncontrolled cooling. *Cryobiology* **1**, 305–311.

IHA, K., ALLI, M., SINGH, R. and BHATTACHARYA, P. (1965) Increasing rice production through the inoculation of *Tolypothrix tenuis*, a nitrogen-fixing blue-green alga. *J. Indian Soc. Sci.* **13**, 161–167.

INGRAM, L. O. and VAN BAALEN, C. (1970) Characteristics of a stable filamentous mutant of a coccoid blue-green alga. *J. Bact.* **102**, 784–789.

IYENGAR, M. O. P. and DESIKACHARY, T. V. (1953) Occurrence of three-pored heterocysts in *Brachytrichia balani*. Curr. Sci. **22**, 180–181.

JACKIM, E. and GENTILE, J. (1968) Toxins of blue-green algae: similarity to saxitoxin. *Science, N.Y.* **162**, 915–916.

JACKSON, W. S. and VOLK, R. J. (1970) Photorespiration. *A. Rev. Pl. Physiol.* **21**, 385–432.

JAKOB, H. (1954) Compatibilities et antagonismes entre algues du sol. *C.r. hebd. Séanc. Acad. Sci., Paris* **238**, 928–930.

JAROSCH, R. (1959) Zur Gleitbewegung der niederen Organismen. *Protoplasma* **50**, 277–289.

JAROSCH, R. (1962) Gliding. *In* "Physiology and Biochemistry of Algae" (R. A Lewin, ed.) pp. 573–581. Academic Press, New York and London.

JENKIN, P. M. (1957) The filter-feeding and food of flamingoes (Phoenicopteri). *Phil. Trans. R. Soc.* B **240**, 401–493.

JENSEN, T. E. (1968) Electron microscopy of polyphosphate bodies in a blue-green alga, *Nostoc pruniforme*. *Arch. Mikrobiol.* **62**, 144–152.

JENSEN, T. E. (1969) Fine structure of developing polyphosphate bodies in a blue-green alga, *Plectonema boryanum*. *Arch. Mikrobiol.* **67**, 328–338.

JENSEN, T. E. and BOWEN, C. C. (1971) Organization of the centroplasm in *Nostoc pruniforme*. *Iowa Acad. Sci. Proc.* **68**, 86–89.

JENSEN, T. E. and BOWEN, C. C. (1970) Cytology of blue-green algae. II. Unusual inclusions in the cytoplasm. *Cytologia* **35**, 132–152.

JENSEN, T. E. and CLARK, R. L. (1969) Cell wall and coat of the developing akinete of a *Cylindrospermum* species. *J. Bact*, **97**, 1494–1495.

JENSEN, T. E. and SICKO, L. M. (1971a) Fine structure of poly-β-hydroxybutyric acid granules in a blue-green alga, *Chlorogloea fritschii*. *J. Bact.* **106**, 683–686.

JENSEN, T. E. and SICKO, L. M. (1971b) The effect of lysozyme on cell wall morphology in a blue-green alga, *Cylindrospermum* sp. *J. gen. Microbiol.* **68**, 71–75.

JEWELL, W. J. and KULASOORIYA, S. A. (1970) The relation of acetylene reduction to heterocyst frequency in blue-green algae. *J. exp. Bot.* **21**, 881–886.

JOHNSON, G. V., MAYEUX, P. A. and EVANS, H. J. (1966) A cobalt requirement for symbiotic growth of *Azolla filiculoides* in the absence of combined nitrogen. *Pl. Physiol., Lancaster* **41**, 852–855.

JONES, D. D., HAUG, A., JOST, M. and GRABER, D. R. (1969) Ultrastructural and conformational changes in gas vacuole membranes isolated from *Microcystis aeruginosa*. *Archs Biochem. Biophys.* **135**, 296–303.

JONES. D. D. and JOST, M. (1970) Isolation and chemical characterization of gas vacuole membranes from *Microcystis aeruginosa* Kuetz. emend. Elenkin. *Arch. Mikrobiol.* **70**, 43–64.

JONES, D. D. and JOST, M. (1971) Characterization of the protein from gas vacuole membranes of the blue-green alga, *Microcystis aeruginosa*. *Planta* **100**, 277–287.

JONES, J. (1930) An investigation into the bacterial associations of some Cyanophyceae. *Ann. Bot.* **44**, 721–740.

JONES, K. and STEWART, W. D. P. (1969a) Nitrogen turnover in marine and brackish habitats. III. The production of extracellular nitrogen by *Calothrix scopulorum*. *J. mar. biol. Ass. U.K.* **49**, 475–488.

JONES, K. and STEWART, W. D. P. (1969b) Nitrogen turnover in marine and brackish habitats. IV. Uptake of the extracellular products of the nitrogen-fixing alga *Calothrix scopulorum*. *J. mar. biol. Ass. U.K.* **49**, 701–716.

JONES, L. W. and MYERS, J. (1964) Enhancement in the blue-green alga, *Anacystis nidulans*. *Pl. Physiol., Lancaster* **39**, 938–946.

JONES, L. W. and MYERS, J. (1965) Pigment variations in *Anacystis nidulans* induced by light of selected wavelengths. *J. Phycol.* **1**, 7–14.

JOST, M. (1965) Die Ultrastruktur von *Oscillatoria rubescens* D.C. *Arch. Mikrobiol.* **50**, 211–245.

JOST, M. and JONES, D. D. (1970) Morphological parameters and macromolecular organization of gas vacuole membranes of *Microcystis aeruginosa* Kuetz. emend. Elenkin. *Can. J. Microbiol.* **16**, 159–164.

JOST, M. and MATILE, P. (1966) Zur Charakterisierung der Gasvakuolen der Blaualge *Oscillatoria rubescens*. *Arch. Mikrobiol.* **53**, 50–58.

JOST, M. and ZEHNDER, A. (1966) Die Gasvakuolen der Blaualge *Microcystis aeruginosa*. *Schweiz. Z. Hydrol.* **28**, 1–3.

JURGENSEN, M. F. and DAVEY, C. B. (1968) Nitrogen-fixing blue-green algae in acid forest and nursery soils. *Can. J. Microbiol.* **14**, 1179–1183.

KAJIYAMA, S., MATSUKI, T. and NOSOH, Y. (1969) Separation of the nitrogenase system of *Azotobacter* into three components and purification of one of the components. *Biochem. biophys. Res. Commun.* **37**, 711–717.

KALE, S. R. and TALPASAYI, E. R. S. (1969) Heterocysts: a review. *Indian Biologist* **1**, 19–29.

KANDLER, O. (1961) Verteilung von C^{14} nach Photosynthese in $C^{14}O_2$ von *Anacystis nidulans*. *Naturwissenschaften* **48**, 604.

KANN, E. (1959) Die eulitorale Algenzone in Traumsee (Oberösterreich). *Arch. Hydrobiol.* **55**, 129–192.

KANN, E. (1966) Der Algenaufwuchs in einigen Bächen Österreichs. *Verh. int. Ver. theor. angew. Limnol.* **16**, 646–654.

KANTZ, T. and BOLD, H. C. (1969) "Phycological studies. IX. Morphological and taxonomic investigations of *Nostoc* and *Anabaena* in culture." University of Texas Publication no. 6924.

KARRER, P. and RUTSCHMANN, J. (1944) Beitrag zur Kenntnis der Carotinoide aus *Oscillatoria rubescens*. *Helv. chim. Acta* **27**, 1691–1695.

KATOH, S. (1959) Studies on the algal cytochrome of *c*-type. *J. Biochem., Tokyo* **46**, 629–632.

KAUSHIK, M. and KUMAR, H. D. (1970) The effect of light on growth and development of two nitrogen-fixing blue-green algae. *Arch. Mikrobiol.* **74**, 52–57.

KELLENBERGER, E., RYTER, A. and SÉCHAUD, J. (1958) Electron microscope study of DNA-containing plasma. II. Vegetative and mature phage DNA as compared with normal bacterial nucleoids in different physiological states. *J. biophys. biochem. Cytol.* **4**, 671–678.

KENYON, C. N. (1972) Fatty acid composition of unicellular strains of blue-green algae. *J. Bact.* **109**, 827–834.

KENYON, C. N., RIPPKA, R. and STANIER, R. Y. (1972) Fatty acid composition and physiological properties of some filamentous blue-green algae. *Arch. Mikrobiol.* **83**, 216–236.

KENYON, C. N. and STANIER, R. Y. (1970) Possible evolutionary significance of polyunsaturated fatty acids in blue-green algae. *Nature, Lond.* **227**, 1164–1166.

KHOJA, T. and WHITTON, B. A. (1971) Heterotrophic growth of blue-green algae. *Arch. Mikrobiol.* **79**, 280–282.

KINDEL, P. and GIBBS, M. (1963) Distribution of carbon-14 in polysaccharide after photosynthesis in carbon dioxide labelled with carbon-14 by *Anacystis nidulans*. *Nature, Lond.* **200**, 260–261.

KIYOHARA, T., FUJITA, Y., HATTORI, A. and WATANABE, A. (1960) Heterotrophic culture of a blue-green alga, *Tolypothrix tenuis*. *J. gen. appl. Microbiol. Tokyo* **6**, 176–182.

KIYOHARA, T., FUJITA, Y., HATTORI, A. and WATANABE, A. (1962) Effect of light on glucose assimilation in *Tolypothrix tenuis*. *J. gen. appl. Microbiol., Tokyo* **8**, 165–168.

KLEBAHN, H. (1895) Gasvakuolen, ein Bestandteil der Zellen der wasserblütebildenden Phycochromaceen. *Flora, Jena* **80**, 241–282.

KLEBAHN, H. (1922) Neue Untersuchungen über die Gasrakuolen. *Jb. wiss. Bot.* **61**, 535–589.

KLEIN, G. (1915) Zur Chemie der Zellhaut der Cyanophyceen. *Sber. Akad. Wiss. Wien* **121**, 529–545.

KLEIN, R. M. (1970) Relationships between blue-green and red algae. *In* "Phylogenesis and Morphogenesis in the Algae" (J. F. Fredrick and R.M. Klein, eds.) pp. 623–633. *Ann. N.Y. Acad. Sci.* **175**.

KLUCAS, R. V. (1967) Ph.D. Thesis, University of Wisconsin.

KNOBLOCH, K. (1966) Photosynthetische Sulfid-Oxydation grüner Pflanzen. I. *Planta* **70**, 73–86.

KOBAYASHI, M., TAKAHASHI, E. and KAWAGUCHI, K. (1967) Distribution of nitrogen-fixing microorganisms in paddy soils of Southeast Asia. *Soil Sci.* **104**, 113–118.

KOHL, F. G. (1903) "Über die Organisation und Physiologie der Cyanophyceenzelle und die mitotische Teilung ihres Kernes". Gustav Fischer, Jena.

KOLKWITZ, R. (1928) Über Gasvakuolen bei Bakterien. *Ber. dt. bot. Ges.* **46**, 29–34.

KORSCHIKOFF, A. A. (1924) Protistologische Beobachtungen. I. *Cyanophora paradoxa. Arch. russ. Protistenk.* **3**, 57–74.

KRASILNIKOV, N. A. (1949) Does *Azotobacter* occur in lichens? *Mikrobiologiya* **18**, 3–6.

KRATZ, W. A. and MYERS, J. (1955a) Nutrition and growth of several blue-green algae. *Am. J. Bot.* **42**, 282–287.

KRATZ, W. A. and MYERS, J. (1955b) Photosynthesis and respiration of three blue-green algae. *Pl. Physiol., Lancaster* **30**, 275–280.

KRAUS, M. P. (1966) Preparation of pure blue-green algae. *Nature, Lond.* **211**, 310.

KRAUSS, R. W. (1960) Transaction of 1960 Seminar on Algae and Metropolitan Wastes. Robert A. Taft Sanitary Engineering Centre. Technical Report W61–3, p. 40. U.S. Public Health Service, Cincinatti, Ohio.

KRAUSS, R. V. (1962) Inhibitors. *In* "Physiology and Biochemistry of Algae" (R. A. Lewin, ed.) pp. 673–685. Academic Press, New York and London.

KRINSKY, N. I. (1966) The role of carotenoid pigments as protective agents against photosensitized oxidations in chloroplasts. *In* "Biochemistry of Chloroplasts" (T. W. Goodwin, ed.) Vol. 1, pp. 423–430. Academic Press, London and New York.

KUBÍN, Š. (1959) Catalase activity in thermal blue-green algae in relation to temperature. *Biologia Pl., Prague* **1**, 3–8.

KUCHKAROVA, M. A. (1962) The algal flora of rice fields in the Tashkent region. *Uzb. biol. Zh.* **1**, 35–38.

KUHL, A. (1968) Phosphate metabolism of green algae. *In* "Algae, Man and Environment" (D. F. Jackson, ed.) pp. 37–52. Syracuse University Press, Syracuse, N.Y.

KULASOORIYA, S. A. (1971) "Studies on heterocyst differentiation and nitrogen fixation in blue-green algae". Ph.D. Thesis, University of London.

KULASOORIYA, S. A., LANG, N. J. and FAY, P. (1972) The heterocysts of blue-green algae. III. Differentiation and nitrogenase activity. *Proc. R. Soc.* B **181**, 199–209.

KUMAR, H. D. (1962) Apparent genetic recombination in a blue-green alga. *Nature, Lond.* **196**, 1121–1122.

KUMAR, H. D. (1963) Effects of radiations on blue-green algae. I. The production and characterization of a strain of *Anacystis nidulans* resistant to ultraviolet radiation. *Ann. Bot.* **27**, 723–733.

KUMAR, H. D. (1964) Streptomycin- and penicillin-induced inhibition of growth and pigment production in blue-green algae and production of strains of *Anacystis nidulans* resistant to these antibiotics. *J. exp. Bot.* **15**, 232–250.

KUMAR, H. D. (1968) Inhibitory action of the antibiotics mitomycin C and neomycin on the blue-green alga *Anacystis nidulans*. *Flora, Jena* **159**, 437–444.

KUMAR, H. D., SINGH, H. N. and PRAKASH, G. (1967) The effect of proflavine on different strains of the blue-green alga *Anacystis nidulans*. *Pl. Cell Physiol., Tokyo* **8**, 171–179.

KUNISAWA, R. and COHEN-BAZIRE, G. (1970) Mutations of *Anacystis nidulans* that affect cell division. *Arch. Mikrobiol.* **71**, 49–59.

KYLIN, H. (1943) Zur Biochemie der Cyanophyceen. *Kgl. Fysiograf. Sällskap. Lund Förh.* **13**, no. 7, 1–14.

LAMI, R. (1958) Extension de deux Chlorophycees dans la Manche accidentale. *Revue algol.* **4**, 61–63.

LAMONT, H. C. (1969a) Shear-orientated microfibrils in the mucilaginous investments of two motile oscillatoriacean blue-green algae. *J. Bact.* **97**, 350–361.

LAMONT, H. C. (1969b) Sacrificial cell death and trichome breakage in an oscillatoriacean blue-green alga: the role of murein. *Arch. Mikrobiol.* **69**, 237–259.

LANG, N. J. (1965) Electron microscopic study of heterocyst development in *Anabaena azollae* Strasburger. *J. Phycol.* **1**, 127–134.

LANG, N. J. (1968) The fine structure of blue-green algae. *A. Rev. Microbiol.* **22**, 15–46.

LANG, N. J. and Fay, P. (1971) The heterocysts of blue-green algae. II. Details of ultrastructure. *Proc. R. Soc.* B **178**, 193–203.

LANG, N. J. and FISHER, K. A. (1969) Variation in the fixation image of "structured granules" in *Anabaena*. *Arch. Mikrobiol.* **67**, 173–181.

LANG, N. J. and RAE, P. M. M. (1967) Structures in a blue-green alga resembling prolamellar bodies. *Protoplasma* **64**, 67–74.

LANG, N. J., SIMON, R. D. and WOLK, C. P. (1972) Correspondence of cyanophycin granules with structured granules in *Anabaena cylindrica*. *Arch. Mikrobiol.* **83**, 313–320.

LANGE, W. (1970) Cyanophyta-bacteria systems: effects of added carbon compounds or phosphate on algal growth at low nutrient concentrations. *J. Phycol.* **6**, 230–234.

LAPORTE, G. and POURROIT, R. (1967) Fixation de l'azote atmospherique par les algues Cyanophycées. *Revue Ecol. Biol. Sol.* **4**, 81–112.

LARSEN, H., OMANG, S. and STEENSLAND, H. (1967) On the gas vacuoles of halobacteria. *Arch. Mikrobiol.* **59**, 197–203.

LAUTERBORN, R. (1895) Protozoenstudien. II. *Paulinella chromatophora* nov. gen. nov. spec. *Z. wiss. Zool.* **59**, 537–544.

LAZAROFF, N. (1966) Photoinduction and photoreversal of the Nostocacean developmental cycle. *J. Phycol.* **2**, 7–17.

LAZAROFF, N. and VISHNIAC, W. (1961) The effect of light on the developmental cycle of *Nostoc muscorum*, a filamentous blue-green alga. *J. gen. Microbiol.* **25**, 365–374.

LAZAROFF, N. and VISHNIAC, W. (1962) The participation of filament anastomosis in the developmental cycle of *Nostoc muscorum*, a blue-green alga. *J. gen. Microbiol.* **28**, 203–210.

Lazaroff, N. and Vishniac, W. (1964) The relationship of cellular differentiation to colonial morphogenesis of the blue-green alga, *Nostoc muscorum*. *J. gen. Microbiol.* **35**, 447–457.

Leach, C. K. and Carr, N. G. (1968) Reduced nicotinamide-adenine dinucleotide phosphate oxidase in the autotrophic blue-green alga *Anabaena variabilis*. *Biochem. J.* **109**, 4P–5P.

Leach, C. K. and Carr, N. G. (1969) Oxidative phosphorylation in an extract of *Anabaena variabilis*. *Biochem. J.* **112**, 125–126.

Leach, C. K. and Carr, N. G. (1970) Electron transport and oxidative phosphorylation in the blue-green alga *Anabaena variabilis*. *J. gen. Microbiol.* **64**, 55–70.

Leach, C. K. and Carr, N. G. (1971) Pyruvate: ferredoxin oxidoreductase and its activation by ATP in the blue-green alga *Anabaena variabilis*. *Biochim. biophys. Acta* **245**, 165–174.

Leak, L. V. (1965) Electron microscopic autoradiography. Incorporation of H^3-thymidine in a blue-green alga, *Anabaena* sp. *J. ultrastruct. Res.* **12**, 135–146.

Leak, L. V. (1967a) Studies on the preservation and organization of DNA-containing regions in a blue-green alga, a cytochemical and ultrastructural study. *J. ultrastruct. Res.* **20**, 190–205.

Leak, L. V. (1967b) Fine structure of the mucilaginous sheath of *Anabaena* sp. *J. ultrastruct. Res.* **21**, 61–74.

Leak, L. V. and Wilson, G. B. (1960) The distribution of chromatin in the blue-green alga *Anabaena variabilis* Kütz. *Can. J. Genet. Cytol.* **2**, 320–324.

Leak, L. V. and Wilson, G. B. (1965) Electron microscopic observations on a blue-green alga, *Anabaena* sp. *Can. J. Genet. Cytol.* **7**, 237–249.

Le Campion-Alsumard, T. (1970) Cyanophycées marines endolithes colonisant les surfaces rocheuses denudées (Etages supralittoral et mediolittoral de la région de Marseille). *Schweiz. Z. Hydrol.* **32**, 552–558.

Lefèvre, M. (1964) Extracellular products of algae. *In* "Algae and Man" (D. F. Jackson, ed.) pp. 337–367. Plenum Press, New York.

Lefort, N. (1960a) Structure inframicroscopique du chromatoplasma de quelques Cyanophycées. *C. r. hebd. Séanc. Acad. Sci., Paris* **250**, 1525–1527.

Lefort, M. (1960b) Nouvelles recherches sur l'infrastructure de chromatoplasma des Cyanophycées. *C. r. hebd. Séanc. Acad. Sci., Paris* **251**, 3046–3048.

Lefort, M. (1965) Sur le chromatoplasma d'une cyanophycée endosymbiotique: *Glaucocystis nostochinearum* Itzigs. *C. r. hebd. Séanc. Acad. Sci., Paris* **261**, 223–236.

Lehman, H. and Jost, M. (1971) Kinetics of the assembly of gas vacuoles in the blue-green alga *Microcystis aeruginosa* Kuetz. emend. Elenkin. *Arch. Microbiol.* **79**, 59–68.

Lemmermann, E. (1910) "Kryptogamenflora der Mark Brandenburg. Algen 1." Verlag von Gebrüder Borntraeger, Leipzig.

Lester, R. L. and Crane, F. L. (1959) The natural occurrence of coenzyme Q and related compounds. *J. biol. Chem.* **234**, 2169–2175.

LEVIN, E. Y. and BLOCH, Y. (1964) Absence of sterols in blue-green algae. *Nature, Lond.* **202**, 90–91.

LEVINE, R. P. (1968) Genetic dissection of photosynthesis. *Science, N.Y.* **162**, 768–771.

LEWIN, R. A. (1959) The isolation of algae. *Revue algol.* **3**, 181–197.

LEWIN, R. A. (1960) A Spirochaete phage. *Nature, Lond.* **186**, 901–902.

LEWIN, R. A. (1966) Apochlorotic cyanophytes and their allies. *In* "Cultures and Collections of Algae" (A. Watanabe and A. Hattori, eds.) pp. 25–29. Japanese Society of Plant Physiologists, Tokyo.

LEWIN, R. A. (1969) The classification of flexibacteria. *J. gen. Microbiol.* **58**, 189–206.

LEWIN, R. A., CROTHERS, D. M., CORRELL, D. L. and REIMANN, B. E. (1964) A phage infecting *Saprospira grandis. Can. J. Microbiol.* **10**, 75–78.

LEX, M., SILVESTER, W. and STEWART, W. D. P. (1972) Photorespiration and nitrogenase activity in the blue-green alga, *Anabaena cylindrica. Proc. R. Soc.* B **180**, 87–102.

LEX, M. and STEWART, W. D. P. (1973) Algal nitrogenase, reductant pools and Photosystem I activity. *Biochim. biophys. Acta* **292**, 436–443

LEY, S. H. (1959) The effect of nitrogen-fixing blue-green algae on the yields of rice plant. *Acta hydrobiol. sin.* **4**, 440–444.

LIGHTBODY, J. J. and KROGMANN, D. W. (1967) Isolation and properties of plastocyanin from *Anabaena variabilis. Biochim. biophys. Acta* **131**, 508–515.

LINKO, P., HOLM-HANSEN, O., BASSHAM, J. A. and CALVIN, M. (1957) Formation of radioactive citrulline during photosynthetic $C^{14}O_2$ fixation by blue-green algae. *J. exp. Bot.* **8**, 147–156.

LINDSTEDT, A. (1943) Die Flora der marinen Cyanophyceen der schwedischen Westküste. Håkan Ohlssons Buchdrukerei, Lund.

LOENING, U. E. (1968) Molecular weights of ribosomal RNA in relation to evolution. *J. molec. Biol.* **38,** 355–365.

LORENZEN, H. and VENKATARAMAN, G. S. (1969) Synchronous cell division in *Anacystis nidulans* Richter. *Arch. Mikrobiol.* **67**, 252–255.

LOTSY, J. P. (1907) "Vorträge über botanische Stammesgeschichte". Gustav Fischer, Jena.

LUFT, J. H. (1956) Permanganate—a new fixative for electron microscopy. *J. biophys. biochem. Cytol.* **2**, 799–802.

LUFTIG, R. and HASELKORN, R. (1967) Morphology of a virus of blue-green algae and properties of its deoxyribonucleic acid. *J. Virol.* **1**, 334–361.

LUND, J. W. G. (1947) Observations on soil algae II. *New Phytol.* **46**, 35–60.

LUND, J. W. G. (1957) Fungal diseases of plankton algae. *In* "Biological Aspects of the Transmission of Disease" (C. Horton-Smith, ed.) pp. 19–23. Oliver and Boyd, Edinburgh and London.

LUND, J. W. G. (1964) Primary production and periodicity of phytoplankton. *Verh. int. Ver. theor. angew. Limnol.* **15**, 37–56.

LUND, J. W. G. (1967) Soil algae. In "Soil Biology" (A. Burges and F. Raw, eds.) pp. 129–148. Academic Press, London and New York.

MacRae, I. C. and Castro, T. F. (1967) Nitrogen fixation in some tropical rice soils. *Soil Sci.* **103**, 277–280.

Maertens, H. (1914) Das Wachstum von Blaualgen in mineralischen Nährlösungen. *Beitr. Biol. Pfl.* **12**, 439–496.

Magee, W. E and Burris, R. H. (1954) Fixation of N_2 and utilization of combined nitrogen by *Nostoc muscorum. Am. J. Bot.* **11**, 777–782.

Marker, A. F. H. (1965) Extracellular carbohydrate liberation in the flagellates *Isochrysis galbana* and *Prymnesium parvum. J. mar. biol. Ass. U.K.* **45**, 755–772.

Marler, J. E and Van Baalen, C. (1965) Role of H_2O_2 in single-cell growth of the blue-green alga, *Anacystis nidulans. J. Phycol.* **1**, 180–185.

Marré, E. (1962) Temperature. *In* "Physiology and Biochemistry of Algae" (R. A. Lewin, ed.) pp. 541–550. Academic Press, New York and London.

Martinez Nadal, N. G. (1971) Sterols of *Spirulina maxima. Phytochemistry* **10**, 2537–2538.

Marty, F. and Busson, F. (1970) Données cytologiques sur deux Cyanophycées: *Spirulina platensis* (Gom.) Geitler et *Spirulina geitleri* J. de Toni. *Schweiz. Z. Hydrol.* **32**, 559–565.

Materassi, R. and Balloni, W. (1965) Quelques observations sur la présence de micro-organismes autotrophes fixateurs d'azote dans les rivières. *Annls Inst. Pasteur, Paris* **3**, Suppl. 218–223.

Mayland, H. F. and McIntosh, T. H. (1966) Availability of biologically fixed nitrogen-15 to higher plants. *Nature, Lond.* **209**, 421–422.

Mayland, H. F., McIntosh, T. H. and Fuller, W. H. (1966) Fixation of isotopic nitrogen on a semi-arid soil by algal crust organisms. *Soil Sci. Soc. Am. Proc.* **30**, 56–60.

McCarty, R. E. and Coleman, C. H. (1969) The uncoupling of photophosphorylation by valinomycin and ammonium chloride. *J. biol. Chem.* **244**, 4292–4298.

McDaniel, H. R., Middlebrook, J. B. and Bowman, R. O. (1962) Isolation of pure cultures of algae from contaminated cultures. *Appl. Microbiol.* **10**, 223.

McLachlan, J. and Gorham, P. R. (1961) Growth of *Microcystis aeruginosa* Kütz in a precipitate-free medium buffered with TRIS. *Can. J. Microbiol.* **7**, 869–882.

McLachlan, J. and Gorham, P. R. (1962) Effects of pH and nitrogen sources on growth of *Microcystis aeruginosa* Kütz. *Can. J. Microbiol.* **8**, 1–11.

McLeod, G. C., Curby, W. A. and Bibblis, F. (1962) The study of the physiological characteristics of *Trichodesmium thiebautii.* A.E.C. Report Contrib. AT(30–1)2646. Bermuda Biol. Station.

Meeuse, B. J. D. (1962) Storage products. *In* "Physiology and Biochemistry of Algae" (R. A. Lewin ed.) pp. 289–313. Academic Press, New York and London.

Menke, W. (1961) Über das Lamellarsystem des Chromatoplasmas von Cyanophyceen. *Z. Naturf.* **16**, 543–546

Mereschkowsky, C. (1905) Über Natur and Ursprung der Chromatophoren in Pflanzenreiche. *Biol. Zbl.* **25**, 593–604.

METZNER, I. (1955) Zur Chemie und zum submikroskopischen Auflbau der Zell-wände, Scheiden und Gallerten von Cyanophyceen. *Arch. Mikrobiol.* **22**, 45–77.

MICKELSON, J. C., DAVIS, E. B. and TISCHER, R. G. (1967) The effect of various nitrogen sources upon heterocyst formation in *Anabaena flos-aquae* A–37. *J. exp. Bot.* **18**, 397–405.

MILLBANK, J. W. (1972) Nitrogen metabolism in lichens. IV. The nitrogenase activity of the *Nostoc* phycobiont in *Peltigera canina*. *New Phytol.* **71**, 1–10.

MILLBANK, J. W. and KERSHAW, K. A. (1969) Nitrogen metabolism in lichens. I. Nitrogen fixation in the cephalodia of *Peltigera aphthosa*. *New Phytol.* **68**, 721–729.

MILLER, A. G., CHANG, K. H. and COLEMAN, B. (1971) The uptake and oxidation of glycolic acid by blue-green algae. *J. Phycol.* **7**, 97–100.

MILLER, M. M. and LANG, N. J. (1968) The fine structure of akinete formation and germination in *Cylindrospermum*. *Arch. Mikrobiol.* **60**, 303–313.

MILLER, S. L. (1953) Production of amino acids under possible primitive earth conditions. *Science, N.Y.* **117**, 528–529.

MISHUSTIN, E. N. and SHIL'NIKOVA, V. K. (1971) "Biological Fixation of Atmospheric Nitrogen". MacMillan Press, London.

MITRA, A. K. (1950) Two new algae from Indian soils. *Ann. Bot.* **14**, 457–464.

MITRA, A. K. (1951) The algal flora of certain Indian soils. *Indian J. agric. Sci.* **21**, 357–373.

MITRA, A. K. (1961) Some aspects of fixation of elementary nitrogen by blue-green algae in the soil. *Proc. natn. Acad. Sci. India* A **31**, 98–99.

MITSUI, A. and ARNON, D. I. (1971) Crystalline ferredoxin from a blue-green alga, *Nostoc* sp. *Phycologia Pl.* **25**, 135–140.

MOEWUS, L. (1953) About the occurrence of fresh water algae in the semi-desert around Broken Hill. *Bot. Notiser* **114**, 399–416.

MOIKEHA, S. N. and CHU, G. W. (1971) Dermatitis-producing alga *Lyngbya majuscula* Gomont in Hawaii. II. Biological properties of the toxic factor. *J. Phycol.* **7**, 8–13.

MOIKEHA, S. N., CHU, G. W. and BERGER, L. R. (1971) Dermatitis-producing alga *Lyngbya majuscula* Gomont in Hawaii. I. Isolation and chemical characterization of the toxic factor. *J. Phycol.* **7**, 4–8.

MOORE, B. G. and TISCHER, R. G. (1965) Biosynthesis of extracellular polysaccharides by the blue-green alga *Anabaena flos-aquae*. *Can. J. Microbiol.* **11**, 877–885.

MOORE, R. T. and MCALEAR, J. H. (1960) Fine structure of Mycota. 2. Demonstration of the haustoria of lichens. *Mycologia* **52**, 805–807.

MOSS, B. (1968) The chlorophyll *a* content of some benthic algal communities. *Arch. Hydrobiol.* **65**, 51–62.

MOSS, B. and MOSS, J. (1969) Aspects of the limnology of an endorheic African lake (L. Chilwa, Malawi). *Ecology* **50**, 109–118.

MOYSE, A. and GUYON, D. (1963) Effet de la temperature sur l'efficacite de la phycocyanine et de la chlorophylle chez *Aphanocapsa* (Cyanophycée). *In* "Microalgae and Photosynthetic Bacteria" (Jap. Soc. Pl. Physiologists, ed.) pp. 253–270. University of Tokyo Press, Tokyo.

MUNK, W. H. and RILEY, G. A. (1952) Absorption of nutrients by aquatic plants. *J. mar. Res.* **11**, 215–240.

MUZAFAROV, A. M. (1953) Importance of blue-green algae in the fixation of atmospheric nitrogen. *Trudy Bot. Akad. Nauk. UzbSSR* (2), 3–11.

MYER, F. H. (1966) Mycorrhiza and other plant symbioses. *In* "Symbioses" (S. M. Henry, ed.) pp. 171–244. Academic Press, New York.

MYERS, J. and CLARK, L. B. (1944) Culture conditions and the development of the photosynthetic mechanism. II. An apparatus for the continuous culture of *Chlorella. J. gen. Physiol.* **28**, 103–112.

MYERS, J. and KRATZ, W. A. (1955) Relations between pigment content and photosynthetic characteristics in a blue-green alga. *J. gen. Physiol.* **39**, 11–22.

NADSON, G. (1900) Die perforierenden (kalkbohrenden) Algen und ihre Bedeutung in der Natur. *Script. Bot. Hort. Univ. Petropolitani* **18**, 1–40.

NADSON, G. (1932) Contribution à l'étude des algues perforantes. *Bull. Acad. Sci. URSS, Cl. sci. mat. nat.* **7**, 833–855.

NAKAMURA, H. (1938) Über die Kohlenäsureassimilation bei niederen Algen in Anwesenheit des Schwefelwasserstoffes. *Acta phytochim., Tokyo* **10**, 271–281.

NASH, A. (1938) "The Cyanophyceae of the thermal regions of Yellowstone National Park, U.S.A., and of Rotorua and Whakarewarewa, New Zealand, with some ecological data." Ph.D. Thesis, University of Minnesota, Minneapolis.

NAUWERCK, A. (1963) Die Beziehungen zwischen Zooplankton und Phytoplankton im See Erken. *Symb. bot. upsal.* **17**, 1–163.

NAWAWY, A. S., LOTEI, M. and FAHMY, M. (1958). Studies on the ability of some blue-green algae to fix atmospheric nitrogen and their effect on growth and yield of paddy soils. *Agric. Res. Rev.* **36**, 308–320.

NEILSON, A., RIPPKA, R. and KUNISAWA, R. (1971) Heterocyst formation and nitrogenase synthesis in *Anabaena* sp. *Arch. Mikrobiol.* **76**, 139–150.

NEUMANN, J., OGAWA, T. and VERNON, L. P. (1970) Increased rate of cyclic photophosphorylation in preparations from *Anabaena variabilis* cells grown in the presence of diphenylamine. *FEBS Letters* **10**, 253–256.

Nichols, B. W. (1970) Comparative lipid biochemistry of photosynthetic organisms. *In* "Phytochemical Phylogeny" (J. B. Harbourne, ed.) pp. 105–118. Academic Press, London.

NICHOLS, B. W. and WOOD, B. J. B. (1968) New glycolipid specific to nitrogen-fixing blue-green algae. *Nature, Lond.* **217**, 767–768.

NIENBURG, W. (1916) Die Perzeption des Lichtreizes bei den Oscillarien und ihre Reaktionen auf Intensitätsschwankungen. *Z. Bot.* **8**, 161–193.

NIKLITSCHEK, A. (1934) Das Problem der Oscillatorien-Bewegung. I. Die Bewegungserscheinungen der Oscillatorien. *Beih. bot. Zbl.* A **52**, 205–257.

NIKLOWITZ, W. and DREWS, G. (1956) Beiträge zur Cytologie de Blaualgen. I. Untersuchungen zur Substruktur von *Phormidium uncinatum* Gom. *Arch. Mikrobiol.* **24**, 134–146.

NIKLOWITZ, W. and DREWS, G. (1957) Beiträge zur Cytologie der Blaualgen. IV. Vergleichende elektronenmikroskopische Untersuchungen zur Substruktur einiger Hormogonales. *Arch. Mikrobiol.* **27**, 150–165.

Norris, L., Norris, R. E. and Calvin M. (1955) A survey of the rates and products of short-term photosynthesis in plants of nine phyla. *J. exp. Bot.* **6**, 64–74.

Nultsch, W. (1961) Der Einfluss des Lichtes auf die Bewegung der Cyanophyceen. I. Phototopotaxis von *Phormidium autumnale. Planta* **56**, 632–647.

Nultsch, W. (1962a) Der Einfluss des Lichtes auf die Bewegung der Cyanophyceen. II. Photokinesis bei *Phormidium autumnale. Planta* **57**, 613–623.

Nultsch, W. (1962b) Der Einfluss des Lichtes auf die Bewegung der Cyanophyceen. III. Photophobotaxis von *Phormidium uncinatum. Planta* **58**, 647–663.

Odintzova, S. V. (1941) Role of blue-green algae in nitrate formation in deserts. *Dokl.* (*Proc.*) *Akad. Sci. USSR* **32**, 578–580.

Oehler, J. H. and Schopf, J. W. (1971) Artificial microfossils: experimental studies of permineralization of blue-green algae in silica. *Science, N.Y.* **174**, 1229–1231.

Ogawa, R. E. and Carr, J. F. (1969) The influence of nitrogen on heterocyst production in blue-green algae. *Limnol. Oceanogr.* **14**, 342–351.

Ogawa, T., Kanai, R. and Shibata, K. (1968) Distribution of carotenoids in the two photochemical systems of higher plants and algae. *In* "Comparative Biochemistry and Biophysics of Photosynthesis" (K. Shibata, A. Takamiya, A. T. Jagendorf and R. B. Fuller, eds.) pp. 22–35. University of Tokyo Press, Tokyo and University Park Press, State College, Pennsylvania.

Ogawa, T. and Vernon, L. P. (1969) A fraction from *Anabaena variabilis* enriched in the reaction center chlorophyll P700. *Biochim. biophys. Acta* **180**, 334–346.

Ogawa, T. and Vernon, L. P. (1970) Properties of partially purified photosynthetic reaction centers from *Scenedesmus* mutant 6E and *Anabaena variabilis* grown in the presence of diphenylamine. *Biochim. biophys. Acta* **197**, 292–301.

Ogawa, T., Vernon, L. P. and Mollenhauer, H. H. (1969) Properties and structure of fractions prepared from *Anabaena variabilis* by the action of Triton X–100. *Biochim. biophys. Acta* **172**, 216–229.

Ohmori, K. and Hattori, A. (1970) Induction of nitrate and nitrite reductases in *Anabaena cylindrica. Pl. Cell Physiol., Tokyo* **11**, 873–878.

Olson, J. M. (1970) The evolution of photosynthesis. *Science, N.Y.* **168**, 438–446.

Oparin, A. J. (1957) "The Origin of Life on the Earth". Oliver and Boyd, Edinburgh.

Oppenheim, J. and Marcus, L. (1970) Correlation of ultrastructure in *Azotobacter vinelandii* with nitrogen source for growth. *J. Bact.* **101**, 286–291.

Öquist, G. (1971) Changes in pigment composition and photosynthesis induced by iron deficiency in the blue-green alga *Anacystis nidulans. Physiologia Pl.* **25**, 188–191.

Oró, J. and Nooner, D. W. (1967) Aliphatic hydrocarbons in Pre-Cambrian rocks. *Nature, Lond.* **213**, 1082–1085.

Padan, E., Raboy, B. and Shilo, M. (1971) Endogenous dark respiration of the blue-green alga *Plectonema boryanum. J. Bact.* **106**, 45–50.

PADAN, E. and SHILO, M. (1968) Distribution of cyanophages in natural habitats. *Verh. int. Ver. theor. angew. Limnol.* **17**, 747–751.

PADAN, E., SHILO, M. and KISLEV, N. (1967) Isolation of "cyanophages" from freshwater ponds and their interaction with *Plectonema boryanum. Virology* **32**, 234–246.

PAN, P. (1972) Growth of a photoautotroph, *Plectonema boryanum*, in the dark on glucose, *Can. J. Microbiol.* **18**, 275–288.

PANDEY, D. C. (1965) A study of the algae from paddy soils of Ballia and Ghazipur districts of Uttar Pradesh, India. I. Cultural and ecological considerations. *Nova Hedwigia* **9**, 299–334.

PANKOW, H. (1964) Bemerkung über die Schädlichkeit von Blaualgenwasserblüten für Tiere. *Naturwissenschaften* **51**, 146–147.

PANKOW, H. and MAERTENS, B. (1964) Über *Nostoc sphaericum* Vauch. *Arch. Mikrobiol.* **48**, 203–212.

PANKRATOVA, E. M. and VAKHRUSHEV, A. S. (1969) Assimilation by higher plants of atmospheric nitrogen fixed by blue-green algae. *Mikrobiologiya* **38**, 928–931.

PANKRATZ, H. S. and BOWEN, C. C. (1963) Cytology of blue-green algae. I. The cells of *Symploca muscorum. Am. J. Bot.* **50**, 387–399.

PANOSYAN, A. K. and NIKOGOSYAN, K. (1966) The presence of nitrogen fixers in lichens. *Biol. Zh. Arm.* **19**, 3–11.

PAPAGEORGIOU, G. and GOVINDJEE (1967) Oxygen evolution from lyophilized *Anacystis* with carbon dioxide as oxidant. *Biochim. biophys. Acta* **131**, 173–178.

PARK, R. B. and BIGGINS, J. (1964) Quantasome: size and composition. *Science, N.Y.* **144**, 1009–1010.

PARKINSON, D., GRAY, T. R. G. and WILLIAMS, S. T. (1972) "Methods for studying the ecology of soil microorganisms." International Biological Program Handbook no. 19. Blackwell, Oxford.

PASCHER, A. (1914a) Über Flagellaten und Algen. *Ber. dt. bot. Ges.* **32**, 136–160.

PASCHER, A. (1914b) Über Symbiosen von Spaltpilzen und Flagellaten mit Blaualgen. *Ber. dt. bot. Ges.* **32**, 339–352.

PASCHER, A. (1929a) Über einige Endosymbiosen von Blaualgen in Einzellern. *Jb. wiss. Bot.* **71**, 386–462.

PASCHER, A. (1929b) Über die Natur der blaugrünen Chromatophoren des Rhizopoden *Paulinella chromatophora. Zool. Anz.* **81**, 189–194.

PATTNAIK, H. (1966) Studies on nitrogen fixation by *Westiellopsis prolifica* Janet. *Ann. Bot.* **30**, 231–238.

PAUL, E. A., MYERS, R. J. K. and RICE, W. A. (1971) Nitrogen fixation in grassland and associated cultivated ecosystems. *In* "Biological Nitrogen Fixation in Natural and Agricultural Habitats". (T. A. Lie and E. G. Mulder, eds.) pp. 495–507. *Pl. Soil*, special volume.

PAYNE, P. I. and DYER, T. A. (1972) Plant 5·8 S RNA is a component of 80 S but not 70 S ribosomes. *Nature, Lond.* **235**, 145–147.

PEARCE, J. and CARR, N. G. (1967) The metabolism of acetate by the blue-green algae, *Anabaena variabilis* and *Anacystis nidulans. J. gen. Microbiol.* **49**, 301–313.

PEARCE, J. and CARR, N. G. (1969) The incorporation and metabolism of glucose by *Anabaena variabilis. J. gen. Microbiol.* **54**, 451–462.

PEARCE, J., LEACH, C. K. and CARR, N. G. (1969) The incomplete tricarboxylic acid cycle in the blue-green alga *Anabaena variabilis. J. gen. Microbiol.* **55**, 371–378.

PEARSALL, W. H. (1932) Phytoplankton in the English lakes. II. The composition of the phytoplankton in relation to dissolved substances. *J. Ecol.* **20**. 241–262.

PEARSON, J. E. and KINGSBURY, J. (1966) Culturally induced variation in four morphologically diverse blue-green algae. *Am. J. Bot.* **53**, 192–200.

PEARY, J. A. and CASTENHOLZ, R. W. (1964) Temperature strains of a thermophilic blue-green alga. *Nature, Lond.* **202**, 720–721.

PEARY, J. A. and GORHAM, P. R. (1966) Influence of light and temperature on growth and toxin production by *Anabaena flos-aquae* NRC-44-1. *J. Phycol.* **2** (Suppl.), 3–4.

PEAT, A. and WHITTON, B. A. (1967) Environmental effects on the structure of the blue-green alga, *Chlorogloea fritschii. Arch. Mikrobiol.* **57**, 155–180.

PEAT, A. and WHITTON, B. A. (1968) Vegetative cell structure in *Anabaenopsis* sp. *Arch. Mikrobiol.* **63**, 170–176.

PECK, H. D. (1968) Energy-coupling mechanism in chemolithotrophic bacteria. *A. Rev. Microbiol.* **22**, 489–518.

PELROY, R. A. and BASSHAM, A. L. (1972) Photosynthetic and dark carbon metabolism in unicellular blue-green algae. *Arch. Mikrobiol.* **86**, 25–38.

PELROY, R. A., RIPPKA, R. and STANIER, R. Y. (1972) Metabolism of glucose by unicellular blue-green algae. *Arch. Mikrobiol.* **87**, 303–322.

PERCIVAL, E. and MCDOWELL, R. H. (1967) "Chemistry and Enzymology of Marine Algal Polysaccharides". Academic Press, London and New York.

PERMINOVA, G. N. (1964) Effect of blue-green algae on the development of micro-organisms in the soil. *Mikrobiologiya* **33**, 472–476.

PETERSEN, J. B. (1935) Studies on the biology and taxonomy of soil algae. *Dansk. bot. Ark.* **8**, 1–180.

PETRACK, B. and LIPMANN, F. (1961) Photophosphorylation and photohydrolysis in cell-free preparations of blue-green algae. *In* "Light and Life" (W. D. McElroy and B. Glass, eds.) pp. 621–630. John Hopkins Press, Baltimore.

PETTER, H. F. M. (1932) Over roode en andere bakterien van gezouten visch. Doctoral Thesis, University of Utrecht, Netherlands.

PFLEIDERER, W. (1964) Recent developments in the chemistry of pteridines. *Angew. Chem.* (intern. edit.) **3**, 114–132.

PFLUG, H. D. (1967) Structured organic remains from the Fig Tree Series (Precambrian) of the Barberton Mountain Land (South Africa). *Rev. Palaeobot. Palynol.* **5**, 9–29.

PFLUG, H. D., MEINEL, W., NEUMANN, K. H. and MEINEL, M. (1969) Entwicklungstendenzen des frühen Lebens auf der Erde. *Naturwissenschaften* **56**, 10–14.

PICKEN, L. E. R. (1936) Mechanical factors in the distribution of a blue-green alga, *Rivularia haematites. New Phytol.* **35**, 221–228.

PIERSON, B. K. and CASTENHOLZ, R. W. (1971) Bacteriochlorophylls in gliding prokaryotes from hot springs. *Nature, Lond.* **233**, 25–27.

PIGOTT, G. H. and CARR, N. G. (1972) Homology between nucleic acids of blue-green-algae and chloroplasts of *Euglena gracilis. Science, N.Y.* **175**, 1259–1261.

PIKÁLEK, P. (1967) Attempt to find genetic recombination in *Anacystis nidulans. Nature. Lond.* **215**, 666–667.

PINTER, I. J. and PROVASOLI, L. (1958) Artificial cultivation of a red-pigmented marine blue-green alga, *Phormidium persicinum. J. gen. Microbiol.* **18**, 190–197.

PIRIE, N. W. (1969) "Food Resources: Conventional and Novel". Penguin Books, Harmondsworth.

PLESSL, A. (1963) Über die Beziehung von Haustorientypus und Organisationshöhe bei Flechten. *Öst. bot. Z.* **110**, 193–269.

POLJANSKY, G. and PETRUSCHEWSKY, G. (1929) Zur Frage über die Struktur der Cyanophyceenzelle. *Arch. Protistenk.* **67**, 11–45.

POWRIE, J. K. (1964) The effect of cobalt on the growth of young lucerne on siliceous sand. *Pl. Soil* **21**, 81–93.

PRELL, H. (1921) Zur Theorie der sekretorischen Ortsbewegung. II. Die Bewegung der Gregarines. *Arch. Protistenk.* **42**, 157–175.

PRESCOTT, G. W. (1951) "Algae of the Western Great Lakes Area". Cranbrook Institute of Science, Bulletin no. 30.

PRICE, J. H. (1971) The shallow sublittoral marine ecology of Aldabra. *Phil. Trans. R. Soc.* B **260**, 123–172.

PRINGSHEIM, E. G. (1914) Kulturversuche mit chlorophyllführenden Mikroorganismen. III. Zur Physiologie der Schizophyceen. *Beitr. Biol. Pfl.* **12**, 49–108.

PRINGSHEIM, E. G. (1946) "Pure Cultures of Algae". Univeristy Press, Cambridge.

PRINGSHEIM, E. G. (1949) The relationship between bacteria and Myxophyceae. *Bact. Rev.* **13**, 47–98.

PRINGSHEIM, E. G. (1963) "Farblose Algen. Ein Beitrag zur Evolutionsforschung". Gustav Fischer, Stuttgart, Germany.

PRINGSHEIM, E. G. (1964) Heterotrophism and species concepts in *Beggiatoa. Am. J. Bot.* **51**, 898–913.

PRINGSHEIM, E. G. (1966) The nature of pseudovacuoles in Cyanophyceae. *Nature, Lond.* **210**, 549–550.

PROCTOR, V. W. (1957a) Some controlling factors in the distribution of *Haematococcus pluvialis. Ecology* **38**, 457–462.

PROCTOR, V. W. (1957b) Studies of algal antibiosis using *Haematococcus* and *Chlamydomonas. Limnol. Oceanogr.* **2**, 125–139.

PROVASOLI, L. and Pintner, I. J. (1953) Ecological implications of *in vitro* nutritional reqirements of algal flagellates. *Ann. N.Y. Acad. Sci.* **56**, 839–851.

PROWSE, G. A. and TALLING, J. F. (1958) The seasonal growth and succession of plankton algae in the White Nile. *Limnol. Oceanogr.* **3**, 222–238.

RABINOWITCH, E. I. (1956) "Photosynthesis and Related Processes". Vol. 2. Interscience Publ., New York.

RABINOWITCH, E. and GOVINDJEE (1969) "Photosynthesis". John Wiley and Sons, New York and London.

RAMAMURTHY, V. D. (1970) Experimental study relating to red tide. *Mar. Biol.* **5**, 203–204.

RAMAMURTHY, V. D. and KRISHNAMURTHY, S. (1968) Nitrogen fixation by the blue-green alga *Trichodesmium erythraeum* (Ehr.) *Curr. Sci.* **37**, 21–22.

REINKE, J. (1879) Zwei parasitische Algen. *Bot. Ztg.* **37**, 473–478.

REITZ, R. C. and HAMILTON, J. G. (1964) The isolation and identification of two sterols from two species of blue-green algae. *Comp. Biochem. Physiol.* **25**, 401–416.

REYNOLDS, C. S. (1967) The breaking of the Shropshire meres. *Shropshire Conservation Trust Bull.* **10**, 9–14.

REYNOLDS, C. S. (1971) The ecology of the planktonic blue-green algae in the North Shropshire meres. *Fld Stud.* **3**, 409–432.

RICHTER, G. (1961) Die Auswirkung von Mangan-Mangel auf Wachstum und Photosynthese bei der Blaualge *Anacystis nidulans*. *Planta* **57**, 202–214.

RIPPKA, R., NEILSON, A., KUNISAWA, R. and COHEN-BAZIRE, G. (1971) Nitrogen fixation by unicellular blue-green algae. *Arch. Mikrobiol.* **76**, 341–348.

RIS, H. and SINGH, R. N. (1961) Electron microscope studies on blue-green algae. *J. biophys. biochem. Cytol.* **9**, 63–80.

ROBINSON, B. L. and MILLER, J. H. (1970) Photomorphogenesis in the blue-green alga *Nostoc commune* 584. *Physiologia Pl.* **23**, 461–472.

ROBINSON, D. G. and PRESTON, R. D. (1971a) Studies on the fine structure of *Glaucocystis nostochinearum* Itzigs. *J. exp. Bot.* **22**, 635–643.

ROBINSON, D. G. and PRESTON, R. D. (1971b) Studies on the fine structure of *Glaucocystis nostochinearum* Itzigs. II. Membrane morphology and taxonomy. *Br. phycol. J.* **6**, 113–128.

ROGERS, R. W., LANGE, R. T. and NICHOLAS, D. J. T. (1966) Nitrogen fixation by lichens of arid soil crusts. *Nature, Lond.* **209**, 96–97.

ROTH, A. G. (1797–1806) "Catalecta Botanica". Vol. 1–3. Leipzig.

ROUND, F. E. (1957) Studies on bottom-living algae in some lakes of the English Lake District. III. The distribution of the sediments of algal groups other than the Bacillariophyceae. *J. Ecol.* **45**, 649–664.

ROUND, F. E. (1961) Studies on the bottom-living algae in some lakes of the English Lake District. V. The seasonal cycles of the Cyanophyceae. *J. Ecol.* **49**, 31–38.

ROUND, F. E. (1965) "The Biology of the Algae". Edward Arnold, London.

RUBENCHIK, L. I., BERSHOVA, O. I., NOVIKOVA, N. S. and KOPTYEVA, Z. P. (1966) Lysis of the blue-green alga *Microcystis pulverea*. (Ukranian) *Mikrobiol. Zh. Akad. Nauk. Ukr. RSR* **28**, 88–91.

RURIANSKI, H. J., RANDLES, J. and HOCH, G. E. (1970) A comparative study of photosynthetic electron transport in algal cells and spinach chloroplasts. *Biochim. biophys. Acta* **205**, 254–262.

RUSNESS, D. and BURRIS, R. H. (1970) Acetylene reduction (nitrogen fixation) in Wisconsin lakes. *Limnol. Oceanogr.* **15**, 808–813.

RUTTEN, M. G. (1962) "The Geological Aspects of the Origin of Life on Earth". Elsevier, Amsterdam.

RZOSKA, J. BROOK, A. J. and PROWSE, G. A. (1955) Seasonal plankton development in the White and Blue Nile near Khartoum. *Verh. int. Ver. theor. angew. Limnol.* **12**, 327–334.

SACHS, J. (1874) "Lehrbuch der Botanik". 4th ed. Leipzig.

SAFFERMAN, R. S., DIENER, T. O., DESJARDINS, P. R. and MORRIS, M. E. (1972) Isolation and characterization of AS-1, a phycovirus infecting the blue-green alga *Anacystis nidulans* and *Synechococcus cedrorum*. *Virology* **47**, 105–113.

SAFFERMAN, R. S. and MORRIS, M. E. (1963) Algal virus: isolation. *Science, N.Y.* **140**, 679–680.

SAFFERMAN, R. S. and MORRIS, M. E. (1964) Growth characteristics of the blue-green algal virus LPP-1. *J. Bact.* **88**, 771–775.

SAFFERMAN, R. S. and Morris, M. E. (1967) Observations on the occurrence, distribution and seasonal incidence of blue-green algal viruses. *Appl. Microbiol.* **15**, 1219–1222.

SAFFERMAN, R. S., SCHNEIDER, I. R., STEERE, R. L., Morris, M. E. and DIENER, T. O. (1969) Phycovirus SM-1: a virus infecting unicellular blue-green algae. *Virology* **37**, 386–395.

SAGAN, L. (1967) On the origin of mitosing cells. *J. theor. Biol.* **14**, 225–274.

SARÀ, M. (1971) Ultrastructural aspects of the symbiosis between two species of the genus *Aphanocapsa* (Cyanophyceae) and *Ircinia variabilis* (Demospongiae). *Mar. Biol.* **11**, 214–221.

Sauvageau, C. (1897) Sur le *Nostoc punctiforme*. *Annls. Sci. nat.* (*Bot.*) VIII, **3**, 366–378.

SAWYER, P. J., GENTILE, J. H. and SASNER, J. J. (1968) Demonstration of a toxin from *Aphanizomenon flos-aquae* (L.) Ralfs. *Can. J. Microbiol.* **14**, 1199–1204.

SCHAEDE, R. (1951) Über die Blaualgensymbiose von *Gunnera*. *Planta* **39**, 154-170.

SCHELL, P. M. and ALEXANDER, V. (1970) Improved incubation and gas sampling techniques for nitrogen fixation studies. *Limnol. Oceanogr.* **15**, 961–962.

SCHLICHTING, H. E. (1969) The importance of airborne algae and protozoa. *J. Air Poll. Control Ass.* **19**, 946–951.

SCHLOESING, T. and LAURENT, E. (1892) Sur la fixation de l'azote libre par les plantes. *Annls Inst. Pasteur, Paris* **6**, 824–840.

SCHNEIDER, K. C., BRADBEER, C., SINGH, R. N., WANG, L. C., WILSON, P. W. and BURRIS, R. H. (1960) Nitrogen fixation by cell-free preparations from microorganisms. *Proc. natn. Acad. Sci. USA* **46**, 726–733.

SCHNEPF, E. (1964) Zur Feinstruktur von *Geosiphon pyriforme*. Ein Versuch zur Deutung cytoplasmatischer Membranen und Kompartimente. *Arch. Mikrobiol.* **49**, 112–131.

SCHNEPF, E. (1965) Struktur der Zellwände und Cellulosefibrillen bei *Glaucocystis*. *Planta* **67**, 213–224.

SCHNEPF, E. and KOCH, W. (1966) Golgi-Apparat und Wasserausscheidung bei *Glaucocystis*. *Z. Pflanzenphysiol.* **55**, 97–109.

SCHNEPF, E., KOCH, W. and DEICHGRÄBER, G. (1966) Zur Cytologie und taxonomischer Einordnung von *Glaucocystis*. *Arch. Mikrobiol.* **55**, 149–174.

SCHÖLLHORN, R. and BURRIS, R. H. (1966) Study of intermediates in nitrogen fixation. *Fedn Proc. Fedn Am. Socs exp. Biol.* **25**, 710.

SCHÖLLHORN, R. and BURRIS, R. H. (1967) Reduction of azide by the N_2-fixing enzyme system. *Proc. natn. Acad. Sci. U.S.A.* **57**, 1317–1323.

SCHOPF, J. W. (1970) Electron microscopy of organically preserved Precambrian fossils. *J. Palaeontol.* **44**, 1–6.

SCHOPF, J. W. (1973) "The Age of Blue-green Algae". (In press.)

SCHOPF, J. W. and BARGHOORN, E. S. (1967) Algae-like fossils from the early Precambrian of South Africa. *Science, N.Y.* **156**, 508–512.

SCHOPF, J. W. and BARGHOORN, E. S. (1969) Microorganisms from the late Precambrian of South Africa. *J. Palaeontol.* **43**, 111–118.

SCHRAM, B. L. and KROES, H. H. (1971) Structure of phycocyanobilin. *Eur. J. Biochem.* **19**, 581–594.

SCHULZ, G. (1955) Bewegungsstudien sowie elektronenmikroskopische Membranuntersuchungen an Cyanophyceen. *Arch. Mikrobiol.* **21**, 335–370.

SCHWABE, G. H. (1966) Ökologischer Charakter und System der Cyanophyten. *Verh. int. Ver. theor. angew. Limnol.* **16**, 1541–1548.

SCHWABE, G. H. and EL AYOUTY, E. (1966) Über die hormogonale Blaualgen aus indischen Böden. *Nova Hedwigia* **10**, 527–536.

SCHWIMMER, D. and SCHWIMMER, M. (1964) Algae and medicine. *In* "Algae and Man", (D. F. Jackson, ed.) pp. 368–412. Plenum Press, New York.

SCHWIMMER, D. and SCHWIMMER, M. (1968) Medical aspects of phycology. *In* "Algae, Man and Environment" (D. F. Jackson, ed.) pp. 279–358. Syracuse University Publ., Syracuse, N.Y.

SCOTT, G. D. (1956) Further investigations of some lichens for fixation of nitrogen. *New Phytol.* **55**, 111–116.

SCOTT, W. E. and FAY, P. (1972) Phosphorylation and amination in heterocysts of *Anabaena variabilis*. *Br. Phycol. J.* **7**, 283–284.

SDOBNIKOVA, N. V. (1958) On the algal flora of the takyrs in the northern part of the Turansk Plain. *Bot. Zh. SSSR* **43**, 1675–1681. (Russian.)

SETCHELL, W. A. (1903) The upper temperature limits of life. *Science, N.Y.* **17**, 934–937.

ŠETLÍCK, T., ŠUST, V. and MÁLEK, T. (1970) Dual purpose open circulation units for large scale culture of algae in temperate zones. I. Basic design considerations and scheme of a pilot plant. *Algol. Studies* (*Trebon*) **1**, 111–164.

SHATKIN, A. J. (1960) A chlorophyll-containing cell-fraction from the blue-green alga *Anabaena variabilis*. *J. biophys. biochem. Cytol.* **7**, 583–584.

SHESTAKOV, S. V. and KHYEN, N. T. (1970) Evidence for genetic transformation in blue-green alga *Anacystis nidulans*. *Molec. gen. Genetics* **107**, 372–375.

SHIELDS, L. M. and DROUET, F. (1962) Distribution of terrestrial algae within the Nevada test site. *Am. J. Bot.* **49**, 547–554.

SHIELDS, L. M. and DURREL, L. W. (1964) Algae in relation to soil fertility. *Bot. Rev.* **30**, 92–128.

SHILO, M. (1966) Predatory bacteria. *Science J.* (Sept. 1966) 33–38.

SHILO, M. (1970) Lysis of blue-green algae by Myxobacter. *J. Bact.* **104**, 453–461.

SHTINA, E. A. (1959) Algae of sod-podzolic soils of the Kirov district. *Trudy bot. Inst. Akad. Nauk. SSSR.*, Ser. 2, *Sporovye Rasteniya* **12**, 36–141.

SHTINA, E. A. (1961) Experience in the cultivation of algae as fertilizer. *In* "Proceedings of the All-Union Conference on the Cultivation of Unicellular Algae", pp. 64–65. Leningrad. (Russian.)

SHTINA, E. A. (1963) Nitrogen fixation by blue-green algae. *Uspekhi sorr Biologii* **56**, 284–299. (Russian.)

SHTINA, E. A. (1964a) The role of the blue-green algae in processes of soil formation. *In* "Biology of the Cyanophyta" (V. D. Fedorov and M. M. Telitchenko, eds.) pp. 66–79. University Press, Moscow. (Russian.)

SHTINA, E. A. (1964b) Role of blue-green algae in the accumulation of nitrogen in the soil. *Agrokhimiya* **4**, 77–83. (Russian.)

SHTINA, E. A. (1965a) The role of the blue-green algae in processes of soil formation. *In* "Biology of the Cyanophyta" (V. D. Federov and M. M. Telitchenko, eds.) pp. 66–79. University Press, Moscow. (Russian.)

SHTINA, E. A. (1965b) Features of the communities of algae in deep chernozems of the Central Chernozem Reserve. *Trudy Tsentral'no Chernozemnogo Zapovednika* **9**, 146–155. (Russian.)

SHTINA, E. A. and ROIZIN, M. B. (1966) The algae of the podzolic soils of Khibin. *Bot. Zh.* **51**, 509–519.

SIEBOLD, VON C.T. (1849) Über einzellige Pflanzen und Tiere. *Z. wiss. Zool.* A **1**, 270–294.

SIEGELMAN, H. W., TURNER, B. C. and HENDRICKS, S. B. (1966) The chromophore of phytochrome. *Pl. Physiol., Lancaster* **41**, 1289–1292.

SILVESTER, W. B. and SMITH, D. R. (1969) Nitrogen fixation by *Gunnera-Nostoc* symbiosis. *Nature, Lond.* **224,** 1231.

SIMON, R. D. (1971) Cyanophycin granules from the blue-green alga *Anabaena cylindrica*: a reserve material consisting of copolymers of aspartic acid and arginine. *Proc. natn. Acad. Sci. U.S.A.* **68**, 265–267.

SINGH, H. N. (1968) Effect of acriflavine on ultra-violet sensitivity of normal, ultra-violet sensitive and ultra-violet resistant strains of a blue-green alga, *Anacystis nidulans*. *Radiat. Bot.* **8**, 355–361.

SINGH, H. N. and SHRIVASTAVA, B. S. (1968) Studies on morphogenesis in a blue-green alga. I. Effect of inorganic nitrogen sources on developmental morphology of *Anabaena doliolum*. *Can. J. Microbiol.* **14**, 1341–1346.

SINGH, R. N. (1950) Reclamation of 'Usar' lands in India through blue-green algae. *Nature, Lond.* **165**, 325–326.

SINGH, R. N. (1955) Limnological relations of Indian inland waters with special reference to water blooms. *Verh. int. Ver. theor. angew. Limnol.* **12**, 831–836.

SINGH, R. N. (1961) "Role of Blue-green Algae in Nitrogen Economy of Indian Agriculture". Indian Council of Agricultural Research, New Delhi.

SINGH, R. N. and SINGH, P. K. (1967) Isolation of cyanophages from India. *Nature, Lond.* **216**, 1020–1021.

SINGH, R. N. and SINHA, R. (1965) Genetic recombination in a blue-green alga, *Cylindrospermum majus* Kuetz. *Nature, Lond.* **207**, 782–783.

SINGH, R. N. and SUBBARAMAIAH, K. (1970) Effects of chemicals on *Fischerella muscicola* (Thuret) Gom. *Can. J. Microbiol.* **16**, 193–199.

SINGH, R. N. and TIWARI, D. N. (1969) Induction by ultraviolet irradiation of mutation in the blue-green alga *Nostoc linckia* (Roth) Born. et Flah. *Nature, Lond.* **221**, 62–64.

SINGH, R. N. and TIWARI, D. N. (1970) Frequent heterocyst germination in the blue-green alga *Gloeotrichia ghosei* Singh. *J. Phycol.* **6**, 172–176.

SIRENKO, L. A., STETSENKO, N. M., ARENDARCHUK, V. V. and KUZ'MENKO, M. I. (1968) Role of oxygen conditions in the vital activity of certain blue-green algae. *Mikrobiologiya* **37**, 199–202.

SKUJA, H. (1948) Taxonomie des Phytoplanktons einiger Seen in Uppland Schweden. *Symb. bot. upsal.* **9**, 1–399.

SMILLIE, R. M. (1965) Isolation of two proteins with chloroplast ferredoxin activity from a blue-green alga. *Biochem. biophys. Res. Commun.* **20**, 621–629.

SMITH, A. J., LONDON, J. and STANIER, R. Y. (1967) Biochemical basis of obligate autotrophy in blue-green algae and thiobacilli. *J. Bact.* **94**, 972–983.

SMITH, D. C. and DREW E. A. (1965) Studies on the physiology of lichens. V. Translocation from the algal layer to the medulla in *Peltigera polydactyla*. *New Phytol.* **64**, 195–200.

SMITH, D. C., MUSCATINE, L. and LEWIS, D. (1970) Carbohydrate movement from autotrophs to heterotrophs in parasitic and mutualistic symbiosis. *Biol.* Rev. **44**, 17–90.

SMITH, G. M. (1938) "Cryptogamic Botany". Vol. 1. McGraw-Hill Book Co., New York.

SMITH, G. M. (ed.) (1961) "Manual of Phycology". Chronica Botanica Co., Waltham, Massachusetts.

SMITH, J. H. C. and FRENCH, C. S. (1963) The major and accessory pigments in photosynthesis. *A. Rev. Pl. Physiol.* **14**, 181–224.

SMITH, K. M., BROWN, R. M. and WALNE, P. L. (1967) Ultrastructural and time-lapse studies on the replication cycle of the blue-green algal virus LPP-1. *Virology* **31**, 329–337.

SMITH, R. V. and EVANS, M. C. W. (1970) Soluble nitrogenase from vegetative cells of the blue-green alga *Anabaena cylindrica*. *Nature. Lond.* **225**, 1253–1254.

SMITH, R. V. and EVANS, M. C. W. (1971) Nitrogenase activity in cell-free extracts of the blue-green alga, *Anabaena cylindrica*. *J. Bact.* **105**, 913–917.

SMITH, R. V., NOY, R. J. and EVANS, M. C. W. (1971) Physiological electron donor systems to the nitrogenase of the blue-green alga *Anabaena cylindrica*. *Biochim. biophys. Acta* **253**, 104–109.

SMITH, R. V. and PEAT, A. (1967a) Comparative structure of the gas vacuoles of blue-green algae. *Arch. Mikrobiol.* **57**, 111–122.

SMITH, R. V. and PEAT, A. (1967b) Growth and gas vacuole development in vegetative cells of *Anabaena flos-aquae*. *Arch. Mikrobiol.* **58**, 117–126.

SMITH, R. V., PEAT, A. and BAILEY, C. J. (1969) The isolation and characterization of gas cylinder membranes and α-granules from *Anabaena flos-aquae* D 124. *Arch. Mikrobiol.* **65**, 87–97.

SMITH, R. V., TELFER, A. and EVANS, M. C. W. (1971) Complementary functioning of nitrogenase components from a blue-green alga and a photosynthetic bacterium. *J. Bact.* **107**, 574–575.

SORIANO, E. and LEWIN, R. A. (1965) Gliding microbes: some taxonomic reconsiderations. *Antonie van Leeuwenhoek* **31**, 66–80.

SOROKIN, C. (1959) Tabular comparative data for the low- and high-temperature strains of *Chlorella*. *Nature, Lond.* **184**, 613–614.

SPEARING, J. K. (1937) Cytological studies of the Myxophyceae. *Arch. Protistenk.* **89**, 209–278.

SRIVASTAVA, B. S. (1969) Ultra-violet induced mutations to growth factor requirement and penicillin resistance in a blue-green alga. *Arch. Mikrobiol.* **66**, 234–238.

STACEY, K. A. (1956) "Light-scattering in Physical Chemistry". Butterworths, London.

STADELMANN, E. J. (1962) Permeability. *In* "Physiology and Biochemistry of Algae" (R. A. Lewin, ed.) pp. 493–528. Academic Press, New York and London.

STANIER, R. Y., KUNISAWA, R., MANDEL, M. and COHEN-BAZIRE, G. (1971) Purification and properties of unicellular blue-green algae (order Chroococcales). *Bact. Rev.* **35**, 171–205.

STANIER, R. Y. and VAN NIEL, C. B. (1962) The concept of a bacterium. *Arch. Mikrobiol.* **42**, 17–35.

STARR, T. J. (1956) Relative amounts of vitamin B_{12} in detritus from oceanic and estuarine environments near Lapelo Islands, Georgia. *Ecology* **37**, 658–664.

STAUB, R. (1961) Ernährungsphysiologisch-autökologische Untersuchungen an der planktischen Blaualge *Oscillatoria rubescens* DC. *Schweiz. Z. Hydrol.* **23**, 82–198a.

STEIN, J. R. (1963) Morphological variation of a *Tolypothrix* in culture. *Br. phycol. Bull.* **2**, 206–209.

STEPHENSON, T. A. and STEPHENSON, A. (1949) The universal features of zonation between tidemarks on rocky coasts. *J. Ecol.* **37**, 289–305.

STEVENS, S. E. and VAN BAALEN, C. (1969) N-methyl-N'-nitro-N-nitrosoguanidine as a mutagen for blue-green algae: evidence for repair. *J. Phycol.* **5**, 136–139.

STEVENS, S. E. and VAN BAALEN, C. (1970) Growth characteristics of selected mutants of a coccoid blue-green alga. *Arch. Mikrobiol.* **72**, 1–8.

STEWART, J. R. and BROWN, R. M. (1969) Cytophaga that kills and lyses algae. *Science, N.Y.* **164**, 1523–1524.

STEWART, J. R. and BROWN, R. M. (1971) Algicidal non fruiting myxobacteria with high G + C ratios. *Arch. Mikrobiol.* **80**, 170–190.

STEWART, K. W. and SCHLICHTING, H. E. (1966) Dispersal of algae and protozoa by selected aquatic insects. *J. Ecol.* **54**, 551–562.

STEWART, W. D. P. (1962) Fixation of elemental nitrogen by marine blue-green algae. *Ann. Bot.* **26**, 439–445.

STEWART, W. D. P. (1964a) Nitrogen fixation by Myxophyceae from marine environments. *J. gen. Microbiol.* **36**, 415–422.

STEWART, W. D. P. (1964b) The effect of available nitrate and ammonium nitrogen on the growth of two nitrogen-fixing blue-green algae. *J. exp. Bot.* **15**, 138–145.

STEWART, W. D. P. (1965) Nitrogen turnover in marine and brackish habitats. I. Nitrogen fixation. *Ann. Bot.* **20**, 229–239.

STEWART, W. D. P. (1966) "Nitrogen Fixation in Plants". Athlone Press, London.

STEWART, W. D. P. (1967a) Nitrogen turnover in marine and brackish habitats. II. Use of ^{15}N in measuring nitrogen fixation in the field. *Ann. Bot.* **31**, 385–407.

STEWART, W. D. P. (1967b) Transfer of biologically fixed nitrogen in a sand dune slack region. *Nature, Lond.* **214**, 603–604.

STEWART, W. D. P. (1967c) Nitrogen-fixing plants. *Science, N.Y.* **158**, 1426–1432.

STEWART, W. D. P. (1968) Nitrogen input into aquatic ecosystems. *In* "Algae, Man and Environment" (D. F. Jackson, ed.) pp. 53–72. University Press, Syracuse, N.Y.

STEWART, W. D. P. (1969) Biological and ecological aspects of nitrogen fixation by free-living micro-organisms. *Proc. R. Soc.* B **172**, 367–388.

STEWART, W. D. P. (1970a) Algal fixation of atmospheric nitrogen. *Pl. Soil* **32**, 555–588.

STEWART, W. D. P. (1970b) Nitrogen fixation by blue-green algae in Yellowstone thermal areas. *Phycologia* **9**, 261–268.

STEWART, W. D. P. (1971a) Nitrogen fixation in the sea. *In* "Fertility of the Sea" (J. D. Costlow, ed.) pp. 537–564. Gordon and Breach, London.

STEWART, W. D. P. (1971b) Physiological studies on nitrogen-fixing blue-green algae. *In* "Biological Nitrogen Fixation in Natural and Agricultural Habitats" (T. A. Lie and E. G. Mulder, eds.) pp. 377–391. *Pl. Soil*, special volume.

STEWART, W. D. P. (1972a) Algal metabolism and water pollution in the Tay Region. *Proc. R. Soc. Edinb.* **71**, 209–224.

STEWART, W. D. P. (1972b) Heterocysts of blue-green algae. *In* "Taxonomy and Biology of Blue-green Algae" (T.V. Desikachary, ed.) pp. 227–235. University of Madras.

STEWART, W. D. P. (1973a) Blue-green algae. *In* "Nitrogen Fixation" (R. W. F. Hardy, ed.) Vol. 2, Wiley-Interscience, New York. (In press.)

STEWART, W. D. P. (1973b) Nitrogen fixation. *In* "The Biology of Blue-green Algae" (N. G. Carr and B. A. Whitton, eds.) Blackwells, Oxford.

STEWART, W. D. P. (1973c) Nitrogen fixation by photosynthetic micro-organisms. *A. Rev. Microbiol.* (In press.)

STEWART, W. D. P. (1973d) Blue-green algae. *In* "Biological Nitrogen Fixation" (A. Quispel, ed.) North Holland Publishing Co., Amsterdam. (In press.)

STEWART, W. D. P. and ALEXANDER, G. (1971) Phosphorus availability and nitrogenase activity in aquatic blue-green algae. *Freshwater Biol.* **1**, 389–404.

STEWART, W. D. P. and Boalch, G. T. (1973) Occurrence of *Trichodesmium thiebautii* in the south west English Channel. *Br. Phycol. J.* (In press.)

STEWART, W. D. P., FITZGERALD, G. P. and BURRIS, R. H. (1967) *In situ* studies on N_2 fixation using the acetylene reduction technique. *Proc. natn. Acad. Sci. U.S.A.* **58**, 2071–2078.

STEWART, W. D. P., FITZGERALD, G. P. and BURRIS, R. H. (1968) Acetylene reduction by nitrogen-fixing blue-green algae. *Arch. Mikrobiol.* **62**, 336–348.

STEWART, W. D. P., FITZGERALD, G. P. and BURRIS, R. H. (1970) Acetylene reduction assay for determination of phosphorus availability in Wisconsin lakes. *Proc. natn. Acad. Sci. U.S.A.* **66**, 1104–1111.

STEWART, W. D. P., HAYSTEAD, A. and PEARSON, H. W. (1969) Nitrogenase activity in heterocysts of blue-green algae. *Nature, Lond.* **224**, 226–228.

STEWART, W. D. P. and LEX, M. (1970) Nitrogenase activity in the blue-green alga, *Plectonema boryanum* strain 594. *Arch. Mikrobiol.* **73**, 250–260.

STEWART, W. D. P., MAGUE, T., FITZGERALD, G. P. and BURRIS, R. H. (1971) Nitrogenase activity in Wisconsin lakes of differing degrees of eutrophication. *New Phytol.* **70**, 497–509.

STEWART, W. D. P. and PEARSON, H. W. (1970) Effects of aerobic and anaerobic conditions on growth and metabolism of blue-green algae. *Proc. R. Soc.* B **175**, 293–311.

STEWART, W. D. P. and PUGH, G. F. J. (1963) Blue-green algae of a developing salt marsh. *J. mar. biol. Ass. U.K.* **43**, 309–317.

STITZENBERGER, E. (1860) "Dr. L. Rabenhorsts Algen Sachsens etc., Systematisch Geordnet mit Zugrundlegung Eines Neuen Systems". Dresden.

STOECKENIUS, W. and KUNAU, W. H. (1968) Further characterization of particulate fractions from lysed cell envelopes of *Halobacterium halobium* and isolation of gas vacuole membranes. *J. Cell Biol.* **38**, 336–357.

STRANSKY, H. and HAGER, A. (1970) Das Carotinoidmuster und die Verbreitung des lichtinduzierten Xanthophyllcyclus in verschiedenen Algenklassen. IV. Cyanophyceae und Rhodophyceae. *Arch. Mikrobiol.* **72**, 84–96.

STRODTMANN, S. (1895) Die Ursache des Schwebevermögens bei den Cyanophyceen. *Biol. Zbl.* **15**, 113–115.

SUBRAHMANYAN, R., RELWANI, L. L. and MANNA, G. B. (1965) Nitrogen enrichment of rice soils by blue-green algae and its effect on the yield of paddy. *Proc. natn. Acad. Sci., India* **35**, 382–386.

SUNG-HONG HIEN (1957) Use of *Azolla* in fertilizing rice fields in the Vietnamese Democratic Republic. (Quoted by E. N. Mishustin and V. K. Shil'nikova, 1971, *in* "Biological Fixation of Atmospheric Nitrogen". Macmillan, London.).

SUSOR, W. A., DUANE, W. C. and KROGMANN, D. W. (1964) Studies on photosynthesis using cell-free preparations of blue-green algae. *Rec. Chem. Prog.* **25**, 197–208.

SUSOR, W. A. and KROGMANN, D. W. (1964) Hill activity in cell-free preparations of a blue-green alga. *Biochim. biophys. Acta* **88**, 11–19.

Susor, W. A. and Krogmann, D. W. (1966) Triphosphopyridine nucleotide photoreduction with cell-free preparations of *Anabaena variabilis*. *Biochim. biophys. Acta* **120**, 65–72.

Taha, M. S. (1963) Isolation of some nitrogen-fixing blue-green algae from the rice fields of Egypt in pure culture. *Mikrobiologiya* **32**, 421–425.

Taha, M. S. (1964) Effect of nitrogen compounds on the growth of blue-green algae and the fixation of molecular nitrogen by them. *Mikrobiologiya* **33**, 352–358.

Takesige, T. (1937) Die Bedeutung der Symbiose zwischen einigen endophytischen Blaualgen und ihren Wirtspflanzen. *Bot. Mag., Tokyo* **51**, 514–524.

Talling, J. F. (1961) Photosynthesis under natural conditions. *A. Rev. Pl. Physiol.* **12**, 133–154.

Talling, J. and Rzoska, J. (1967) The development of plankton in relation to hydrological regime in the Blue Nile. *J. Ecol.* **55**, 637–662.

Talpasayi, E. R. S. (1963) Polyphosphate containing particles of blue-green algae. *Cytologia* **28**, 76–80.

Talpasayi, E. R. S. (1967) Localization of ascorbic acid in heterocysts of blue-green algae. *Curr. Sci.* **36**, 190–191.

Talpasayi, E. R. S. and Bahal, M. R. (1967) Cellular differentiation in *Anabaena cylindrica*. *Z. Pflanzenphysiol.* **56**, 100–101.

Talpasayi, E. R. S. and Kale, K. S. (1967) Induction of heterocysts in the blue-green alga *Anabaena ambigua*. *Curr. Sci.* **36**, 218–219.

Taylor, M. M. and Storck, R. (1964) Uniqueness of bacterial ribosomes. *Proc. natn. Acad. Sci. U.S.A.* **52**, 958–965.

Tchan, Y. T. (1953) Study of soil algae. I. Fluorescence microscopy for the study of soil algae. *Proc. Linn. Soc. N.S.W.* **77**, 265–269.

Tchan, Y. T. and Beadle, N. C. W. (1955) Nitrogen economy in semi-arid plant communities. II. The non-symbiotic nitrogen-fixing organisms. *Proc. Linn. Soc. N.S.W.* **80**, 97–104.

Thomas, J. (1970) Absence of the pigments of photosystem II of photosynthesis in heterocysts of a blue-green alga. *Nature, Lond.* **228**, 181–183.

Thomas, J., David, K. A. V. and Gopal-Ayengar, A. R. (1972) On the biology of heterocysts in blue-green algae. *In* "Taxonomy and Biology of Blue-green Algae" (T. V. Desikaehary, ed.) pp. 218–226. University of Madras.

Thornber, J. P. (1969) Comparison of a chlorophyll *a*-protein complex isolated from a blue-green alga with chlorophyll-protein complexes obtained from green bacteria and higher plants. *Biochim. biophys. Acta* **172**, 230–241.

Thuret, G. (1875) Essai de classification des Nostochinées. *Annls Sci. nat.* (*Bot.*) **6**, 372-382.

Thurston, E. L. and Ingram, L. O. (1971) Morphology and fine structure of *Fischerella ambigua*. *J. Phycol.* **7**, 203–210.

Tiffany, L. H. (1951) Ecology of freshwater algae. *In* "Manual of Phycology" (Smith, G. M., ed.) pp. 293–311. Chronica Botanica, Waltham, Massachusetts.

Tilden, J. E. (1933) A classification of the algae based on evolutionary development, with special reference to pigmentation. *Bot. Gaz.* **95**, 59–77.

TILDEN, J. E. (1935) "The Algae and their Life Relations". University of Minnesota Press.

TISCHER, I. (1957) Untersuchungen über die granulären Einschlüsse und das Reduktions-Oxydations-Vermögen der Cyanophyceen. *Arch. Mikrobiol.* **27**, 400–428.

TISCHER, J. (1938) Über die Polyenpigmente der Blaualge *Aphanizomenon flos-aquae. Hoppe-Seyler's Z. physiol. Chem.* **251**, 109–128.

TISCHER, J. (1958) Carotinoide der Süsswasseralgen. XI. Über die Carotinoide aus *Oscillatoria amoena. Hoppe-Seyler's Z. physiol. Chem.* **311**, 140–147.

TISCHER, R. G. (1965) Pure culture of *Anabaena flos-aquae* A-37. *Nature, Lond.* **205**, 419–420.

TREUB, M. (1888) Notice sur la nouvelle flora de Krakatau. *Annls Jard. bot. Buitenz.* **7**, 221–223.

TROITZKAYA, O. V. (1924) Quelques observations biométriques sur les hétérocystes de *Anabaena scheremetievi* Elenk. au point de vue de leur signification biologique. *Arch. Russ. Protistol.* **3**, 117–135.

TSCHERMAK, E. (1941) Untersuchungen über die Beziehungen von Pilz und Alge im Flechtenthalus. *Öst. bot. Z.* **90**, 233–307.

TSUSUÉ, Y. and FUJITA, Y. (1964) Mono- and oligo-saccharides in the blue-green alga *Tolypothrix tenuis. J. gen. appl. Microbiol., Tokyo* **10**, 283–294.

TSUSUÉ, Y. and YAMAKAWA, T. (1965) Chemical structure of oligosaccharides in the blue-green alga *Tolypothrix tenuis. J. Biochem., Tokyo* **58**, 587–594.

TUFFERY, A. A. (1969) Light and electron microscopy of the sheath of a blue-green alga. *J. gen. Microbiol.* **57**, 41–50.

TYAGI, V. V. S. (1973) Effects of some metabolic inhibitors on heterocyst formation in the blue-green alga, *Anabaena doliolum. Ann. Bot.* **37**, 361–368.

ULLRICH, H. (1926) Über die Bewegung von *Beggiatoa mirabilis* und *Oscillatoria jenensis. Planta* **2**, 295–324.

UREY, H. C. (1952) "The Planets: Their Origin and Development". Yale University Press, New Haven, Connecticut.

UTTERBACK, C. L., PHIFER, L. D. and ROBINSON, R. J. (1942) Some chemical, physical and optical properties of Crater Lake. *Ecology* **23**, 97–103.

VAIDYA, S. M. W. (1964) "Ecology of Algae in Indian Soils with Special Reference to the Light Factor." Ph.D. Thesis, University of London.

VAN BAALEN, C. (1961) Vitamin B_{12} requirement of a marine blue-green alga. *Science, N.Y.* **133**, 1922.

VAN BAALEN, C. (1962) Studies on marine blue-green algae. *Botanica mar.* **4**, 129–139.

VAN BAALEN, C. (1965a) Quantitative surface plating of coccoid blue-green algae. *J. Phycol.* **1**, 19–22.

VAN BAALEN, C. (1965b) Aldolase in blue-green algae. *Nature, Lond.* **206**, 193-195.

VAN BAALEN, C. (1965c) The photooxidation of uric acid by *Anacystis nidulans. Pl. Physiol., Lancaster* **40**, 368–371.

VAN BAALEN, C. (1968) The effects of ultra-violet irradiation on a coccoid blue-green alga: survival, photosynthesis and photoreactivation. *Pl. Physiol., Lancaster* **43**, 1689–1695.

Van Baalen, C. and Brown, R. M. (1969) The ultrastructure of the marine blue-green alga, *Trichodesmium erythraeum*, with special reference to the cell wall, gas vacuoles and cylindrical bodies. *Arch. Mikrobiol.* **69**, 79–91.

Van Baalen, C., Hoare, D. S. and Brandt, E. (1971) Heterotrophic growth of blue-green algae in dim light. *J. Bact.* **105**, 685–689.

Van Baalen, C. and Marler, J. E. (1963) Characteristics of marine blue-green algae with uric acid as nitrogen source. *J. gen. Microbiol.* **32**, 457–463.

Vance, B. D. (1965) Composition and succession of cyanophycean water blooms. *J. Phycol.* **1**, 81–86.

Vance, B. D. (1966) Sensitivity of *Microcystis aeruginosa* and other blue-green algae and associated bacteria to selected antibiotics. *J. Phycol.* **2**, 125–128.

Vance, B. D. and Ward, H. B. (1969) Preparation of metabolically active protoplasts of blue-green algae. *J. Phycol.* **5**, 1–3.

Van Gorkom, H. J. and Donze, M. (1971) Localization of nitrogen fixation in *Anabaena*. *Nature, Lond.* **234**, 231–232.

Van Niel, C. B. and Stanier, R. Y. (1959) Bacteria. *In* "Freshwater Biology" (W. T. Edmonson, ed.) pp. 16–46. 2nd ed. John Wiley and Sons, New York and London.

Vasconcelos, A. C. L. and Bogorad, L. (1971) Proteins of cytoplasmic, chloroplast and mitochondrial ribosomes of some plants. *Biochim. biophys. Acta*, **228**, 492–502.

Vaucher, J. P. (1803) "Histoire des Conferves d'eau douce". Genève.

Venkataraman, G. S. (1957) A note on the occurrence of three-pored heterocysts in *Mastigocladus laminosus* Cohn. *Curr. Sci.* **26**, 254.

Venkataraman, G. S. (1961a) Studies on nitrogen fixation by *Cylindrospermum sphaerica* Prasad under various conditions. *Proc. natn. Acad. Sci. India* **31**, 100–104.

Venkataraman, G. S. (1961b) Nitrogen fixation by *Stigonema dendroideum*. *Indian J. agric. Sci.* **31**, 213–215.

Venkataraman, G. S. (1962) Studies on nitrogen fixation by *Anabaena azollae*. *Indian J. agric. Sci.* **32**, 22–24.

Venkataraman, G. S. (1969) "The Cultivation of Algae". Indian Council for Agricultural Research, New Delhi.

Venkataraman, G. S. and Lorenzen, H. (1969) Biochemical studies on *Anacystis nidulans* during its synchronous growth. *Arch. Mikrobiol.* **69**, 34–39.

Venkataraman, G. S. and Neelakantan, S. (1967) Effect of the cellular constituents of the nitrogen-fixing blue-green alga, *Cylindrospermum muscicola*, on the root growth of rice plants. *J. gen. appl. Microbiol.* **13**, 53–62.

Volk, S. L. and Phinney, H. K. (1968) Mineral requirements for the growth of *Anabaena spiroides* in vitro. *Can. J. Bot.* **46**, 619–630.

Vollenweider, R. A. (1968) "Scientific Fundamentals of the Eutrophication of Lakes and Flowing Waters with Particular Reference to Nitrogen and Phosphorus as Factors in Eutrophication". O.C.D.E., Directorate for Scientific Affairs, Paris.

Vologdin, A. G. and Kordé, K. B. (1965) Several species of ancient Cyanophyta and their coenoses. (Russian) *Dokl. Akad. Nauk. SSSR* **164**, 429–432.

WAALAND, J. R. (1969) "The Ultrastructure and Development of Gas Vacuoles" Ph.D. Thesis, University of California, Berkeley.

WAALAND, J. R. and BRANTON, D. (1969) Gas vacuole development in a blue-green alga. *Science, N.Y.* **163**, 1339–1341.

WAALAND, J. R. and WAALAND, S. D. (1970) Light induced gas vacuoles and their effect on light absorption by the blue-green alga *Nostoc muscorum*. *J. Phycol.* **6** (Suppl.), 7.

WAALAND, J. R., WAALAND, S. D. and BRANTON, D. (1971) Gas vacuoles. Light shielding in blue-green algae. *J. Cell Biol.* **48**, 212–215.

WALSBY, A. E. (1965) Biochemical studies on the extracellular polypeptides of *Anabaena cylindrica* Lemm. *Br. phycol. Bull.* **2**, 514–515.

WALSBY, A. E. (1967a) A new culture flask. *Biotechnol. Bioengng* **9,** 443–447.

WALSBY, A. E. (1967b) Mucilage secretion and the movements of blue-green algae. *Br. phycol. Bull.* **3**, 412.

WALSBY, A. E. (1968a) An alga's buoyancy bags. *New Scientist* **40**, 436–437.

WALSBY, A. E. (1968b) Mucilage secretion and the movements of blue-green algae. *Protoplasma* **65**, 223–238.

WALSBY, A. E. (1969) The permeability of blue-green algal gas-vacuole membranes to gas. *Proc. R. Soc.* B **173**, 235–255.

WALSBY, A. E. (1970) "Studies on the Physiology of Blue-green Algae" Ph.D. Thesis, University of London.

WALSBY, A. E. (1970) The nuisance algae: curiosities in the biology of planktonic blue-green algae. *Water Treatment and Examination* **19**, 359–373.

WALSBY, A. E. (1971) The pressure relationships of gas vacuoles. *Proc. R. Soc.* B **178**, 301–326.

WALSBY, A. E. (1972a) Gas filled structures providing buoyancy in photosynthetic organisms. *In* "The Effect of Pressure on Organisms" (M. A. Sleigh and A. G. MacDonald, eds.) *26th Symp. Soc. exp. Biol.*, pp. 233–250. University Press, Cambridge.

WALSBY, A. E. (1972b) Structure and function of gas vacuoles. *Bact. Rev.* **36**, 1–32.

WALSBY, A. E. and BUCKLAND, B. (1969) Isolation and purification of intact gas vesicles from a blue-green alga. *Nature, Lond.* **224**, 716–717.

WALSBY, A. E. and EICHELBERGER, H. H. (1968) The fine structure of gas-vacuoles released from cells of the blue-green alga *Anabaena flos-aquae*. *Arch. Mikrobiol.* **60**, 76–83.

WALSBY, A. E. and NICHOLS, B. W. (1969) Lipid composition of heterocysts. *Nature, Lond.* **221**, 673–674.

WALTON, J. (1953) "An Introduction to the Study of Fossil Plants." 2nd ed. Adam and Charles Black, London.

WATANABE, K. (1924) Studien über die Koralloide von *Cycas revoluta*. *Bot. Mag., Tokyo* **38**, 165–187.

WATANABE, A. (1951) Production in cultural solution of some amino acids by the atmospheric nitrogen-fixing blue-green algae. *Archs Biochem. Biophys.* **34**, 50–55.

WATANABE, A. (1959) Distribution of nitrogen-fixing blue-green algae in various areas of South and East Asia. *J. gen. appl. Microbiol., Tokyo* **5**, 21–29.

WATANABE, A. (1961) Collection and cultivation of nitrogen-fixing blue-green algae and their effect on growth and crop yield of rice plants. *Stud. Rokugawa Inst. Tokyo* **9**, 162–166.

WATANABE, A. (1966) The blue-green algae as the nitrogen fixators. *Proc. 9th Intern. Congr. Microbiol.* pp. 77–85.

WATANABE, A. (1970) Studies on application of Cyanophyta in Japan. *Schweiz. Z. Hydrol.* **32**, 566–569.

WATANABE, A., HATTORI, A., FUJITA, Y. and KIYOHARA, T. (1959) Large scale culture of a blue-green alga, *Tolypothrix tenuis*, utilizing hot spring and natural gas as heat and carbon dioxide sources. *J. gen. appl. Microbiol., Tokyo* **5**, 51–57.

WATANABE, A. and KIYOHARA, T. (1963) Symbiotic blue-green algae of lichens, liverworts and cycads. *In* "Studies on Microalgae and Photosynthetic Bacteria" (Japanese Society of Plant Physiologists, eds.) pp. 189–196. University of Tokyo Press.

WATANABE, A., NISHIGAKI, S. and KONISHI, C. (1951) Effect of nitrogen-fixing blue-green algae on the growth of rice plants. *Nature, Lond.* **168**, 748–749.

WATANABE, A. and YAMAMOTO, Y. (1967) Heterotrophic nitrogen fixation by the blue-green alga *Anabaenopsis circularis*. *Nature, London.* **214**, 738.

WATANABE, A. and YAMAMOTO, Y. (1971) Algal nitrogen fixation in the tropics. *In* "Biological Nitrogen Fixation in Natural and Agricultural Habitats" (T. A. Lie and E. G. Mulder, eds.) pp. 403–413. *Pl. Soil*, special volume.

WEBBER, E. (1967) Blue-green algae from a Massachusetts salt marsh. *Bull. Torey bot. Club* **94**, 99–106.

WEBER, H. L. and BOCK, A. (1968) Comparative studies on the regulation of DAHP synthetase activity in blue-green and green algae. *Arch. Mikrobiol.* **61**, 159–168.

WEBSTER, D. A. and HACKETT, D. P. (1964) The respiratory chain in colorless algae. *Pl. Physiol., Lancaster* Suppl. **39**, LIX.

WEBSTER, D. A. and HACKETT, D. P. (1965) Respiratory chain of colourless algae. I. Chlorophyta and Euglenophyta. *Pl. Physiol., Lancaster* **40**, 1091–1100.

WEBSTER, D. A. and HACKETT, D. P. (1966) Respiratory chain of colourless algae. II. Cyanophyta. *Pl. Physiol., Lancaster* **41**, 599–605.

WEBSTER, G. C. and FRENKEL, A. W. (1953) Some respiratory characteristics of the blue-green alga, *Anabaena*. *Pl. Physiol., Lancaster* **28**, 63–69.

WEIBULL, C. (1960) Movement. *In* "The Bacteria" (I. C. Gunsalus and R. Y. Stanier, eds.) Vol. 1, pp. 153–205. Academic Press, New York.

WEISE, G., DREWS, G., JANN, B. and JANN, K. (1970) Identification and analysis of a lipopolysaccharide in cell walls of the blue-green alga *Anacystis nidulans*. *Arch. Mikrobiol.* **71**, 89–98.

WERBIN, H. and RUPERT, C. S. (1968) Presence of photoreactivating enzyme in blue-green algal cells. *Photochem. Photobiol.* **7**, 225–230.

WEST, G. S. and FRITSCH, F. E. (1927) "A Treatise on the British Fresh-water Algae". University Press, Cambridge.

WHATLEY, F. R. and LOSADA, M. (1964) The photochemical reaction of photosynthesis. *In* "Photophysiology" (A. C. Giese, ed.) Vol. 1 pp. 111–154. Academic Press, London.

WHITTON, B. A. (1962) Effect of deep-freeze treatment on blue-green algal cultures. *Br. phycol. Bull.* **2**, 177–178.

WHITTON, B. A. (1965) Extracellular products of blue-green algae. *J. gen. Microbiol.* **40**, 1–11.

WHITTON, B. A. (1967a) Phosphate accumulation by colonies of *Nostoc*. *Pl. Cell Physiol., Tokyo* **8**, 293–296.

WHITTON, B. A. (1967b) Studies on the toxicity of polymyxin B to blue-green algae. *Can. J. Microbiol.* **13**, 987–993.

WHITTON, B. A. (1969) The taxonomy of blue-green algae. *Br. phycol. J.* **4**, 121–123.

WHITTON, B. A. (1971) Terrestrial and freshwater algae of Aldabra. *Phil. Trans. R. Soc.* Ser. B **260**, 249–255.

WHITTON, B. A. and MACARTHUR, K. (1967) The action of two toxic quinones on *Anacystis nidulans*. *Arch. Mikrobiol.* **57**, 147–154.

WHITTON, B. A. and PEAT, A. (1967) Heterocyst structure in *Chlorogloca fritschii*. *Arch. Mikrobiol.* **58**, 324–338.

WHITTON, B. A. and PEAT, A. (1969) On *Oscillatoria redekei* Van Goor. *Arch. Mikrobiol.* **68**, 362–376.

WIERINGA, K. T. (1968) A new method for obtaining bacteria-free cultures of blue-green algae. *Antonie van Leeuwenhoek* **34**, 54–56.

WIESSNER, W. (1970) Photometabolism of organic substrates. *In* "Photobiology of Microorganisms" (P. Halldal, ed.) pp. 95–133. John Wiley and Sons, London.

WILCOX, M. (1970) One-dimensional pattern found in blue-green algae. *Nature, Lond.* **228**, 686–687.

WILDEMANN, L. (1934) Weiterer Beitrag für Symbiose von *Azolla* und *Anabaena*. Dissertation, Münster University, Germany.

WILDON, D. C. (1965) "Ultrastructure and Respiratory Function in Blue-green Algae". Ph.D. Thesis, University of Sydney, Australia.

WILDON, D. C. and MERCER, F. V. (1963a) The ultrastructure of the vegetative cell of blue-green algae. *Aust. J. biol. Sci.* **16**, 585–596.

WILDON, D. C. and MERCER, F. V. (1963b) The ultrastructure of the heterocyst and akinete of the blue-green algae. *Arch. Mikrobiol.* **47**, 19–31.

WILDON, D. C. and REES, T. (1965) Metabolism of glucose-C^{14} by *Anabaena cylindrica*. *Pl. Physiol., Lancaster* **40**, 332–335.

WILLARD, J. M., SCHULMAN, M. and GIBBS, M. (1965) Aldolase in *Anacystis nidulans* and *Rhodopseudomonas spheroides*. *Nature, Lond.* **206**, 195.

WILLIAMS, A. E. and BURRIS, R. H. (1952) Nitrogen fixation in blue-green algae and their nitrogenous composition. *Am. J. Bot.* **39**, 340–342.

WILSON, A. T. (1965) Escape of algae from frozen lakes and ponds. *Ecology* **46**, 376–378.

WILSON, P. W. (1951) Biological nitrogen fixation. *In* "Bacterial Physiology "(C. H. Werkman and P. W. Wilson, eds.) pp. 467–499. Academic Press, New York.

WILSON, P. W. (1958) Asymbiotic nitrogen fixation. *In* "Encyclopaedia of Plant Physiology" (W. Ruhland, ed.) Vol. 8, pp. 9–47. Springer Verlag, Berlin.

WINKENBACH, F., WOLK, C. P. and JOST, M. (1972) Lipids of membranes and of the cell envelope in heterocysts of a blue-green alga. *Planta* **107**, 69–80.

WINTER, G. (1935) Über die Assimilation des Luftstickstoffs durch endophytische Blaualgen. *Beitr. Biol. Pfl.* **23**, 295–335.

WIRTH, T. L. and DUNST, R. C. (1967) "Limnological changes resulting from artificial destratification and aeration of an impoundment". Wisconsin Conservation Department, Research and Planning Division. Progress report no. 22.

WOLFE, M. (1954a) The effect of molybdenum upon the nitrogen metabolism of *Anabaena cylindrica*. I. A study of the molybdenum requirement for nitrogen fixation and for nitrate and ammonia assimilation. *Ann. Bot.* **18**, 299–308.

WOLFE, M. (1954b) The effect of molybdenum upon the nitrogen metabolism of *Anabaena cylindrica*. II. A more detailed study of the action of molybdenum in nitrate reduction. *Ann. Bot.* **18**, 309–325.

WOLK, C. P. (1965a) Heterocyst germination under defined conditions. *Nature, Lond.* **205**, 201–202.

WOLK, C. P. (1965b) Control of sporulation in a blue-green alga. *Devl. Biol.* **12**, 15–35.

WOLK, C. P. (1966) Evidence of a role of heterocysts in the sporulation of a blue-green alga. *Am. J. Bot.* **53**, 260–262.

WOLK, C. P. (1967) Physiological basis of the pattern of vegetative growth of a blue-green alga. *Proc. natn. Acad. Sci. U.S.A.* **57**, 1246–1251.

WOLK, C. P. (1968) Movement of carbon from vegetative cells to heterocysts in *Anabaena cylindrica*. *J. Bact.* **96**, 2138–2143.

WOLK, C. P. (1970) Aspects of the development of a blue-green alga. *Ann. N.Y. Acad. Sci.* **175**, 641–647.

WOLK, C. P. and SIMON, R. D. (1969) Pigments and lipids of heterocysts. *Planta* **86**, 92–97.

WOLK, C. P. and WOJCIUCH, E. (1971a) Photoreduction of acetylene by heterocysts. *Planta* **97**, 126–134.

WOLK, C. P. and WOJCIUCH, E. (1971b) Biphasic time course of solubilization of nitrogenase during cavitation of aerobically grown *Anabaena cylindrica*. *J. Phycol.* **7**, 339–344.

WOOD, E. J. F. (1965) "Marine Microbial Ecology". Chapman and Hall, London.

WOODCOCK, C. L. F. and BOGORAD, L. (1971) Nucleic acids and information processing in chloroplasts. *In* "Structure and Function of Chloroplasts" (M. Gibbs, ed.) pp. 89–128. Springer Verlag, Berlin.

WRIGHT, W. D. (1964) Structural colours of biological material. *In* "Colour and Life" (W. B. Broughton, ed.) pp. 1–11. *12th Symp. Inst. Biol.* Institute of Biology, London.

WU, B., HANDY, M. K. and HOWE, H. B. (1968) Antimicrobial activity of a myxobacterium against blue-green algae. *Bact. Proc.* 48.

Wyatt, J. T., Lawley, G. G. and Barnes, R. D. (1971) Blue-green algal response to some organic nitrogen substrates. *Naturwissenschaften* **58**, 570.

Wyatt, J. T. and Silvey, J. K. G. (1969) Nitrogen fixation by *Gloeocapsa*. *Science. N.Y.* **165**, 908–909.

Yamanaka, T., Takenami, S., Wada, K. and Okunuki, K. (1969) Purification and some properties of ferredoxin derived from the blue-green algae, *Anacystis nidulans*. *Biochim. biophys. Acta* **180**, 196–198.

Yoshida, T., Roncal, R. A. and Bautista, E. M. (1973) Nitrogen fixation in a Philippine soil. *Proc. 2nd Asian Soil Conf.*, Indonesia. (In press.)

Zaneveld, J. S. (1969) Cyanophytan mat communities in some melt water lakes at Ross Island, Antarctica. *Proc. K. med. Akad. Wet.* C **72**, 299–305.

Zehnder, A. and Hughes, E. O. (1958) The antialgal activity of actidione. *Can. J. Microbiol.* **4**, 399–408.

Zelitch, I., Rosenblum, E. D. Burris, R. H. and Wilson, P. W. (1951) Isolation of the key intermediate in biological nitrogen fixation by *Clostridium*. *J. biol. Chem*, **191**, 295–298.

Zimmerman, U. (1969) Ökologische und physiologische Untersuchungen an der planktonischen Blaualge *Oscillatoria rubescens* D.C. unter besonderer Berücksichtigung von Licht und Temperatur. *Schweiz. Z. Hydrol.* **31**, 1–58.

Index

B

C

N

P

S

T

U